CHEMILUMINESCENCE IN ANALYTICAL CHEMISTRY

Chemiluminescence in Analytical Chemistry

EDITED BY

Ana M. García-Campaña
University of Granada
Granada, Spain

Willy R. G. Baeyens
Ghent University
Ghent, Belgium

CRC Press
Taylor & Francis Group
Boca Raton London New York

CRC Press is an imprint of the
Taylor & Francis Group, an **informa** business

CRC Press
Taylor & Francis Group
6000 Broken Sound Parkway NW, Suite 300
Boca Raton, FL 33487-2742

First issued in paperback 2019

© 2001 by Taylor & Francis Group, LLC
CRC Press is an imprint of Taylor & Francis Group, an Informa business

No claim to original U.S. Government works

ISBN-13: 978-0-8247-0464-3 (hbk)
ISBN-13: 978-0-367-39748-7 (pbk)

Visit the Taylor & Francis Web site at
http://www.taylorandfrancis.com

and the CRC Press Web site at
http://www.crcpress.com

Preface

For more than 30 years, the phenomenon of luminescence—originally a curiosity in the physical laboratory—has been the basis of a well-established and widely applied spectrometric branch of analytical chemistry. Specifically, chemiluminescence (CL)-based analysis is growing rapidly, offering a simple, low-cost, and sensitive means of measuring a variety of compounds. Owing to elegant new instrumentation and, especially, to new techniques, some of which are entirely new and some borrowed from other disciplines, CL and bioluminescence (BL) can now be routinely applied to solve diverse qualitative and quantitative analytical problems.

Although luminescence phenomena date back beyond 300 B.C., the development of CL and BL analytical applications is relatively recent. Simple measuring devices and the high versatility for the determination of a wide variety of species have enabled CL-based detection to develop into a highly sensitive and most useful analytical technique. The first application of CL as an analytical tool was carried out in the early 1950s, employing several substances such as luminol, lophine, and lucigenin as volumetric indicators. Investigations on the potential of CL for analytical routine applications date from the 1970s for gas-phase and from the 1980s for liquid-phase reactions. In trace analysis for inorganic compounds, CL is one of the most sensitive techniques, compared to atomic absorption spectrometry (AAS), inductively coupled plasma-optical emission spectrometry (ICP-OES), and inductively coupled plasma-mass spectrometry (ICP-MS). Together with classical CL reactions, new strategies have been proposed, considering not only the effect of inorganic ions as oxidants, reductants, catalysts, or inhibitors but also the use of coupling reactions, time-resolved techniques, and solid-surface analysis. Also, in organic analysis the number of reactions producing CL cited in the literature is increasing annually. For example, the inherent power of applying the peroxyoxalate CL system to a vast number of natively fluorescing species or fluorophores formed after chemical derivatization broadens the scope of this relatively new detection technique. In drug analysis, CL has become a powerful tool in recent years, due to the discovery of new CL systems based on the direct oxidation of molecules with different common oxidants in acid or alkaline media.

Since the discovery in 1947 of the essential role of ATP in the BL reactions by which fireflies produce light, simple and very sensitive methods for its determination have been applied in such areas as medicine, biology, agriculture, industry, and environmental sciences. In the past few years, BL applications have increased, mainly in the biomedical field, owing to the further development of gene technology and the use of different new methods to study BL at the molecular level. As an example, CL precursors have been used from the 1970s to the present as sensitive substitute labels for isotopic labeling, replacing radioisotopes and providing a new strategy, considerably better in terms of sensitivity and safety, in immunoassay. In this sense, increasing interest has been focused on CL products for life sciences research. For example, isoluminol derivatives and acridinium esters have proved to be successful in the development of commercial kits in clinical diagnostics. In the 1980s, the discovery of the light-yield enhancement when firefly luciferase was accidentally added to a mixture of horseradish peroxidase, luminol, and hydrogen peroxide marked the beginning of a very successful analytical era for immunoassay and diverse blotting applications (protein, DNA, and RNA). More recently, a new technology using novel acridan esters as chemiluminogenic signal reagents has demonstrated its suitability in immunoassay.

The characteristics of CL emission make this phenomenon suitable as a detecting tool in flow injection, gas, and column liquid chromatographic separating systems. Continuous-flow CL-based detection of several analytes has been widely applied by several groups for the determination of diverse biological and pharmaceutical compounds. In combination with HPLC separations, several CL reactions have been used, including peroxyoxalates, firefly luciferase, lucigenin, and luminol, the peroxyoxalate reaction being most commonly used for postcolumn detection in conventional and microcolumn LC setups. Applications in analytical research, biotechnology, and quality control areas are currently being amply described.

A recent trend in analytical chemistry involves the application of CL as a detection system in combination with capillary electrophoresis as prior separation methodology, providing excellent analytical sensitivity and selectivity and allowing the resolution and quantification of various analytes in relatively complex mixtures. Until the 1990s, chemiluminometric detection was not applied after capillary electrophoretic separation, but fast developments from some important research groups have been noticed in the past few years; hence, further developments are expected.

Immobilization techniques have been applied in the preparation of immobilized CL reagents, with specific advantages such as reusability, improved stability, and increased efficiency. These strategies have been applied in the development of CL sensors, which today constitute the most important tools in analytical chemistry because of the high sensitivity offered. Optical fibers have been used to transfer light in order to improve the quality of detection, and new types of flow-through cells have been introduced in the construction of CL sensors. Also, selectivity has been considerably improved by the utilization of enzymatic or antigen–antibody reactions.

It is clear that the need for improving detection technology is related to the general trend in analytical chemistry to miniaturize, and thus reduce, waste volumes of organic solvents in separational setups and, by using more aqueous systems, study smaller samples at increasingly lower concentrations. As the CL technique may provide solutions for these specific challenges, the instrumentation for CL measurements and the coupling with a selective physical or chemical interface to achieve selective measurements are likewise being explored. In this way, disadvantages of direct CL-based techniques (e.g., lack of selectivity, sensitivity to various physicochemical factors) are avoided. As an example, in recent years a CL-based detection system using electrophoretically mediated microanalysis (EMMA) has been described, allowing the detection of enzymes at the *zeptomole* level in both open tubular capillaries and channels in microfabricated devices.

The degree of scientific interest toward the application of CL in the various disciplines of analytical chemistry may be illustrated by the growing position that is being attributed to this physicochemical phenomenon in the luminescence-based analytical symposia that have been organized over the globe since the early 1980s, series that appear about to receive increasing interest by the scientific community in the decade to come. Moreover, in the past two decades the number of published papers in prestigious analytical journals and in related dedicated journals such as the *Journal of Biological and Chemical Luminescence* has considerably grown.

All these considerations encouraged us to produce a multiauthored book focussing on the importance and versatility of CL in the actual scientific context through the different perspectives related to its potential as an analytical technique. Our aim was to provide the reader with a wide overview of chemical reactions producing light, with emphasis on the analytical uses of the phenomenon and its recent applications, in a style accessible to readers at various levels (researchers, industrial workers, undergraduates, and graduates, as well as Ph.D. students). With this purpose, we have organized the available information on the various aspects of CL into different chapters, each produced by authors with recognized international expertise in the specific areas. In our modest opinion, a comprehensive volume was built up in this way, useful to students at the various university levels; chemists; pharmacists; biologists; medical doctors; technicians in food, clinical, toxicological, and environmental disciplines; quality control managers—primarily in chemical analytical laboratories—and, in general, researchers applying luminescence-based techniques.

The selection of essential topics and expert authors was not an easy task. We tried to include the most representative applications of CL and BL in analytical chemistry. The contributors were invited to elaborate on the subjects according to their knowledge and experience in the field, and we think we have succeeded in unifying the contents of the overall volume. We heartily thank the contributing authors for agreeing to collaborate on this project; their efforts led to the comprehensive structure of this book.

Apart from an overview on the historical evolution of luminescence phenomena, and more specifically of CL and BL, the volume treats the physicochemical nature of these reactions, the basic principles, the evolution in instrumentation—from the use of simple PMTs to the implementation of CCD cameras and the development of imaging technology—and general applications in organic and inorganic analysis, considering the use of organized media so as to enhance sensitivity. Different analytical CL approaches related to the intrinsic kinetic nature of CL emission and specific analytical topics such as the recently applied electrogenerated CL, the relative unknown possibilities offered by photosensitized CL used in medical and industrial routine analysis, and the wide uses of CL detection in the gas phase—mainly in atmospheric research—have been included.

Optimization and applications of CL detection in flow injection and liquid chromatographic analysis and the relatively new use of CL in capillary electrophoresis are extensively described. Particular interest is attached to the universally applied peroxyoxalate CL reactions, as well as to the applications of new acridan esters in immunoassay. Obviously, the related applications of BL and CL imaging techniques in analytical chemistry, and the increasing importance of these techniques in DNA analysis—including the recent strategies in the development of CL sensors—are also presented.

It is our wish to encourage the analytical community to discover more about this most exciting analytical technique and to consider it a powerful alternative in the resolution of a variety of analytical challenges.

Ana M. García-Campaña
Willy R. G. Baeyens

Contents

vii

Contents

Contributors

Hassan Y. Aboul-Enein   Pharmaceutical Analysis Laboratory, Department of Biological and Medical Research, King Faisal Specialist Hospital and Research Centre, Riyadh, Saudi Arabia

Willy R. G. Baeyens   Department of Pharmaceutical Analysis, Ghent University, Ghent, Belgium

Mario Baraldini   Institute of Chemical Sciences, University of Bologna, Bologna, Italy

John W. Birks   Department of Chemistry and Biochemistry and Cooperative Institute for Research in Environmental Sciences, University of Colorado, Boulder, Colorado

Luca Bolelli   Institute of Chemical Sciences, University of Bologna, Bologna, Italy

Richard Bos   School of Biological and Chemical Sciences, Deakin University, Geelong, Victoria, Australia

James E. Boulter   Department of Chemistry and Biochemistry, University of Colorado, Boulder, Colorado

Antony C. Calokerinos Laboratory of Analytical Chemistry, Department of Chemistry, University of Athens, Athens, Greece

Elida Nora Ferri Institute of Chemical Sciences, University of Bologna, Bologna, Italy

Fabiana Fini Institute of Chemical Sciences, University of Bologna, Bologna, Italy

Ana M. García-Campaña Department of Analytical Chemistry, University of Granada, Granada, Spain

Stefano Girotti Institute of Chemical Sciences, University of Bologna, Bologna, Italy

Massimo Guardigli Department of Pharmaceutical Sciences, University of Bologna, Bologna, Italy

Norberto A. Guzman Department of Bioanalytical Drug Metabolism, The R.W. Johnson Pharmaceutical Research Institute, Raritan, New Jersey

Knut Irgum Department of Analytical Chemistry, Umeå University, Umeå, Sweden

Marjorie Jacquemijns National Institute of Public Health and the Environment, Bilthoven, The Netherlands

Tobias Jonsson Department of Analytical Chemistry, Umeå University, Umeå, Sweden

Masaaki Kai Department of Medicinal Chemistry, School of Pharmaceutical Sciences, Nagasaki University, Nagasaki, Japan

Andrew W. Knight Department of Instrumentation and Analytical Science, University of Manchester Institute of Science and Technology, Manchester, England

Naotaka Kuroda Department of Analytical Chemistry, School of Pharmaceutical Sciences, Nagasaki University, Nagasaki, Japan

Dan A. Lerner Department of Physical Chemistry, Ecole Nationale Supérieure de Chimie, Montpellier, France

Gudrun Lewin   Research Institute for Antioxidant Therapy, Berlin, Germany

Mara Mirasoli   Department of Pharmaceutical Sciences, University of Bologna, Bologna, Italy

Monica Musiani   Division of Microbiology, Department of Clinical and Experimental Medicine, University of Bologna, Bologna, Italy

Kenichiro Nakashima   Department of Analytical Research for Pharmacoinformatics, Graduate School of Pharmaceutical Sciences, Nagasaki University, Nagasaki, Japan

Kazuko Ohta   School of Pharmaceutical Sciences, Nagasaki University, Nagasaki, Japan

Leonidas P. Palilis   Laboratory of Analytical Chemistry, Department of Chemistry, University of Athens, Athens, Greece

Patrizia Pasini   Department of Pharmaceutical Sciences, University of Bologna, Bologna, Italy

Dolores Pérez-Bendito   Department of Analytical Chemistry, University of Córdoba, Córdoba, Spain

Einar Pontén   Department of Analytical Chemistry, Umeå University, Umeå, Sweden

Igor Popov   Research Institute for Antioxidant Therapy, Berlin, Germany

Yener Rakicioğlu   Department of Chemistry, Istanbul Technical University, Istanbul, Turkey

Aldo Roda   Division of Analytical Chemistry, Department of Pharmaceutical Sciences, University of Bologna, Bologna, Italy

Manuel Román-Ceba   Department of Analytical Chemistry, University of Granada, Granada, Spain

Carmela Russo   Department of Pharmaceutical Sciences, University of Bologna, Bologna, Italy

José Juan Santana Rodríguez   Department of Chemistry, University of Las Palmas de G.C., Las Palmas de G.C., Spain

Joanna M. Schulman   Department of Botany, University of Florida, Gainesville, Florida

Stephen G. Schulman   Department of Medicinal Chemistry, College of Pharmacy, University of Florida, Gainesville, Florida

Gloria Sermasi   Institute of Chemical Sciences, University of Bologna, Bologna, Italy

Manuel Silva   Department of Analytical Chemistry, University of Córdoba, Córdoba, Spain

Raluca-Ioana Stefan   Department of Chemistry, University of Pretoria, Pretoria, South Africa

Malin Stigbrand   Department of Analytical Chemistry, Umeå University, Umeå, Sweden

Jacobus F. van Staden   Department of Chemistry, University of Pretoria, Pretoria, South Africa

Xinrong Zhang   Department of Chemistry, Tsinghua University, Beijing, P. R. China

Gijsbert Zomer   National Institute of Public Health and the Environment, Bilthoven, The Netherlands

Chemiluminescence in Analytical Chemistry

1

Historical Evolution of Chemiluminescence

Ana M. García-Campaña and Manuel Román-Ceba
University of Granada, Granada, Spain

Willy R. G. Baeyens
Ghent University, Ghent, Belgium

1

1. INTRODUCTION TO THE DISCOVERY AND THE DEVELOPMENT OF LUMINESCENCE

It has been known for centuries that many compounds emit visible radiation when they are exposed to sunlight. Luminescence phenomena, such as the aurora borealis, phosphorescence of the sea, luminous animals and insects, phosphorescent wood, etc., have fascinated man since antiquity, being reflected in the early scientific literature. Aristotle (384–322 B.C.) appears to be one of the first philosophers to recognize "cold light" in dead fish, fungi, and the luminous secretion of the cuttlefish [1].

These luminescence phenomena have been known since ancient times; according to the legend, about 1000 B.C., a Chinese emperor possessed a magic paint on which the image of an ox appeared at sunset. The chemical composition of the paint used was not known. This is the first known case of a man-made substance capable of storing daylight for later recovery [2].

Also, early written references citing luminescence phenomena appeared in the Chinese literature of around 1500–1000 B.C., describing glowworms and fireflies [3]. In fact, all of these first observations were related mainly to living organisms that emit light such as the fireflies, luminous bacteria and protozoa, the sea pansy, the marine fireworm, unicellular organisms such as the dinoflagellates, etc. Harvey describes in his book an interesting chapter on the many early approaches to explain luminescence, as a matter of fact an established modern scientific approach on the subject, from the seventeenth century [3].

Francis Bacon reports in 1605 the different kinds of luminescence in relation to its origins and writes: "sugar shineth only while it is scraping; and salt water while it is in dashing; glowworms have their shining while they live, or a little after; only scales of fish putrefied seem to be of the same nature with shining wood: and it is true, that all putrefaction has with it an inward motion, as well as fire or light" [3].

The first example of luminescence emission from solids, of which written documents exist, date from the Italian Renaissance, originating from the accidental discovery around the year 1600 (1602 or 1603) by a Bolonian shoemaker and alchemist, called Vincencio Casciarolo or Casciarolus. He melted heavy bricks, close to his house, hoping to extract precious metals from them.

These bricks, after calcination with carbon and exposure to daylight, emitted a reddish glittering in the dark. These "Bolonian stones," also named "moonstones," particularly those from the Monte Paterno, remain among the most famous ones and were the subject of scientific interest during the next two centuries; they were termed "phosphor" (Greek: "light bearer"). They are considered the first inorganic artificial "phosphors" [2–4]. The first natural phosphor was diamant, whose luminescence was cited by Cellini in 1568 [5].

The discovery of the ''Bolonian stones'' attracted the interest of Galileo (1564–1642) and colleagues from that period who stated that a ''phosphor'' does not emit luminescence before having been exposed to natural light, in the way that light enters the stone, like a sponge taking up water, then producing light emission, so that the ''phosphor'' behavior implicated some time during which light remained for a while in the substance material [6]. In 1652 the Italian mathematician Zucchi described that the Bolonian stone emitted more intensely once exposed to brilliant light and that the color of the emitted light did not change when the stone was exposed to white light, green, yellow, or red light. He concluded that the light was not simply absorbed and emitted in an unchanged form, like a sponge, but that on the contrary, during the excitation process reactions occurred with some substance (''spiritus'') present in the brick, and that when illumination ceased, the light produced by the substance gradually diminished [3].

Originally, the Bolonian stone, as mentioned, was considered a ''light sponge,'' but clearly this term was poetic rather than exact. In this poetic way, as a matter of fact, the poet Goethe described the stone. When light of different colors impinged upon this stone, some red light was produced when exciting with blue light, red light as such being unable to do so. The light emitted by the stone in the dark did not appear to be similar to the illuminating light applied, which made clear that the Bolonian stone in fact did not act as a sponge that *post factum* emitted the absorbed light [7].

The delayed light emission as observed from the Bolonian stone is now classified as phosphorescence. We know now that these stones contain barium sulfate with traces of bismuth and manganese, and that the corresponding reducing process concerns the transformation of sulfate into sulfur. It is now well known that alkaline earth metal sulfates emit phosphorescence that strongly increases when traces of heavy metals are present. The so-called inorganic multi-component compounds ''phosphor'' and ''crystallophosphor'' are in fact polycrystalline substances containing traces of some ionic activators of luminescence.

The term ''phosphor'' obviously is employed as well for the chemical element discovered by a Hamburg alchemist, Hennig Brand, in 1669, recognized to be the first scientist discovering a chemical element. In fact Brand, searching for the philosopher's stone, distilled a mixture of sand and evaporated urine and obtained a product that was capable of shining in the dark. They called it ''Brand's phosphor'' to distinguish it from other luminescent materials also termed ''phosphor'' [8]. Brand called the product obtained ''miraculous light'' [3]. The element P, in its white or yellow allotropic variety, emits light in the dark, but this is *not* a photoluminescent (phosphorescent) phenomenon but a *chemiluminescence* produced by the reaction of this element with oxygen, in a humid environment [9].

The first example of protein luminescence was made by Beccari in 1746, who detected a visible, blue phosphorescence proceeding from frozen hands when entering a dark room after exposure to sunlight [10].

During the seventeenth and eighteenth centuries, numerous "phosphors" were discovered, but little progress in their characterization occurred. An attempt to classify luminescent phenomena, indicated by the general term "phosphorescence," appears at the end of the eighteenth century. The Encyclopedia of Diderot and Alembert, in its Geneva edition of 1778–1779, mentions six classes of "phosphors" differentiating the slow oxidation of the metalloid from: physiological phenomena (fireflies, glowworms, mosquitos from the Venice lagoon, flies from the Antilles, sting of irritated vipers); electric phenomena (diamond, strongly rubbed tissues and clothes, Hauxbee globe); mechanical effects (friction of sugar or cadmia, metals trapped in steel or iron); physical phenomena (Bolonian stone and spar exposed to sunlight); biological phenomena (will-o'-the wisp) [4].

The first observation of fluorescence in solution occurred in 1565 by the Spanish physician and botanist Nicolás Monardes, who noticed a blue tint in the water contained in a recipient fabricated with a specimen of wood called "lignum nephriticum" [3, 11]. It was known in 1570 that the blue coloration that is produced by white light from the aqueous extract of the "lignum nephriticum" or "peregrinum" disappeared in acid medium. In 1615 a similar behavior was observed from the rind of *Aesculos hippocastanum* in aqueous medium [4]. When placed in water, the rind of chestnut produces a colorless liquid with bluish reflections; today it is known that this originates from aesculin fluorescence [12].

The luminescent properties of the aqueous extracts of some wood specimens were of interest to many scientists of the seventeenth and the beginning of the eighteenth century. Athanasius Kircher, who investigated aqueous extracts of wood, postulated in 1646 that the observed color depended on the intensity of the ambient light [3]. Robert Boyle (1627–1691), in 1664, Isaac Newton (1662–1727), and Robert Hooke, in 1678, disagreed with Kircher, who indicated that the color of the wood extract depended on the angle of observation, being yellow for the transmitted light and blue for the reflected light [6].

In the year 1704, in an optical treatise Newton stated that the tincture of *lignum nephriticum* showed a variable color depending on the position of the sun and of the incident light: yellow by transparency and blue from a lateral view [4]. Although Kircher is generally recognized as the discoverer of the fluorescence phenomenon in solution, it was Boyle who was the first to describe some of the most important characteristics of fluorescence from organic solutions. Boyle, after carrying out various extractions of wood, obtained a fluorescent solution and thought this to be an "essential salt" present in the wood, responsible for the observed luminescence. All cited scientists describe the production of a certain "reflection" phenomenon, without clearly providing any differentiation between the terms "emission" and "reflection" [3]. Hooke, in 1665, mentions that because of internal vibrations, some matrices emit light [3]. In 1718 Newton produces a report agreeing with Hooke's hypothesis, stating that the incandescence of luminous bodies—hot or cold—originates from vibrational movements of their particles [13].

Although numerous materials and fluorescing solutions were described in the seventeenth and eighteenth centuries, and in spite of the fact that since around 1860 mineralogists started the use of fluorescence for detection of mineral deposits, little progress was observed concerning the explanation of the phenomenon, and it was only around the mid-nineteenth century that important achievements were made in the study and understanding of luminescence phenomena.

In 1833 David Brewster (1781–1868) describes the red fluorescence from an alcoholic extract from green leaves (i.e., chlorophyll) and the fluorescence from fluorspar crystals, but he considered the effect to be caused by "dispersed" light, rather than by emitted light. John Herschel (1792–1871), in 1845, uses a prism to obtain crude spectral analysis of the fluorescence from quinine solutions. Apparently he did not realize that the emitted light had a longer wavelength. He observed that the solution emitted a noticeable luminescent radiation when exposed to sunlight. The solution of quinine sulfate was colorless when observed by transparency and bluish white when examined from a certain angle. He postulated that the blue light was produced at the surface of the liquid and called this phenomenon "dispersion epipolique." He had already suggested in 1825 that "dispersed" light might be employed for the detection of small quantities of some compounds. His compatriot Brewster, who studied the dissolution of quinine and of aesculine, mentioned that the "dispersion" occurred much more internally than superficially. Hence, the concepts of absorption and of emission as well as the phenomenon of fluorescence had not been established yet [5, 14]. It is important to note that the ideas of light absorption and emission were suggested much earlier for the "phosphors" (seventeenth century) than for fluorescent materials (nineteenth century), probably because understanding of the phosphorescence emission phenomenon occurred only later [5].

Sir George Gabriel Stokes (1820–1903), physicist and professor of mathematics at Cambridge, bears the merit of having established the theoretical principles of fluorescence, in an important publication in the journal *Philosophical Transactions* that appeared in the year 1852. By means of a setup of prisms he obtained a solar spectrum that he utilized to illuminate a tube containing a solution of quinine sulfate in the way that the red, yellow, green, etc. light passed the solution. When coming close to the violet or further spectral zones, a blue shining was progressively produced by the solution. It is extraordinary, described Stokes, to see how the tube is illuminated instantaneously by the "invisible rays." These rays are what is today called ultraviolet radiation. He stated that the blue light in fact was made by the material starting from other radiations that were absorbed by the liquid, in the way that light production was more or less important depending on the way the irradiation penetrated, which in fact had already been observed many years before from inorganic "phosphors." He demonstrated that there was no dispersion, superficial or internal, rejecting at the same time the term "dispersion" suggesting to replace the latter by "fluorescence," as derived from a certain variety of fluorspar (fluorite) that showed blue reflections, similar to what was observed from solutions of quinine sulfate [12, 14, 15].

Numerous substances that produced fluorescence were examined by Stokes: plant extracts (e.g., chestnut rind, chlorophyll in water), glass, paper, animal material, uranium compounds, etc., and he pointed out that "the rays produced by the fluorescence process were much more "refrangible" than the rays initiating them."

Yellow light is much more deflected by a prism than blue light, as a result of which more pronounced yellow fluorescence is induced by the blue, but never the other way round. Hence violet and ultraviolet radiation are most active in many cases of fluorescence [12].

In the previously mentioned publication by Stokes, apart from introduction of the term "fluorescence," the concept of fluorescence being emission of light was proposed, being the first to clearly define fluorescence as a process of emission. He worked out the technique for observing fluorescence using filters of various colors, one to allow the exciting light to impinge on the compound, and one to observe the emitted fluorescence, and he developed the physical statement that is actually known as the "Stokes law"; namely the wavelength of the emitted light is higher than that of the exciting light. It is worth mentioning that the Stokes law from 1852 is valid, for example, for the phosphorescence from the Bolonian stone as well as for the fluorescence of solutions of quinine sulfate, and that, for the first time, two types of (photo)luminescent phenomena were comprised, being phosphorescence and fluorescence, until then considered independent [15, 16].

Stokes observed that the fluorescent emissions from certain crystals were polarized, although he did not detect polarized fluorescence emerging from solutions [17].

In a later work, Stokes established the relationship between the intensity of fluorescence and the concentration, pointing out that the emission intensity depended on the concentration of the sample (analyte), but that attenuation of the signal occurred at higher concentrations as well as in the presence of foreign substances. He actually was the first to propose, in 1864, the application of fluorescence as an analytical tool, based on its sensitivity, on the occasion of a conference given previously in the Chemical Society and the Royal Institution, and entitled "On the Application of the Optical Properties to the Detection and Discrimination of Organic Substances" [5].

In 1867 Goppelsröder introduced the term "Fluoreszenzanalyse" (analysis by fluorescence or fluorimetry) and proposed the first fluorimetric analysis in history: the determination of Al(III) by the fluorescence of its chelate with morin [5]. In 1889 he proposed the capillary analytical technique using ultraviolet light. This technique, which was frequently used in paper and thin-layer chromatography, was again put into practice by Danckwortt and Pfau in 1928 [4].

By the end of the nineteenth century around 600 fluorescent compounds had been identified [3], including fluorescein (A. von Baeyer, 1871), eosine (H. Garo, 1874), and polycyclic aromatic hydrocarbons (C. Liebermann, 1880) [5]. Although it is generally accepted that fluorescence markers are relatively new analytical benefits, it is surprising to note that their chemical synthesis is rather old, such as the fluorescein reported by Baeyer, the 2,5-diphenyloxazole by Fisher in 1896, and the fluorene by Berthelot in 1867 [18].

In 1888, Walter studied the quenching of fluorescence, by the concentration effect, of fluorescein solutions. Nicols and Merrit observed in 1907, in solutions of eosine and resorufine, the symmetry existing between their absorption and fluorescence spectra. In 1910, Ley and Engelhardt determined the fluorescence quantum yield of various benzene derivatives, values that were still referred to until recent years [18]. The works by Lehmann and Wood, around 1910, marked the beginning of analysis based on fluorescence [4].

Edmond Becquerel (1820–1891) was the nineteenth-century scientist who studied the phosphorescence phenomenon most intensely. Continuing Stokes's research, he determined the excitation and emission spectra of diverse "phosphors," determined the influence of temperature and other parameters, and measured the time between excitation and emission of phosphorescence and the duration time of this same phenomenon. For this purpose he constructed in 1858 the first "phosphoroscope," with which he was capable of measuring lifetimes as short as 10^{-4} s. It was known that lifetimes considerably varied from one compound to the other, and he demonstrated in this sense that the phosphorescence of Iceland spar stayed visible for some seconds after irradiation, while that of the potassium platinum cyanide ended after 3.10^{-4} s. In 1861 Becquerel established an exponential law for the decay of phosphorescence, and postulated two different types of decay kinetics, i.e., exponential and hyperbolic, attributing them to monomolecular or bimolecular decay mechanisms. Becquerel criticized the use of the term fluorescence, a term introduced by Stokes, instead of employing the term phosphorescence, already assigned for this use [17, 19, 20]. His son, Henri Becquerel (1852–1908), is assigned a special position in history because of his accidental discovery of radioactivity in 1896, when studying the luminescence of some uranium salts [17].

The term "luminescence" (Latin "lucifer," meaning "light carrier," cf. emission of cold light) was introduced in 1888 by Eilhardt Wiedemann to distinguish between the light proceeding from the thermal excitation of substances, and the emission of light by molecules excited by other means without increasing their kinetic energy. He stated that the phosphorescence and fluorescence phenomena indicate that compounds are capable of emitting light without necessarily having been heated up. In other words, Wiedemann characterized luminescence by the fact that this "cold light" does not obey Kirchoff's laws of thermal absorption and emission by black bodies. He observed phosphorescence from colored aniline derivatives in "solid" solutions and gelatines. He mentioned that double salts of platinum emitted polarized light when excited by cathode rays, and specified that the luminescence could be initiated by various types of excitation, proposing that along with these modes of excitation, six different classes of luminescence are to be considered. Based on the mechanisms of excitation, which are much better understood nowadays, the present-day classification of luminescence phenomena is essentially the same as proposed by Wiedemann [5, 17].

In fact, an important advance in the phosphorescence theory was realized by Wiedemann in 1889, stating that a "phosphor" exists in two forms, a stable one, A, and an unstable one, B. Light absorption brings along conversion of form A to B, which then returns to A emitting light. This hypothesis was in agreement with the exponential decay law as postulated years before by Becquerel, but who did not provide any information about the nature of both forms [5].

In 1935, after studying the luminescence of various colorants, Jablonski suggested the "electronic energy diagram" of the singlet and triplet states to explain the luminescence processes of excitation and emission. The proposed diagram of molecular electronic energy levels formed the basis of the theoretical interpretation of all luminescent phenomena [21].

Spectroscopists interested in elucidation of the molecular energy schemes studied the phosphorescence emission of over 200 compounds, of which 90 were tabulated by Lewis and Kasha in 1944. They classified phosphorescing substances in two classes, based on the mechanism of phosphorescence production. The first group comprises minerals or crystals named "phosphors," where the individual molecule is not phosphorescent as such, but emits a shining associated with the presence of some impurity localized in the crystal. This type of phosphorescence cannot be attributed to a concrete substance. The second type of phosphorescence emission is attributed to a specific molecular species, being a pure substance in crystalline form, adsorbed on a suitable surface or dissolved in a specific rigid medium [22].

Lewis and Kasha identified phosphorescence as a forbidden transition from the triplet to the lowest singlet state, and suggested the phosphorescence emission spectrum as an analytical tool to assess molecular identification. Each phosphorescence phenomenon is unique with respect to frequency, lifetime, quantum yield, and vibrational pattern (band spacing). Moreover, as phosphorescence characteristics are a unique property of each organic species, they suggested application of this phenomenon for identification of mixtures [22]. Hence, the development of phosphorescence as analytical technique was showing up in the analytical investigations describing metastable phosphorescence or triplet-state emission together with a description of the paramagnetic nature of this luminescence phenomenon [22–25].

James Dewar observed in 1894 phosphorescence from frozen solutions utilizing liquid air [5]. Jean Becquerel discovers in 1907 that samples frozen at liquid air temperatures considerably narrow the spectral shape and increased information is obtained from the luminescence spectra [26].

Lewis et al. stated in 1941 that when a liquid is frozen the phosphorescence intensity increases along with an increase of viscosity [27]. Some of the works on phosphorescence were done, before 1940, in frozen aqueous solutions, and during the fifties for studying compounds of biochemical importance [28, 29]. However, it was demonstrated quite rapidly that it was more advantageous to use organic solvents to create transparent, rigid glassy matrices to measure phosphorescence in. In fact, the requirement of freezing conditions for phosphorescence measurements represented a major disadvantage in the use of this luminescence-based analytical technique. Later developments applied the technique of room temperature phosphorescence (RTP). As a matter of fact, Schmidt already observed in 1896 phosphorescence from colorants, adsorbed on solid gels. Without any doubt, historically this is the first observation of RTP [30].

It is worth noting some historical aspects in relation to the instrumentation for observing phosphorescence. Harvey describes in his book that pinhole and the prism setup from Newton were used by Zanotti (1748) and Dessaignes (1811) to study inorganic phosphors, and by Priestley (1767) for the observation of electroluminescence [3]. None of them were capable of obtaining a spectrum utilizing Newton's apparatus; that is, improved instrumentation was required for further spectroscopic developments. Of practical use for the observation of luminescence were the spectroscopes from Willaston (1802) and Frauenhofer (1814) [13].

Before the nineteenth century and during part of the latter, the observations of fluorescence occurred visually and in the course of this century photographic observation is being proposed. However, for measuring the intensity of fluorescence, until around 1930 methods based on visual comparison were used. Desha realizes in 1920 the first quantitative measurement of fluorescence, using a nephelometer with variable optical pathway (similar to the Duboscq colorimeter), and in a later work he states that at low concentrations, the fluorescence intensity increases linearly with the concentration [31]. Utilizing an instrument of the visual comparison type—a prism spectrometer—Bayle, Fabre, and George measure in 1925, employing a tedious procedure, the fluorescence emission spectra of a great number of drugs [5]. Gaviola mentions in 1927 the construction of the first phase fluorimeter, based on the phenomenon that the intensity of the exciting source radiation modulates sinusoidally. With this apparatus he measured the lifetime of rhodamine B (2 ns) and of fluorescein (4.5 ns), values that are still accepted [32].

The first photoelectric fluorimeter was described by Jette and West in 1928. The instrument, which used two photoemissive cells, was employed for studying the quantitative effects of electrolytes upon the fluorescence of a series of substances, including quinine sulfate [5]. In 1935, Cohen provides a review of the first photoelectric fluorimeters developed until then and describes his own apparatus using a very simple scheme. With the latter he obtained a typical analytical calibration curve, thus confirming the findings of Desha [33]. The sensitivity of these photoelectric instruments was limited, and as a result utilization of the photomultiplier tube, invented by Zworykin and Rajchman in 1939 [34], was an important step forward in the development of suitable and more sensitive fluorometers. The pulse fluorimeter, which can be used for direct measurements of fluorescence decay times and polarization, was developed around 1950, and was initiated by the commercialization of an adequate photomultiplier [35].

The first complete commercial spectrofluorimeter was manufactured by the American Instrument Company (Aminco), based on a design published by Bowman, Caulfield, and Udenfriend in 1955. The appearance of this commercial model instrument was of utmost importance for spectrofluorimetric investigations by numerous chemists, biologists, and biochemists [5].

In spite of the suggestions made by Lewis and Kasha in 1944, the analytical applications of phosphorescence appeared in the next decade only, when suitable phosphoroscopes were available, employing practical modifications of the Becquerel phosphoroscope, whose scheme can be found in the monograph by Bernard [20]. Keirs et al. created and utilized a resolution phosphoroscope, based on suggestions by their laboratory companion M. Kasha, which allowed them to distinguish the phosphorescence emission from the exciting light, and the simple resolution of mixtures of phosphorescent compounds based on their lifetimes. In 1957 they published the first work on quantitative phosphorescence, stressing two important aspects: obtaining phosphorimetric analytical curves of diverse organic molecules and the quantitative analysis of mixtures of two or three components by means of selective excitation, phosphoroscopic resolution, and simultaneous equations. They could phosphoroscopically and spectroscopically resolve mixtures of benzaldehyde, benzophenone, and 4-nitrobiphenyl; the phosphoroscopic resolution of mixtures of acetophenone and benzophenone; and the determination of mixtures of diphenylamine and triphenylamine by means of selective excitation. As solvent they used EPA (a mixture of ethyl ether, isopentane, and ethyl alcohol in a volume ratio of $5:5:2$) at liquid nitrogen temperatures, behaving as a clear and transparent glass [36].

In 1962, Parker and Hatchard described a photoelectric spectrometer for phosphorescence measurements with which they were capable of obtaining phosphorescence spectra, and of determining lifetimes and quantum efficiencies of a large number of organic compounds. This work stimulated intensely the interest in the phosphorimetry of diverse chemical analytes [5], and one year later, Winefordner and Latz proposed a phosphorimetric method for determination of aspirin in blood [37]. The development of a phosphorimetric method for aspirin in blood was the first application to a "real" sample and it contributed very much to the further acceptance of phosphorimetry [5]. Since then, phosphorimetry has been developing as a full analytical technique, which, when compared to fluorimetry, often is more sensitive for specific organic molecules and sometimes provides complementary information about the structure, reactivity, and surrounding requirements.

However, since the second half of the eighties, practically no more phosphorescence appears, at least in the analytical literature for quantitative estimations, being nearly completely substituted by sophisticated fluorescence and laser-induced fluorescence methods, mostly applied as detection tools in diverse flowing streams.

2. THE DISCOVERY OF BIOLUMINESCENCE AND CHEMILUMINESCENCE PHENOMENA

The significance of oxygen in bioluminescence (BL) was first established by Robert Boyle in 1669, who carried out experiments with shining wood, fish, and flesh and found that the emitted light was largely reduced and in some cases disappeared on removal of air [38, 39]. Although Boyle was not aware of the existence of oxygen, as it was discovered independently by Scheele and Priestley over 100 years later, this was the first experimental demonstration that oxygen, or one of its derivatives, is required in all known bioluminescent reactions and most artificial organic CL. By that period, it was not realized that living organisms were responsible for the shining of wood and flesh. In fact, proof that the glowing or shining of the latter was caused by a luminous fungus and luminous bacteria, respectively, was first reported by Heller in 1843. The requirements for oxygen in the bioluminescent reactions can be explained by the very high affinity for oxygen of some enzymes involved in this kind of luminescent processes.

Explanation of the mechanisms of BL systems begins in 1821, when Macaire suggested that the source of light in the glowworm might be some organic compound, rather than the inorganic phosphor, as previously assumed. In his studies, he observed that all chemical reagents that caused albumin to coagulate also extinguish the glowworm's light and concluded that the luminous material was composed mainly of albumin and required oxygen [40]. Following the study of living organisms that emit light, Pasteur reported in 1864 the spectrum of light from the tropical luminous beetle *Phyrophorus* as continuous, without a dark or light band [3]. In 1885, Dubois stimulated the interest in BL by carrying out a series of experiments using these luminous beetles and the luminous rock-boring bivalve *Pholas* [41–43]. He obtained a cold-water and hot-water extract form

Phyrophorus, which, when mixed together, reacted to produce light. He showed that luminescence was the result of a chemical reaction that requires a heat-stable factor, named *luciferine*, and a heat-labile factor, named *luciferase*. He was able to demonstrate that both compounds comprised an enzyme-substrate system, which required the presence of oxygen. From this period on, in which lumines-cence bacteria were used analytically for the first time by Beijerinck to detect small amounts of oxygen, BL reactions have been widely studied in this kind of organism, mainly by Harvey. He traveled around the world observing, collecting, and describing bioluminescent organisms and his publications still provide the most comprehensive description to date of the distribution of luminescence in nature [3, 44, 45].

3. STUDY OF THE CHEMILUMINESCENT SYSTEMS

In the mid-nineteenth century the chemiluminogenic capacity of simple organic compounds was discovered. By 1880, Radziszewski elaborated a long list includ-ing synthetic chemiluminescing organic compounds and compounds of biological origin, such as terpenes, cholates, and fatty acids and in the same year he was able to obtain the first CL spectrum of a synthetic organic compound.

In his paper dating from 1877, Radziszewski reported for the first time on the CL exhibited by the synthetic organic compound lophine (2,4,5-triphenylimi-dazole). He found that lophine emitted green light when it reacted with oxygen in the presence of strong base [46]. In the same year, Eder accidentally observed the luminescence of alkaline pyrogallol when it was employed as a developer for photographic plates [47] (Fig. 1).

The term "chemiluminescence" was not introduced until 1888, when Wie-demann defined the term "luminescence." He was able to classify luminescence phenomena of six different kinds, according to the manner of excitation: *photolu-minescence*, caused by the absorption of light, *electroluminescence*, produced in

(A)   (B)

Figure 1 Chemical structure of (A) lophine (2,4,5-triphenylimidazol) and (B) pyro-gallol.

gases by an electric discharge; *thermoluminescence*, produced by slight heating; *triboluminescence*, as a result of friction; *cristalloluminescence*, as a result of crystallization; and *chemiluminescence*, caused by a chemical reaction [48].

Many more CL reactions were discovered during the early twentieth century. In 1905, Trauzt described in an extensive report the known examples of CL and systematically reported the luminescent properties of the reactions from several hundred organic compounds with various oxidants. In this study, some of the earliest investigations of the spectral distribution of the emitted light were performed and the emission was attributed to some form of "activated" oxygen [49]. In the same period, Wedekind reported the first luminescent assay with a Grignard reagent. He described the brilliant green emission observed when an ether solution of phenylmagnesium bromide (Fig. 2) or iodide reacted with chloropicrin [50]. Following these studies, Hezcko reported that Grignard reagents emitted visible light in the presence of oxygen [51]. He carried out his luminescent experiment during a lecture demonstration in front of a large auditorium. Some years later, Dufford, Evans, and co-workers systematically investigated the CL properties of a large number of Grignard compounds to establish a relation between the intensity of light emission and the chemical structure [52–54]. Also, in 1927 light emitted during electrolysis was observed by Dufford et al. [55] for solutions of Grignard compounds in anhydrous ether.

In 1912 Delepine observed light generated in the gaseous phase from the vapors of some phosphorus-sulfur compounds in the presence of oxygen [56]. Two years later, Bancroft published a paper on the luminescence generated at mercury and other electrodes in the electrolysis of halides [57].

The chemistry of siloxenes and their light emission properties were studied by Kautsky et al. [58–60]. These complex silicon compounds are highly polymerized solids that have a permutoid structure, forming an isolated network that has a thickness the size of a molecule and where all reacting groups are quantitatively accessible for external agents (Fig. 3). These substances were first prepared more than a century ago by Wöhler, using calcium silicide and concentrated hydrochloric acid [61]. They emitted a bright light when a suspension of the siloxene in dilute acid was treated with strong oxidizing agents such as permanganate, ceric compounds, or nitric acid, the color and intensity of the emission varying strikingly with time.

Figure 2 A Grignard compound: *p*-chlorophenylmagnesium bromide.

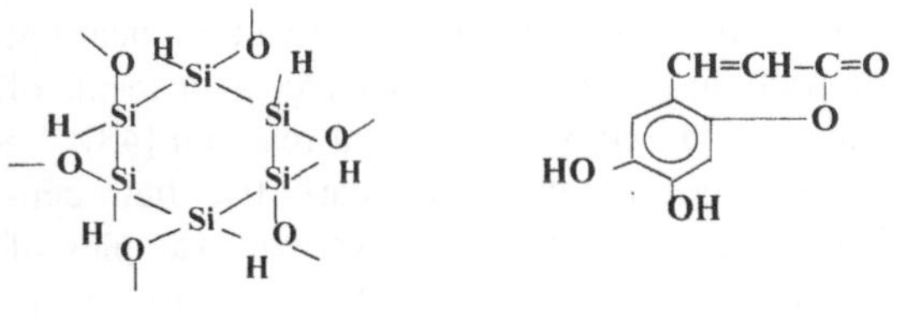

(A)　　　　　　　　　　　　　　　(B)

Figure 3　Chemical structure of (A) siloxene and (B) aesculin.

Mallet reported in 1927 that the intensity of light emitted in the reaction of hydrogen peroxide and the hypochlorite ion was enhanced when eosin, fluorescein, anthracene, quinine sulfate, or aesculin (Fig. 3) was added to the reaction medium [62].

Although the synthetic substance luminol was discovered in the mid-nineteenth century, it was not until 1928 that it was reported by Albrecht, who described the intense luminescence associated with the alkaline oxidation of luminol (5-amino-2,3-dihydro-1,4-phthalazinedione) and other *N,N*-diacylhydrazides [63] (Fig. 4). Soon after, Harvey [64] observed the light emitted during its electrolysis in alkaline solution at the anode. In 1934, the name luminol was given to this compound [65] and in 1936, confirmation of previously reported findings was made; it was reported that the reactions with hematin were the most intense [66]. The first proposal for the use of luminol in medicolegal investigations as a presumptive test for blood was reported by Specht in 1937 [67] and studied and confirmed in 1939 [68]. Most notable was the finding that dried, decomposed, and generally older bloodstains produced a much more brilliant and longer-lasting reaction with luminol than did fresh blood. Applying fresh luminol spraying—after allowing the previous applications to dry—can reactivate the luminescence; hematin can be detected in a dilution of $1:10^8$. Luminol is best employed to

(A)　　　　　　　　　　　　　(B)

Figure 4　Chemical structure of (A) luminol (5-amino-2,3- dihydro-1,4-phthalazinedione) and (B) isoluminol.

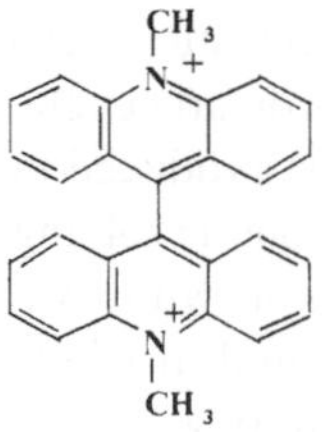

Figure 5 Chemical structure of lucigenin (10,10′- dimethyl-9,9′-biscridinium (nitrate)).

detect trace quantities of blood that are not visible to the naked eye, e.g., areas intentionally wiped clean of blood, washed clothes, dark surfaces, cracks and crevices, plumbing segments, and large areas to be screened, being used in criminal investigations [69]. Since then, several luminol derivatives have been synthesized, the largest CL quantum yield being shown from a benzoperylene derivative [70]. Also, isoluminol derivatives such as aminobutylethylisoluminol (ABEI) were synthesized in 1978 by Schroeder et al. [71] and subsequently widely applied in analytical chemistry.

One of the more efficient CL substances, lucigenin (10,10′-dimethyl-9,9′-biscridinium nitrate), was discovered by Gleu and Petsch in 1935 (Fig. 5). They observed an intense green emission when lucigenin was oxidized in an alkaline medium [72]. Other acridinium derivatives were shown to produce CL emission upon hydrogen peroxide oxidation of aqueous alkaline solutions. The main reaction product was N-methylacridone, acting as an active intermediate in the mechanism proposed by Rauhut et al. [73, 74] (Fig. 6).

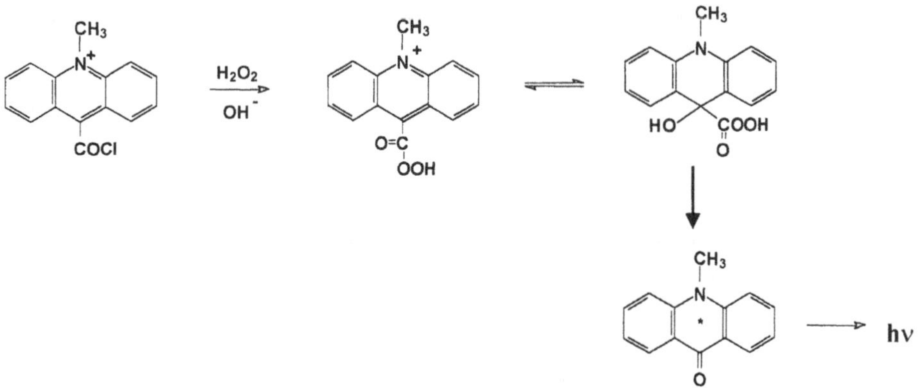

Figure 6 Chemiluminescent mechanism of acridinium salts.

Some of the investigations carried out in the first half of the twentieth century were related to CL associated with thermal decomposition of aromatic cyclic peroxides [75, 76] and the extremely low-level ultraviolet emission produced in different reaction systems such as neutralization and redox reactions involving oxidants (permanganate, halogens, and chromic acid in combination with oxalates, glucose, or bisulfite) [77]. In this period some papers appeared in which the bright luminescence emitted when alkali metals were exposed to oxygen was reported. The phenomenon was described for derivatives of zinc [78], boron [79], and sodium, potassium, and aluminum [80].

In 1950, Pruett et al. synthesized tetrakis(dimethylamino)ethylene, a clear, slightly yellow and mobile liquid. The authors observed a prolonged bright blue-green luminescence when the compound was exposed to oxygen or air in protic solvents [81]. The mechanism for this CL reaction, as proposed by Fletcher and Heller in 1967 [82], is shown in Figure 7.

In the sixties, with the development of instrumentation and the use of more sensitive photomultiplier tubes (PMT), the range of CL reactions studied was extended. Vasil'ev's group studied intensively the low emission produced in the autoxidation of a variety of hydrocarbons and noted that the addition of certain fluorescent molecules considerably enhanced the luminescence intensity [83, 84]. These mechanisms refer to the term "sensitized" CL. Chandross in 1963 [85] and McKeown and Waters in 1964 [86] observed the visible light emitted during the reaction of hydrogen peroxide with oxalyl chloride or certain nitriles. Hercules [87] and Visco and Chandross [88] independently reported the visible production of light generated in the vicinity of the cathode when a series of highly condensed aromatic hydrocarbons were electrolyzed in acetonitrile or dimethylformamide with tetraethylammonium salts employed as supporting electrolytes. This was the first time that the phenomenon of electrogenerated CL (ECL) was investigated in detail. Chandross and Sonntag [89] found a similar behavior when chemically produced aromatic radical anions were reacted with electron acceptors such as 9,10-dichloro-9,10-diphenyl-9,10-dihydroanthracene, benzoyl peroxide, oxalyl chloride, mercuric chloride, and aluminum chloride. Throughout the 1960s and 1970s there was much interest in these phenomena for studying new compounds, the different mechanisms, and the nature of the emitting state. In particular, polyaromatic hydrocarbons and their derivatives, ruthenium, osmium, platinum, palladium, and other transition metal complexes, and molybdenum and tungsten clusters have been studied in relation to their photochemical and electrochemical properties. However, only in the 1990s was application of ECL in analytical chemistry fully exploited (Chapter 9).

In 1961, Ashby reported the weak light emission produced from several polymers such as nylon, when heated [90]. The phenomenon was termed oxyluminescence because it was caused by oxidative processes and required the presence

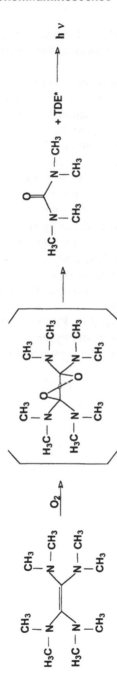

Figure 7 Chemiluminescent reaction of tetrakis(dimethylamino) ethylene (TDE).

of oxygen. This property was employed for determining the stability of a polymer toward oxidative degradation for estimating the ability of several compounds to act as antioxidants.

The CL behavior of 1,2-dioxetanes, bearing a four-membered ring, was studied by McCapra [91] in 1968. He showed that these compounds were easily converted to an excited product by heat and produced light due to compound cleavage to form two carbonyl compounds, one of them simultaneously being electronically excited and producing emission of light (Fig. 8A). This explanation supported the assumption that the CL reaction of lophine or indole includes a 1,2-dioxetane as an active intermediate. Kopecky et al. synthesized 3,3,4-tri-methyl-1,2-dioxetane for the first time, and observed that the CL emission was strongly enhanced by the addition of a fluorophore [92]. On the other hand, Rich-ardson et al. proposed another mechanism by which dioxetane is decomposed by heat to an excited carbonyl compound via a biradical form (Fig. 8B) [93, 94]. Because of the weak emission by these compounds, analytical applications were not so common although Hummelen et al. synthesized stable 1,2-dioxetanes as a label for thermoluminescence immunoassay of proteins [95, 96]. Some of these derivatives are shown in Figure 9.

Indole derivatives were studied by Philbrook et al. in 1965, showing the CL emission produced in the presence of oxygen and under strong alkaline condi-tions [97], following the reaction scheme depicted in Figure 10 for skatole (3-methylindole).

In 1965, Rauhut et al. [73] reviewed the oxalyl chloride CL system and showed that oxalyl esters could be used for this system instead of oxalyl chloride. Since then, they synthesized a number of oxalates including oxamides and estab-lished a new, potent luminescent system, namely the peroxyoxalate CL (PO CL) system. Much work has been carried out to synthesize suitable oxalic compounds. The first study dealing with different reagents was published in 1967 by Rauhut et al. [98] for the American Cyanamid Company with the purpose of developing

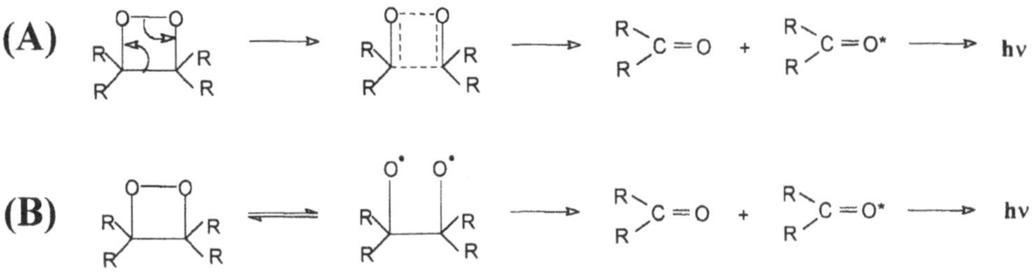

Figure 8 Chemiluminescent mechanism for 1,2-dioxetanes: (A) a concerted decomposi-tion process; (B) a two-step biradical process.

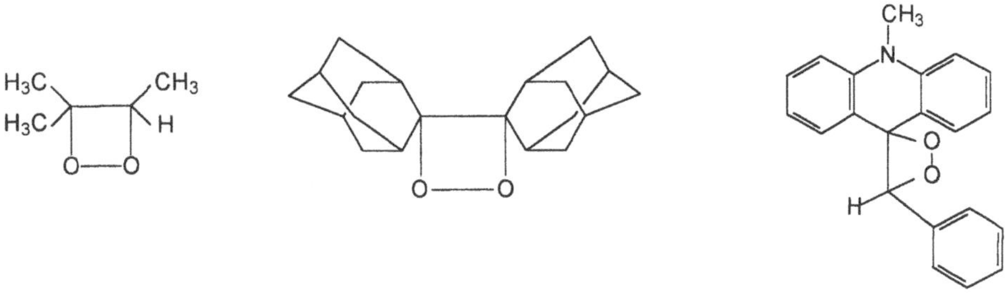

Figure 9   Some dioxetane derivatives.

reagents suitable for different kinds of emergency lights, where a high quantum yield in combination with a long duration of emission were considered optimal properties. They found that phenyl esters with strongly electron-withdrawing substituents were most efficient, whereas aryl oxalates with electron-donating or weak electron-withdrawing groups were capable of producing only low CL emission. The most efficient oxalate ester appeared to be *bis*(2,4-dinitrophenyl)oxalate (DNPO), being one of the most widely used esters together with *bis*(2,4,6-trichlorophenyl)oxalate (TCPO). Maulding et al. compared the reactivity of 19 oxamides in the peroxyoxalate reaction and found that the strongest intensity of all oxamides tested was obtained from 1,1′-oxalylbisbenzimidazole [99]. The PO CL system is subject to catalysis by weak bases (e.g., amines) and to inhibition by organic acids. Sherman et al. [100] used the TCPO system to determine hydrogen peroxide and aromatic hydrocarbon fluorescers in a static system; when metal

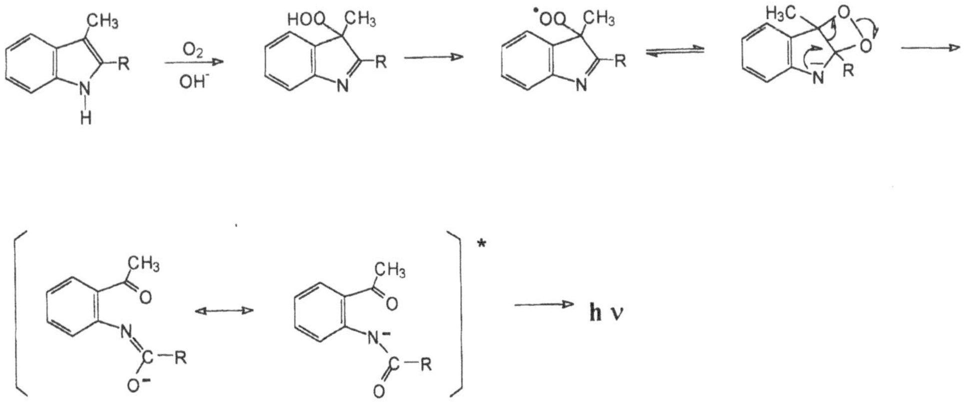

Figure 10   Chemiluminescent mechanism for 3-methylindole (skatole).

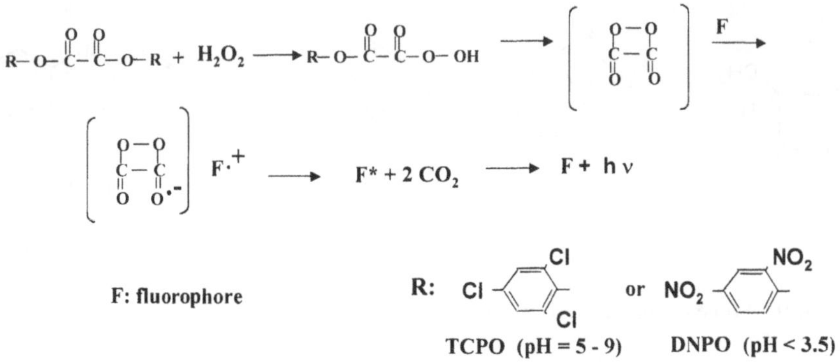

F: fluorophore

R: TCPO (pH = 5 - 9)   or   DNPO (pH < 3.5)

Figure 11 Proposed mechanism for the PO CL reaction.

chelates are employed as fluorescers, trace metals can be determined. In 1977, Curtis and Seitz [101, 102] applied the PO CL reaction to the detection of fluorescers separated by thin-layer chromatography (TLC). Dansyl derivatives separated by TLC could be detected by successive spraying with solutions of TCPO and hydrogen peroxide in dioxane. The suggested method, comparable to conventional fluorescence detection, had the advantages that it did not require excitation source and could be used to excite the plate uniformly. However, in our opinion, further TLC applications of analytical CL reactions were rarely published thereafter.

From the first papers produced by Rauhut et al. [98, 103] numerous mechanisms for these reactions have been reported to explain the suitability of peroxyoxalate reactions to easily excite oxidizable fluorescers down to 280 nm, although efficiency decreases markedly in the UV region. A widely accepted mechanism

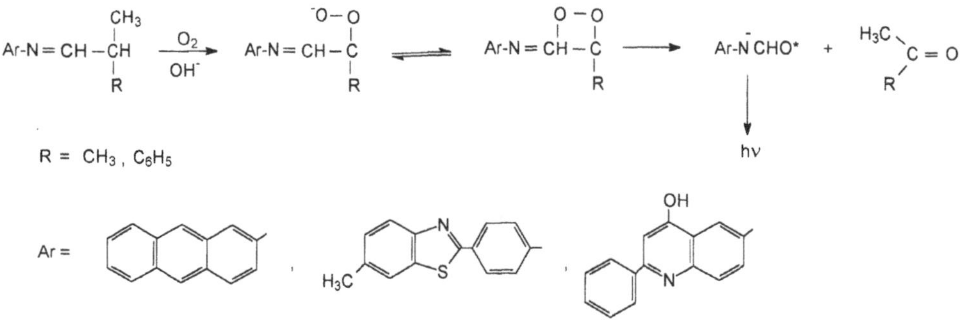

$R = CH_3, C_6H_5$

Ar =

Figure 12 Proposed chemiluminescent mechanism for Schiff bases.

is termed ''chemically initiated electron exchange luminescence'' (CIEEL) [104], which is based on the formation of an intermediate of the reaction of 1,2-dioxe-tanedione, which forms a charge-transfer complex with a fluorophore that donates one electron to the intermediate. This electron is transferred back to the fluoro-phore at a higher energy level, resulting in an excited fluorophore (Fig. 11). The energy content of the intermediates has been determined to be 105 kcal/mol^{-1}, which corresponds to an excitation wavelength around 280 nm [105]. The light emitted corresponds to the first singlet excited state of the fluorophore. Different ''key intermediates'' for this reaction have been proposed. For example, Catherall et al. proposed substituted 1,2-dioxetanedione as a key intermediate alternative to 1,2-dioxetane [106, 107]. This assumption was supported by the results of analysis by computer simulation by Givens's group, who studied the complex role of some base catalysts in this mechanism [108, 109]. The mechanism and application of the PO CL system are discussed in Chapter 7.

In 1976, McCapra and Burford [110] studied CL reactions of Schiff bases, proposing the mechanism shown in Figure 12. Although the efficiency of the CL reaction is high, a strong base is needed; furthermore, the reaction takes place

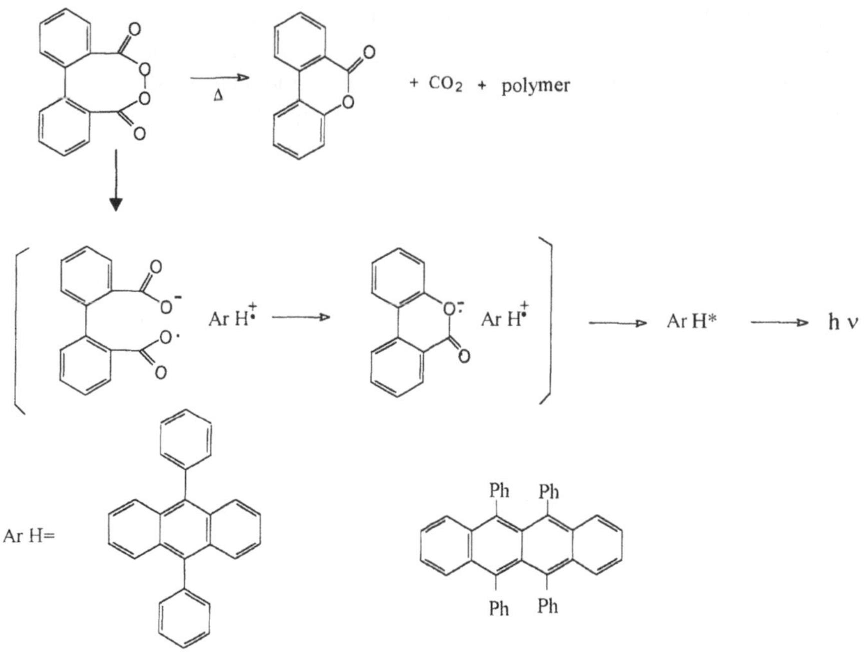

Figure 13 Chemiluminescent reaction of diphenoyl peroxide based on CIEEL mechanism.

only in aprotic or anhydrous solvents, which restricts analytical applications of this system.

In 1977, Koo and Schuster studied the CL emission produced when diphenoyl peroxide was decomposed at 24°C in dichloromethane in the dark producing benzocoumarin and polymeric peroxide [111, 112]. No CL emission was observed directly as benzocoumarin is nonfluorescent; however, in the presence of aromatic hydrocarbons light was produced because of the fluorescence of these hydrocarbons. The explanation of this phenomenon was based on the abovementioned CIEEL: the aromatic hydrocarbons, which have a low oxidation potential, transfer one electron to diphenoyl peroxide to form a charge-transfer complex, from which benzocoumarin and the corresponding hydrocarbon in the excited state are produced (Fig. 13).

4. THE FIRST ANALYTICAL USES OF BIOLUMINESCENCE AND CHEMILUMINESCENCE

Investigations of CL for analytical use began around 1970 [113]. In 1974 Isacsson and Wettermark presented an extensive review covering the general field of analytical methods based on the recording of CL [114]. The applications included gas-phase, solid-state, and liquid-phase analysis as well as special applications (identification of bloodstains in forensic chemistry, the analysis of microorganisms, and the CL of organic compounds induced by ozone). In another review article, Seitz and Neary reported in the same year on the advantages of CL and BL for chemical analysis, in relation to the extreme sensitivity and simple instrumentation. They described a small number of CL and BL analytical uses because of lack of available reactions [115].

Burdo and Seitz reported in 1975 the mechanism of the formation of a cobalt peroxide complex as the important intermediate leading to luminescence in the cobalt catalysis of the luminol CL reaction [116]. Delumyea and Hartkopf reported metal catalysis of the luminol reaction in chromatographic solvent systems in 1976 [117], while Yurow and Sass [118] reported on the structure-CL correlation for various organic compounds in the luminol-peroxide reaction.

Routine application of CL as an analytical tool dates from around the 1980s for the liquid phase. In 1978, Paul focused attention on recent advances in CL analysis in solution, stressing the high sensitivities that are possible and the use of rather inexpensive equipment [119].

4.1 Chemiluminescence in the Gas Phase

The development of CL methods for determining components of a gas largely originated from the need to determine atmospheric pollutants. In 1965, a hydrogen-rich flame photometric detector was developed, in which a strong emission was produced in the presence of volatile compounds of phosphorus or sulfur because of a reduction reaction in the flame [120, 121]. The detector was sensitive down to ppb levels of phosphorus compounds and less than ppm levels for sulfur compounds. In the case of sulfur dioxide and other sulfur compounds the emission was due to the electronically excited sulfur produced [122,123]:

$$SO_2 + 2H_2 \rightarrow S + 2H_2O$$
$$S + S \rightarrow S_2{}^*$$
$$S_2{}^* \rightarrow S_2 + h\nu \ (300\text{--}425 \ nm)$$

Also, a CL reaction between sulfur dioxide and oxygen atoms was studied by Mulcahy and Williams [124] and proposed for the analysis of sulfur dioxide with a sensitivity of 0.001 ppm. However, a disadvantage of this analysis lies in the difficulty for finding a stable source of oxygen atoms.

$$O^\bullet + O^\bullet + SO_2 \rightarrow O_2 + SO_2{}^*$$
$$SO_2{}^* \rightarrow SO_2 + h\nu \ (280 \ nm)$$

The oxidation of phosphorus by molecular oxygen occurs with vapor just above the solid to give a green emission. The mechanism of the reaction is not known [125] though the emitting species have been identified as $(PO)_2$ and HPO [126]. In 1965 Nederbragt et al. proposed a method for analysis of ozone based on a CL reaction in the presence of ethylene. Light intensity was proportional to the concentration of ozone and was emitted between 300 and 600 nm [127]. Using this method, monitors for field use as well as ozone detectors were constructed [128].

Some methods were proposed for the determination of nitrogen oxides based on the reaction with ozone [128, 129]:

$$NO + O_3 \rightarrow NO_2{}^* + O_2$$
$$NO_2{}^* \rightarrow NO_2 + h\nu \ (600\text{--}2800 \ nm),$$

with atomic oxygen [128]:

$$NO + O^\bullet \rightarrow NO_2{}^* \rightarrow NO_2 + h\nu \ (450\text{--}1800 \ nm)$$
$$NO_2 + O^\bullet \rightarrow NO + O_2$$

or with hydrogen atoms [130]:

$$NO_2 + H^\bullet \rightarrow NO + {}^\bullet OH$$
$$H^\bullet + NO + M \rightarrow HNO^* + M$$
$$HNO^* \rightarrow HNO + h\nu \ (650\text{--}760 \ nm)$$

Applying these processes, several commercial instruments were developed for pollutant monitoring with sensitivities at the ppb level.

Bowman and Alexander described the CL emission induced by ozone from a variety of organic compounds when absorbed on a silica gel surface and dissolved in an organic solvent [131]. The intensity of emission was proportional to the quantity of analyte in the ng range, offering a sensitivity comparable to fluorescence-based methods. In this sense, Regener proposed a very sensitive and selective method for the determination of ozone based on the CL induced when rhodamine B, adsorbed on silica gel, was exposed to ozone [132]. The system was improved by Hodgeson et al. eliminating the humidity by treating the gel surface with a hydrophobic agent [133]. This procedure has been used for the study of vertical distribution of ozone in the atmosphere using a rocket-borne probe.

The characteristics of this kind of CL emission, design of reactors, CL reactions in gas phase, and applications as detection technique in gas chromatography (GC) and atmospheric research are extensively described in Chapter 13.

4.2 Chemiluminescent Systems as Indicators in Titrations

For analysis in solutions, the most frequently used CL reaction is alkaline oxidation of luminol and lucigenin in the presence of hydrogen peroxide as oxidant, although sodium hypochlorite, sodium perborate, or potassium ferricyanide may also be used. CL reactions involving alkaline oxidation have been used to indicate acid-base, precipitation, redox, or complexometric titration endpoints either by the appearance or the quenching of CL when an excess of titrant is present [114, 134]. An example of these mechanisms is shown in Figure 14.

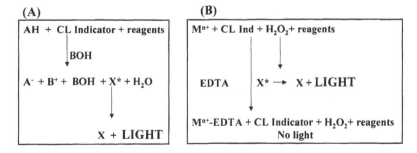

Figure 14 Some examples of endpoint determination in titrations using chemiluminescent indicators. (A) Acid-base titration: the endpoint is detected by the emission of light; (B) complexometric titration: the endpoint is detected by disappearance of light. M, metal acting as a catalyst; X^*, excited state from the CL precursor acting as indicator.

The systems luminol-H_2O_2-catalyst [135, 136], lucigenin-H_2O_2 [137, 138], lophine-H_2O_2-catalyst [137, 139], and pyrogallol [140, 141] were used as acid-base indicators based on the fact that these substances emit light only in an alkaline medium, allowing the titration of bases as well as acids when including hydrogen peroxide in the indicator system. Lucigenin can be considered the most satisfying indicator because no catalyst is required. These systems find useful application in the determination of acidity of dark-colored or turbid solutions such as red wine, fruit juices, milk, mustard, etc. and in the determination of the acid and saponification numbers of fats and oils.

Luminol, lucigenin [142–145], and siloxene were found suitable for redox titrations. In these systems, an oxidizing agent is used as titrant, and the titration is carried out until light emission begins. In some cases a reducing agent is used as titrant, the titration being carried out until light emission decreases. The use of luminol as redox indicator involves the application of hypochlorite or hypobromite as titrant in alkaline medium. The endpoint is observed when oxidization of luminol starts and light emission appears. In the case of lucigenin, the CL emission is produced when this reagent is oxidized by hydrogen peroxide in alkaline solution, which allows its use as indicator. Titrations with hydrogen peroxide can be carried out, and light is produced only when hydrogen peroxide is in excess. Siloxene can also be used as redox indicator in the direct titration of several inorganic species employing different oxidizing agents such as MnO_4^-, Ce(IV), $Cr_2O_7^{2-}$, and VO_3^- [142, 146–149]. Kenny and Kurtz found that a potential of about $+1.17$ V had to be reached before the indicator emits sufficient light to be detected [147]. The reaction is instantaneous in the presence of a small excess of oxidant and no catalyst is required. Indirect methods using siloxene as indicator are based on the reduction of I_2 or IO_3^- with zinc to iodide, on the precipitation of Ag(I) as AgI by adding an excess of iodine, which is then titrated in the presence of the precipitate, or on the addition of an excess of Fe(II) for the analysis of VO_3^-. Some applications of these systems are shown in Table 1.

Precipitation titrations were developed using lucigenin in the presence of hydrogen peroxide as an adsorption indicator for the argentometric determination of I^- in the presence of Cl^- and Br^- [150]. The indicator, as a positive species, was absorbed onto the precipitated silver iodide, which is negatively charged with I^- and the luminescence disappears. At the endpoint, I^- is desorbed from silver iodine and the solution emits light. Siloxene was also employed for the determination of Pb(II) by formation of a precipitate with potassium chromate solution [151]. The endpoint was observed by the emission of siloxene when a small excess of oxidant is present. The method was also used for indirect determination of sulfate by precipitating lead sulfate and titrating the excess of Pb(II) [152] and for determination of Cd(II) by precipitation with hexacyanoferrate(III) [153]. Using this methodology, potassium was determined by precipitation with a known quantity of a standard sodium tetraphenylborate solution, of which the

Table 1   Chemiluminescent Systems in Redox Titrations

Indicator	Species	Titrating reagent
Luminol	OCl^-, OBr^-	As_2O_3
	As(III), Sb(III), $S_2O_3^{2-}$, SO_3^{2-}, S^{2-}, CN^-, SCN^-	NaOBr
	As(III), Sb(III), $S_2O_3^{2-}$	NaOCl
	As(III), SO_3^{2-}	I_2
Lucigenin	$Fe(CN)_6^{4-}$, As(III), OCl^-, OBr^-, Cr(III)	H_2O_2
	OCl^-, OBr^-	$N_2H_4 \cdot H_2SO_4$
Siloxene	I^-, Fe(II), Sn(II), Mo(III), Ti(III), As(III), Pb(II),	MnO_4^-
	$C_2O_4^{2-}$, H_2O_2, $I_2^{(a)}$ $IO_3^{-(a,b)}$, $Ag^{+(c)}$, $VO_3^{-(d)}$	
	Fe(II), As(III), I^-, $C_2O_4^{2-}$, H_2O_2	Ce(IV)
	Fe(II), Tl(I)	$Cr_2O_7^{2-}$
	Fe(II)	VO_3^-

[a] After reduction with Zn to iodide.
[b] After addition of an excess of iodide solution of known concentration.
[c] AgI is precipitated by adding an excess of iodide, which is then titrated in the presence of the precipitate.
[d] After adding Fe(II) in excess.

excess was back-titrated with a standard solution of lucigenin [154]. Luminol in the presence of H_2O_2 and Co(II) as a catalyst was used for analysis of aluminum in colored samples by titrating 8-hydroxyquinoline [155]. The endpoint is detected by the decrease of CL emission due to hydroxyquinoline removing Co(II) from its association with luminol.

In complexometric titrations with EDTA, luminol and lucigenin have been employed as indicators based on the fact that complexing the ions of a catalytic heavy metal can cause changes in the emission from these reactions in alkaline solution and in the presence of oxidizing agents. In the case of luminol, if the metals are titrated with EDTA, the endpoint is observed by the disappearance of the light originating from luminol. On the contrary, using lucigenin the CL emission indicates the end of the process. Ca(II), Sr(II), and Ba(II) were determined by direct titration using luminol [115] and Cu(II) [115], Cd(II), Zn(II), and Ni(II) using luminol or lucigenin [156]. Employing the indirect method, Pb(II) and Hg(II) were determined by adding EDTA in excess and back-titration with Cu(II) solution [150].

4.3   Application to Firefly Luciferase Reaction in the Analysis of ATP

Although already in 1947 McElroy discovered that adenosine triphosphate (ATP) and magnesium were essential for the BL reactions by which fireflies produce light [157], the reaction mechanisms and kinetics were only reported much later [158]. Because of the importance of ATP in these mechanisms, the BL system was used as standard procedure for an extremely sensitive and selective determination of the cited molecule. The detection limit for the firefly assay method is about 10^{-11} mol/L ATP, allowing the detection of single cells in certain cases [159]. Since ATP is present in all organisms with a fairly constant intercellular concentration, the ATP concentration is a useful index of total biomass and has been applied as such to monitor bacteria in food, biomass in wastewater, and sewage sludge [160]. The high sensitivity of this assay allows the use of the firefly assay as a tool for detection of life on other planets. Table 2 shows some applications of the analysis of ATP.

4.4 The Application of Chemiluminescence in Immunoassay

Since the development of radioimmunoassay (RIA), many assays that rely on the specificity of the antigen-antibody binding reaction have been developed because of their inherent sensitivity and specificity. A typical competitive binding

Table 2 Some Applications in the Analysis of ATP

Direct determinations	
Applications	Field
Monitoring of fermentation rates	Process control in food, beverage, and drug industries
Detection of bacterial contamination	Quality control in food, beverage, drug, and cosmetic industries
Measuring biomass	Treatments of wastewaters, oceanography studies
Detecting the presence of life	Extraterrestrial investigations
Control of cell viability	Medical
Measuring infections caused by bacteria	Medical
Indirect determinations	
Species analyzed	Importance
Creatinine phosphate (CP + ADP $\rightarrow$ creatinine + ATP)	Energy reservoir for muscle activity
AMP	Mediator of hormone activity
Dissolved O_2	Measurement of water quality Medical
Inorganic pyrophosphate	Starting material for several biological compounds

assay for the determination of an antigen (Ag) can be represented by the following scheme, where unlabeled and labeled antigen (Ag*) compete for sites on the antibody (Ab):

$$Ag + Ag* + Ab \rightarrow AgAb + Ag*Ab$$

After equilibration, the amount of bound and free-labeled antigen can be measured, and a calibration curve can be used to determine the analyte. Radioactive labels have been extensively used because of the sensitivity of the measurement; however, they have several disadvantages, such as the waste disposal problem and the unstable nature of reagents. CL tags were therefore considered attractive alternatives due to their low (excellent) detectability, which was not fully provided by most fluorescent labels.

In 1976, Schroeder et al. developed a CL immunoassay as alternative to RIA in the analysis of very low levels of biological substances [161]. They used isoluminol derivatives as fluorescent labels and specific protein bindings were monitored allowing the assay of biotin and thyroxin. Campbell and McCapra's groups also studied the potential of acridinium esters as labels [162], being successfully used to achieve detection limits in the femtomole ranges. The phenyl carboxylate ester group was attached to the amine group on the antigen through a N-succinimide funtional group on the ester [163].

The PO CL detection system was also applied in immunoassay to measure the CL of fluorescent labels, but solvent restrictions significantly affected the precision. Arakawa et al. [164], in 1982, developed a sensitive, selective CL enzyme immunoassay of 17 α-hydroprogesterone using the glucose oxidase and TCPO-8-anilinonaphthalene-1-sulfonic acid system. Grayeski and Seitz used PO CL of rhodamine B isothiocyanate-labeled analytes for batch analysis and also as an alternative to radioisotope immunoassays [165]. Chapter 18 discusses practical aspects of enzyme immunoassay, the mechanism of enhanced CL using acridan esters as chemoluminogenic reagents, and its synthesis and recent applications.

4.5 Chemiluminescence in Flow Analysis and Immobilization Techniques

The flow-cell design was introduced by Stieg and Nieman [166] in 1978 for analytical uses of CL. Burguera and Townshend [167] used the CL emission produced by the oxidation of alkylamines by benzoyl peroxide to determine aliphatic secondary and tertiary amines in chloroform or acetone. They tested various coiled flow cells for monitoring the CL emission produced by the cobalt-catalyzed oxidation of luminol by hydrogen peroxide and the fluorescein-sensitized oxidation of sulfide by sodium hypochlorite [168]. Rule and Seitz [169] reported one of the first applications of flow injection analysis (FIA) in the CL detection of peroxide with luminol in the presence of a copper ion catalyst. They envisaged routine applications in clinical analysis via coupling reactions that generate hydrogen peroxide. Chapter 12 reports the basic principles of FIA, instrumentation, recent FIA versions, and an extensive overview of the last analytical applications, mainly in drug analysis.

Immobilization techniques based on chemical or physical processes were introduced from the seventies onward to prepare immobilized luminescent reagents with specific advantages such as reusability, improved stability, and increased efficiency. Examples of CL reagents are luminol using a solid-phase indicator for hydrogen peroxide in gas mixtures utilizing paper impregnated with luminol [170], luminol-impregnated membranes coupled with peroxidase-glucose oxidase–impregnated paper pads and the instant photographic detection for the analysis of glucose [171], or isoluminol immobilized on sepharose via a disulfide linkage for the analysis of serum cholinesterase [172]. This enzyme produces a thiol by enzymatic cleavage of an acetylcholine thioester; the thiol reacts with immobilized isoluminol and a thiol-disulfide interchange releases soluble isoluminol, which is determined chemiluminescently. Luminol and 9-carboxy-m-chlorophenolate acridane derivatives of sepharose have also been prepared and used in thiol-disulfide exchange reactions for the assay of picomole amounts of thiol [173].

Chemical immobilization procedures of bioluminescent enzymes such as firefly luciferase and bacterial luciferase-NAD(P)H:FMN oxidoreductase to glass beads or rods [174, 175], sepharose particles [176], and cellophane films [177] have produced active immobilized enzymes. Picomole-femtomole amounts of ATP or NAD(P)H could be detected using immobilized firefly luciferase or bacterial luciferase-oxidoreductase, respectively.

4.6 First Uses of Chemiluminescence as a Detection Technique in HPLC

Promising perspectives introduced the first applications of CL reactions as detection mode in liquid chromatography (LC), combining the high efficiency in separation and the low detection limits inherent to several CL systems. The luminol CL system was applied as postcolumn detection mode in HPLC by Kawasaki et al. [178]. They used labeled amines and carboxylic acids with an isoluminol derivative prior to separation by HPLC; postcolumn reaction with hydrogen peroxide and potassium ferricyanide enables the application of CL detection. The postcolumn CL reaction of lucigenin with glucose [179], heparin [180], cortisol, and carboxylic acid p-nitrophenacyl esters [181] allowed sensitive detection of these analytes.

Outstanding research—among others—on the use of aryl oxalates for CL detection in high-performance liquid chromatography (HPLC) was introduced in the 1980s by Imai's group from Tokyo University. They determined fluorescers or fluorophore-labeled compounds by HPLC and CL detection [182], using TCPO in ethyl acetate, and hydrogen peroxide in acetone, mixed with the column eluate using separate pumps and a suitable mixing vessel. Dansyl amino acids (alanine, glutamic acid, methionine, and norleucine) were separated on a μ-Bondapak C18 column and fmol limits could be reached. The experiments in static systems demonstrated that the CL intensity decays apparently according to a first-order rate equation; so the measurements have to be carried out fast. Also, the authors state that care should be taken during CL detection in TLC experiments. In flow systems, on the other hand, reproducible peak heights that are proportional to the concentration of the fluorescers are obtained since the detector, which is a fluorescence spectrophotometer with the light source switched off, measures the CL intensity during a fixed stage of the reaction. An analogous TCPO–hydrogen peroxide system was used for determination of fluorescamine-labeled catecholamines in urine [183], providing a 25-fmol detection limit, which is about 20 times lower than that using conventional fluorescence detection. Several aryl oxalates were evaluated for the hydrogen peroxide–induced CL detection of various fluorescers for use in HPLC postcolumn reactors [184]. Different oxalate esters were synthesized for use in the PO-CL reaction and their characteristics were studied in relation to the introduction of alkoxy side chains in the phenyl ring and the solubility effects [185, 186].

The same group reported in 1986 a sensitive and selective HPLC method employing CL detection utilizing immobilized enzymes for simultaneous determination of acetylcholine and choline [187]. Both compounds were separated on a reversed-phase column, passed through an immobilized enzyme column (acetylcholine esterase and choline oxidase), and converted to hydrogen peroxide, which was subsequently detected by the PO-CL reaction. In this period, other advances in this area were carried out such as the combination of solid-state PO CL detection and postcolumn chemical reaction systems in LC [188] or the development of a new low-dispersion system for narrow-bore LC [189].

Features of CL detection in HPLC, CL reactions used, construction of devices, optimization of experimental conditions, and recent applications are discussed in Chapter 14.

4.7 Application of Chemiluminescence in DNA Analysis

In 1987, CL started to be applied in DNA hybridization assays as an alternative to the use of radioactive tags. These assays are based on the specificity of a binding process: that of DNA strands for each other. An unknown DNA can be identified with the Southern blot method in which the strands of the analyte are separated and allowed to interact with labeled probe DNA strands on nitrocellulose filter paper. If the label on the probe is detected, the DNA can be identified and, in some cases, quantitated. Conventionally, radioactive tags were used because of the sensitivity required for quantitation. However, horseradish peroxidase (HRP) can be used as a DNA probe to be detected with the luminol CL reaction, using photographic detection. This alternative avoids the disadvantages from the use of radioactive material and the long time required by these processes; the endpoint is reached rapidly and levels as low as 1 pg of DNA can be easily detected. The CL assay was commercialized in 1985 [190].

In Chapter 19, basic aspects and recent trends in CL-based DNA analysis are discussed.

4.8 Chemiluminescence Sensors

In 1981 Freeman and Seitz reported on the first optical sensors to detect oxygen in the gas and liquid phases using the tetrakis(dimethylamino)ethylene oxidation reaction [191]. The reagent was held behind a Teflon poly(tetrafluoroethylene)-membrane through which oxygen diffused while light was transmitted to the photomultiplier detector. Detection limits and responses were similar to those of the oxygen electrode but the interferences from several species were removed because they do not affect the CL reaction. Aizawa et al. developed a sensor for analysis of hydrogen peroxide with HRP immobilized on a photodiode, which was immersed in an alkaline solution containing luminol and the analyte [192]. Also, glucose was determined with a bienzyme sensor containing both HRP and glucose oxidase.

Chapter 20 includes an extensive review of the CL sensors for determination of analytes in air, vapors, or liquids using different immobilization modes, its application to different analytes, based on enzyme or nonenzyme reactions, as well as CL immunosensors and DNA sensors.

4.9 Chemiluminescence Produced by Direct Oxidation

Until the end of the 1980s, CL analytical determinations involved interaction of the analyte with well-established CL reactions, often as a catalyst or as a sensitizer, or after a sequence of other reactions. However, Townshend's group discovered new CL reactions by testing the analyte with a wide range of oxidants (and reductants) under a large range of reaction conditions [193]. The typical tested oxidants were H_2O_2, ClO^-, Ce(IV), IO_4^-, MnO_4^- (in acid and alkaline medium), and Br_2, allowing the possibility of adding catalysts and/or carrying out the reactions at different pH values. There are some guidelines for predicting which analytes are likely to generate CL. For example, if oxidation of the molecule is known to give a fluorescent product, or if the analyte itself has a structure that might lead to fluorescence behavior, there is a possibility that oxidation of the analyte will give CL. Some of the first applications were the CL determination of morphine [194], buprenorphine hydrochloride [195], and the benzodiazepine

Table 3   CL Determinations Based on Direct Oxidation

Analyte	Oxidant system	Detection limit (M)	Ref.
Sulfide	ClO^-	1×10^{-7}	197
	H_2O_2-peroxidase	1.5×10^{-9}	198
Adrenaline	H_2O_2-OH^-	6×10^{-7}	199
Humic acids	MnO_4^--OH^-	0.7 mg/l	200
Paracetamol	Ce(IV)	4×10^{-7}	201
Quinones	H_2O_2-OH^-	1×10^{-5}	202
Hydrazine	ClO^-	ca. 10^{-8}	203, 204
Morphine	MnO_4^--H^+	1×10^{-10}	194
Naphthols	MnO_4^--H^+	5×10^{-7a}	205
Tetracycline	Br_2-OH^-	4×10^{-5}	206
	Dibromodimethyl hydantoin	1×10^{-6}	207
Sulphenyl esters	Iodosobenzene	5×10^{-4}	208
Loprazolam	MnO_4^--H^+	7×10^{-6}	209
Streptomycin	MnO_4^--H^+	ca. 10^{-5}	210
Sulfite	MnO_4^--H^+	1×10^{-6a}	211
	MnO_4^--H^+	1.5×10^{-8b}	212

[a] With sensitizer.
[b] With 3-cyclohexylaminopropanesulfonic acid.

loprazolam [196]. In these cases, the sample containing the drug was injected into a phosphoric acid stream that was subsequently merged with the oxidant stream, offering an extremely simple, economic, and effective experimental procedure. Since then, a wide range of such reactions have been reported mainly in the analysis of drugs [194, 197–212]. Table 3 shows some of the first direct CL determinations.

5. CONCLUSIONS

General books [213–217], chapters [218], and reviews were published in the 1980s reporting the suitability of CL and BL in chemical analysis [219–222], the specific analytical applications of BL [223], the CL detection systems in the gas phase [224], in chromatography [225, 226], the use of different chemiluminescent tags in immunoassay, and applications in clinical chemistry [227–232] as well as the applications of CL reactions in biomedical analysis [233].

In the 1990s, CL detection was coupled as a detection system in the by then recently introduced separation technique in routine analysis, capillary electrophoresis (CE). Hara's group reported the first application in 1991 [234], using a homemade CE-CL device for the determination of proteins using the PO-CL reaction. Since then, more investigations have been carried out to improve the detection devices and their coupling to the flowing stream as well as to optimize the different experimental factors influencing both detection and separation processes.

It is clear from the literature, and as stressed on various occasions, that CL measuring systems applied to detection in flowing streams and immunoassays are of growing importance, based on the advantages including low detection limits and the relatively simple instrumentation. Also, CL-based detection appears to be of utmost importance in the area of CE micromachining in which microchannel-based mixing of reagents and the creation of reaction chambers open new analytical perspectives. It should be emphasized as well that the fast evolution of immobilization techniques (e.g., employing beds of CL reagents or enzymes) improves the application of CL detection, among others, in separational streams and in the biosensing area.

REFERENCES

1. Aristotle. De anima, De sensu, Historia animalium, Meteorologica. Oxford; 1923.
2. R Bernard. La Luminescence. Paris: Presse Universitaire de France, 1961, p 5.
3. EN Harvey. A History of Luminescence from the Earliest Times until 1900. Philadelphia: American Philosophical Society, 1957.
4. M Pesez. Cronache Chim 28:12–27, 1970.
5. TC O'Haver. J Chem Educ 55:423–428, 1978.
6. MC Goldberg, ER Wedner. In: MC Goldberg, ed. Luminescence Application in Biological, Chemical, Environmental and Hydrological Sciences. Washington, DC: American Chemical Society, 1989, chapter 1, p 1.
7. R Bernard, La Luminescence. Paris: Presse Universitaire de France, 1961, p 6.
8. JA Babor, J Ibarz Aznárez. Química General Moderna. Barcelona: Manuel Marín y cia, 7th ed., 1962, p 635.
9. R Bernard. La Luminescence. Paris: Presse Universitaire de France, 1961, p 19.
10. JB Beccari. Phil Trans Roy Soc 44:81, 1746.
11. A Fernández-Gutierrez, A Muñoz de la Peña. In: SG Schulman, ed. Molecular Luminescence Spectrometry. Methods and Applications: Part I. Vol 77. New York: Wiley, 1985, p 372.
12. R Bernard. La Luminescence. Paris: Presse Universitaire de France, 1961, p 8.
13. MC Goldberg, ER Wedner. In: MC Goldberg, ed. Luminescence Application in Biological, Chemical, Environmental and Hydrological Sciences. Washington, DC: American Chemical Society, 1989, chapter 1, p 3.
14. R Bernard. La Luminescence. Paris: Presse Universitaire de France, 1961, p 7.
15. GG Stokes. Phil Trans 142:463, 1852.
16. R Bernard. La Luminescence. Paris: Presse Universitaire de France, 1961, p 9.
17. MC Goldberg, ER Wedner. In: MC Goldberg, ed. Luminescence Application in

Biological, Chemical, Environmental and Hydrological Sciences. Washington, DC: American Chemical Society, 1989, chapter 1, p 4.

18. B del Castillo, MA Martin. In: O Vals, B del Castillo, eds. Técnicas Instrumentales en Farmacia y Ciencias de la Salud. Barcelona: Piros, 1998, p 191.

19. E Becquerel, La Lumière, ses Causes et ses Effets. Paris: Didot, 1867.

20. R Bernard. La Luminescence. Paris: Presse Universitaire de France, 1961, pp 9–11.

21. A Jablonski. Z Physik 94:38–46, 53–65, 1935.

22. GN Lewis, M Kasha. J Am Chem Soc 66:2100, 1944.

23. GN Lewis, M Kasha. J Am Chem Soc 67:1332, 1945.

24. M Kasha. Chem Rev 41:401, 1947.

25. DS McClure. J Chem Phys 17:905, 1949.

26. MC Goldberg, ER Wedner. In: MC Goldberg, ed. Luminescence Application in Biological, Chemical, Environmental and Hydrological Sciences. Washington, DC: American Chemical Society, 1989, chapter 1, p 6.

27. GN Lewis, D Lipkin, TT Magel. J Am Chem Soc 63:3005, 1941.

28. A Szent-Gyorgi. Science 124:124, 873.

29. RA Steel, A Szent-Gyorgi. Proc Natl Acad Sci USA 43:477, 1957.

30. GC Schmidt. Ann Phys (Leipzig) 58:103, 1896.

31. LJ Desha. J Am Chem Soc 42:1350, 1920.

32. E Gaviola. Z Phys 42:853–861, 1927.

33. FA Cohen. Rev Trav Chim 54:133, 1938.

34. FA Zworykin, JA Rajchman. Proc IRE 27:528, 1939.

35. MC Goldberg, ER Wedner. In: MC Goldberg, ed. Luminescence Application in Biological, Chemical, Environmental and Hydrological Sciences. Washington, DC: American Chemical Society, 1989, chapter 1, pp 8–9.

36. RJ Keirs, D Britt, WE Wentworth. Anal Chem 29:202–209, 1957.

37. DJ Winefordner, HW Latz. Anal Chem 35:1517, 1963.

38. R Boyle, Phil Trans Roy Soc 2:211–214, 1668.

39. R Boyle, Phil Trans Roy Soc 2:215–216, 1668.

40. J Macaire. Ann Chim Phys 17:251–267, 1821.

41. R Dubois. CR Scéanc Soc Biol 1:661–664, 1884.

42. R Dubois. CR Scéanc Soc Biol 2:559–562, 1885.

43. R Dubois. CR Scéanc Soc Biol 4:564–565, 1884.

44. EN Harvey. Living Light. Princeton: Princeton University Press, 1940.

45. EN Harvey. Bioluminescence. New York: Academic Press, 1952.

46. B Radziszewski. Berichte d Chemischen Gesellschaft 10:70–75, 1877.

47. JM Eder. Phot Mitt 24:74, 1877.

48. E Wiedemann. Ann Phys Chem 34:446–463, 1888.

49. M Trautz. Z Phys Chem 53:1–111, 1905.

50. E Wedekind. Chem Zentralbl 30:921, 1905.

51. A Hezcko. Chem Zentralbl 35:199, 1911.

52. WV Evans, RT Dufford. J Am Chem Soc 45:278, 1923.

53. RT Dufford, D Nightgale, S Calvert. J Am Chem Soc 45:2058, 1923.

54. WV Evans, EM Diepenhorst. J Am Chem Soc 48:715, 1926.

55. RT Dufford, D Nightingale, LW Gaddum. J Am Chem Soc 49:1858, 1927.

56. M Delepine. Comp Rend 154:1171, 1912.
57. J Bancroft. J Phys Chem 18:762, 1914.
58. H Kautsky, H Zochar. Z Phys 9:267–284, 1922.
59. H Kautsky. Trans Faraday Soc 35:216–219, 1922.
60. H Kautsky, H Thiele. Z Anorg Allgem Chem 144:197–217, 1925.
61. F Wöhler. Ann 127:263, 1863.
62. L Mallet. Comp Rend 185:352–354, 1927.
63. HO Albrecht. Z Phys Chem 136:321–330, 1928.
64. EN Harvey. J Phys Chem 33:1456, 1929.
65. E Huntress, L Stanley, A Parker. J Am Chem Soc 56:241–242, 1934.
66. K Gleu, K Pfannstiel. J Prak Chem 146:137, 1936.
67. W Specht. Angew Chem 50:155–157, 1937.
68. F Proescher, AM Moody. J Lab Clin Med 24:1183–1189, 1939.
69. BJ Culliford. The Examination and Typing of Bloodstains in the Crime Laboratory. Washington, DC: US Govt Printing Office, 1971.
70. CC Wei, EH White. Tetrahedron Lett 39:3559, 1971.
71. HR Schroeder, RC Boguslaski, RJ Carrico, RT Buckler. Methods Enzymol 57:424, 1978.
72. K Gleu and W Petsch. Angew Chem 48:57–59, 1935.
73. MM Rauhut, D Sheehan, RA Clarke, AM Semsel. Photochem Photobiol 4:1097–1110, 1965.
74. MM Rauhut, D Sheehan, RA Clarke, BG Roberts, AM Semsel. J Org Chem 30:3587, 1965.
75. C Dufrassie, L Velluz. Bull Soc Chim France 4:349, 1937.
76. C Dufrassie, L Velluz. Bull Soc Chim France 9:171, 1942.
77. R Audebert. Trans Faraday Soc 35:197, 1939.
78. CH Bamford, DW Newitt. J Chem Soc 688, 1946.
79. CH Bamford, DW Newitt. J Chem Soc 695, 1946.
80. M Dimbat, GA Harlow. Anal Chem 34:450, 1962.
81. RL Pruett, JT Barr, KE Rapp, CT Bahner, JD Gibson, RH Lafferty. J Am Chem Soc 72:3646, 1950.
82. AN Fletcher, CA Heller. J Phys Chem 71:1507, 1967.
83. VL Shaliapintokh, RF Vasil'ev, ON Karpukhine, ML Postnikov, LA Kibalko. J Chim Phys 57:1112, 1960.
84. RF Vasil'ev. Nature 196:668, 1962.
85. EA Chandross. Tetrahedron Lett 761–765, 1963.
86. E McKeown, WA Waters. Nature 203:1063, 1964.
87. DM Hercules. Science 145:808, 1964.
88. RE Visco, EA Chandross. J Am Chem Soc 86:5350, 1964.
89. EA Chandross, FI Sonntag. J Am Chem Soc 86:3179, 1964.
90. GE Ashby. Polym Sci 50:99, 1961.
91. F McCapra. J Chem Soc Chem Commun 155, 1968.
92. KR Kopecky, JH Van de Sande, C Mumford. Can J Chem 46:25, 1968.
93. WH Richardson, MB Yelvington, HE O'Neal. J Am Chem Soc 94:1619, 1972.
94. WH Richardson, HE O'Neal. J Am Chem Soc 94:8665, 1972.
95. JC Hummelen, TM Luider, H Wynberg. Methods Enzymol 133:531, 1986.

96.   JC Hummelen, TM Luider, H Wynberg. Pure Appl Chem 59:639, 1987.
97.   GF Philbrook, JB Qyers, JR Totter. Photochem Photobiol 4:869, 1965.
98.   MM Rauhut, LJ Bollyky, BG Robert, M Loy, RH Whitman, AV Iannotta, AM Semsel, RA Clarke. J Am Chem Soc 89:6515–6522, 1967.
99.   DR Maulding, RA Clarke, BG Robert, MM Rauhut. J Org Chem 33:250–254, 1968.
100. PA Sherman, J Holzbecher, DE Ryan. Anal Chim Acta 97:21–27, 1978.
101. TG Curtis, WR Seitz. J Chromatogr 134:343–350, 1977.
102. TG Curtis, WR Seitz. J Chromatogr 134:513–516, 1977.
103. MM Rauhut, BG Robert, AM Semsel. J Am Chem Soc 88:3604–3617, 1966.
104. GB Schuster. Acc Chem Res 12:366–373, 1979.
105. P Lechtken, NJ Turro. Mol Photochem 6:95–99, 1974.
106. CLR Catherall, TF Palmer, RB Cundall. J Chem Soc Faraday Trans 80:823–834, 1984.
107. CLR Catherall, TF Palmer, RB Cundall. J Chem Soc Faraday Trans 80:837–849, 1984.
108. FL Alvárez, JP Parekh, B Matusewski, RS Givens, T Higuchi, RL Schowen. J Am Chem Soc 108:6435–6437, 1986.
109. N Hanaoka, RS Givens, RL Schowen, T Kuwana. Anal Chem 60:2193–2197, 1988.
110. F McCapra, A Burford. J Chem Soc Chem Commun 607, 1976.
111. JY Koo, GB Schuster. J Am Chem Soc 99:6107, 1977.
112. JY Koo, GB Schuster. J Am Chem Soc 100:4496, 1978.
113. JH Hass. J Chem Ed 44:396–402, 1967.
114. U Isacsson, G Wettermark. Anal Chim Acta 68:339–362, 1974.
115. WR Seitz, MP Neary. Anal Chem 46:188A–202A, 1974.
116. TG Burdo and WR Seitz. Anal Chem 47:1639–1643, 1975.
117. R Delumyea, AV Hartkopf. Anal Chem 48:1402–1405, 1976.
118. HW Yurow, S Sass. Anal Chim Acta 88:389–394, 1977.
119. DB Paul. Talanta 25:377–382, 1978.
120. WL Crider. Anal Chem 37:1770, 1965.
121. SS Brody, JE Chaney. J Gas Chromatogr 4:42, 1966.
122. RK Stevens, AE O'Keeffe, GC Ortman. Environ Sci Technol 3:652, 1969.
123. RK Stevens, JD Mulik, AE O'Keeffe, KJ Krost. Anal Chem 43:827, 1971.
124. MFR Mulcahy, DJ Williams. Chem Phys Lett 7:455, 1970.
125. IM Campbell, DL Baulch. Gas Kin Energy Trans 3:42, 1978.
126. RL Vanzee, AV Khan. J Am Chem Soc 96:6805, 1974.
127. GW Nederbragt, A Van der Horst, J Van Durn. Nature 206:87, 1965.
128. RK Stevens, JA Hodgeson. Anal Chem 45:443A, 1973.
129. A Fontijn, AJ Sabadell, RJ Ronco. Anal Chem 42:575, 1970.
130. KJ Krost, JA Hodgeson, RK Stevens. Publication Preprint. Research Triangle Park, NC: Environmental Protection Agency, 1972.
131. RL Bowman, N Alexander. Science 154:1454, 1966.
132. VH Regener. J Geophys Res 69:3795, 1964.
133. JA Hodgeson, KJ Krost, AE O'Keeffe, RK Stevens. Anal Chem 42:1795, 1970.
134. A Fernández-Gutierrez, A Muñoz de la Peña. In: SG Schulman, ed. Molecular

Luminescence Spectrometry. Methods and Applications: Part I. Vol 77. New York: Wiley, 1985, pp 484–487.
135. F Kenny, RB Burtz. Anal Chem 23:339, 1951.
136. F Kenny, RB Burtz. Anal Chem 24:1218, 1952.
137. L Erdey. Indust Chem 33:459, 523, 575, 1957.
138. L Erdey. Acta Chim Acad Sci Hung 3:81, 1953.
139. L Erdey, I Buzas. Anal Chim Acta 15:322, 1956.
140. J Slawinski. Rocz Glebozn 18:191, 1967.
141. D Slawinska, S Slawinski. Zeszyty Nauk Wyz Szk Rol Szc No 10:163, 1963.
142. L Erdey. Magyar Kem Lapja 13:7, 1958.
143. L Erdey, I Buzas. Acta Chim Acad Sci Hung 6:77, 93, 115, 123, 1955.
144. L Erdey, I Buzas. Magyar Tudomayos Akad Kem Tudomanyok Osztalayyynak Kolemenyei 5:279, 293, 313, 321, 325, 1954.
145. U Fritsche, U Mihm, A Koenig. Mikrochim Acta II:85, 1980.
146. L Erdey, I Buzas, L Polos. Z Anal Chem 169:187, 1959.
147. F Kenny, RB Kurtz. Anal Chem 22:693, 1950.
148. I Buzas, L Erdey. Talanta 10:467, 1963.
149. FF Grigorenko, LI Dubovenko, GI Kovalenko. Zavod Lab 39:133, 1973.
150. L Erdey, I Buzas. Anal Chim Acta 22:524, 1960.
151. F Kenny, RB Kurtz. Anal Chem 25:1550, 1953.
152. F Kenny, RB Kurtz, I Beck, I Lukosevicius. Anal Chem 29:543, 1957.
153. F Kenny, RB Kurtz, AC Vandenoever, CJ Sanders, CA Novarro, LE Menzel, R Kukla, KM McKenny. Anal Chem 36:529, 1964.
154. I Sarudi. Z Anal Chem 260:114, 1972.
155. AK Babko, NM Lukovskaya. Ukr Khim Zh 35:1060, 1969.
156. L Erdey, O Weber, I Buzas. Talanta 17:1221, 1970.
157. WD McElroy. Proc Natl Acad Sci USA 33:342, 1947.
158. M De Luca, WD McElroy. In: M De Luca, ed. Methods in Enzymology. New York: Academic Press, 1978, p. 3.
159. G Wettermark, H Stymne, SE Brolin, B Peterson. Anal Biochem 63:293, 1975.
160. A Lundin. In: LJ Kricka, TJN Carter, ed. Clinical and Biochemical Luminescence. New York: Marcel Dekker, 1982, pp 43–74.
161. HR Schroeder, PO Vogelhut, RJ Carrico, RT Buckler. Anal Chem 48:1933–1937, 1976.
162. I Weeks, I Beheshti, F McCapra, AK Campbell, JS Woodhead. Clin Chem 29:1474–1479, 1983.
163. I Weeks, JS Woodhead. Clin Chim Acta 141:275–280, 1984.
164. H Arakawa, M Maeda, A Tsuji. Chem Pharm Bull 30:3036–3039, 1982.
165. ML Grayeski, WR Seitz. Anal Biochem 136:277–282, 1984.
166. S Stieg, TA Nieman. Anal Chem 50:401–404, 1978.
167. JL Burguera, A Townshend. Talanta 26:795–798, 1979.
168. JL Burguera, A Townshend. Anal Chim Acta 114:209–214, 1980.
169. G Rule, WR Seitz. Clin Chem 25:1635–1638, 1979.
170. I Kh Agranov, LV Reiman. Zh Anal Khim 34:1533, 1979; Chem Abstr 92:33267, 1980.
171. TJN Carter, TP Whitehead, LJ Kricka. Talanta 29:529, 1982.

172. RD Lippman. In: MA De Luca, WD McElroy, ed. Bioluminescence and Chemilu-
      minescence. New York: Academic Press, 1981, p. 633.
173. RD Lippman. Anal Chim Acta 116:181, 1980.
174. E Jablonski, M DeLuca. Proc Natl Acad Sci USA 73:3848, 1976.
175. E Jablonski, M DeLuca. Methods Enzymol 57:202, 1978.
176. J Ford, M DeLuca. Anal Biochem 110:43, 1981.
177. NN Ugarova, LY Brovko, EI Beliaieva. Enzyme Microb Technol 5:60, 1983.
178. T Kawasaki, M Maeda, A Tsuji. J Chromatogr 328:121–126, 1985.
179. RL Veazey, T Nieman. J Chromatogr 200:153–162, 1980.
180. RA Steen, T Nieman. Anal Chim Acta 155:123–129, 1983.
181. M Maeda, A Tsuji. J Chromatogr 352:213–220, 1986.
182. S Kobayashi, K Imai. Anal Chem 52:424–427, 1980.
183. S Kobayashi, J Sekino, K Honda, K Imai. Anal Biochem 112:99–104, 1981.
184. K Honda, K Miyaguchi, K Imai. Anal Chim Acta 177:103–110, 1985.
185. K Imai, H Nawa, M Tanaka, H Ogata. Analyst 111:209–211, 1986.
186. K Nakashima, K Maki, S Akiyama, WH Wang, Y Tsukamoto, K Imai. Analyst
      114:1413–1416, 1989.
187. K Honda, K Miyaguchi, H Nishino, H Tanaka, T Yao, K Imai. Anal Biochem 153:
      50–53, 1986.
188. JR Poulsen, JW Birks, P Van Zoonen, C Gooijer, NH Velthorst, RW Frei. Chro-
      matographia 21:587–595, 1986.
189. GJ De Jong, N Lammers, FJ Spruit, C Dewaele, M Verzele. Anal Chem 59:1458–
      1461, 1987.
190. JA Matthews, A Batki, C Hynds, LJ Kricka. Anal Biochem 151:205–209, 1985.
191. TM Freeman, WR Seitz. Anal Chem 53:98–102, 1981.
192. M Aizawa, Y Ikariyama, H Kuno. Anal Lett. 17:555–564, 1984.
193. RW Abbott, A Townshend. Anal Proc 23:25–26, 1986.
194. RW Abbott, A Townshend, R Gill. Analyst 111:635–640, 1986.
195. AA Alwarthan, A Townshend. Anal Chim Acta 185:329–353, 1986.
196. ARJ Andrews, A Townshend. Anal Chim Acta 26:368–369, 1989.
197. D Klockow, J Teckentrup. Talanta 23:889, 1976.
198. JL Burguera, A Townshend. Talanta 27:309, 1980.
199. T Nakagama, M Yamada, S Susuki. Anal Chim Acta 217:305, 1987.
200. DF Marino, JD Ingle Jr. Anal Chim Acta 124:23, 1981.
201. II Koukli, AC Calokerinos, TP Hadjiioannou. Analyst 114:711, 1989.
202. JL Burguera, M Burguera. Talanta 31:1027, 1984.
203. AR Wheatley, PhD thesis, University of Hull, 1983.
204. AT Faizullah, A Townshend. Anal Proc 22:15, 1985.
205. SA Al-Tamrah, A Townshend. Anal Chim Acta 202:247, 1987.
206. AA Alwarthan, A Townshend. Anal Chim Acta 205:261, 1988.
207. T Owa, T Masujima, H Yoshida, H Imai. Bunseki Kagaku 33:570, 1984.
208. JS Lancaster, PJ Worsfold. Anal Proc 26:19, 1989.
209. ARJ Andrews, A Townshend. Anal Chim Acta 227:65, 1989.
210. RW Abbott. PhD thesis, University of Hull, 1986.
211. M Yamada, T Nakada, S Suzuki. Anal Chim Acta 147:401, 1983.
212. SA Al-Tamrah, A Townshend, AR Wheatley. Analyst 112:883, 1987.

213. MA DeLuca. Bioluminescence and Chemiluminescence. Orlando: Academic Press, 1978.
214. JG Burr, ed. Chemi- and Bioluminescence. New York: Marcel Dekker, 1985.
215. K Van Dyke, ed. Bioluminescence and Chemiluminescence. Instruments and Applications. Cleveland: CRC Press, Vol I and II, 1985.
216. MA De Luca, WD McElroy, eds. Bioluminescence and Chemiluminescence. Part B. Orlando: Academic Press, 1986.
217. AK Campbell. Chemiluminescence. Principles and Applications in Biology and Medecin. Chichester: Ellis Horwood/VCH, 1988.
218. JD Ingle, SR Crouch. Spectrochemical Analysis. Englewood Cliffs, NJ: Prentice-Hall, 1988, pp 478–485.
219. JL Burguera, M Burguera, A Townshend. Acta Cient Venezolana 32:115–122, 1981.
220. LJ Kricka. GHG Thorpe. Analyst 108:1274–1296, 1983.
221. JL Bernal, V Cerdá. Quím Anal 4:205–235, 1983.
222. ML Grayeski. Anal Chem 59:1243 A-1256 A, 1987.
223. M DeLuca, LJ Kricka, WD McElroy. Trends Anal Chem 1:225–228, 1982.
224. A Fontijn, ed. Gas-Phase Chemiluminescence and Chemi-Ionization. New York: Elsevier, 1985.
225. K Imai. Methods Enzymol 133:435–449, 1986.
226. JW Birks, ed. Chemiluminescence and Photochemical Reaction Detection in Chromatography. New York: VCH, 1989.
227. HR Schroeder, RC Boguslaski, RJ Carrico, RT Buckler. Methods Enzymol 57: 424–445, 1978.
228. HR Schroeder. Trends Anal Chem 15:352–354, 1982.
229. W Klingler, CL Strasburger, WG Wood. Trends Anal Chem 6:132–136, 1983.
230. GJR Barnard, JB Kim, JL Williams, WP Collins. In: K Van Dycke, ed. Bioluminescence and Chemiluminescence Vol 1. West Palm Beach: CRC Press, 1985, pp. 151–183.
231. I Weeks, M Sturgess, RC Brown, JS Woodhead. Methods Enzymol 133:366–387, 1987.
232. A Tsuji, M Maeda, H Arakawa. In: K Van Dycke, ed. Bioluminescence and Chemiluminescence Vol 1. West Palm Beach: CRC Press, 1985, pp. 185–202.
233. WRG Baeyens, K Nakashima, K Imai, B Lin Ling, Y Tsukamoto. J Pharm Biomed Anal 7:407–412, 1989.
234. T Hara, S Okamura, J Kato, J Yokogi, R Kakajima. Anal Sci 7: 261–264, 1991.

2

Chemiluminescence-Based Analysis

An Introduction to Principles, Instrumentation, and Applications

Ana M. García-Campaña
University of Granada, Granada, Spain

Willy R. G. Baeyens
Ghent University, Ghent, Belgium

Xinrong Zhang
Tsinghua University, Beijing, P. R. China

1. INTRODUCTION

Chemiluminescence (CL) is defined as the emission of electromagnetic radiation (usually in the visible or near-infrared region) produced by a chemical reaction.

Table 1   Classification of Luminescence Phenomena

Produced from irradiation
A. Photoluminescence: An excited state is produced by the absorption of ultraviolet, visible, or near-infrared radiation
   Fluorescence: Short-lived emission from a singlet electronically excited state
   Phosphorescence: Long-lived emission from a triplet electronically excited state
B. Cathodoluminescence: Emission produced from irradiation of β-particles
C. Anodoluminescence: Emission produced from irradiation of α-particles
D. Radioluminescence: Emission produced from irradiation of γ-particles or X-rays
Produced from heating
A. Candoluminescence: Emission from incandescent solids
B. Thermoluminescence: Emission from solids and crystals on mild heating
C. Pyroluminescence: Emission from metal atoms in flames
Produced from structural rearrangements in solids
A. Triboluminescence: Emission from shaking, rubbing, or crushing crystals
B. Crystalloluminescence: Emission from crystallization
C. Lyoluminescence: Emission from dissolving crystals
Produced from electrical phenomena
A. Electroluminescence: Emission from electrical discharges
B. Galvanoluminescence: Emission during electrolysis
C. Sonoluminescence: Emission from exposure to ultrasonic sound waves in solution
D. Piezoluminescence: Emission from frictional charges separation at the crystal surface
Produced from chemical reactions
A. Bioluminescence: Emission from living organisms or biological systems
B. Chemiluminescence: Emission from a chemical reaction
   Electrochemiluminescence: Emission occurring in solution, from an electronically excited state produced by high-energy electron transfer reactions
   Electrogenerated chemiluminescence: Emission produced at an electrode surface
   Oxyluminescence: Emission from polymers caused by oxidative processes (presence of oxygen is required)

When this emission originates from living organisms or from chemical systems derived from them, it is named bioluminescence (BL). Both phenomena are luminescence processes that have been traditionally distinguished from related emissions by a prefix that identifies the energy source responsible for the initiation of emission of electromagnetic radiation. Based on Wiedemann's classification, which was discussed in Chapter 1, contemporary luminescence processes have been added to the list of luminescence phenomena, as can be seen in Table 1.

In CL, reactions generally yield one of the reaction products in an electronic excited state producing light on falling to the ground state. As can be seen in Figure 1, the process of light emission in CL is the same as in photoluminescence, except for the excitation process. In fluorescence and phosphorescence the electronically excited state is produced by absorption of ultraviolet or visible light, returning to the ground state (S_0) from the lowest singlet excited state (S_1) or from the triplet excited state (T_1) (Fig. 1). A more extensive discussion of these processes has been included in Chapter 3.

Because the emission intensity is a function of the concentration of the chemical species involved in the CL reaction, measurement of emission intensities can be used for analytical purposes. An advantage of CL techniques is that it is possible to employ rather simple basic instrumentation, as the optical system requires no external light source. CL is often described as a dark-field technique: the absence of strong background light levels, such as found in spectrophotometry and fluorimetry, reduces noise signals and leads to improved detection limits. Instrumentation for CL measurements ranges from simple to very complex, that is, allowing the use of some fluorometers by turning off the excitation source, or applying the facilities offered by more sophisticated systems.

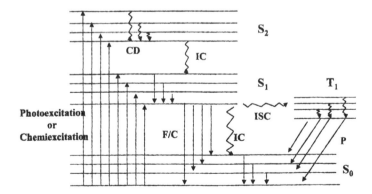

Figure 1 Jablonski diagram showing energy levels and transitions: F, fluorescence; C, chemiluminescence; P, phosphorescence; CD, collisional deactivation; IC, internal conversion; ISC, intersystem crossing; S_0, ground singlet state; S_1, S_2, excited singlet states; T_1, excited triplet state.

However, some limitations must be considered in CL analysis, such as the dependence of the CL emission on several environmental factors that therefore must be controlled, the lack of selectivity because a CL reagent is not limited to just one unique analyte, and finally, like other mass flow detection approaches, since CL emission is not constant but varies with time (light flash composed of a signal increase after reagent mixing, passing through a maximum, then declining to the baseline), and this emission-versus-time profile can vary widely in different CL systems, care must be taken to detect the signal in flowing streams at strictly defined periods.

In this introductory chapter, the basic principles of CL will be presented, with a brief introduction to the essential instrumentation as well as some general aspects showing that this technique is suitable as a detection mode for analytical purposes. Details on each specific topic can be found later in this book.

2.   GENERAL PRINCIPLES

2.1   Mechanisms of Chemiluminescence Reactions

In general, a chemiluminescent reaction can be generated by two basic mechanisms (Fig. 2). In a direct reaction, two reagents, usually a substrate and an oxidant in the presence of some cofactors, react to form a product or intermediate, sometimes in the presence of a catalyst. Then some fraction of the product or intermediate will be formed in an electronically excited state, which can subse-

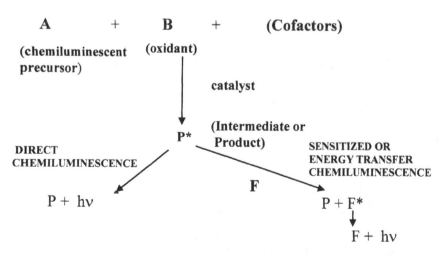

Figure 2   Types of CL reactions. P, product; F, fluorescing substance.

quently relax to the ground state with emission of a photon. The substrate is the CL precursor, which is converted into the electronically excited molecule, responsible for light emission or acting as the energy transfer donor in indirect CL. The catalyst, enzyme or metal ions, reduces the activation energy and provides an adequate environment for producing high CL efficiency out of the process. Cofactors sometimes are necessary to convert one or more of the substrates into a form capable of reacting and interacting with the catalyst, or to provide an efficient leaving group if bond cleavage is required to produce the excited emitter. On the contrary, indirect or sensitized CL is based on a process of transfer of energy of the excited specie to a fluorophore. This process makes it possible for those molecules that are unable to be directly involved in CL reactions to transfer their excess of energy to a fluorophore that in turn is excited, releasing to its ground state with photon emission. All of these paths lead to a great variety of practical uses of CL in solid, gas, and liquid phases.

2.2 Requirements for Chemiluminescence Emission

For a chemical reaction to produce light, it should meet some essential requirements:

1. The reaction must be exothermic to produce sufficient energy to form the electronically excited state. To predict whether the CL reaction will occur or not, it is possible to use the free energy (ΔG):

$$\Delta G = \Delta H - T\Delta S \tag{1}$$

where T is temperature, ΔH is enthalpy, and ΔS is entropy. Enthalpy is the actual energy source for generating the electronically excited state product. To initiate the chemical reaction the activation energy (ΔH_A) is absorbed. The available energy to produce the excited state (ΔE_{EX}) must be the difference between the reaction energy and the activation energy. If this difference is equal to or greater than the energy required for generating the excited state ΔE_{EX}, the CL process will be produced:

$$\text{Energy available} = \Delta H_A - \Delta H_R \geq \Delta E_{EX} \tag{2}$$

In many CL reactions, the entropy change is small, so ΔG and ΔH are very similar in magnitude and the energetic requirements can be established in terms of ΔG (Kcal.mol^{-1}). In this sense, for CL to occur the reaction must be sufficiently exothermic such that:

$$-\Delta G \geq \frac{hc}{\lambda_{ex}} = \frac{2.86.\ 10^4}{\lambda_{ex}} \tag{3}$$

where λ_{ex} is the long-wavelength limit (nanometers) for excitation of the luminescent species. Because most of the CL reactions produce photons in the range 400 (violet)–750 (red) nm, the creation of the electronically excited state and the generation of CL in the visible region require around 40–70 Kcal.mol^{-1}. This exothermic condition is associated with redox reactions using oxygen and hydrogen peroxide or similar potential oxidants.

2. The reaction pathway must be favorable to channel the energy for the formation of an electronically excited state. In case the chemical energy is lost as heat, e.g., via vibrational and rotational energy ways, the reaction will not be chemiluminescent.

3. Photon emission must be a favorable deactivation process of the excited product in relation to other competitive nonradiative processes that may appear in low proportion (Fig. 3). In the case of sensitized CL, both the efficiency of energy transfer from the excited species to the fluorophore and the fluorescence efficiency of the latter must be important.

In all the luminescent processes, the intensity of the produced emission depends on the efficiency of generating molecules in the excited state, which is represented by the quantum efficiency (quantum yield) and the rate of the reaction. In the case of CL reactions, the intensity can be expressed as:

$$I_{CL} = \phi_{CL} \frac{-dA}{dt}$$

where I_{CL} is the CL emission intensity (photons/seconds), ϕ_{CL} is the CL quantum yield, and $(-dA/dt)$ is the rate at which the CL precursor A is consumed. Higher values of quantum yields are usually associated with BL reactions whereas in most of the CL reactions used for analytical purposes, ϕ_{CL} ranges from 0.001 to

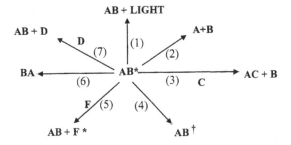

Figure 3 Different processes for losing energy from the excited state: (1) direct CL; (2) molecular dissociation; (3) chemical reaction with other species; (4) intramolecular energy transfer; (5) intermolecular energy transfer (in case of a fluorophore, indirect CL); (6) isomerization; (7) physical quenching. (Adapted from Ref. 1.)

0.1, and even very inefficient systems with much lower quantum yields can be used in analysis based on the almost complete absence of background emission. This is the case with ultraweak CL reactions, in which ϕ_{CL} can be less than 0.001 (often 10^{-3}–10^{-8} and sometimes 10^{-9}–10^{-15}). Ultraweak CL is produced from oxidative reactions in living cells and the emitted signals are mostly 10^3–10^6 times less intense than those from luminous organisms. The main characteristic of ultraweak CL emission is that it is invisible to the naked eye. This kind of CL includes a group of CL reactions that, at least in living cells, involve a number of oxygen intermediates that play an important role in certain types of cell activation, in the defense system of the body, and in ischemic heart diseases. It has been detected from a wide variety of intact organs, isolated cells, and tissue homogenates from vertebrates, invertebrates, plants, and from several reactions in vitro. Ultraweak CL is associated with some important cellular functions, such as mitochondrial respiration, photosynthesis, cell division, or phagocytosis, among others [1].

2.3 Factors Influencing Chemiluminescence Emission

Because of the cited dependence of CL intensity upon various parameters, CL measurements are strongly modified by experimental factors that affect quantum yield and rate of reaction, such as:

> The chemical structure of the CL precursor, not only the central portion containing the electronically excited group, but also the side chain
>
> The nature and concentration of other substrates affecting the CL pathway and favoring other nonradiative competition processes
>
> The selected catalyst
>
> The presence of metal ions, especially transition metals involved in the processing of the oxidant
>
> Temperature
>
> pH and ionic strength
>
> The hydrophobicity of the solvent and solution composition [as example, the ϕ_{CL} of luminol oxidized in dimethylsulfoxide (DMSO) is 0.05 compared with 0.01 in water, the colors being blue-violet (425 nm) and blue-green (480–502 nm), respectively]
>
> The presence of energy transfer acceptors

2.4 Characteristics of Chemiluminescence as Analytical Technique

Because the reaction rate is a function of the chemical concentration, CL techniques are suitable for quantitative analysis. The usefulness of CL systems in analytical chemistry is based on some special characteristics:

1. As the technique simultaneously comprises kinetic and luminescence features, it provides high sensitivity and wide dynamic ranges. Excellent detection limits, in the range of femtomoles, can be reached if ϕ_{CL} is high enough. As an example, in the gas phase, typical detection limits are 10 pmol NO using ozone, and 0.1 pmol for sulfur compounds using a hydrogen flame followed by ozone; in the liquid phase down to 1 fmol of the fluorophore may be detected using the peroxyoxalate CL system, and 0.1 fmol peroxidase when using luminol. In comparison to other spectrometric techniques, in general CL is approximately 10^5 times more sensitive than absorption spectrometry and at least 10^3 times more sensitive than fluorimetry.
2. Compared to photoluminescence processes, no external light source is required, which offers some advantages such as the absence of scattering or background photoluminescence signals, the absence of problems related to instability of the external source, reduction of interferences due to a nonselective excitation process, and simple instrumentation.
3. Selectivity and linearity are most dependent on the reaction and reaction conditions chosen. As for photoluminescence processes, absorption or emission radiation by the analyte, products, or concomitants can cause nonlinearity or spectral interferences.
4. The technique is versatile for determination of a wide variety of species that can participate in the CL process, such as: CL substrates or CL precursors responsible for the excited state; the necessary reagent for the CL reaction (usually an oxidant); some species that affect the rate or efficiency of the CL reaction: activators such as catalysts (enzymes or metal ions) or inhibitors such as reductants that inhibit the CL emission; fluorophores in the case of sensitized CL; some species that are not directly involved in the CL reaction but that can react with other reagents in coupled reactions to generate a product that is a reactant in the CL reaction; species that can be derivatized with some CL precursors or fluorophores, being determined by direct or sensitized CL.
5. Depending on the nature of the analyte and the CL reaction, the increase or decrease of CL intensity will be directly related to the analyte concentration.
6. CL reactions can be coupled as a detection technique in chromatography, capillary electrophoresis, or immunoassay, providing qualitative and/or quantitative information of a large variety of species in the gas and liquid phases.

As an empirical rule, CL behavior may be predicted if a compound or its derivatization product has fluorescence properties. It is possible that oxidation of such a species may produce CL, but there are many exceptions to this general rule.

3. BASIC INSTRUMENTATION

Measurement of light emitted from chemical or biochemical reactions is related to the concentration of participant species: the rate of light output is directly related to the amount of light emitted and, accordingly, proportional to the concentration of the specific species. For this reason, the measurement of light is an indicator of the amount of analyte and the basic instrumentation for these measurements is termed a luminometer. One of the most important advantages of CL as analytical technique is the simplicity of this instrumentation, which includes as principal components: a reaction cell, a lighttight housing, a device for introducing and mixing the reagents and/or the sample, a light detector, and an acquisition and signal-processing system (Fig. 4). In the housing a sample chamber holds the reaction cell (a test tube, a microplate, a flow cell, etc.) to present the luminescencing sample to the detector. This chamber must be sealed from ambient light to minimize potential interferences and it must be positioned as close to the detector as possible to maximize optical efficiency. Additional mirrors, lenses, and other devices can be used to improve the angle of collection and to obtain high optical efficiency, which is desirable for an optimum signal-to-noise ratio, allowing rapid and precise measurements. The function of the housing is to maintain the mixture of sample and CL reagents at an adequate temperature and in the complete dark to isolate the external light from the detector. Usually no monochromator is required because the CL intensity arises from

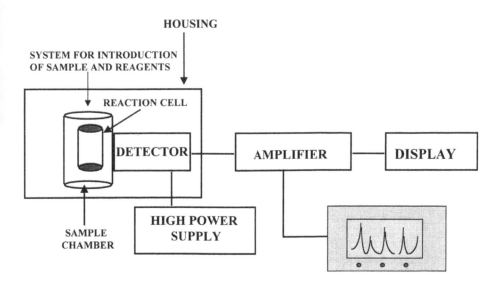

Figure 4 Schematic configuration of a basic luminometer.

one species and other substances rarely affect the spectral distribution of the light emitted.

In CL techniques, after mixing of the reagents and the sample, the CL reaction starts and the emission intensity is being produced, decreasing once the reactants are consumed. This implies a transient character of the CL emission, the time scale of which depends on the specific reaction, and which can range from a short flash to a continuous glow. This fact is crucial for selection of the most convenient systems of reagent addition.

3.1 Introduction of Sample and Reagents

Depending on the configuration of the device and the method for sample and reagents introduction, it is possible to classify the systems into static (batch or discrete sampling instrument) or flowing stream, both using continuous-flow or stopped-flow systems.

In static systems, discrete portions of the CL reagent and the sample are mixed rapidly in the reaction cell, often at a controlled temperature. Usually the final reagent that initiates the CL reaction is added with a syringe or using an automatic injector to provide a more reproducible injection speed and volume and a synchronization of data collection in relation to the initiation of the reaction. Mixing of sample and reagents is provided by the force of injection, although a magnetic stirring bar and stirrer can be used. In this case, the whole CL intensity-versus-reaction-time profile is monitored. Depending on the situation, the analytical signal can be taken as the maximum CL intensity (peak height), the signal after some fixed delay time from the point of mixing, the integral of the signal over some time period, or the integral of the entire peak (full area) (Fig. 5). All of these are correlated with the analyte concentration. This approach is used in selective reactions in the solution phase showing high quantum yields or long-time emission, such as BL reactions, CL immunoassay, or nucleic acid assays.

Continuous-flow CL systems are used in gas-phase and solution-phase reactions. The sample and the CL reagents are continuously pumped and mixed together using a merging zones flow-injection manifold and sent to a flow cell or mixed and the emission is observed in an integral reactor flow cell. The signal is observed when the cell is totally filled with the reaction mixture, at a fixed period after mixing. In this case, it is possible to obtain a constant and reproducible signal, easier to measure, which represents the integrated output over the residence time of the reaction mixture in the cell. Optimum sensitivity is achieved by adjusting the flow rate, the observation cell volume, and transfer channel volumes, with the aim of the observed portion of emission profile to occur at the maximum of the CL intensity-versus-time profile. Disadvantages of this approach include the high reagent consumption and the fact that no kinetic information is

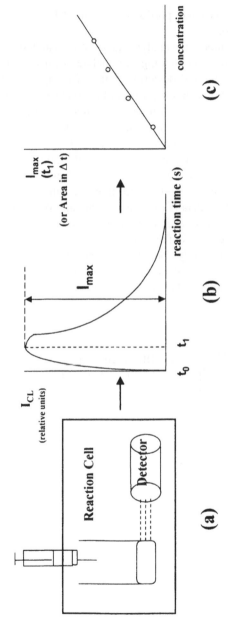

Figure 5 Basic steps in a CL process: (a) the sample and reagent(s) are introduced in the reaction cell and the final reagent is injected to initiate the CL emission, then light is monitored by the detector; (b) curve showing CL intensity as a function of time after reagent mixing to initiate the reaction (the decay of the signal is due to the consumption of reagents and changes in the CL quantum efficiency with time); (c) a calibration function is established in relation to increasing analyte concentrations.

obtained and that only a very small portion of the total CL intensity is measured for slow reactions (Fig. 6).

In the flow injection mode, the analyte carrier stream and the reagent stream flow continuously; the sample containing the analyte is injected and the carrier transports it into the reaction and detection zones. The signal is acquired as a narrow peak, which corresponds to the portion of analyte passing though the cell. Depending on the situation, it is possible to inject not only the sample, but also other CL reagents, e.g., in the case of more expensive chemicals. This methodology has the advantage of low consumption of analyte or CL reagents and the smaller volume of the cell, so as to prevent excessive dilution; the signal is highly reproducible and precise because it corresponds to the height of the peak; moreover it is possible—in given cases—to adapt online chemical treatments to improve selectivity. Chapter 12 details extensively this approach in CL analysis.

Stopped-flow approaches have interesting features for CL monitoring in very fast CL reactions. In this case, the sample and reagent are efficiently mixed in a reaction chamber and rapidly forced to an observation cell, in which the flow is violently stopped, allowing monitoring of the full intensity-time curve and allowing kinetics measurements (Fig. 7), which can be related to the analyte

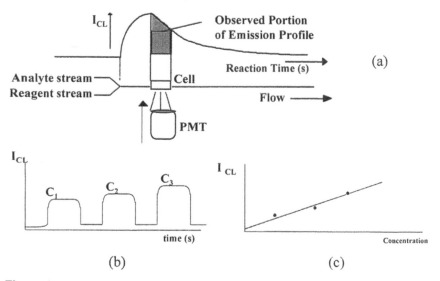

Figure 6   (a) Schematic CL device for continuous flow measurements; (b) CL emission intensity-versus-reaction time for different increasing analyte concentrations; (c) calibration function. (Adapted from Ref. 9.)

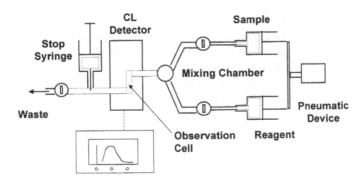

Figure 7 Schematic diagram for stopped-flow CL measurements.

concentration with a higher precision and selectivity than using peak height or areas. This analytical methodology is considered in Chapter 8.

3.2 Reaction Cell

Depending on the CL reaction that is carried out in the liquid, solid, or gas phases, it is possible to use different reaction cells, the essential requirement for their design being that the maximum intensity must be reached while the stream containing the analyte and the CL reagent is in front of the detector. It is a very critical variable because in both static and flow methods the magnitude of the CL intensity is proportional to the volume of the cell. Usually it is possible to apply any material that transmits light in the visible range and that is compatible with the sample, such as glass, quartz, or even acrylic or other plastics. Polystyrene is not recommended because it can build up static electricity, which may contribute to elevated background levels. More transparent cell materials allow more light to reach the detector, resulting in better detection limits.

In the case of flow CL measurements in the liquid phase, the cell design most frequently used is a long spiral cell (about 1 cm), which is located as close to the detector as possible. The large volume is required due to the need to collect a greater amount of emitted light and because the thin size required (50–100 μm) minimizes band broadening and provides a thin optical path to reduce inner filtering effects of the CL emission. With the purpose of minimizing sample dispersion, usually the injection valve and the cell must be placed as close to each other as possible.

In heterogeneous systems, the reagent can be immobilized on a solid surface either covalently, behind a membrane, or by absorption, and it can interact with the analyte by diffusion or convection. In gas-phase CL reactions, the sample and the CL reagent flow inside a reactor through calibrated leaks and the CL emission is produced and monitored by the detector through an optical window. Commercial and homemade luminometers for liquid-, solid-, and gas-phase analysis use various cell configurations. Detailed descriptions of reaction cells for different CL applications are included in Chapters 12, 13, 14, 15, and 20.

3.3   Detector

According to Campbell [1] there are four basic requirements for the detector, which constitutes the critical part of the luminometer:

1. It must be able to detect a light signal over several orders of magnitude of intensity (from just a few photons per second to tens of millions photons per second).
2. It must be sensitive at least over the spectral range 400–600 nm, and ideally over the complete range of the visible spectrum (i.e., 380–750 nm) and even over the UV and IR regions. When its sensitivity varies with wavelength, as is usually the case, it must be possible to correct the data in this respect.
3. The signal output presented by the detector should be directly related to the light intensity reaching it, ideally linearly over the entire sensitivity range. The signal from the detector should be produced in a form that can be easily displayed, recorded, and analyzed.
4. The speed of response of the detector must be much faster than the rate of the CL reaction, if the signal output is not to be a distorted version of the true signal.

Light is detected by photosensitive devices through the generation of photocurrent. Photomultiplier tubes (PMT) are the detection devices more frequently used in CL because of their high gains making them suitable for low light-level detection [2, 3]. However, they must be operated with very stable power supplies to keep the total gain constant. There are two different configurations for PMTs depending on their position in relation to the reaction cell: either to the side ("side-on" configuration) or underneath the reaction cell ("end-on" configuration). Side-on PMTs are usually more economical than end-on models and their vertical configuration takes up less space than the end-on versions. However, end-on PMTs have a large photocathode area and offer a greater uniformity of light collection and of the response [4].

PMTs contain a photosensitive cathode and a collection anode that are separated by electrical electrodes called dynodes, which provide electron multiplication or gain. The cathode is biased negatively by 400–2500 V with respect to the anode. An incident photon ejected by the photocathode strikes the first dynode and releases two to five secondary electrons, which are accelerated by the field between the first and second dynodes and strike the next dynode with sufficient energy to release other electrons and so on. Since each dynode down the chain is biased approximately 100 V more positive than the preceding dynode, this multiplication process continues until the anode is reached. In this sense, the first-incident photon produces an amplified cascade of electrons very quickly and essentially free of externally generated electrical noise. The gain (the average number of electrons per anode pulse) depends on the power supply voltage, being usually between 10^4 and 10^7. A schematic diagram of a PMT is shown in Figure 8.

Also, in the absence of light stimulation, all practical light detectors produce an unwanted output signal termed "background," which is referred to as "dark counts" or "dark current," depending on the application. At moderate bias voltages, dark current can be due primarily to thermal emission of electrons at the photocathode and from the dynodes. This undesirable signal can be reduced by cooling the PMT using temperatures of -60 to $0°C$.

Most of the commercial and homemade luminometers use PMTs with two different modes of operation in relation to their signal processing and readout design: photon counting or current measuring. The most common method of operation is current measuring and it consists of measuring the average current that results from the arrival of many anodic pulses. Then it is possible to obtain an average photoanodic current that is obtained by integration using electronics with a time constant much longer than the pulse duration. However, in the photon-

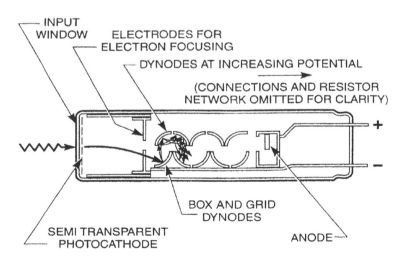

INPUT
WINDOW   ELECTRODES FOR
        ELECTRON FOCUSING

        DYNODES AT INCREASING POTENTIAL

        (CONNECTIONS AND RESISTOR
        NETWORK OMITTED FOR CLARITY)

        +

        −

        BOX AND GRID
        DYNODES

SEMI TRANSPARENT
PHOTOCATHODE   ANODE

Figure 8 Schematic diagram of a photomultiplier tube. (Reproduced by permission, from the Book of Photon Tools, Oriel, Stafford, CT.)

counting mode, the number of anode pulses per unit time or per unit event is counted. Both photon-counting and current-measuring instruments have a response that is proportional to the sample light output; however, the first mode reads in ''photons per seconds'' while the second one reads in light arbitrary units, usually referred to as relative light units. The signal recovery performance offered by photon counting has been shown to be superior to the current-detection method over the entire dynamic range of PMT operation. In photon-counting terms this corresponds to rates of a few photons per second to $\sim 10^8$/s, equivalent to ~ 1 pA to ~ 100 μA in output current. A recent paper discusses the advantages of photon-counting mode versus current-measuring mode, which can be summarized as follows [5]:

> The gain process is noisy and introduces a factor into current-detection measurements.
>
> The dark current in current detection always exceeds the dark-current equivalent of dark counts in photon counting because of the leakage. Also, no component of the dark current can be eliminated by discrimination as in photon counting.
>
> It is less sensitive to temperature effects, aging, high voltage stability, rate effects, magnetic effects, and microphonics.
>
> The dynamic range is superior at fixed gains.
>
> The photon-counting method preserves the temporal structure of the signal.

On the other hand, the photon-counting mode has some disadvantages, such as the limited dynamic ranges (while very low light levels can be measured, the maximum measurable light level is relatively low), they are not able to count fast enough to respond to the high light levels and short duration of fast reactions, and they are less sensitive to quenching effects (e.g., methods based on the signal decrease when analyte concentration increases).

Liquid scintillation counters are highly efficient for low CL intensities and consist of two photon-counting channels provided with a variable discriminator. The sample is placed between these two detectors to ensure a high optical efficiency. The discriminator is adjusted to allow photon impacts to be transmitted and small background noise pulses to be rejected. As disadvantages they suffer from saturation errors and provide nonlinear relationships between the CL intensity and the total counts.

Other detection modes in bright CL or BL reactions are multichannel detectors, which provide simultaneous detection of the dispersed radiation and produce a permanent image of a wide area. Photographic films or plates are emulsions that contain silver halide crystals in which incident photons produce stable clusters of silver atoms within the crystals. Internal amplification is provided in the development process by an electron donor that reduces the remaining silver ions to silver atoms within the exposed crystals. A complexing agent is used to remove the unexposed silver ions. The photographic detection is an integrating process in that the output (density of silver) is a result of the cumulative effect of the entire incident radiation during the exposing time. The image is constituted by many thousands of grains on each square centimeter of film, only 10–100 photons being necessary to produce a developable grain. The main limitations are: nonlinearity of the response over the wide range of light intensities required to observe CL, restricted dynamic range reciprocity failure, and adjacent effects. More efficient imaging devices for CL are vidicon cameras and charged coupled devices (CCDs), which have recently been incorporated in CL instrumentation. These kinds of detectors are extensively discussed in Chapter 4 and applications of CL imaging are included in Chapter 16.

3.4 Signal Conditioning, Manipulation, and Readout

The photoanodic current from the PMT is first converted to voltage with an operational amplifier in the current-to-voltage configuration. Often, after further voltage amplification the readout signal from the analog transducers is converted to digital because the digital domain allows the signal to be treated with highly accurate digital methods or with software in a microcomputer. Only in the case of the photon-counting mode, the PMT output is directly digital because the registered quantity is the total number of accumulated counts over some boundary conditions. The increasing use of computer-controlled CL instruments allows easy keyboard control of instrumental parameters, and the analytical response is stored in memory for later manipulation. Even specific software has been designed for specialized applications such as kinetics studies and measurements, flow systems CL studies, CL detectors coupled to HPLC or GC, etc.

4. MAIN CHEMILUMINESCENCE APPLICATIONS

In the last few years, interest in the use of CL systems in analytical chemistry has been growing exponentially, mainly in gas and liquid phases; however, CL applications in the solid phase are more limited. Table 2 shows some of the applications of the more widely used CL systems that will be considered in detail in this volume.

Because weak CL emission often is produced from the oxidation of many solid organic compounds, the measurement of this light emission may be used as an indicator of changes in materials composition due to oxidation processes, and for evaluating stabilizers intended to prevent or retard these oxidative alterations [6]. Some examples of materials than can be characterized by CL emission are the polymers that are degraded by weathering, exposure to heat, or exposure to ionizing radiation, or food components that suffer flavor alterations. In this

Table 2 Common Applications of the Main CL Systems

Phase	Reagent	Analyte
Gas	Ethylene	O_3
Gas	O_3	Hydrocarbons
		NO
Gas	O_3 after conversion of the analyte to NO	Nitrosamines
		Total nitrogen
Gas	H_2 flame	Sulfur compounds
Gas	H_2 flame followed by O_3	Sulfur compounds
Liquid	Luminol and derivatives	Metal ions and complexes (Co(II), Cu(II), Fe(III), Zn(II), Cd(II), Mn(II), Cr(III), Cr(IV), Pt(IV), ClO$^-$, Fe(CN)$_6^{3-}$)
		Heme compounds
		Peroxidases
		Oxidants (H_2O_2, O_2, I_2, etc.)
		Inhibitors (Ag(I), Ce(IV), Ti(IV), V(V), etc.)
		Substances easily oxidized and indirectly determined (ascorbic, acid, carboxylic acids, amines, etc.)
		Substances converted into H_2O_2 (glucose, etc.)
		Substances labeled with luminol and derivatives
Liquid	Acridinium esters	Substances labeled with acridinium esters
		Ions (Ag(I), Bi(III), Pb(II), Co(II), Cr(III), Cu(II), Fe(III), Mn(II), etc.)
		Oxidants (H_2O_2, O_2, etc.)
		Substances converted into H_2O_2
		Reducing compounds (Cr(II), Fe(II), Mo(V), ascorbic acid, tetracyclines, sugars, etc.)
Liquid	Ru(bpy)$_3^{3+}$	Aliphatic amines
Liquid	Peroxyoxalates	Oxidants (H_2O_2, etc.)
		Fluorophores (polycyclic aromatic hydrocarbons, etc.)
		Derivatized compounds with fluorophores (amino acids, steroids, aliphatic amines, carboxylic acids, catecholamines, etc.)
Liquid	Direct oxidation with MnO$_4^-$, ClO$^-$, Ce(IV), IO$_4^-$, etc.	Different molecules (usually in pharmaceutical applications)
Solid	O_2 plus heat	Polymers

sense, polyolefins, polyamides, rubber, epoxies, and lubricating or edible oils can be characterized in this way. The methodology used in these cases is based on monitoring of the CL intensity-versus-time curve when the solid is submitted to a controlled high temperature in front of the detector. The shape of this curve is characteristic for the material and can help to identify its composition and characterize its degradation process.

In the past, general chapters and reviews have been published, related to the characteristics of CL as analytical technique [7–9], mainly in the liquid phase [10–14], and its use as detection mode in flowing streams and immunoassay [15–17]. Two extensive reviews reported on the specific application of CL reactions according to the nature of the analyte (inorganic species, enzymes and nucleotides, acids and amines, carbohydrates, steroids, polycyclic aromatic compounds, and drugs) and covering the literature from 1983 to 1991 [18] and from 1991 to mid-1995 [19].

Specific reviews, books, and chapters reported the applications of BL [20] and CL as a highly sensitive method of detection in the flame [21], in FIA [22, 23], in LC [22, 24, 25], in GC [26], in CE [27–30], and in immunoassays [31, 32]. Application of the most commonly used CL system for postcolumn detection in conventional and in microcolumn LC, the peroxyoxalate-CL reaction, has been revised [33] and the mechanism of this reaction has been critically reviewed and reevaluated [34].

Advances have been achieved in recent years, such as the use of CL reagents as labels to derivatize and sensitively determine analytes containing amine, carboxyl, hydroxy, thiol, and other functional groups and their application in HPLC and CE [35, 36], the synthesis and application of new acridinium esters [37], the development of enhanced CL detection of horseradish peroxidase (HRP) labels [38], the use of immobilization techniques for developing CL-based sensors [39–42], some developments of luminol-based CL in relation to its application to time-resolved or solid-surface analysis [43], and the analytical application of electrogenerated CL (ECL) [44–47], among others.

The number of reactions producing CL as cited in the literature of the last decade is high. Extensive reviews have been reported on analytical applications in different disciplines such as medical and biochemical [48–52], food [53, 54], environmental, and toxicological [55] analysis.

In the gas and liquid phases, very well-established CL reactions exist that have been chronologically introduced in Chapter 1, together with their mechanisms; they will be treated in different chapters of this book. Particularly, some chapters include descriptions of the CL systems and applications in the liquid phase in organic and inorganic analysis (Chapters 5 and 6, respectively), for BL systems (Chapter 10); applications derived from the use of organized media (Chapter 11); the specific study of the mechanism and applications of a widely applied CL system based on the reaction of peroxyoxalates (Chapter 7); kinetics considerations in CL analysis (Chapter 8); and the recent applications of ECL (Chapter 9).

As far as the gas phase is concerned, the CL systems that occur at room temperature or in the presence of a flame are extensively described in Chapter 13, together with the applications of these reactions in gas chromatography (GC).

Also, specific chapters deal with the use of CL reactions as detection mode in FIA (Chapter 12), in separational techniques, such as liquid chromatography (LC) (Chapter 14) or capillary electrophoresis (CE) (Chapter 15), in immunoassay (Chapter 18), and in the development of sensors (Chapter 20). The recent use of this technique for the analysis of DNA (Chapter 19) and a photosensitized CL mode for medical routine and industrial applications (Chapter 17) are also considered in this book.

It is clear that the areas of CL and BL, from both the analytical and biochemical/biological points of view, have been the subject of various international series of symposia that have been organized since the early 1970s, and the importance of which apparently is increasing for the next decade, as can be noticed from the calendar of events published by reputed international journals. An exciting analytical future is hence to be expected.

5.   CHEMILUMINESCENCE AND BIOLUMINESCENCE ON THE INTERNET

Recently Stroebel et al. published the first survey of Internet sites relating to various aspects of luminescence [56]. In their classification they cite relevant companies and existing products, general websites, images, movies, and demonstrations for educational purposes. Below are summarized some of these useful websites to which others have been added.

5.1   General Websites

Analytical Chemiluminescence. Naval Air Warfare Center Weapons Division NA
         WCWPNS, China Lake. Chemiluminescence history.
    http://www.nawcwpns.navy.mil/clmf/chemilume.html
Chemiluminescence.
    http://fischer.union.edu/chem20/experiments/chemiluminescence.htlm
Chemiluminescence.
    http://www.deakin.edu.au/~swlewis/cl.htm
Chemiluminescence and laser-induced fluorescence spectra of lightsticks.
    http://www.cs.moravian.edu/chemistry/lightstick/1_st_scheme.html
Chemiluminescence definitions and primer.
    http://www.shsu.edu/~chm_tgc/chemilumdir/Define.html

Chemiluminescence detector.
  http://www.techlab.de/produkte/c1-2.html
Chemiluminescence-light without heat.
  http://scifun.chem.wisc.edu/homeexpts/chemilum.html
Chemiluminescence of tris ruthenium(ll) ion.
  http://chemed.chem.purdue.edu/~genchem/demosheets/5.12.html
Chemiluminescence reaction.
  http://learn.chem.vt.edu/user/long/demo/luminol.html
Chemiluminescence sites of interest.
  http://www.shsu.edu/~chm_tgc/chemilumdir/chemisites.html
Chemiluminescence spectroscopy.
  http://www.scimedia.com/chem-ed/spec/molec/chemilum.htm
Luminol/hydrogen peroxide reaction.
  http://www.pky.ufl.edu/homepages/faculty/eg/demo.html
Bioluminescence production of light by living organisms resulting from the con-
    version of chemical energy to light energy.
  http://www.encyclopedia.com/printable/01486-a.html
The Bioluminescence Web Page.
  http://lifesci.ucsb.edu/~biolum/index.shtml

5.2 Images, Movies, and Demonstrations

Chemiluminescence movie.
  http://www.shsu.edu/~chm-tgc/chemilumdir/movie.html
Chemiluminescence: firefly reaction Quick-Time video.
  http://www.newlisbon.k12.wi.us/reactions/home.html%234
Chemiluminescence: glowing tornados.
  http:1/learn.chem.vt.edu/user/long/demo/chemiluminescence.html
Chemiluminescence: the Cyalume lightstick (demonstration).
  http://chemed.chem.prudue.edu/~genchem/demosheets/5.8.html
The Chemiluminescence Home Page.
  http://www.shsu.edu/~chm_tqc/chemilumdir/chemiluminescence2.html
Demonstration of chemiluminescence. Luminol chemical concept.
  http://chemed.chem.purdue.edu/~genchem/demosheets/5.9.html
Chemiluminescence experiments.
  http://library.advanced.org/3310/nographics/experiments/lumine.html

5.3 Companies, Instruments, and Products

Andor technology
  http://www.andor-tech.com/

Argus Laboratories Ltd. Chemiluminescent and bioluminescent products
    Berthold. Luminescence assays
    http://www.berthold-online.com/
http://www.net-escape.co.uk/business/argus/
Biosynth. Bioluminescent products list, bioluminescent substrates, and related
    products.
    http://www.biosynth.inter.net/page14.html
Chemiluminescence
    http://www.etd.ameslab.gov/etd/technologies/projectss/chemilum.html
Hamamatsu. Photomultipliers and photon countings for bioluminescence and
    chemiluminescence.
    http://www.hamamatsu.com/
Kodak. Chemiluminescence imaging
    http://www.nenlifesci.com/
Lumigen, Inc.
    http://www.lumigen.com/
McPherson, Inc. Model 660 HPLC Chemiluminescence detector.
    http://www.mcphersoninc.com/hplcdetectors/mode1660description.htm
Ocean Optics. Optical accessories
    http://www. OceanOptics.com/
Oriel. Photomultipliers and electronic and optical devices
    http://www.oriel.com/
Photometrics. High-performance CCD imaging for bioluminescence.
    http://www.photomet.com/images/im.biolum.html
Photomultipliers for bioluminescence and chemiluminescence.
    http://www.electron-tubes.co.uk/pmts/pmtchem.html
Photosensitized chemiluminescence
    http://www.riat.com/
Photoreactors
    http://www.bioanalytical.com/
Sievers Instruments. Gas-phase chemiluminescence detector
    http://www.SieverInst.com/
Tropix, Inc.
    http://www.tropix.com/
Turner Designs TD-20/20 luminometer.
    http://www.seoulin.co.kr/protocol/turner/chem.htm

It is clear that when surfing on the Internet applying uniquely the keyword
"chemiluminescence," many important sites show up. These include principles
and applications, spectroscopy, potentials in liquid chromatography, university
training courses, products, demonstrations, etc.

The journal *Luminescence (The Journal of Biological and Chemical Luminescence)* includes a section featuring new products, people, and conferences, regular literature reviews, and abstracts from Conference Proceedings (http://www.interscience.wiley.com/jpages/0884-3996/). Interesting also is the website of the International Society for Bioluminescence and Chemiluminescence (http://www.unibo.it/isbc/).

REFERENCES

1. AK Campbell. Chemiluminescence. Principles and Applications in Biology and Medicine. Chichester: Ellis Horwood/VCH, 1988.
2. JD Ingle Jr, SR Crouch. Spectrochemical Analysis. Englewood Cliffs, NJ: Prentice-Hall, 1988, pp. 106–117.
3. GK Turner. In: K Van Dyke, ed. Bioluminescence and Chemiluminescence: Instruments and Applications. Vol. I. Boca Raton, FL: CRC Press, 1985, pp. 43–78.
4. ZB Drozdowicz, ed. The Book of Photon Tools. Oriel Instruments, USA, pp 6–7; 6–65.
5. RJ Ellis, AG Wright. Luminescence 14:11–18, 1999.
6. TA Nieman. In: A Townshend, ed. Encyclopedia of Analytical Science. Vol. 1. London: Academic Press, 1995, pp 611–612.
7. K Nakashima, K Imai. In: SG Schulman, ed. Molecular Luminescence Spectrometry, part 3. New York, Wiley, 1993, pp 1–23.
8. TA Nieman. In: A Townshend, ed. Encyclopedia of Analytical Science. Vol. 1. London: Academic Press, 1995, pp 608–612.
9. TA Nieman. In F Settle, ed. Handbook of Instrumental Techniques for Analytical Chemistry. New York: Marcel Dekker, 1997, pp 541–559.
10. A Townshend. Analyst 115:495–500, 1990.
11. JS Lancaster. Endeavour, New Series. Vol 16. Great Britain: Pergamon Press, 1992, pp 194–200.
12. MJ Pringle. In: Recent Advances in Clinical Chemistry. New York: Academic Press, 1993, Vol. 30, pp 89–183.
13. TA Nieman. In: A Townshend, ed. Encyclopedia of Analytical Science. Vol. 1. London: Academic Press, 1995, pp 613–621.
14. MG Sanders, KN Andrew, PJ Worsfold. Anal Commun 34:13H–14H, 1997.
15. LP Palillis, AC Calokerinos, WRG Baeyens, Y Zhao, K Imai. Biomed Chromatogr 11:85–86, 1997.
16. WRG Baeyens, SG Schulman, AC Calokerinos, Y Zhao, AM García-Campaña, K Nakashima, D De Keukeleire, J Pharm Biomed Anal 17:941–953, 1998.
17. AM García-Campaña, WRG Baeyens, XR Zhang, E Smet, G Van der Weken, K Nakashima, AC Calokerinos. Biomed Chromatogr 14:166–172, 2000.
18. K Robards, PJ Worsfold. Anal Chim Acta 266:147–173, 1992.
19. AR Bowie, G Sander, PJ Worsfold. J Biolum Chemilum 11:61–90, 1996.
20. RA LaRosa, ed. Bioluminescence Methods and Protocols. (Methods in Molecular Biology, Vol. 102). New Jersey: The Humana Press, 1998.
21. DA Stiles, AC Calokerinos, A Townshend. Flame Chemiluminescence Analysis by Molecular Emission Cavity Detection. New York: Wiley, 1994.
22. SW Lewis, D Price, PJ Worsfold. J Biolum Chemilumin 8:183–199, 1993.

23. TA Nieman. In: WRG Baeyens, D De Keukeleire, K Korkidis, eds. Luminescence Techniques in Chemical and Biochemical Analysis. New York: Marcel Dekker, 1991, pp 523–561.
24. B Yan, SW Lewis, PJ Worsfold, JS Lancaster, A Gachanja. Anal Chim Acta 250: 145–155, 1991.
25. DS Hage. In: G Patonay, ed. HPLC Detection, Newer Methods. New York: VCH, 1992, pp. 57–75.
26. JS Lancaster. In: A Townshend, ed. Encyclopedia of Analytical Science. Vol. 1. London: Academic Press, 1995, pp 621–626.
27. WRG Baeyens, B Lin Ling, K Imai, AC Calokerinos, SG Schulman. J Microcol Sep 6: 195–206, 1994.
28. AM García-Campaña, WRG Baeyens, Y Zhao. Anal Chem 69:83A–88A, 1997.
29. TD Staller, MJ Sepaniak. Electrophoresis 18:2291–2296, 1997.
30. AM García-Campaña, WRG Baeyens, NA Guzman. Biomed Chromatogr 12:172–176: 1998.
31. EHJM Jansen, G Zomer, CH Van Peteghem, G. Zomer. In: WRG Baeyens, D De Keukeleire, K Korkidis, eds. Luminescence Techniques in Chemical and Biochemical Analysis. New York: Marcel Dekker, 1991, pp 477–504.
32. I Weeks. Chemiluminescence Immunoassay. Amsterdam: Elsevier, 1992.
33. PJM Kwakman, UATh Brinkman. Anal Chim Acta 266:175–192, 1992.
34. G Orozs, RS Givens, RL Schowen. Crit Rev Anal Chem 26:1–27, 1996.
35. R Zhu, WTh Kok. J Pharm Biomed Anal 17:985–999, 1998.
36. N Kuroda, K Nakashima. In: T Toyo'oka. Modern Derivatization Methods for Separation Sciences. New York: Wiley, 1999, pp 168–189.
37. G Zomer, JFC Stavenuiter, RH Van Den Berg, EHJM Jansen. In: WRG Baeyens, D De Keukeleire, K Korkidis, eds. Luminescence Techniques in Chemical and Biochemical Analysis. New York: Marcel Dekker, 1991, pp 505–521.
38. LJ Kricka, RAW Stott, GHG Thorpe. In: WRG Baeyens, D De Keukeleire, K Korkidis, eds. Luminescence Techniques in Chemical and Biochemical Analysis. New York: Marcel Dekker, 1991, pp 599–635.
39. PR Coulet, LJ Blum. Trends Anal Chem 11:57–61, 1992.
40. S Girotti, EN Ferri, S Ghini, F Fini, M Musiani, G Carrea, A Roda, P Rauch. Chemicke List 91:477–482, 1997.
41. LJ Blum. Bio- and Chemiluminescent Sensors. Singapore: World Scientific, 1997.
42. XR Zhang, WRG Baeyens, AM García-Campaña, J Ouyang. Trends Anal Chem 18: 384–391, 1999.
43. XR Zhang, WRG Baeyens, AM García-Campaña. Biomed Chromatogr 13:169–170, 1999.
44. J Kankare. In: A Townshend, ed. Encyclopedia of Analytical Science. Vol. 1. London: Academic Press, 1995, pp 626–632.
45. AW Knight, GM Greenway. Analyst 119:879–890, 1994.
46. AW Knight, GM Greenway. Analyst 120:1077–1082, 1995.
47. AW Knight. Trends Anal Chem 18:47–62, 1999.
48. J Martínez Calatayud, S Laredo Ortíz. Cienc Pharm 4:190–201, 1994.
49. AC Calokerinos, NT Deftereos, WRG Baeyens. J Pharm Biomed Anal 13:1063–1071, 1995.

50. J Martínez Calatayud. Flow Injection Analysis of Pharmaceuticals. Automation in the Laboratory. London: Taylor & Francis, 1996, pp. 183–186.
51. C Dodeigne, L Thunus, R Lejeune. Talanta 51:415–439, 2000.
52. A Roda, P Pasini, M Guardigli, M Baraldini, M Musiani, M Mirasoli. Fresenius J Anal Chem 366:752–759, 2000.
53. MJ Navas, AM Jiménez. Food Chem 55:7–15, 1996.
54. S Girotti, E Ferri, S Ghini, A Roda, P Pasini, G Carrea, R Bovara, S Lodi, G Lasi, J Navarro, P Raush. Quím Anal 16:S111–S117, 1997.
55. AM Jiménez, MJ Navas. Crit Rev Anal Chem 27:291–305, 1997.
56. J Stroebel, LJ Kricka, PE Stanley. Luminescence 14:63–66, 1999.

3
The Nature of Chemiluminescent Reactions

Stephen G. Schulman and Joanna M. Schulman
University of Florida, Gainesville, Florida

Yener Rakicioğlu
Istanbul Technical University, Istanbul, Turkey

1. INTRODUCTION

Chemiluminescence (CL) is the emission of light by molecules that are electronically excited by virtue of their precursors' participation in a highly exergonic chemical reaction, almost invariably an oxidation reaction. Such reactions occur in vitro and in vivo and in the latter case they are mediated by enzymes and the resulting chemiluminescence is called bioluminescence. Chemiluminescence is actually fluorescence (spin-allowed, radiative deactivation of the lowest excited singlet state) from the electronically excited product of the chemical reaction. It differs from the process normally called fluorescence (photofluorescence) only in that in fluorescence, the light-emitting molecules are electronically excited by the absorption of light whereas in CL, no external light source is employed. Although, in principle, CL could arise from the lowest triplet state as a phosphorescent emission, the long inherent lifetimes of triplet states and their near-total quenching by photoreaction in fluid solutions has, to our knowledge, precluded such a phenomenon from ever having been observed.

In this chapter will be considered:

1. The means by which the light-emitting molecule is formed in its emissive state
2. The nonradiative processes immediately leading to and competing with light emission
3. The kinetics of CL reactions

This discussion will be confined to strictly chemical reactions and will not include electrogenerated chemiluminescence.

2. CHEMICAL CONSIDERATIONS

A sine qua non for CL to occur is that the precursor(s) of the light-emitting species must participate in a reaction that releases a considerable amount of energy. For visible emission (say 400–750 nm in wavelength) 40–70 kcal/mol is required. Actually, the enthalpies of reaction have to be slightly greater than this owing to energy losses by thermal relaxation in the ground (after emission) and lowest excited singlet (after excitation) states of the fluorophore. Normally, only certain oxidation-reduction reactions generate this much energy. In addition, at least some of the energy produced must be channeled into a reaction pathway, in which at least one of the upper vibrational levels of the reactants, probably corresponding to the transition state of the initial reaction, has the same energy and a comparable geometrical structure as an upper vibrational level of the lowest excited singlet state of a potentially emissive product of the reaction. Higher excited singlet states, although of consequence in the generation of photofluorescence, need not be considered here because their energies are so high, relative to the thermal energies of the reactants, that they would be impossible to populate, even in a very exergonic chemical reaction. The geometric identity or near-identity of the interconverting species ensures that most of the energy of activation is channeled into free energy of activation rather than being wasted as entropy of activation.

Invariably, there will also be a pathway for "dark" reaction. The dark reaction may lead directly to the ground, electronic-state species that also results from the fluorescence of the excited product, but this is not necessarily so. It is possible to have a competing dark reaction that leads directly to ground-state products different from that produced by fluorescence. If ΔH_a is the sum of the enthalpies of activation for all dark reaction competing with the chemiluminescent pathway, whose enthalpy of activation is ΔH_a^*, CL will be probable when $\Delta H_a^* < \Delta H_a$. This occurs when the lowest excited singlet state of the flurophore has the same geometrical configuration as the ground electronic state of the reactants at lower energy than when the dark products in the ground electronic state have the same configuration as the reactants in their ground electronic state. This is illustrated in Figure 1. Often, catalysts, which are usually transition metal

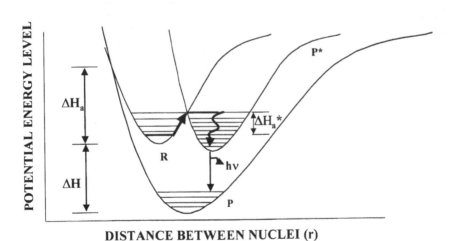

DISTANCE BETWEEN NUCLEI (r)

Figure 1 Relative positions of the potential energy (E) surfaces of the electronic states involved in a hypothetical chemiluminescent reaction as a function of internuclear separation (r). P and P* represent the ground and lowest electronically excited singlet states of the product of the reaction, respectively. R represents the ground electronic state of the reactant. ΔH is the enthalpy of the "dark" reaction while ΔH_a is its enthalpy of activation. ΔH_a^* is the enthalpy of activation of the photoreaction. hv denotes the emission of chemiluminescence.

ions in chemiluminescent reactions and enzymes in bioluminescent reactions, are employed to make light-emitting reactions proceed at a convenient rate. This is accomplished by the catalyst forming a transient complex with the reagents, whose ΔH_a^* is lower than that of the uncatalyzed reaction. The catalyst is usually a co-oxidant; i.e., it assists electron transfer. The catalysis of the luminol–hydrogen peroxide reaction by Co(II) and of the luciferin-oxygen reaction by the enzyme luciferase are important examples.

3. PHOTOPHYSICAL CONSIDERATIONS

Subsequent to the formation of the potentially CL molecule in its lowest excited singlet state, a series of events carry the excited molecule down to its ground electronic state. Since the electronically excited molecule is initially also vibrationally excited, rapid, stepwise (10^{-13}–10^{-12} s) thermal deactivation of the excited molecule, called thermal or vibrational relaxation, in which the molecule loses vibrational energy by inelastic collisions with the solvent, occurs. The process of vibrational relaxation carries the molecule to the lowest vibrational level of the lowest excited singlet state. Certain molecules may return radiationlessly all the way to the ground electronic state in the same time frame by thermal deactivation, a process known as internal conversion. However, in some molecules, for a variety of reasons, the return from the lowest vibrational level of the lowest excited singlet state to the lowest vibrational level of the ground state by internal conversion and vibrational relaxation is forbidden (i.e., of low probability or long duration). In these molecules return to the ground electronic state occurs by one of two alternative pathways, the simpler being the direct emission of ultraviolet or visible radiation whose frequency or wavelength is governed by the energy gap between the lowest excited singlet state and the ground electronic state. The radiative transition between excited and ground states of the same spin multiplicity occurs in a time frame of 10^{-11}–10^{-7} s after excitation, and is called fluorescence. Owing to the fact that the ground electronic state of a molecule has several vibrational levels associated with it, fluorescence emission does not occur at a single wavelength, but rather over a range of wavelengths corresponding to several vibrational transitions as components of a single electronic transition.

Several processes may compete with fluorescence for deactivation of the lowest excited singlet state. As a result only a fraction of the molecules formed in the lowest excited singlet state, ϕ_f, actually fluoresce. ϕ_f is called the quantum yield or fluorescence efficiency. It is usually a fraction but may be unity in some exceptional cases and is related to the probabilities (rate constants) of fluorescence (k_f) and competitive processes (k_d) by

$$\phi_f = \frac{k_f}{k_f + k_d} \tag{1}$$

Thus, the greater the numbers or rates of processes competing with fluorescence for deactivation of the lowest excited singlet state, the lower the value of ϕ_f. The quantum yield of fluorescence is important in determining how intense chemiluminescence can be for a particular reaction.

Another important property of fluorescing molecules is the lifetime of the lowest excited singlet state (τ_f). If the mean rate of fluorescence is the number of fluorescence events per unit of time, the mean lifetime of the excited state is the reciprocal rate, or the mean time per fluorescence event. The quantum yield of fluorescence and the lifetime of the excited state are related by

$$\tau_f = \phi_f \tau_N \tag{2}$$

where τ_N is called the natural lifetime of the excited state and represents the lifetime the fluorescing molecule would have if fluorescence was the sole pathway for deactivating the lowest excited singlet state. While τ_f and ϕ_f are determined largely by the kinds and rates of processes competing with fluorescence, τ_N is a function only of molecular structure.

The second pathway for deactivation of the lowest excited singlet state competes temporally with fluorescence and is called intersystem crossing from the lowest excited singlet to the lowest triplet state. Intersystem crossing entails a change in spin angular momentum, which, classically, violates the law of conservation of angular momentum. Although it is about a million times less probable than the corresponding singlet-singlet radiationless process, internal conversion, its rate is comparable to that of the spin allowed radiative transition (fluorescence). Subsequent to intersystem crossing, the molecules populating the lowest triplet state undergo vibrational relaxation to the lowest vibrational level of the lowest triplet state. Molecules in the lowest triplet state can return to the ground state radiationlessly by triplet-singlet intersystem crossing or by the emission of light. The emission of light accompanying the transition from the triplet state to the ground singlet state is also a forbidden transition, is characterized by a long duration (10^{-4}–10 s), and is called phosphorescence. Because triplet states are so long-lived, radiationless chemical and physical processes in fluid solution compete effectively with phosphorescence for deactivation of the lowest excited triplet state. Except for the shorter-lived phosphorescences, collisional deactivation by solvent molecules, quenching by paramagnetic species (e.g., oxygen), photochemical reactions, and certain other processes preclude the observation of phosphorescence in fluid media unless the potentially phosphorescent molecules are protected from their environment by inclusion in micelles. Owing to the need for diffusion of the reactants in a chemiluminescent reaction into contact with each other, CL requires a fluid environment for its observation. Hence, phosphorescence has never been observed in a CL reaction. Although CL has a long duration (several seconds to several hours), this is a function of the slow reaction kinetics rather than the decay characteristics of the fluorescence. Fluorescent decay lasts only nanoseconds.

4. STRUCTURAL AND ENVIRONMENTAL CONSIDERATIONS

Molecules that enter into CL reactions are generally reduced species that can be easily oxidized. Molecules containing amino and hydroxy groups fall into this category, as do polycyclic aromatic ring systems. Under strongly alkaline conditions, hydroxy groups and even some arylamino or arylamido groups can be deprotonated making them even more susceptible to oxidation. On the other hand, electron-withdrawing groups tend to stabilize electronic charge and make oxidation more difficult in molecules to which they are affixed.

From the chemical point of view, the solvent in which the CL experiment is carried out can have a dramatic influence on the efficiency of the CL reaction as solvation can alter the shapes, the depths, and the densities of the vibrational states of the potential surfaces representing the ground states of products and reactants and the lowest excited singlet state of the potential fluorophore. The alteration of the intersections of these potential energy surfaces can affect the enthalpies of reaction and the enthalpies of activation for dark and lumigenic reactions. In some cases, these changes will favor CL (if ΔH_a^* decreases relative to ΔH_a) and in some cases, they will make it thermodynamically unfavorable for CL to occur.

Other influences of the structure and the environment are manifest in the rates of processes competing with fluorescence for deactivation of the lowest excited singlet state. These are the processes and properties that influence the fluorescence process and will be discussed briefly here.

5. THE INFLUENCE OF MOLECULAR STRUCTURE ON FLUORESCENCE

Among other factors, the quantum yield of fluorescence determines the intensity of light emission in a CL. This, as well as the position in the spectrum occupied by the fluorescence band, is largely a function of the molecular structure.

Fluorescence is most often observed in highly conjugated or aromatic organic molecules with rigid molecular skeletons. The less vibrational and rotational freedom that the molecules has, the greater is the probability that the energy gap between the lowest excited singlet state and the ground electronic state will be large so that fluorescence will predominate over nonradiative deactivation and assure a high fluorescence efficiency.

Aromatic molecules containing freely rotating substituents usually tend to fluoresce less intensely than those without these substituents. This results from the introduction into each electronic state of rotational and vibrational substates by the exocyclic substituents. At very low temperatures, the quantum yields of fluorescence tend to be greater than at ambient temperatures as a result of restricted vibrational and rotational freedom and, consequently, a lower efficiency of internal conversion. It is, thus, likely that if appropriate solvents that remain fluid at low temperatures can be used to conduct CL experiments, higher light yields will result. There is a tendency for many substituents to increase quantum yields of fluorescence. This arises from the increased rate of radiative decay accompanying extension of an aromatic system by substitution with a strongly interacting group such as $-NH_2$ or $-OH$. However, with certain substituents, the fluorescence quantum yields of aromatic molecules are diminished. This is especially so in the case of substitution by heavy atoms such as bromium and iodine, six-membered heterocyclic rings (e.g., quinoline), and other groups having sp^2 hybridized nonbonding electrons. Each of these substituents has the ability to cause mixing of the spin and orbital electronic motions of the aromatic system. Spin-orbital coupling destroys the concept of molecular spin as a well-defined property of the molecule and thereby enhances the probability or rate of singlet-triplet intersystem crossing. This process favors population of the lowest triplet state at the expense of the lowest excited singlet state and thus decreases the fluorescence quantum yield. Consequently nitro compounds, bromo and iodo derivatives, aldehydes, ketones, and N-heterocyclics tend to fluoresce very weakly or not at all and are not likely to function as the fluorophores in CL reactions.

Molecular structure can have a profound effect on the position in the spectrum where fluorescence occurs, as well as on its intensity. It can be shown by quantum mechanics that the more extended a conjugated system is, the smaller will be the separation in energy between the ground state and the lowest excited singlet state. This is evident in the fact that benzene, naphthalene, and anthracene, having one, two, and three rings, fluoresce maximally at 262 nm, 320 nm, and 379 nm, respectively.

Similarly, the affixment of conjugate substituents onto aromatic systems extends the conjugation of the latter and causes fluorescence maxima of the substituted derivatives to lie at wavelengths longer than those of the parent compound. Hence, the fluorescence of aniline lies at 340 nm while that of the parent hydrocarbon benzene lies at 262 nm.

6. THE INFLUENCE OF THE ENVIRONMENT ON THE FLUORESCENCE SPECTRUM

6.1 Solvent Effects

Although the capacity for fluorescence is primarily a function of molecular structure, the solvent of a potentially fluorescing molecule can have a dramatic effect on the fluorescence.

Solvent interactions with solute molecules are mainly electrostatic and it is usually the differences between the electrostatic stabilization energies of ground and excited states that contribute to the relative intensities and spectral positions of fluorescence in different solvents.

The changes in π electron distribution that take place upon transition from the lowest excited singlet state to the ground state during fluorescence cause changes in the dipolar and hydrogen-bonding properties of the solute. If the solute is more polar in the excited state than in the ground state, fluorescence will occur at longer wavelengths in a polar solvent than in a nonpolar solvent because the more polar solvent will stabilize the excited state relative to the ground state. Moreover, the fact that photoluminescence originates from an excited state that is in equilibrium with its solvent cage and terminates in a ground state that is not causes the fluorescence to lie at longer wavelengths the more polar or more strongly hydrogen-bonding the solvent.

In molecules having atoms with trigonally hybridized nonbonded electron pairs (e.g., carbonyl compounds and certain N-heterocyclics), the lowest excited singlet state is formed by promoting a lone-pair electron to a vacant π orbital (this is called an n, π* state). These molecules tend to show very little fluorescence in aprotic solvents such as aliphatic hydrocarbons because the n, π* excited singlet state is efficiently deactivated by intersystem crossing. However, in protic solvents such as water or ethanol, these molecules become fluorescent. This results from the destabilization of the lowest singlet n, π* state by hydrogen bonding. If this interaction is sufficiently strong, the fluorescent π, π* state drops below the n, π* state allowing intense fluorescence. Quinoline, for example, fluoresces in water but not in cyclohexane.

Solvents containing atoms of high atomic number (e.g., alkyl iodides or bromides) also have a substantial effect on the intensity of fluorescence of solute molecules. Atoms of high atomic number in the solvent cage of the solute molecule enhance spin-orbital coupling in the lowest excited singlet state of the solute. This favors the radiationless population of the lowest triplet state at the expense of the lowest excited singlet state. Thus in heavy-atom solvents, all other things being equal, fluorescence is always less intense than in solvents of low molecular weight.

6.2 Quenching of Fluorescence

Fluorescence may be decreased or completely eliminated by interactions with other chemical species. This phenomenon is called quenching of fluorescence. Obviously, if the fluorescence of a fluorophore generated in a CL reaction is quenched the observation of chemiluminescence will be precluded.

Two kinds of quenching are distinguished. In static quenching, interaction between the potentially fluorescent molecule and the quencher takes place in the ground state, forming a nonfluorescent complex. The efficiency of quenching is governed by the formation constant of the complex as well as by the concentration of the quencher. The quenching of the fluorescence of salicylic acid by Cu(II) is an example.

In dynamic quenching (or diffusional quenching) the quenching species and the potentially fluorescent molecule react during the lifetime of the excited state of the latter. The efficiency of dynamic quenching depends upon the viscosity of the solution, the lifetime of the excited state (τ_o) of the luminescent species, and the concentration of the quencher [Q]. This is summarized in the Stern-Volmer equation:

$$\phi/\phi_o = \frac{1}{1 + k_Q \tau_o [Q]} \tag{3}$$

where k_Q is the rate constant for encounters between quencher and potentially luminescing species and ϕ and ϕ_o are the quantum yields of fluorescence in the absence and presence of concentration [Q] of the quencher, respectively.

k_Q is typical of diffusion-controlled reactions ($\sim 10^{10}$ M^{-1} s^{-1}) while τ_o for a fluorescent molecule is typically 10^{-8} s or less. Hence, $k_Q \tau_o \leq 10^2$ M^{-1} and for dynamic quenching to be observed (say 1% quenching), [Q] must be greater than or equal to 10^{-4} M.

6.3 Energy Transfer

Energy transfer entails the excitation of a molecule that during the lifetime of the excited state passes its excitation energy to another molecule. The loss of excitation energy from the initial excited species (the donor) results in quenching of the luminescence of the energy donor and may result in luminescence from the energy recipient (acceptor), which becomes excited in the process.

Energy transfer can occur by either of two acceptor-concentration-dependent processes. In the resonance excitation transfer mechanism or dipole (Förster) mechanism, the donor and acceptor molecules are not in contact with one another and may be separated by as much as 10 nm (although transfer distances closer to 1 nm are more common). In the classical sense, the excited energy donor molecule may be thought of as a transmitting antenna that creates an electrical field in its vicinity. Potential acceptor molecules within the range of this electrical field function as receiving antennae and absorb energy from the field resulting in their electronic excitation.

The rate of resonance energy transfer decreases with the sixth power of the distance between the donor and acceptor dipoles according to

$$k_{ET} = \frac{1}{\tau_D} \left(\frac{R_o}{R} \right)^6 \tag{4}$$

where k_{ET} is the rate constant for resonance energy transfer, τ_D is the lifetime of the excited state of the donor molecule, R is the mean distance between the centers of the donor and acceptor dipoles, and R_o is a constant for a given donor-acceptor pair, corresponding to the mean distance between the centers of the donor and acceptor dipoles for which energy transfer from donor to acceptor and fluorescence from the donor are equally probable. Another general requirement for the occurrence of resonance energy transfer is the overlap of the fluorescence spectrum of the donor and the absorption spectrum of the acceptor. Any degree of overlap of these spectra will satisfy the quantization requirements for the energy of the thermally equilibrated donor molecule to promote the acceptor to a vibrational level of its excited singlet state. The greater the degree of overlap of the luminecsence spectrum of the donor and the absorption spectrum of the acceptor, the greater is the probability that energy transfer will take place.

The exchange (Dexter) mechanism of excitation energy transfer is important only when the electron clouds of donor and acceptor are in direct contact. In this circumstance, the highest energy electrons of the donor and acceptor may exchange places. Thus, the optical electron of an excited donor molecule may become part of the electronic structure of an acceptor molecule originally in the ground singlet state while the donor is returned to its ground singlet state by acquiring an electron from the acceptor. Exchange energy transfer is also most efficient when the fluorescence spectrum of the donor overlaps the absorption spectrum of the acceptor. Exchange energy transfer is a diffusion-controlled process (i.e., every collision between donor and acceptor leads to energy transfer) and as such its rate depends upon the viscosity of the medium. Resonance energy transfer, on the other hand, is not diffusion-controlled, does not depend upon solvent viscosity, and may be observed at lower concentrations of acceptor species. Energy transfer from the initially chemiexcited species to a suitable acceptor followed by fluorescence from the acceptor is an important process in chemiluminescence.

7. CHEMILUMINESCENCE REACTION PARAMETERS

CL reactions are commonly divided into two classes. In the type I (direct) reaction the oxidant and reductant interact with rate constant k_r to directly form the excited product whose excited singlet state decays with the first (or pseudofirst)-order rate constant $k_s = k_f + k_d$. In the type II (indirect) reaction the oxidant and reactant interact with the formation of an initially excited product (k_r) followed by the formation of an excited secondary product, either by subsequent chemical reaction or by energy transfer, with rate constant k_A. The secondary product then decays from the lowest excited singlet state with rate constant k_g. Type II reactions are generally denoted as complex or sensitized chemiluminescence.

If ϕ_{CL} is the efficiency of the chemiluminescent reaction, which is the ratio of the number of photons emitted to the number of molecules of reactant reacting *in toto*, it can be defined for a type I reaction as

$$\phi_{CL} = \phi_c \times \phi_E \times \phi_f \tag{5}$$

where ϕ_c, the chemical yield, is the ratio of the number of molecules that react through the chemiluminescent pathway to the total number of molecules reacted; ϕ_E, the excitation yield, is the ratio of the number of molecules that form an electronically excited product to the number of molecules that react through the chemiluminescent pathway; and ϕ_f is the quantum yield of fluorescence of the light-emitting species.

In a type II reaction

$$\phi_{CL} = \phi_c \times \phi_E \times \phi_{ET} \times \phi_f \tag{6}$$

where all symbols have the same meaning as above and ϕ_{ET} is the efficiency of energy transfer from the initially chemiexcited species to the energy-transfer acceptor. ϕ_{CL} is, of course, a function of the chemical and photophysical factors described in Sec. 1–3 such as solvent polarity, reagent concentrations, and molecular structure.

The physical significance of ϕ_{CL} is that under defined experimental conditions it is the constant of proportionality between I_{CL}, the observed intensity of chemiluminescence, and the rate of consumption of the initial luminophore (reactant L); i.e.,

$$I_{CL} = \phi_{CL} \times (-dL/dt) \tag{7}$$

8. KINETIC CONSIDERATIONS

8.1 Type I Chemiluminescence Reaction

If the reactants in a type I chemiluminescence reaction are rapidly mixed they will result in an emission whose intensity I_{CL} can be measured as a function of time. A typical time intensity curve for a CL reaction is shown in Figure 2.

The shape of the curve depends on the kinetics of the reaction. In Figure 2 the mixing process is not rate limiting and Eq. (7) is obeyed (if the mixing rate is rate limiting, the time required to achieve $I_{CL(max)}$ will be much longer). If the reaction is first order in analyte (L) (the oxidizing agent is in excess and $k_f \gg k_r$) then Eq. (7) can be written as:

$$I_{CL} = \phi_{CL} k_r [L] \tag{8}$$

L is the analyte concentration as a function of time ($L_o \times e^{-k_r t}$) and k_r is as defined in Sec. 7. In this case the intensity, as in all kinetic methods of analysis, can be

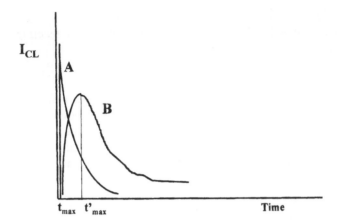

Figure 2   Chemiluminescence intensity (I_{CL}) as a function of time (t) for A, a reaction whose rate is much slower than the time for mixing of the reagents, and B, a reaction whose rate is comparable to the mixing time.

measured at a fixed time after mixing. The maximum emission intensity can be measured and related to concentration. If the total intensity-time curve is integrated as in endpoint or equilibrium analysis, the total light yield is obtained and the resulting integrated intensity will be proportional to the concentration and independent of the reaction rate. It is from the integration over the time course of the reaction that the great sensitivity of CL analysis is derived.

8.2   Type II Chemiluminescence Reaction

A type II CL reaction can be represented as sequential chemical and energy transfer reactions.

$$L + O_x \xrightarrow{k_r} P^* \qquad P^* + A \xrightarrow{k_{ET}} A^* \xrightarrow{k} h\nu + A$$

Here, O_x is the oxidant that reacts with the reductant L, k_r is the pseudo-first-order rate constant causing the rise in P^*, the excited product that is the intermediate in the type II reaction, and k_s is the rate constant for deactivation of A^* [$k_s = k_f + k_d$ in Eq. (1)]. P^* will fall in concentration with rate constant k_{ET} as it transfers electronic excitation to the acceptor A to form the excited acceptor A^*.

If P is in excess two extreme situations can be distinguished:

1.   If $k_r \gg k_{ET}$, the concentration of P^* will not be depleted and I_{CL} reflects [P^*] as a function of time.

2. If $k_r \ll k_{ET}$, then $P*$ is consumed as soon as it is formed and I_{CL} reflects the rate of the oxidation-reduction reaction.

In the case the k_r and k_{ET} are of the same order of magnitude, the intensity time curve can be represented as a biexponential function.

$$-\frac{dL}{dt} = \frac{k_r L_o}{k_{ET} - k_r} \times (e^{-k_r t} - e^{-k_{ET} t}) \qquad (9)$$

Combining Eq. (7) and (9) gives:

$$I_{CL}(t) = \phi_{CL} \times \frac{k_r L_o}{k_{ET} - k_r} \times (e^{-k_r t} - e^{-k_{ET} t}) \qquad (10)$$

which gives the course of the type II chemiluminescent emission.

9. SUMMARY

CL is the emission of light by molecules that are excited by participation in a highly exergonic reaction, usually an oxidation. It can also be defined as the fluorescence of the electronically excited product of a chemical reaction; however, unlike fluorescence, no external light source is used.

The precursors of the light-emitting molecule must participate in a reaction that releases a large amount of energy. For luminescence in the visible spectrum (400–750 nm) 40–70 kcal/mol must be released. Only certain oxidation reactions will generate this much energy. Often, there will also be a "dark" reaction pathway, which may lead directly to the ground-state species, or to ground-state products that differ from those produced by fluorescence.

Subsequent to the formation of a potentially chemiluminescent molecule in its lowest excited state, a series of events carries the molecule down to its ground electronic state. Thermal deactivation of the excited molecule causes the molecule to lose vibrational energy by inelastic collisions with the solvent; this is known as thermal or vibrational relaxation. Certain molecules may return radiationlessly all the way to the ground electronic state in a process called internal conversion. Some molecules cannot return to the ground electronic state by internal conversion or vibrational relaxation. These molecules return to the ground excited state either by the direct emission of ultraviolet or visible radiation (fluorescence), or by intersystem crossing from the lowest excited singlet to the lowest triplet state.

Molecules that are involved in CL reactions are generally reduced species that can be easily oxidized, such as molecules containing amino and hydroxy groups and polycyclic aromatic ring systems. The solvent in which the experiment is carried out has a dramatic effect on the efficiency of the reaction. Solvation can alter the shapes, the depths, and the densities of the vibrational states of the potential surfaces representing the ground states of products and reactants, as well as the lowest excited singlet state of the potential fluorophore. Structure and environment influence the rates of processes competing with fluorescence for deactivation of the lowest excited singlet state.

The quantum yield of fluorescence determines the intensity of light emission in CL reactions. The spectral wavelengths that are fluoresced depend on the structure of the molecule. Fluorescence is most often observed in highly conjugated or aromatic organic molecules with rigid molecular skeletons. Aromatic molecules with freely rotating substituents tend to fluoresce less intensely than those without these substituents. Molecular structure has a large effect on the position in the spectrum where fluorescence occurs, as well as its intensity.

Solvent interactions with solute molecules are mainly electrostatic. The differences in electrostatic stabilization energies of ground and excited states contribute to the relative intensities and spectral positions of fluorescence in different solvents. Changes in dipolar and hydrogen-bonding properties of the solute also affect the wavelength that is fluoresced. Fluorophores in solvents containing atoms of high atomic number fluoresce less intensely than those without heavy atoms, because the high atomic weight enhances the spin-orbital coupling in the lowest excited state of the solute.

Fluorescence may be decreased or completely eliminated by interactions with other chemical species in a process known as quenching of fluorescence. Two kinds of quenching may occur. The first is known as static quenching, where the interaction between the potentially fluorescent molecule and the quencher takes place in the ground state, forming a nonfluorescent complex. Dynamic quenching may also occur when the quenching species and the potentially fluorescent molecule react during the lifetime of the fluorescent molecule.

Energy transfer involves the passing of excitation energy to another molecule by a molecule in its excited state. The loss of excitation energy from the donor species results in quenching of the luminescence of the energy donor, and may result in luminescence from the energy acceptor that has become excited in the process. Energy transfer can occur by either of two processes. The resonance excitation transfer mechanism, or dipole mechanism, involves energy transfer between two molecules that are not in contact with each other, and may be separated by as much as 10 nm. The exchange mechanism, or Dexter mechanism, of excitation energy transfer is important only when the electron clouds of the donor and acceptor molecules are in direct contact. In this case, the highest-energy electrons of the donor and acceptor may exchange places. This is most efficient when the fluorescence spectrum of the donor overlaps the absorption spectrum of the acceptor.

There are two common classes of CL reactions. In the type I reactions the oxidant and reductant interact to directly form the excited product whose excited singlet state decays with the first order rate constant k_s. In the type II reaction the oxidant and reactant interact with the formation of an initially excited product, followed by the formation of an excited secondary product, either by subsequent chemical reaction or energy transfer. The secondary product decays from the lowest excited singlet state with a rate constant k_g. The kinetics of type I and II reactions also differ.

ACKNOWLEDGMENT

The authors are grateful to Mrs. Nancy Rosa and Mrs. Virginia Schulman for technical assistance with the preparation of this manuscript.

REFERENCES

1. KD Gundermann, F McCapra. Chemiluminescence in Organic Chemistry. Heidelberg Springer-Verlag, 1986.
2. F McCapra, KD Perring. In: JG Burr, ed. Chemi- and Bioluminescence; Clinical and Biochemical Analysis. Vol 16. New York: Marcel Dekker, 1985.
3. E Lissi. J Am Chem Soc 98:3387–3388, 1976.
4. WR Seitz. Crit Rev Anal Chem 13:1–58, 1981.
5. AK Campbell. Chemiluminescence, Principles and Applications in Biology and Medicine. Heidelberg: VCH Publishers, 1988.
6. SG Schulman. Fluorescence and Phosphorescence Spectroscopy, Physiochemical Principle and Practice. London: Pergamon Press, 1977.
7. MA DeLuca, ed. Bioluminescence and Chemiluminescence, Methods in Enzymology. Vol 57. New York: Academic Press, 1978.
8. MA DeLuca, WD McElroy, eds. Bioluminescence and Chemiluminescence, part B, Methods in Enzymology. Vol 133. New York: Academic Press, 1978.
9. LJ Kricka. Clin Chem 37:1472–1481, 1991.
10. BM Krasnovitskii, BM Bolotin. Organic Luminescent Materials. Heidelberg: VCH, 1987.
11. K Nakashima, K Imai. In: SG Schulman, ed. Molecular Luminescence Spectroscopy: Methods and Applications: Part 3. New York: Wiley-Interscience, 1993, pp 1–23.

4

Recent Evolution in Instrumentation for Chemiluminescence

Dan A. Lerner
Ecole Nationale Supérieure de Chimie, Montpellier, France

1. INTRODUCTION

In recent years no truly new basic principles have been introduced for the detection of luminescence. However, the technical evolution in the field of microelectronics and optoelectronics, charge coupled device (CCD) detectors, fiberoptics, assembly techniques, and robotics resulted in the introduction on the market of new generations of instruments with increased performance, speed, and ease of handling. In this chapter, some of their typical features will be reviewed. To keep this presentation at a concrete level and to illustrate some specific item, instruments of different makes will be referred to. However, this does not imply they are better than those not cited. It is more a matter of availability of recent documentation at the time of writing. Note that numerical values cited typically relate what can be done today and may vary from one instrument to another from the same company.

Although the present chapter lays emphasis on CCD and CCD-based instruments, it begins with a section devoted to photomultiplier tubes (PMT)-based luminometers.

2. EVOLUTION OF NONIMAGING LUMINOMETERS

2.1 Early Instruments

The last three decades have seen a slow but constant evolution in the conception of instruments for chemiluminescence (CL) or bioluminescence (BL). Initially users had to assemble their own luminometers but performance was not very good. Photomultipliers were used in the current detection mode, early amplifiers were noisy and affected by drift, and signals were fed to standard recorders with a slow response. Many improvements were made progressively and in different ways. At one time, researchers even tried to turn scintillation counters into luminometers, using them in a noncoincident mode to benefit from their photon-counting capability [1].

The problem of quantification of CL is better understood if the time evolution of intensity for a CL or BL reaction is analyzed. This time evolution is typically an asymmetric bell-shaped curve with a rise time that may last from much less than a second to minutes or more depending on the reaction studied. Several methods were devised to extract useful quantitative data from such a curve. The area under the curve may be integrated to define the total light emission (TL). The observation was also made that the peak intensity for CL is related to the overall rate of the light-producing reaction. So the peak height at maximum intensity may be used to quantify light emission. These basic factors, to which must be added the fact that the amount of light emitted in some important reactions is exceedingly small, explain the problems met by early experimentalists. These factors are also the ones that led the way to further improvements.

Quantification of light was first improved by the use of a printed circuit board, which allowed better reproducibility and new types of measurements. Integration of a delayed light reaction curve [2] could be carried between selected time limits (the mixing time of reagents was used to define the origin of the time scale). Besides integrators, various peak detectors were developed. Until the early nineties, many luminometers were still very simple instruments. They consisted in a PMT detector, a chamber for measurements, a sample injector (hand mixing or syringes), and the associated electronics. Direct reading was usually obtained following a simple calibrating step. Luminometers were basically analog instruments even if some used PMT in the photon-counting mode. Batch measurement was standard and typical instruments required a volume of about 1–10 mL in a test tube or a vial to carry out a determination.

2.2 Last Generation Instruments

The present generation of luminometers is composed of computer- or microprocessor-driven digital instruments. Time limits are precisely defined by the user and controlled over a wider range (0.1–200 s or more). Even if an instrument works with the current measuring method, use of high-gain, high-impedance AC amplifiers and of the electronic nulling of the background results in high sensitivity. Cooling the PMT results in an increased dynamic range. Even if all aspects of current measuring versus photon-counting modes of operation are not yet fully understood, the photon-counting mode appears to deliver the best performance at the low intensity end of the dynamic range. Ellis and Wright [3] justify this conclusion after a careful examination of the various sources of noise and of the dark current or dark count contribution.

The commercial availability of ready-to-use photon-counting packages is also a reason for the success of this technology. These assemblies are light, compact, and all elements are fully integrated, including the high-voltage source. They only require a low-voltage input and generally all controls are preset to provide the best performance so that no adjustment is required. This allows the construction of sensitive laboratory-assembled luminometers for low light level. Spectroscopic studies may be carried out with some of these instruments [4]. Furthermore, these photon-counting modules are also incorporated in commercial instruments [5]. PMTs are by far the preferred detectors for CL. Recently, however, avalanche photodiodes (AVP) brought the cost benefit of solid-state instrument together with good overall performance to new luminometers. An interesting comparison of PMT and AVP has recently been published by Fullam [6].

A few words should be said about the existence of PMT-based instruments that are developed to solve specific problems in chemi- or bioluminescence. For instance, marine laboratories have developed and improved over time a range of so-called bathyphotometers for hydrobiophysical measurements (microalgae, zooplankton in the surface waters of the sea) [7].

The high sensitivity and fast counting rate of recent instruments allow a large number of repetitions per sample, which makes possible meaningful statistical analysis. As noted earlier, most luminometers are computer-driven and often benefit from the presence of software. The latter allows choosing procedures, such as setting the parameters for quantification of the light emitted in assays from flash- or glow-type reagents (TL, integrated light emission between two time limits, average emission during a selected interval, peak emission), or data handling such as curve fitting for kinetic enzyme studies. In case no software is available, provisions are usually made to transfer results to a spreadsheet.

With today's instruments the customer may specify an injector or an automatic reagent dispenser, and a choice of sample formats is available: test tube, vial, microplate, Petri dish. The volumes of sample and reagent required to carry out a determination steadily decrease and are typically a few microliters. Sample temperature during the assays is generally controlled.

Flow injection analysis (FIA) is a convenient technique for automatic measurements. It can be easily adapted to measure CL. Typically two or more liquid solutions containing the analyte, reagents, and/or buffers are mixed sequentially in purposely interconnected narrow tubes. A multichannel peristaltic pump insures an even flow. After the last mixing operation the mixture is pushed through a flow cell where the light-emitting reaction occurs. An example of flow cell consists in a quartz coil (typical length 1 m, total volume 50–100 µL) positioned in front of the PMT window. FIA was developed before the advent of robotized systems to handle microplates. This technique owes much to its versatility. One drawback is its higher consumption of reagent, which means often a higher cost per determination.

Despite all these developments, some problems remain. One of them concerns the comparison of results obtained from different instruments. Solutions involving standards are proposed. But most of them suffer from drawbacks: thermal or photochemical instability, multistep preparation, wavelength inadequacy between standard and actual fluorophore emission. Many solutions have been advocated to calibrate luminometers [8].

2.3 Bibliographic Sources for Commercial Instruments

Scientific and technical literature provide a wealth of information on the detailed historical and technical developments of CL. Early reviews describing commercial apparatus appeared as early as 1968 [9]. In this context it is important to cite the recent comprehensive set of survey updates written by Stanley since 1992 (see [5] and references therein). These include not only instruments but reagents or kits for CL or BL. They constitute an excellent source of information about commercial instrumentation.

3. OVERALL IMPROVEMENT IN THE DESIGN OF IMAGING INSTRUMENTS

Until recently, most standard or imaging luminometers were a simple combination of existing parts, such as a dark cabinet, a photon-counting photomultiplier or a CCD camera, a computer with an acquisition board, and image-processing software. Generally, these components had not been specifically selected or tailored in view of their effective integration. Although good performance could be obtained, reproducibility was not always guaranteed and the analysis of samples and treatment of data were neither straightforward nor fast. The concept of high throughput was not yet important.

Starting in the mideighties, an increase in sensitivity and speed as well as the simplicity in the handling of reagents resulted in the fast development of luminometric tests at the expense of isotopic labeling tests. This move was later associated with an increased availability of molecular luminescence probes. In an almost parallel move, pressure grew due to needs originating from gene sequencing and comparative genomic hybridization (CGH), to name only two. As these techniques present strong requirements in terms of speed, sensitivity, and spatial resolution, the quasi-artisan aspect of instrumentation referred to above was bound to disappear. The design of recent instruments insures generally that the best performance from all components will combine to provide the highest sensitivity, reproducibility, speed, and ease of handling.

As a consequence of the optimization of performance, different performing instruments were first designed for one specific task. However, the best instruments from the latest generation correspond to universal instruments that make it possible to run several types of experiments such as fluorimetry, luminometry, and densitometry for instance. This has interesting consequences for users who work in different domains of luminescence for research and development purposes and do not have to buy a whole range of instruments.

However, many recent instruments are still not considered satisfactory, since professional developers in the field of high-throughput screening (HTS) want to use the full performance of the latest generation of robots and computers for automation. This results in new instrumental developments, like the possibility of reading not only 96, but 384 or even 1536 wells plates as well as DNA chips, very rapidly (in a minute or so) and repeatedly without any mechanical failures. Hence, in the eyes of company scientists developing new assays, many present-day instruments still correspond to an intermediate stage of development. For research laboratory scientists, on the other hand, the actual equipment offers excellent performance.

4. USER-ASSEMBLED SYSTEMS ARE STILL COMPETITIVE

As noted above, the number of integrated instruments has grown tremendously. However, the intrinsic quality of equipment that can be considered as subsystems or components for assembling, for instance, an imaging luminometer has also greatly increased. As a consequence, even today, an expert user may assemble a unit to carry out a specific job with good efficiency.

5. CCD CAMERAS: THE HEART (AND EYES) OF IMAGING SYSTEMS

Special CCD cameras are used to acquire high-resolution images for scientific work. They are also called HCCD cameras. They belong to the two following classes: cooled CCD cameras or intensified CCD cameras. Intensified cameras have been mostly used to follow spectroscopic fast events but they are now used in imaging to offer very high sensitivity. Cooled CCD cameras, however, offer better spatial resolution for luminescence work in the field of analysis and a broader combination of spectral ranges and sensitivity.

Among the components or subsystems found in luminescence imaging, CCD cameras constitute the key element. There are two main reasons to use these devices. The first is that they can detect very weak light emission and thus provide high sensitivity. The second reason is that they have an extended linear range for the conversion of light intensity into an electric signal. For the user this means carrying out the quantitative determination of an analyte over an extended range of concentrations.

Of course, other components are necessary to assemble an instrument. However, they are generally less complex and have been optimized more rapidly or their characteristics were not so critical to the overall performance (at least before the CCD reached the present state of development). In contrast, the problems that faced the use of CCD cameras when they appeared on the market some 25 years ago were not fully appreciated at once. These problems have been solved one after another, leading to new hardware features and new concepts. Even now, the conception of a CCD camera for its integration in a specific instrument is not a straightforward process. Without delving into the hardware complexity, it seems interesting to describe—even in a simplified manner—the structure and basic characteristics of modern CCD cameras.

6. A SIMPLE DESCRIPTION OF THE HARDWARE

In a crude description, a CCD chip consists of a mono- or bidimensional array of tiny elementary detectors (the size of which may reach less than 10 μm by 10 μm), formed on a silicon substrate and with an associated electronic circuitry (Fig. 1). Each of these detectors will define a pixel, that is, an elementary constituent of an image. For this reason, such an elementary detector is also often called a pixel and we will use this terminology in the rest of the chapter. For imaging purposes, which concerns the present chapter, only 2D arrays are used, with a pixel size in the vicinity of 50 μm by 50 μm. In short, during acquisition of a signal, each pixel of the array will convert the photons reaching its surface into electrons that drift toward a well (the depletion region) where they are stored. In the following step, called the readout, the trapped electrons will be counted and the count will define the value of the corresponding pixel in the resulting image.

At a more detailed level, one must mention the presence of elements called electrode gates, which overlap the pixel surface. During the readout process, a voltage is applied in sequential steps across the electrodes to shift the charge of each pixel row by row along one dimension of the array until the charge reaches the bottom row, which constitutes the serial or shift register of the chip (Fig. 1). The charge thus collected will then move along the serial register until it reaches

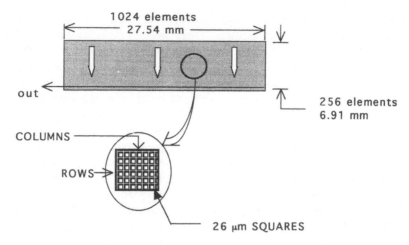

Figure 1 Layout of the pixels of a CCD. (Reproduced by permission, from the Book of Photon Tools, Oriel, Stafford, CT.)

the end of the array and enters a low noise preamplifier (Fig. 2). The preamplified signal is then fed to an analog-to-digital (A/D) converter and the digital signal is sent to the computer. All pixels are treated sequentially.

Readout speed for a typical CCD camera is somewhere between 100 kpixel s^{-1} and 5 Mpixel s^{-1}. Cameras with up to 4 Mpixel chips have been built. How-

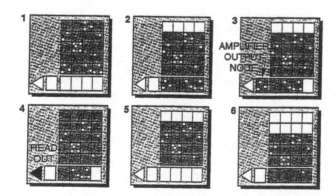

Figure 2 Readout pattern of a two-dimensional CCD. The bottom row corresponds to the shift register. Its output is connected to the preamplifier input. (Reproduced by permission, from the Book of Photon Tools, Oriel, Stafford, CT.)

ever, for imaging very low light levels, slow scan cameras are used that read out at less than 50 kpixel s^{-1}.

A microprocessor set in the camera housing together with the preamplifier and A/D converter controls two clock drivers to precisely time the readout sequences. In some recent cameras the preamplifier is integrated on the chip. The computer sends the commands to control the acquisition step (setting the temperature and exposure to the source of light, acting on the shutter when reading continuous signals) and the readout step (counting the number of electrons stored in each pixel). Exposure and readout are normally triggered as two successive events. Note that in a CCD the process of reading a pixel is destructive.

Strictly speaking, the elements of hardware mentioned above are only those of the chip. A camera is built around a chip, adding ancillary equipment such as an optics or a cooling system. The simple description provided above allows, however, an analysis of the most important features to take into account when selecting or using CCDs.

7. KEY PROPERTIES FOR DEFINING THE PERFORMANCE OF A CCD CAMERA

7.1 Introduction

Some of the parameters defined below are interconnected and it is impossible to introduce all of them one by one independently of the others. The subject is complex and the presentation given here is simplified. More may be learned on the subject from the equipment catalogues of various companies [10–12] or from specialized publications [13, 14].

7.2 Resolution

Spatial resolution is limited by the size of the pixel of the array. Its choice depends on the intended use of the CCD camera (spectroscopy, luminometry, etc.). The pixels of a CCD chip form a flat bidimensional array and as such are conceived to acquire a 2D picture. The consequence is that to keep the benefit of high resolution, the object used to form an image must be as flat as possible (chromatographic plate, blot, microarrays, Petri dish, etc.). However, the associated optics may project a distorted image of the peripheral region of a large object on the chip. As the size of the chip is limited to a few square centimeters, it means that for given optics, the size of the object is limited so as not to degrade performance.

Some objects may have a large size (gels, autoradiographic films, or TLC plates, for instance) and special techniques must be used to obtain a global image while keeping the high resolution constant. The usual way to treat the problem is to scan the object.

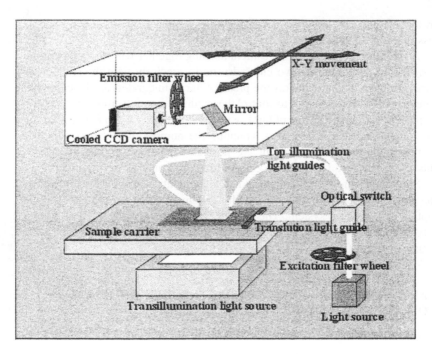

Figure 3 Inside view of the Wallac Arthur multiwavelength imaging system. (Reproduced by permission from EG&G Life Sciences, Evry, France.)

To illustrate the process, we describe the solution provided by the Arthur imaging system from Wallac EGG [15]. It is a general-purpose, CCD-based horizontal scanning imaging system that scans sample areas as large as 23 cm × 28 cm to provide images with a constant resolution down to 50 μm. Both luminescent labels and fluorescent dyes may be used. The general structure of this instrument is shown in Figure 3.

The sample is placed horizontally on a two-axis transport mechanism at a predefined focal plane. A cooled CCD camera scans an area previously defined by the user and takes multiple adjacent frames covering that area. These frames are combined in the computer into one single high-resolution image.

7.3   Quantum Efficiency

Quantum efficiency (QE) is defined as the ratio of the number of photoelectrons created by a given light signal to the number of photons in this light signal. This parameter is wavelength-dependent and is usually available as a graph of QE versus wavelength.

7.4 Dynamic Range

The dynamic range factor is defined as the ratio of the largest signal that can be acquired to the signal corresponding to the limit of detection (LOD). In this definition, the signals correspond to the stored charges and are analogous signals. So the dynamic range is a measure of the range of signals, hence of the range of concentrations of the analyte, that can be measured simultaneously.

When working conditions have been defined for a given experiment, the signal corresponding to the LOD is the signal the level of which is equal to the noise associated with this signal. This definition of the LOD corresponds to a signal-to-noise (S/N) ratio of 1. Note that values of 2 or 3 for S/N are used in different subfields of analytical chemistry to define the LOD. The two most important parameters to set the working conditions are temperature and exposure time.

What is the largest signal that can be measured? It was noted earlier that when a signal is acquired during the exposure time, photoelectrons are generated and stored in each pixel. It turns out that each pixel cannot store more than a specific number of electrons, which is known as the well depth capacity or saturation charge. It is expressed in electrons or charge per pixel. The saturation charge is typically in the range 300,000–500,000 electrons for HCCDs.

The charge corresponding to a given signal varies from 0 (supposing for the moment that one does not have a dark current) to the saturation charge. As noted before, it is an analog signal that is converted after the readout step into a digital binary value by the A/D converter. The total range for the binary values is divided in n levels so that the converted saturation charge corresponds to the maximum value of 2^n. Typical values for n are 12, 14, or 16. In the latter case this corresponds to 65,536 levels (gray levels).

It is clear that the n value should be such that 2^n is at least equal to the dynamic range to benefit from the full performance of the chip.

Note that any signal may be conveniently measured in terms of electrons or of energy per unit surface (J cm^{-2}). To carry the conversion between the two modes, one has to use the wavelength dependence of the QE of the chip and the surface of the pixel.

7.5 Dark Current

All CCD chips have a leakage current, also called dark current, which is of thermal origin as in photomultipliers. As a result of this phenomenon, a charge slowly builds up in each pixel even when the detector is shielded from light.

This means that any signal is contaminated to some amount by the dark current. As the latter is of thermal origin, it varies with temperature. In fact, the variation is exponential and cooling the chip results in reduction of the dark current by a factor of about 3 for each 10°C lowering of temperature. So many CCD devices are cooled to obtain a low dark count value (see Sec. 8.1). This also reduces baseline drift induced by temperature changes.

7.6   Noise

7.6.1   Introduction

In Sec. 7.4, the word *noise* was used. The analysis of noise in CCDs is quite complex and a simplified analysis will be given here. True noise is also called pixel noise and results from the combination of several noise components. The latter, analyzed in the next three sections, does not add together directly to give the total noise [10, 12, 14].

7.6.2   Shot Noise from the Signal

Any light signal is composed of discrete particles, the photons, which are emitted at random. The resulting fluctuation in the number of photons reaching a pixel in a unit of time is transferred to the photoelectron flux generated by absorption. The fluctuation in the associated current is the shot noise. Due to the nature of the photons, the associated shot noise for a flux of N particles is equal to the square root of N. It depends on the exposure time.

7.6.3   Shot Noise from the Dark Current

The electrons associated with the dark current are also released in a statistical manner and are associated with shot noise. The latter is very sensitive to temperature and also depends on the exposure time. Its contribution to the total noise is significant only at low signal levels.

7.6.4   Readout Noise

An electronic noise component is also generated by the transfer of charges and by the preamplifier. For each readout process, one readout noise is generated. This readout noise is not very sensitive to temperature but increases with reading-out speed. Readout noise for a HCCD is about 10 electrons RMS or less.

7.6.5   Fixed Pattern Noise

In the manufacturing process inhomogeneities arise in the silicon substrate and all the pixels in the array are not exactly identical. They may show variations in their response to the same photon flux. Even in the absence of a light signal, differences occur in the dark current in a group of pixels. However, these differences remain constant from frame to frame if temperature is kept constant. As a result, this fixed pattern noise may be corrected for by simple subtraction.

7.7 Binning

Pixel binning is a technique used to increase the S/N ratio and, as a corollary, the dynamic range. It is a specific reading mode of the pixels allowed by the software: in the binning mode, a group of pixels (a superpixel) is read out together as if it were a single large pixel. As one readout noise is generated for each readout, whether it relates to a single pixel or to a superpixel, the S/N ratio is increased as a result of binning. Usually, the software allows the user to control the size of the superpixel. Note, however, that binning results in a loss of spatial resolution so that the user must trade off between dynamic range and spatial resolution.

8. IMPROVING PERFORMANCE

8.1 Dark Current Reduction by Cooling the CCD Chip

Practically, the chip is cooled by means of a cryogenic fluid (often liquid nitrogen) or, more conveniently, by heat flow from the chip to a multistage Peltier thermo-electric device. By cooling at -40 to $-50°C$, dark current may be brought down to less than a hundredth of an electron per pixel per second (e^- pixel^{-1} s^{-1}) allowing a much larger exposure time, hence an increased sensitivity. For instance for the ultimate generation of the ORCA series, which use 1280 by 1024 chips, Hamamatsu claims 3.6 e^- pixel^{-1} h^{-1} [16].

8.2 Effect of Readout Rate on Readout Noise

As noted briefly above, the readout noise level is sharply decreased when reading-out rate is low. Cameras allowing a slow readout are called slow-scan CCD cameras.

8.3 Back-Illuminated CCD Cameras

This class of CCD has a higher quantum yield than a front-illuminated similar CCD. The origin of this improvement is as follows. The silicon substrate for the chip is etched down, on the face opposite the electrodes, to reach a thin layer (~15 μm), and the light illuminates the array from the back, opposite from the usual way (Fig. 4). It was noted in Sec. 5 that electrode gates that are used for the readout process overlap the pixel surface. Since these electrodes have a given thickness and are laid on the front face, photoelectrons have to travel across a

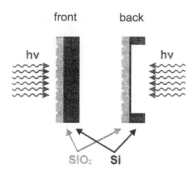

Figure 4   Principle of a back-illuminated thinned CCD chip. (Reproduced by permission from EG&G Life Sciences, Evry, France.)

larger thickness of silicon to reach the well and have a lower probability of reaching it. Furthermore, in a front-face CCD, light must penetrate the surface layer of silicon oxide, which is on top of the array and absorbs UV radiation. For this reason the QE is much higher in a back-thinned chip. It changes from about 40% for a front-face CCD to 70–80%, in the 600-nm region. However, the manufacturing of these chips is delicate and cameras built around a back-illuminated chip are more expensive.

8.4   Other Improvements

Some cameras allow multiple-mode readout. The user may switch modes between ultralow-readout noise for a high dynamic range and high-speed frame capture at reduced resolution. This is the case, for instance, for the Hamamatsu Orca II (high sensitivity at 14 bits vs. 5.3 fps at 12 bits).

QE may be increased selectively in different wavelength ranges. For instance, a layer of a UV-to-visible converter may be deposited on the front of a standard chip to boost sensitivity in the UV region. One example is Metachrome II from Photometrics [17]. This is only one of the solutions and users should carefully specify their requirements to be sure they get exactly what they need. QE may also be increased in the orange-red and NIR region of the spectrum. Special chips are made with an increased thickness to compensate for the low absorption of the material in that region.

Background subtraction is used to improve contrast. In this technique, a background pseudoimage is obtained with the shutter closed and is subtracted from the actual image, which of course must be acquired under the same conditions as the background. Most dark current and fixed pattern noise is removed and sensitivity is enhanced.

In video microscopy, for instance, background is normally subtracted using differential interference contrast (DIC) [18]. This technique, which requires a number of manipulations from the user, may now be automated using a new method called polarization-modulated (PMDIC) [19, 20]. It requires the introduction of a liquid crystal electro-optic modulator and of a software module to handle difference images. PMDIC has been shown to bring improvements in imaging moving cells, which show a low contrast, as well as thick tissue samples.

When reading a gel, a thin-layer chromatography (TLC) plate, or a DNA chip, different spots may give rise to signals that differ greatly in intensity. As a result, one pixel has not yet accumulated a sufficient number of charges to give an acceptable S/N ratio than another one is filled and overflows to neighboring pixels. The latter phenomenon is called blooming. Its consequences are a loss of resolution, errors in appreciation of signal intensity, and loss of time to dissipate the charge to start a new measurement. Some CCD chips are made with an anti-blooming drain attached to each pixel. They avoid the overflow of charges to neighboring pixels.

An operating mode of a CCD called multipinned phase (MPP) allows a hundredfold dark current reduction at room temperature [21].

9. FILL FACTOR AND THE HANDLING OF COLORS

High-resolution CCDs have another interesting property. The fill factor of an imaging detector is the fraction of the array actually used for image acquisition. In these HCCD cameras, the fill factor is 100% as there is no blind zone in the array that would lower the resolution. In color video cameras for production line or mass market, each pixel is a compound unit made of three adjacent pixels, which are respectively sensitive to red, green, and blue (RGB system). Any pure R, G, or B light reaching the array will activate one of the three pixels only, the other two being blind. For any other color, the three pixels will each build a charge and a color will be affected to the set. So the fill factor for this type of camera is $1/3$, with a corresponding decrease in resolution.

To obtain color images with a HCCD camera, filter wheels will be used, in which each filter will be chosen specifically for one fluorescent label, in emission and in excitation. For CL or BL work, one filter wheel is sufficient. For fluorescence, two are necessary (see, for instance, in Fig. 3 the positioning of the filter wheels in the optical path). If, for instance, a three-color experiment is performed, an image will be acquired for each label with the adapted filter(s). The three gray-level images are sequentially acquired. They will then be "colored"

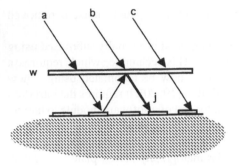

Figure 5 Crosstalk between pixels due to multiple reflections between the window and the surface of the CCD chip. A photon coming from *a* initially should hit the pixel labeled *i*. If it is reflected off the surface, it bounces back toward the window and will be absorbed by pixel *j* as if it were coming from *b*.

individually by the computer in function of the specific filter/dye combination before being combined into a global multicolored image. So the high resolution is preserved and at the same time, each component image may be acquired with an individually optimized exposure time [22].

An interesting feature of the technique is that the photosensitivity of each dye may be taken care of individually.

10. OTHER PROBLEMS LINKED TO HARDWARE OR OPTICAL DESIGN

Only one example will be given. It concerns a specific form of stray light that is observed when the entrance face of the CCD is protected by a transparent window. This situation arises mostly for cooled CCDs. A small fraction of the measured light is reflected off the surface of the CCD chip, hits the window, and bounces back toward another pixel of the CCD chip (Fig. 5). So the reading on the first pixel is lowered and that on the second is increased, the overall effect distorting the true distribution in intensity.

11. CHARGE INJECTION DEVICE

A charge injection device (CID) is very similar to a CCD as it consists also in a 2D grid of pixels on a semiconductor substrate. The main difference between the devices is that the readout process in a CID is a nondestructive readout

(NDRO). It is possible to interrogate any pixel of the array at any time to follow the buildup of the charge and the S/N ratio may be selectively optimized for a pixel by increasing its allotted collection time. However, the surface-mounted complex electronic circuits associated with the NDRO absorb a larger amount of the signal light than the electrode gates of CCD chips. This results in lower quantum yields, lower readout speed, and higher readout noise. So CCDs are preferred for luminescence work.

12. COMMERCIAL INSTRUMENTS

Following are some examples of cameras presently available. Some recent cooled intensified cameras compete with cooled back-illuminated cameras. For instance, Cooke DiCam-PRO [23] is an intensified cooled imaging system, with a 12-bit dynamic range and exposure down to 1.5 ns, conceived for low-level light and high-speed imaging applications. From the designers' data, one may infer that it would have a global performance comparable to that of back-illuminated models such as the PixelVision Bioxight 14 bits modular camera [24]. The latter has a resolution ranging from 512×512 to 4096×4096 pixels. When configured for low-noise operation (10-Mhz system bandwidth), this camera can operate 1024×1024 CCDs at better than 15 e$^-$ rms noise at 10 frames per second (fps). However, comparisons of a limited set of parameters may be misleading and potential users should clearly rank their needs to make the best choice.

Looking at systems that incorporate CCD cameras, the Berthold NightOWL$^{™}$ [25] is a good example for a polyvalent imaging system for detection of ultraweak luminescent and fluorescent signals. Its technology is designed to increase the S/N ratio to the maximum possible, e.g., Peltier/air cooling down to $-70°C$ of a back-thinned CCD. Both microscopic and macroscopic 2D or 3D samples could be measured. The camera is enclosed in a light-tight cabinet and may be moved vertically to fit the CCD detector into the actual sample size. Resolutions up to 80 μm may be reached (Fig. 6).

Inside the dark box main sockets and automation port ensure integration of special probe requirements like incubators. There is provision to insert a fiberoptic-guided light source that accept filters for fluorescent work. The sensitivity of the NightOWL is claimed to enable direct detection of reactive oxygen species, even without enhancers like luminol.

The last generation of instruments make it possible to work faster as they offer an increased S/N ratio for the detectors. This results in shorter integration times. Furthermore, the computer-driven reading of microplates with 384 or 1536 wells rather than 96 increases the speed of analysis. Original and new design of the optical path of light and optimized and computer-driven selection of the optical elements and of the mechanical positioning of filters, detectors, or samples

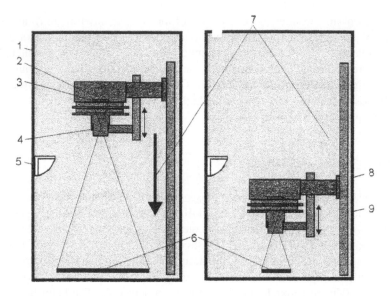

Figure 6   Schematic view of the NightOWL. (Reproduced by permission from Berthold EG&G Life Sciences, Evry, France.)

for a given experiment became important factors. These instruments allow new modes of excitation for fluorescent samples and/or handling of samples with large areas. They are generally built to cover genomics and proteomics applications, such as electrophoresis gels and various microarrays, and to analyze metabolites expressed as cells or organs.

In the field of DNA sequencing instruments, Perkin-Elmer [26] switched from the PMT-based model 373 sequencer to the CCD-based model 377, which allowed the simultaneous discrimination of four or more colors in emission, in a single lane of an electrophoretic gel. The CCD made possible the use of a specific fluorescent label for each type of base and throughput was improved more than four times [27]. Excitation uses an argon laser.

There are, however, many approaches to DNA sequencing, and even some recent instruments such as the sequencer made by LI-COR [28] do not use a CCD. The latter uses instead a very-low-noise silicon avalanche photodetector to excite near infrared-emitting dyes. The source is a laser diode emitting at 785 nm. It is a compact system that can be mounted on a focusing stage with confocal optics and it is meant for small laboratories that do not have HTS requirements.

The Wallac Arthur multiwavelength system, shown earlier (Fig. 3), is an example of a recent polyvalent CCD-based instrument. Three light sources make it possible to perform several types of excitation: top (fluorescence or reflectance), bottom (transmission), and a special edge illumination. The multiwavelength capability is given by six position excitation and emission filter wheels. The basic model uses a Kodak KAF-0400 CCD chip [29]. As a general-purpose imager, it allows measurements on various media, including membranes, blots, TLC plates, exposed autoradiography film, whole body sections, microscopic slides, and all kinds of documents. Kodak distributes CCD chips but also produces systems for CL detection and quantitation of proteins. Its Digital Science Image Station 440CF [30] uses a 752×582 pixel CCD equipped with Peltier cooling. The S/N ratio reaches 1600 for a single frame and rises to 6000 for multiple images. The images generated by the 12-bit camera, which is isolated from the sample to fight contamination and humidity, are linear over 3.0 orders of magnitude [31]. Remember that photographic films have a dynamic range of 1.5–2.0 orders of magnitude. Similar instruments are made by NucleoTech with its NucleoVision 920 instrument [32].

IF CCD-based detectors are the mainstay for luminescence imaging, PMTs are used in many instruments owing to their very high gain and low dark current. Their performance is still being improved in two areas. The first one relates to the nature of the photocathode or dynode design. Such intrinsic improvements lead to the availability of new UV or extended near-infrared PMTs with intrinsically lower noise. The second area of improvement is the integration of a PMT in a convenient module that fulfills a specific function. For example, Hamamatsu produced a side-on cooled PMT, the R6060, with a dark noise 10 times less than for normal PMTs [16]. The R6060-12 version has a built-in heatsink and fan to reduce initial cooling time. This device is convenient for measurements on samples such as living tissues for which it is not possible to control the temperature. Hamamatsu also produces the H7467 series, which is an ultracompact sensor that includes the necessary electronics for photon counting and a high level of functionality. It is an ideal device when a compact unit is required for luminescence work.

The Analyst from LJL Biosystems [33] is specially designed for HTS and runs with robots. It presents analogies to the Arthur described above, but is a different type of instrument as it is not an imager and relies on a PMT to detect light emission. It is dedicated to reading microplates. It allows homogeneous assays using time-resolved fluorescence and fluorescence polarization. Its specially designed read-head, which limits crosstalk between adjacent wells, and a selected PMT make it an efficient tool in luminescence. Each element is optimized for maximum reliability. It is claimed that it delivers the same performance for 384- as for 96-well microplates due to a specific focusing system. It has a fully automated optical system, which reconfigures the system when changing the method of analysis to eliminate downtime.

Other manufacturers of PMT-based luminometers designed to read microplates include Anthos [34] and Tecan [35], which manufacture a wide range of instruments as well as adapted robotic processors and manipulators. The Lumi-count from Packard [36] is also a plate luminometer designed to detect kinetic-based CL. It also uses a PMT, which may run in a mode called digital photon integration in which high voltage, hence gain, is controlled to optimize the S/N ratio for each assay.

13. SOFTWARE

All recent instruments are controlled by personal computers and come with software applications. For luminescence work, the intrinsic performance and ease of operation of the instrument are the most important features. However, the full potential of a good instrument may be partly offset by an inadequate or insufficient software package.

In the field of imaging, present-day software should use mouse-driven or menu-selectable commands to control the camera and allow real-time image capture. The latter feature allows the users to monitor the image buildup and trigger the capture at the best moment. Temperature control and exposure times should be under software control.

For all instruments, the software should allow users to program a number of buttons for frequent assays, to run different types of calculation (examples: background subtraction, statistical analysis, curve fitting, and/or storage of raw data for quality control). Almost all instrument producers include a software with their instruments. The latter display specific features that are added to the basic modules: binning for CCDs in the X and Y directions, negative as well as positive viewing of images, image analysis module, band or lane detection, colony counting, etc. Many software packages dedicated to specific applications and running independently from a given instrument have appeared on the market. A typical example of this type of software is Kalcium PC [37]. It is an integrated package for image analysis of the cellular dynamics of calcium ions.

14. CONCLUSIONS

We have given a brief overview of the present state of instrumentation for CL-based analysis. Technological improvements have resulted in ultrasensitive, reliable, and user-friendly system components and instruments. As will be appreciated, the range of instruments offered has become very broad and the purpose of this chapter was not to be comprehensive and to describe all existing instruments or techniques. Readers interested in more details should read technical journals dedicated to clinical, chemical, or biotechnological laboratories or request leaflets about specific instruments. Note that a host of instruments for very specific applications appeared recently on the market.

To illustrate this point and to close the chapter, three examples will be given. A compact analyzer, the FastPack from Qualisys [38], was designed for use at the point of care, at a doctor's office or in a small laboratory to carry a prostate-specific antigen test (PSA). It is based on chemiluminescent technology and is as accurate as a large laboratory instrument. Vilber Lourmat [39] released the Photo-Print, a photodocumentation system for users interested in the printing-out of photos. It does not require a PC for the printout but it is equipped with a software and floppy disk drive for transfer to a PC if the user wishes to do so. Last, but not least, specific DNA chip readers are available, such as the ChipReader made by Virtek Vision [40]. Its modular design provides flexibility to read microarrays in any format, detecting up to five fluorescent dyes. It reads very quickly due to its high-speed scanning mechanism.

REFERENCES

1. SP Colowick, NO Kaplan. In: MA DeLuca, ed. Methods Enzymol, vol LVII. Biolu-minescence and Chemiluminescence, section IX. Instrumentation and Methods. New York: Academic Press, 1978, pp 529–559.
2. R Johnson, JH Gentile, S Cheer. Anal Biochem 60:115, 1974.
3. RJ Ellis, AG Wright. Luminescence 14:11–18, 1999.
4. P Helle, F Brau, JP Steghens, JC Bernengo. In: A Roda, M Pazzagli, LJ Kricka, PE Stanley, eds. Bioluminescence and Chemiluminescence: Perspectives for the 21st Century. Proceedings of the 10th International Symposium. Chichester: Wiley, 1998, pp 195–198.
5. PE Stanley. J Biolumin Chemilumin 12:61–78, 1997.
6. PS Fullam. In: A Roda, M Pazzagli, LJ Kricka, PE Stanley, eds. Bioluminescence and Chemiluminescence: Perspectives for the 21st Century. Proceedings of the 10th International Symposium. Chichester: Wiley, 1998, pp 207–210.
7. D Lapota, S Paden, D Duckworth, DE Rosenberg, JF Case. In: AK Campbell, LJ Kricka, PE Stanley, eds. Bioluminescence and Chemiluminescence: Fundamentals and Applied Aspects. Chichester: Wiley, 1994, pp 127–130.
8. DH Leaback. In: AA Szalay, LJ Kricka, PE Stanley, eds. Biolumin Chemilumin Status Report. Chichester: Wiley, 1993, pp 33–37.
9. EW Chapelle, GV Levin. Biochem Med 2:49, 1968.
10. The Book of Photon Tools. Oriel Instrument, Stratford, CT, 1999.
11. Guide for Spectroscopy. ISA Jobin Yvon Spex, 1995.
12. The Photonics Handbook. Book 3, Laurin Publishing Co, 1996.
13. EL Dereniak, DG Crowe. Optical Radiation Detectors. Wiley Series in Pure and Applied Optics. Chichester: Wiley, 1984.
14. GC Holst. CCD Arrays, Cameras and Displays. Co-published by SPIE and JCD Publishing Co. SPIE Press, Vol. PM57, 1998.

15. EGG, 9238 Gaither Road, Gaithersburg, MD.
16. Hamamatsu Corporation, Bridgewater, NJ.
17. Photometrics, Tucson, AZ.
18. ED Salmon, P Tran. In: G Sluder and DE Wolf, eds. Video Microscopy. New York: Academic Press, 1998, Chapter 9.
19. G Holzwarth, B Moomaw. Application Report. Hamamatsu Euronews 99 1:13, 1999.
20. G Holzwarth, SC Webb, DJ Kubinski, NS Allen. J Microsc 188:249–254, 1997.
21. Janesick, Proc SPIE 1071:153–155, 1996.
22. SJ Sternberg. J NIH Res 5:79–83, 1993.
23. Cooke Corporation, Tonawanda, NY.
24. Pixel Vision, Beaverton, OR.
25. Wallac Oy, EGG Life Sciences, Turku, Finland.
26. Perkin-Elmer Corporation, Applied Biosystems Division, Foster City, CA.
27. MD O'Neill. Laser Focus World 135–142, 1995.
28. LI-COR, Lincoln, NE.
29. Kodak Digital Science, Rochester, NY.
30. The IS440CF is exclusively distributed by NEN Life Science Product Inc., Boston, MA.
31. PC Mayer, JP Masucci. Intl Biotech Lab 1, April 1999, p. 12.
32. Nucleo Tech Corp, Hayward, CA.
33. LJL Biosystems, Sunnyvale, CA.
34. Anthos Labtec Instruments, Wals/Salzburg, Austria.
35. Tecan, Austria AG, Groedig, Austria.
36. Packard Instruments Corporation, Meriden, CT.
37. Kinetic Imaging, Liverpool, UK.
38. Qualisys Diagnostic Inc, Minneapolis, MN.
39. Vilber Lourmat, Marne la Vallée, France.
40. Virtek Vision Inc, Woburn, MA.

5
Applications of Chemiluminescence in Organic Analysis

Yener Rakicioğlu
Istanbul Technical University, Istanbul, Turkey

Joanna M. Schulman and Stephen G. Schulman
University of Florida, Gainesville, Florida

1. INTRODUCTION

Intense and analytically useful direct chemiluminescence (CL) has been observed from a rather limited group of organic compounds. These include diacylhydrazides, indoles, acridines and acridans, polydimethylaminoethylenes, anthracenes, and aroyl peroxides. A substantial number of other kinds of compounds, when oxidized, give rise directly to very weak or ultraweak CL, which, if not analytically useful in and of itself, may provide excited states of lifetime sufficiently long as to sensitize intense fluorescence in an acceptor fluorophore by energy transfer. In addition, a number of compounds, such as dioxetanes, organic oxalates, and oxamides, that are not necessarily chemiluminescent themselves can be thermally oxidized to electronically excited intermediates (aldehydes and ketones) whose lifetimes in the excited state are sufficient for transfer of their excitation energy to a suitable acceptor fluorophore to occur. These will be the subjects of this chapter.

Considerably more intense CL may be observed from a limited number of naturally occurring substances of biological origin (e.g., luciferin) when they are oxidized in vivo with the intermediacy of enzymes. These "bioluminescent" reactions are interesting and of considerable analytical utility as they can often be conducted under in vitro conditions. Gas-phase reactions also will be considered elsewhere in this book and do not have a wide range of applicability in organic analysis. For these reasons, they too will not be discussed in this chapter. Nor will electrogenerated CL, which entails electron transfer reactions occurring through the agency of electrode surfaces rather than oxidizing or reducing chemical species.

2. CHEMILUMINESCENT ORGANIC MOLECULES OF HISTORICAL SIGNIFICANCE

CL emissions can be characterized by four parameters, including color, intensity, rate of production, and decay of intensity. The properties of several organic, chemiluminescent reactions known to produce emissions of light are shown in Table 1.

2.1 Lophine and Other Indoles

Although bioluminescence had been observed in nature for centuries, that synthetic organic compounds could produce light was first established by Radziszewski [1, 2] in 1877 using lophine (2,4,5-triphenylimidazole) (Fig. 1).

Lophine emits yellow CL upon oxidation by molecular oxygen in alkaline solution. The oxidation is believed to produce a free radical [3], which is further oxidized to a hydroperoxide, which is the light-emitting species [4–6]. A number of chemiluminescent derivatives of lophine have been synthesized and have been shown to exhibit varying efficiencies of CL. Lophine has been used in the analysis of metal ions such as Co^{2+} that catalyze the chemiluminescent reaction between it and hydrogen peroxide [7]. It has also been used as a chemiluminescent indicator in titrimetry [8].

Table 1 Properties of Some Organic Chemiluminescent Reactions

Reaction	Color (λ_{max})	Quantum yield
Luminol oxidation in DMSO	Blue-green 480–502 nm	0.05
Luminol oxidation in aqueous alkali	Blue 425 nm	0.01
Lucigenin oxidation in alkaline H_2O_2	Blue-green 440 nm	0.016
Lophine oxidation in alcoholic NaOH	Yellow 525 nm	
Pyrogallol in alkaline H_2O_2	Reddish-pink	
ATP-dependent oxidation of D-luciferin with firefly luciferase		
pH 8.6	Yellow-green 560 nm	0.88
pH 7.0	Red 615 nm	
Aliphatic aldehyde, reduced flavin mononucleotide oxidation with marine bacteria	Green 490 nm	
Peroxyoxalate [bis(2,4,6-trichlorophenyl) oxalate] oxidation using 9,10-diphenylanthracene as the fluorophore	Blue	0.07–0.50

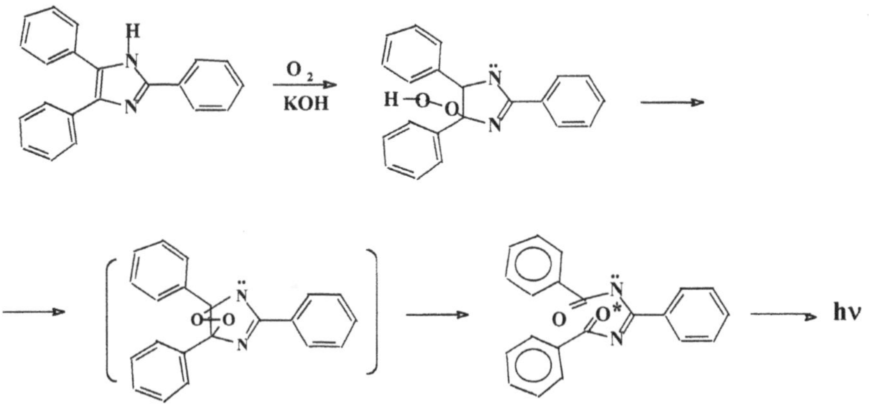

Figure 1 Mechanism for CL emission of lophine (2,4,5-triphenylimidazole).

A number of other indole derivatives have also been observed to yield CL subsequent to oxidation. Among these is skatole (3-methylindole), which emits light after oxidation by O_2 under strongly alkaline conditions [9].

2.2 Luminol and Related Acylhydrazides

Albrecht [10] was the first to report the CL of luminol (5-amino-2,3-dihydro-1,4-phthalazinedione), in 1928. The chemiluminescent reaction involves the oxidation of luminol (usually by H_2O_2) and often occurs in the presence of a catalyst (or co-oxidant) such as $Fe(CN)_6^{-3}$, Cu(II), or Co(II)) (Fig. 2). The light emission, which is blue in water and yellow-green in DMSO, is identical with the fluorescence of the 3-aminophthalate oxidation product [11].

The oxidation of luminol in basic solution is one of the best known and most efficient chemiluminescent reactions, having a quantum yield of CL of about 0.01 in water and 0.05 in DMSO.

White et al. [11, 12] showed that the production of CL from luminol follows from the formation of a dinegative ion of luminol, which reacts with oxygen or an alternative oxidizing agent to yield an excited singlet state of the aminophthalate ion that is responsible for the emission of light. Table 2 shows several oxidizing systems that react with luminol to generate light [13].

The luminol reaction occurs under a wide variety of conditions. Specific analysis using luminol requires that the chemistry be controlled so that the CL intensity is proportional to the concentration of the species of interest.

In the early studies on luminol and related hydrazides the systems used were composed of either sodium or potassium hydroxide, as base, hydrogen peroxide as the oxidizing agent (more recently molecular oxygen, hypochlorite, iodide, and permanganate have also been used), and some type of initiator or activator. This initiator was frequently hypochlorite, persulfate, a transition metal

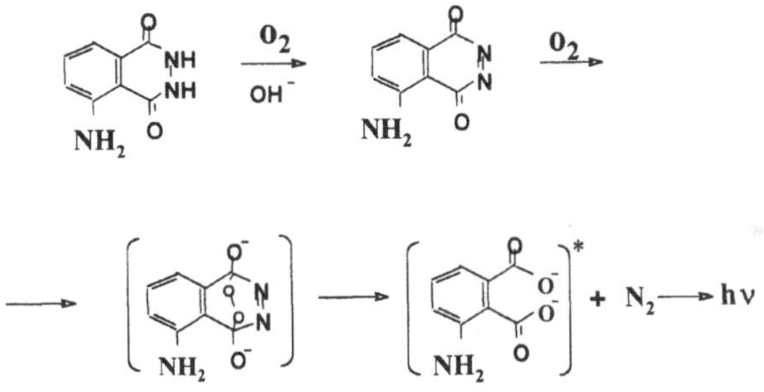

Figure 2 Oxidation reaction of luminol to produce light.

Table 2 Oxidizing Systems that React
with Luminol to Emit Light

Oxidant	Catalyst/(Co)oxidant
H_2O_2	$H_2S_2O_8$
	Peroxidase
	Ferricyanide
	Heme compounds
	Transition metal ions: (Co^{2+}, Cu^{2+}, Cr^{3+}, Ni^{2+}, Fe^{2+}, VO^{2+})
	Hypochlorite
O_2	Ferricyanide
	Fe^{2+}
OCl^-	
I_2	
MnO_4	
NO_2	

complex such as ferricyanide or heme, or a metal ion such as Co(II), Cu(II), Cr(III), Ni(II), Fe(II), or VO^{2+}. Lower alcohols as well as mixtures of water and other water-soluble organics [14] have been used as solvents in place of water. In place of the activating agent, pulse radiolysis [15, 16] and sonic waves have also been used [17]. No activating agent is necessary in some aprotic media, only oxygen and a strong base.

Drew et al. [18–20] showed that alterations to the heterocyclic part of the hydrazide system effectively blocked the chemiluminescent reaction. *O*- and *N*-methyl analogs, the *N*-amino imide, and the corresponding aminoquinazoline-2,3-dione were not chemiluminescent under the reaction conditions used in the oxidation.

The unsubstituted phthalic acid hydrazide and several nonaromatic cyclic hydrazides such as maleic acid hydrazide or succinic acid hydrazide are either nonchemiluminescent or show extremely weak CL. However, the 6-amino isomer of luminol, which is called isoluminol, is chemiluminescent to about the same extent as is luminol. Isoluminol has been used in many chemiluminescent studies, and because the amino group is less sterically hindered than that of luminol, it is probably derivatized for chemiluminescent labeling far more often than is luminol (Fig. 3).

The amino group can be diazotized for coupling to various substrates, as is done in CL immunoassay, without loss of the chemiluminescent properties of the cyclic hydrazide.

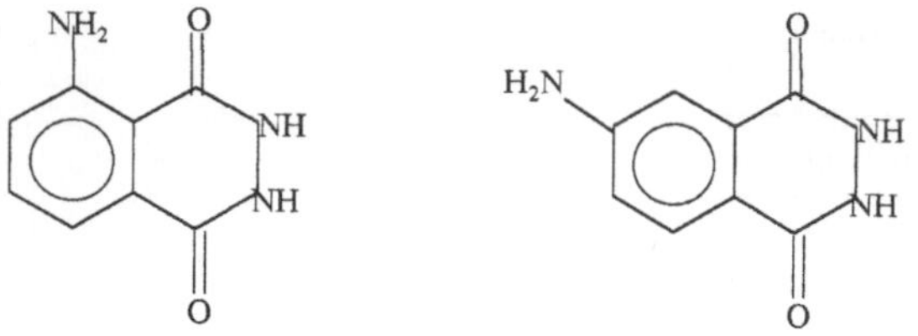

Luminol   **Isoluminol**

Figure 3   Structures of luminol and isoluminol.

Gundermann et al. [21, 22] have extended Drew's work and showed that alkylation of the amino group enhances the efficiency, provided that the steric bulk of the alkyl groups does not interfere with the planarity of the nitrogen and the ring. Later work [23] showed that the chemiluminescent efficiency increased with the size of the aromatic ring.

The luminol reaction has been used for the determination of oxidizing agents such as hydrogen peroxide, for enzymes such as peroxidase and xanthine oxidase, and for metal ions such as copper or cobalt that catalyze this CL reaction [24].

2.3   Peroxyoxalate Chemiluminescence

The brilliant emissions resulting from the oxidation of certain oxalic acid derivatives, especially in the presence of a variety of fluorophores, are the bases of the most active area of current interest in CL. This group of chemiluminescent reactions has been classified as peroxyoxalate chemistry because it derives from the excited states formed by the decomposition of cyclic peroxides of oxalic acid derivatives called dioxetanes, dioxetanones, and dioxetanediones.

Peroxyoxalate chemistry has gained increasing importance in recent years, as it is the basis for a number of practical chemiluminescent devices that are marketed worldwide. The initial example of peroxyoxalate CL was reported in 1963 by Chandross [25]. He found that when oxalyl chloride, oxamide, or oxalate esters were treated with hydrogen peroxide in the presence of a fluorescent compound such as 9,10-diphenylanthracene (DPA), a bright, short-lived blue emission was produced corresponding to the fluorescence of the hydrocarbon. Thus, the energy produced in the chemical reaction was transferred to the fluorescer, sensitizing the formation of the lowest excited singlet state of the fluorescer that emits a typical fluorescence. The energy transfer step is common to all peroxyoxa-

late chemiluminescent systems in that chemical reactions provide a key intermediate (an electronically excited ketone or aldehyde), which then transfers its energy to a fluorescent substance, thereby generating an excited state. Usually
alkaline conditions are required to activate the reaction. The "color" of the emission is controlled by changing the energy acceptor fluorophore in the reaction
mixture.

The reactions of certain esters of oxalic acid with hydrogen peroxide in
the presence of a fluorescer give bright, long-lasting emission yielding the most
efficient nonenzymatic chemiluminescent systems yet known with CL quantum
yields in the range of 22–27% when the reactions are carried out under optimum
conditions. Two of the most popular oxalyl derivatives in current use are *bis*
(2,4-dinitrophenyl) oxalate (DNPO) and *bis* (2,4,6-trichlorophenyl) oxalate
(TCPO) with TCPO being the more frequently used (Fig. 4, A and B).

The "leaving group" of the oxalic ester has a strong effect on the efficiency
of the peroxyoxalate chemiluminescent system. The electron-attracting power of
the substituents on the phenyl rings of the substituted diphenyl oxalates is important to the overall efficiency of the chemiluminescent reactions. Steric effects

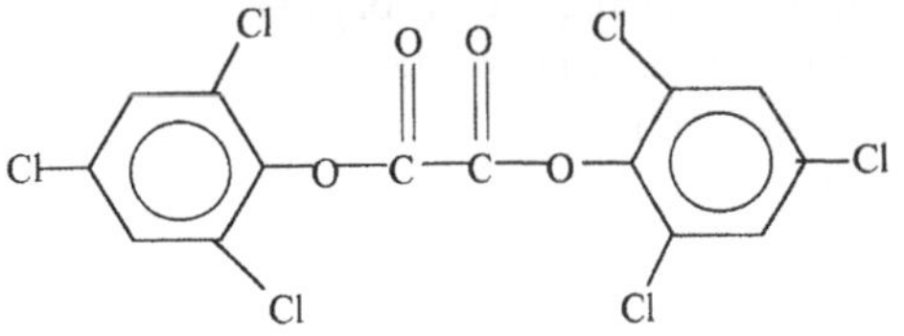

(A) TCPO

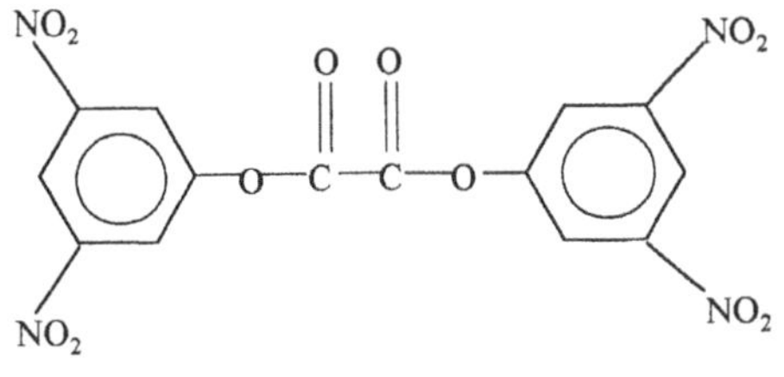

(B) DNPO

Figure 4   Structures of TCPO (A) and DNPO (B).

are also believed to play an important role. Concentration is also significant, as decreasing efficiency is usually observed with increasing concentration of the oxalic ester.

Leaving groups other than phenols were also found to be effective in producing moderately efficient chemiluminescent reactions [26, 27].

The structure of the fluorescent molecule can also contribute substantially to the overall efficiency of a chemiluminescent process. Excitation and fluorescence can be strongly influenced by the structure of the fluorescer.

A wide variety of different classes of fluorescent molecules has been investigated in the peroxyoxalate chemiluminescent systems. Among those screened were fluorescent dyes such as rhodamines and fluoresceins, heterocyclic compounds such as benzoxazoles and benzothiazoles, and a number of polycyclic aromatic hydrocarbons such as anthracenes, tetracenes, and perylenes. The polycyclic aromatic hydrocarbons and some of their amino derivatives appear to be the best acceptors as they combine high fluorescence efficiency with high excitation efficiency in the chemiluminescent reaction [28].

Quenching of the emissions of fluorophores sensitized by dioxetanes has also been used for chemical analysis [29, 30] but this is probably less specific and less sensitive than direct measurement of the sensitized emission of the analyte. Most analytical reactions of the dioxetanes require an organic solvent for optimal solubility and CL yields [31]. This has led to considerable interest in the development of water-soluble oxalate esters [32].

2.4 Lucigenin and Other Acridines

Blue-green CL arising from the oxidation of lucigenin (10,10'-dimethyl-9,9'-bisacridinium dinitrate) by hydrogen peroxide or oxygen in the presence of catalysts such as Co(II), Fe(II), Fe(III), Cu(II), Cr(III), or Ni(II), in alkaline solution, was first observed by Glue and Petsch [33]. The oxidation proceeds through a dioxetane intermediate and yields N-methylacridone as a final decomposition product (Fig. 5). The emission is identical with that of N-methylacridone (10-methylacridan-9-one) and it is the lowest excited singlet state of the latter that is believed to be the result of the lumigenic reactions [34].

Chemiluminescence from lucigenin is observed even without the catalytic transition metal ions but it is more intense when these ions are used. In aqueous or predominately aqueous solutions the CL yield is 0.01–0.02, which makes it a slightly better emitter than luminol [35]. The emission of lucigenin is also catalyzed by Pb(II), Bi(III), Tl(III), and Hg(I) ions, which do not catalyze the CL of luminol [36].

Lucigenin is often used to facilitate the measurement of reactive oxygen species in immunological studies, because it enhances the cellular CL intensity [37]. Both luminol and lucigenin can be used to measure reactive oxygen; how-

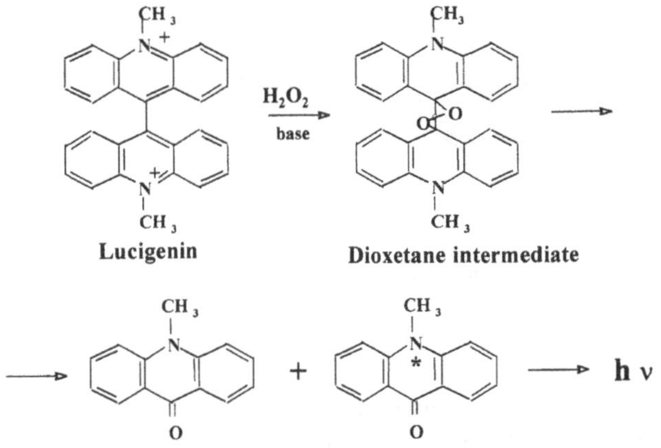

N-methylacridone

Figure 5 Chemiluminescent reaction of lucigenin proceeding via a dioxetane intermediate.

ever, lucigenin is sometimes used in favor of luminol because lucigenin reacts only with O_2^- [38]. Lucigenin-enhanced cellular CL serves as a convenient and sensitive method for investigating factors influencing phagocyte biology and immunology, such as the creation of reactive oxygen species during phagocytosis and other immunopathic responses [37].

Several N-methyl-9-acridinecarboxylic acid derivatives (e.g., 10-methyl-9-acridinecarboxylic chloride and esters derived therefrom [39]) are chemiluminescent in alkaline aqueous solutions (but not in aprotic solvents). The emission is similar to that seen in the CL of lucigenin and the ultimate product of the reaction is N-methylacridone, leading to the conclusion that the lowest excited singlet state of N-methylacridone is the emitting species [40]. In the case of the N-methyl-9-acridinecarboxylates the critical intermediate is believed to be either a linear peroxide [41, 42] or a dioxetanone [43, 44]. Reduced acridines (acridanes) such as N-methyl-9-bis (alkoxy) methylacridan [45] also emit N-methylacridone-like CL when oxidized in alkaline, aqueous solutions. Presumably an early step in the oxidation process aromatizes the acridan ring.

The reaction in Figure 6 shows the CL-producing reaction of an acridinium ester and hydrogen peroxide in the presence of a base. This reaction makes acridinium suitable as a derivatizing agent (tag) for amino acids, peptides, and proteins in capillary electrophoresis. The positive charge provides greater mobility in an applied electric field, and it has high CL efficiency. Its rate can be adjusted for measurements in flowing systems, which require reaction completion

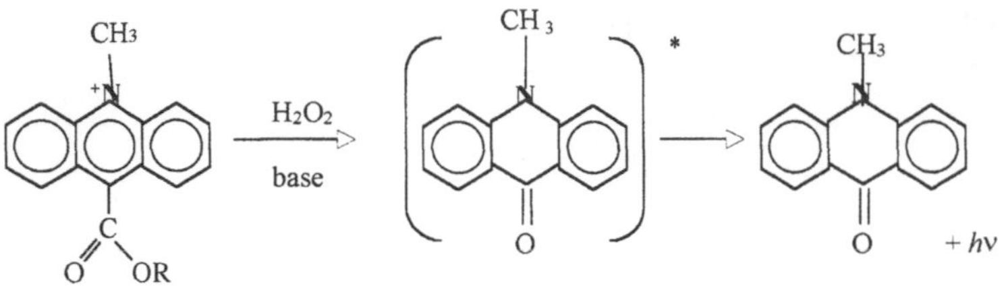

Figure 6 Chemiluminescent reaction of an acridinium ester and hydrogen peroxide in the presence of a base.

in a few seconds to minimize band overlapping. In addition, the acridiniums have been modified to include functional groups suitable for the derivatization of other biomolecules [46].

2.5 Miscellaneous Organic Compounds

A number of organic compounds have demonstrated CL but have not been investigated, in this context, as extensively as the foregoing substances.

2.5.1 Tetrakis (Dimethylamino) Ethylene

This compound chemiluminesces when exposed to air or oxygen. Its CL was first demonstrated by Fletcher and Heller [47, 48] and suggested to occur via formation of a dioxetane by addition of oxygen across the ethylenic double bond. Cleavage of the dioxetane to form excited tetramethylurea results in excitation of the tetrakis (dimethylamino) ethylene, whose CL is in good agreement with the fluorescence spectrum of the parent compound. The reaction has been used for the analysis of oxygen [49, 50].

2.5.2 Diphenoyl Peroxide

Diphenoyl peroxide is the cyclic peroxide of diphenic acid (2,2′-biphenyldicarboxylic acid) (Fig. 7). It undergoes thermal decomposition to form 3,4-benzocou-

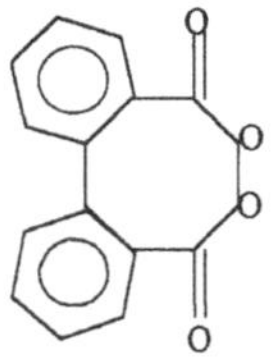

Figure 7 Structure of diphenoyl peroxide.

marin with no emission of light [51, 52]. However, in the presence of some of the larger aromatic hydrocarbons (e.g., 9,10-diphenylanthracene and rubrene) emission corresponding to the fluorescence of the hydrocarbons is observed. In this respect, not surprisingly, diphenoyl peroxide behaves very much like the cyclic peroxides formed by the oxidation of oxalic acid derivatives. The reaction is believed to proceed through a charge-transfer complex formed between the aromatic hydrocarbon and the diphenoyl peroxide, followed by electron transfer from the former to the latter with subsequent decomposition of the complex to benzocoumarin and the excited hydrocarbon. This is called "chemically initiated electron exchange luminescence" (CIEEL).

The oxidation of naphthalic acid (1,8-naphthalenedicarboxylic acid) by peroxide, rather surprisingly, does not proceed by formation of a cyclic peroxide but rather via a dioxirane [53] (a three-membered ring containing a carbon atom and a peroxide group). CL is observed from this reaction.

2.5.3 Schiff Bases

9-Aminoanthracene forms a Schiff base with dimethylacetaldehyde (isobutyraldehyde). This compound can be oxidized by peroxide under basic conditions to form 9-formamidoanthracene and acetone in dimethylformamide as a solvent [54, 55]. CL from this system can be observed in other aprotic solvents as well. A limited amount of work has been done with the CLs of Schiff bases or anthracene derivatives. Presumably, this will change in the future.

3. RECENT DEVELOPMENTS IN ORGANIC CHEMILUMINESCENCE

(E)-2-(phenylsulfonyl)-3-phenyloxaziridine, commonly known as Davis' oxaziridine, and its analogs have been shown to luminesce in the presence of strong bases. Davis' oxaziridine has been widely used in organic syntheses, such as in the hydroxylation of enolates to produce α-hydroxy ketones. Stojanovic and Kishi observed in 1995 that when using more than two equivalents of a strong base, such as lithium diisopropylamide (LDA), Davis' oxaziridine instantly decomposed and no α-hydroxy ketone was produced. However, this reaction was accompanied by an emission of intense yellow light (about 520 nm). They found that not only would LDA produce CL, but tert-butyllithium, n-butyllithium, and potassium hexamethyldisilazanide also cause emission of visible light [56]. In addition to Stojanovic and Kishi's work, other studies suggest that Davis' oxaziridine reacts with LDA through proton- and oxygen-transfer processes [57].

Akhavan-Tafti et al. have developed a new class of peroxidase substrates that produce CL upon enzymatic oxidation. Horseradish peroxidase (HRP) is commonly used as a reporter enzyme in various assays. Previously, the only significant chemiluminescent reagent for detecting HRP was the enhanced luminol detection system. However, the level of concentration at which this enzyme can be detected with CL detection is at least three orders of magnitude lower than that of other marker enzymes. A wide variety of N-alkylacridancarboxylic acid derivatives, including esters, thioesters, and sulfonamides, are effectively oxidized by a peroxidase and a peroxide to enzymatically produce the corresponding chemiluminescent acridinium compound. Use of these substrates provides a more sensitive chemiluminescent detection of HRP and also has a longer signal duration compared to the luminol system [58].

A chemiluminescent assay of β-galactosidase in coliform bacteria, using a phenylgalactose-substituted 1,2-dioxetane derivative as a substrate has been developed by Poucke and Nelis. The assay of β-galactosidase is important in clinical, environmental, food, and molecular microbiology. There are chromogenic and fluorogenic tests that are commonly used to assay β-galactosidase; however, they require visual or spectrophotometric detection of colored reaction products, and therefore have a limited sensitivity. This limited sensitivity is of no concern in a confirmatory test of bacterial isolates because the abundance of available cells allows a rapid cleavage of the substrate. However, when a small number of bacteria have to be detected in a sample, such as in drinking water, an incubation period of at least 24 h is needed to allow sufficient bacterial propagation and enzymatic hydrolysis before any color becomes detectable. Using a chemiluminogenic substrate, such as a phenylgalactose-substituted 1,2-dioxetane derivative, as a substrate would create a substantial gain in sensitivity compared to both chromogenic and fluorogenic tests. This is a powerful substrate for the sensitive assay of bacterial β-galactosidase in samples containing small numbers of cells [59].

Marley and Gaffney have constructed a hydrocarbon analyzer on the basis of the CL reaction with ozone. The detector is designed to operate at various temperatures to take advantage of the different rates of reaction of the hydrocarbon classes with ozone to yield a measure of their atmospheric reactivity. This research was initiated by the growing concern with regard to the atmospheric reactivity of hydrocarbon emissions from fossil-fuel-powered machinery. These hydrocarbons have been increasing the levels of tropospheric ozone and other oxidants. Tropospheric ozone is a concern because of its impact on human health and its effects on plants and crops. Other oxidants can play important roles in the formation of acid rain and other global climate effects. Thus, a need to characterize both the reactivities and concentrations of reactive hydrocarbon emissions exists to better evaluate their atmospheric impact. Marley and Gaffney's chemiluminescent hydrocarbon analyzer represents a new concept in measuring the reactive hydrocarbons in the atmosphere based on the temperature dependences of their reactions with ozone. CL techniques lend themselves readily to real-time

Table 3 Determination of Some Drugs Using Direct Chemiluminescence

Analyte	Method	Limit of detection	Range	Ref.
Tetracycline	Reaction with bromine in alkaline medium, flow injection analysis	19 µg/mL (4×10^{-5} M)	5×10^{-5}–1×10^{-2} M	62
Acetaminophen	Reduction of cerium (IV), flow injection analysis	0.070 µg/mL	1.00–10.0 µg/mL	63
Cortisone	Sensitization of CL of sulfite by cerium (IV), flow injection analysis	0.040 µg/mL	0.100–1.00 µg/mL	64
Hydrocortisone		0.028 µg/mL		
Prednisolone		0.016 µg/mL		
Methylprednisolone		0.021 µg/mL		
Dexamethasone		0.16 µg/mL	0.500–5.00 µg/mL	
Betamethasone		0.30 µg/mL		
Quinine	Sensitization of CL of sulfite by cerium (IV), flow injection analysis	0.64 µg/mL	5.00–500 µg/mL	65
Isoniazid	Oxidation with N-bromosuccinimide	0.024 µg/mL	0.050–20.0 µg/mL	66
Morphine	Reaction with permanganate, flow injection analysis, high-performance liquid chromatography	0.7 pg (2 fmol, 1×10^{-10} M) for flow injection analysis. 25 ng/mL, 50 ng/mL (aqueous solution, biological fluids, for high-performance liquid chromatography	50 ng/mL–500 µg/mL	67, 68
Loprazolam	Reaction with permanganate, flow injection analysis	163 ng (7×10^{-6} mol/50 µL)	1×10^{-5} M–1×10^{-3} M	69
Buprenorphine HCl	Reaction with permanganate, flow injection analysis	5 ng/mL	1×10^{-8}–1×10^{-4} M	70

Table 4 Drugs Analyzed with a Chemiluminescent System

Analyte	Method	Limit of detection	Range	Ref.
Penicillin	Enhancement of luminol chemiluminescent reaction, batch method	100 ng/mol	100 ng/mL–100 μg/mL	71, 72
Cephalothin sodium	Enhancement of luminol chemiluminescent reaction, batch method	—	—	73
Hydrocortisone	Solid-state peroxyoxalate	2 ng/mL	—	74
Digoxin	CL, high-performance	2 ng/mL		
Theophyllin	liquid chromatography	4 ng/mL		

monitoring of atmospheric species, while being selective due to the relatively small number of compound classes that will produce a chemiluminescent reaction with ozone [60].

Chemiluminescence has become a powerful tool for drug determination. It has a wide range of applications, and low detection limits can be measured with a simple, low-cost instrument. The coupling of flow injection with CL has made this technique more popular with wider applications. Sultan and Almuaibed report the CL of medazepam in an oxidation reaction with permanganate in a sulfuric acid medium. A photomultiplier tube was used to detect emitted light. Compounds commonly used in pharmaceutical preparations, such as starch, glucose, maltose, and lactose, had no effect on CL. The coupling of flow injection with CL has allowed the assay of medazepam. Using the simplex optimization for chemiluminescent measurements they were able to determine optimum conditions for the determination of the drug in aqueous solution [61].

Some drugs will emit light under specific reaction conditions. The discovery of these direct chemiluminescent reactions has come about through trial and error. Table 3 lists some of these drugs and the method used for analysis. Other drugs that are not themselves chemiluminescent can be easily analyzed using a chemiluminescent system (Table 4).

4. ULTRAWEAK CHEMILUMINESCENCE

Some organic reactions produce ultraweak chemiluminescence. Grignard reagents [75, 76] and indole derivatives [13], for example, have very low $\phi_{CL}(10^{-5}-10^{-8})$ when they are oxidized. Many cellular systems also produce this dim or low-level chemiluminescence during phagocytosis [77], chemotaxis, and mitosis

Table 5 Ultraweak Chemiluminescence of Drugs

Analyte	Method	LOD	Range	Ref.
Imipramine HCl	Auto-oxidation	—	—	80
Clomipramine HCl	Auto-oxidation	—	—	81
Trimipramine maleate				
Desipramine HCl				
Clocapramine diHCl				
Carpipramine diHCl				
Pepleomycin	Auto-oxidation	—	—	82
Mytomycin C				
Neocarzinostatin	Auto-oxidation	—	—	83

[75, 78, 79]. Several drugs can be analyzed by their ultraweak chemiluminescence. Some examples are given in Table 5.

Other drugs, for example lofepramine HCl, carpipramine, warfarin, and triprolidine HCl, have also shown ultraweak CL intensity [84]. The measurement of ultraweak CL entails integration of the signal over the entire course of the CL reaction to produce analytically useful results. Alternatively, an acceptor fluorophore may be employed for analysis as in the case of dioxetanes. The ultraweak CL is probably generated by the formation of peroxide and hyperperoxide intermediates upon auto-oxidation. Specific functional groups have been tested for ultraweak CL in simple organic compounds [85]. The emissions of alkynes, aliphatic amines, aliphatic aldehydes, epoxides, and peroxides were demonstrated [75, 76]. An ultraweak CL reaction between an amino group and a carbonyl group was also observed [86]. This reaction has been used to quantitate amino acid derivatives.

REFERENCES

1. B Radziszewski. Chem Ber 10:70–75, 1877.
2. B Radziszewski. Liebig's Ann Chem 203:305–336, 1880.
3. T Hayoshi, K Maeda. Bull Chem Soc Jpn 35:2057–2058, 1962.
4. J Sonnenberg, DM White. J Am Chem Soc 86:5685–5686, 1964.
5. EH White, MJC Harding. J Am Chem Soc 86:5686–5687, 1964.
6. EH White, MJC Harding. Photochem Photobiol 4:1129–1155, 1965.
7. DF Marino, JD Ingle Jr. Anal Chem 53:292–294, 1981.
8. GA Shannon. Sch Sci Rev 68:81, 1986.
9. GF Philbrook, JB Oyers, JR Totter. Photochem Photobiol 4:869–876, 1965.
10. HD Albrecht. Z Phys Chem 136:321–330, 1928.

11. EH White, OC Zafiriou, HH Kagi, JHM Hill. J Am Chem Soc 86:940–941, 1964.
12. EH White, MM Bursey. J Am Chem Soc 86:941–942, 1964.
13. KD Gunderman, F McCapra. Chemiluminescence in Organic Chemistry. Heidelberg: Springer-Verlag, 1986.
14. EH White. J Chem Educ 34:275–279, 1957.
15. JH Boxendale. Chem Commun 1470–1471, 1971.
16. G Merenyi, JS Lind. J Am Chem Soc 102:5830–5835, 1980.
17. EN Harvey. J Am Chem Soc 61:2392–2398, 1939.
18. HDK Drew, FH Pearman. J Chem Soc 586–592, 1937.
19. HDK Drew, RF Garwood. J Chem Soc 791–793, 1938.
20. HDK Drew, RF Garwood. J Chem Soc 836–832, 1939.
21. KD Gundermann, M Drawert. Chem Ber 95:2018–2026, 1962.
22. KD Gundermann, W Horstmann, G Bergmann. Liebig's Ann Chem 684:127, 1965.
23. EH White, MM Bursey, DL Rosewell, JHM Hill. J Org Chem 32:1198–1202, 1967.
24. A Ponomarenko. Tr Mosk Obshichestva Ispytaklei Pirody Otd Biol 21:165, 1965.
25. EA Chandross. Tetrahedron Lett 761–765, 1963.
26. MM Rauhut, D Sheehan, RA Clarke, AM Semsel. Photochem Photobiol 4:1097–1110, 1965.
27. LJ Bollyky, BJ Roberts, RH Whitman, JE Lancaster. J Org Chem 34:836–842, 1969.
28. WRG Baeyens, BL Ling, UA Th Brinkman, SG Schulman. J Biolumin Chemilumin 4:484–499, 1989.
29. JK DeVasto, ML Grayeski. Analyst 116:443–447, 1991.
30. P VanZoonen, H Bock, C Gooier, NH Velthorst, RW Frei. Anal Chim Acta 200: 1313–1341, 1987.
31. AL Mohan, JG Burr, eds. Chemi- and Bioluminescence. New York: Marcel Dekker, 1985, pp 245–258.
32. AG Mohan. Investigation of Water Soluble Chemiluminescent Materials. Report of the Federal Highway Administration. FHWA-RD-77-79, 1976.
33. K Glue, W Petsch. Angew Chem 48:57–72, 1935.
34. JR Totter. Photochem Photobiol 22:203–211, 1975.
35. D Betteridge. Anal Chem 50:823A–833A, 1978.
36. K Gundermann, F McCapra. Chemiluminescence in Organic Chemistry. Heidelberg: Springer-Verlag, 1987, p 169.
37. F McCapra, DG Richardson, YC Chang. Accts Chem Res 9:201–208, 1976.
38. P Falck. J Mater Sci 29:4007–4012, 1994.
39. EH White, DF Rosewell, OC Zafiriou. J Org Chem 34:2462–2468, 1969.
40. EH White, OC Zafiriou, HH Kagi, JHM Hill. J Am Chem Soc 86:940–941, 1964.
41. MM Rauhut, D Sheehan, RA Clarke, AM Semsel. Photochem Photobiol 4:1097–1110, 1965.
42. MM Rauhut, D Sheehan, RA Clarke, BG Roberts, AM Semsel. J Org Chem 30: 3587–3592, 1965.
43. F McCapra, DG Richardson, YC Chang. Photochem Photobiol 4:1111–1121, 1965.
44. F McCapra, DG Richardson, YC Chang. Proc Roy Soc Lond B215:247–272, 1982.
45. N Suzuki, M Nakaminami, T Tsukamoto, K Iwasaki, R Izawa. Res Rep Fac Eng Mie Univ 10:41, 1985.

46. WRG Baeyens, BL Ling, K Imai, AC Calokerinos, SG Schulman. J Microcol 6: 195–206, 1994.
47. AN Fletcher, CA Heller. J Chem Phys 71:1507–1518, 1967.
48. CA Heller, RA Hewry, JM Fritsch. In: MJ Cormier, DM Hercules, J Lee, eds. Chemiluminescence and Bioluminescence. New York: Plenum, 1979, p 249.
49. TM Freeman, WR Seitz. Anal Chem 53:98–102, 1981.
50. BF McDonald, WR Seitz. Anal Lett 15:57–66, 1982.
51. JY Koo, GB Schuster. J Am Chem Soc 99:6107–6109, 1977.
52. JY Koo, GB Schuster. J Am Chem Soc 100:4496–4503, 1978.
53. MFD Stainfatt. J Chem Res 5:140–145, 1985.
54. F McCapra, A Burford. J Chem Soc Chem Commun 607–609, 1976.
55. F McCapra, A Burford. J Chem Soc Chem Commun 874–876, 1977.
56. MN Stojanovic, Y Kishi. J Am Chem Soc 117:9921–9922, 1995.
57. WW Zajac, TR Walters, MG Darcy. J Org Chem 53:5856–5860, 1988.
58. H Akhavan-Tafti, R DeSilva, Z Arghavani, RA Eickholt, RS Handley, BA Schoen-felner, K Sugioka, Y Sugioka, AP Schaap. J Org Chem 63:930–937, 1998.
59. SO Van Poucke, HJ Nelis. Appl Environ Microbiol 61:4505–4509, 1995.
60. NA Marley, JS Gaffney. Atmospher Environ 32:1435–1444, 1998.
61. SM Sultan, AM Almuaibed. Fresenius J Anal Chem 362:167–169, 1998.
62. AA Alwarthan, A Townshend. Anal Chim Acta 205:261–265, 1988.
63. II Koukli, AC Calokerinos, T Hadjiioannou. Analyst 114:711–714, 1989.
64. II Koukli, AC Calokerinos. Analyst 115:1553–1557, 1990.
65. II Koukli, AC Calokerinos. Anal Chim Acta 236:463–468, 1991.
66. SA Halvatzis, MM Timotheou-Potamiu, AC Calokerinos. Analyst 113:1229–1234, 1988.
67. RW Abott, A Townshend. Anal Proc 23:25–26, 1988.
68. RW Abott, A Townshend, R Gill. Analyst 112:397–406, 1987.
69. ARJ Andrews, A Townshend. Anal Chim Acta 227:65–71, 1989.
70. AA Alwarthan, A Townshend. Anal Chim Acta 185:329–333, 1986.
71. DS Milbrath. In: M DeLuca, ed. Bioluminescence and Chemiluminescence. New York: Academic Press, 1978, pp 515–518.
72. S Chen, G Yan, MA Schwartz, JH Perrin, SG Schulman. J Pharm Sci 80:1017–1019, 1991.
73. SG Schulman, JH Perrin, GF Yan, S Chen. Anal Chim Acta 255:383–385, 1991.
74. I Aichinger, G Gubitz, JW Birks. J Chrom 523:163–172, 1990.
75. AK Campbell. Chemiluminescence: Principles and Applications in Biology and Medicine. Heidelberg: VCH, 1988.
76. F McCapra, KD Perring. In: JG Burr, ed. Clinical and Biochemical Analysis, vol 16. New York: Marcel Dekker, 1985, pp 259–320.
77. MA Trush, ME Wilson, K VanDyke. In: MA DeLuca, ed. Methods in Enzymology, vol LVII. New York: Academic Press, 1987, pp 462–494.
78. LJ Kricka. Clin Chem 37:1472–1481, 1991.
79. D Slawinska, J Slawinski. In: JG Burr, ed. Clinical and Biochemical Analysis, vol 16. New York: Marcel Dekker, 1985, pp 553–601.
80. H Sato, K Edo, M Mizugaki. Chem Pharm Bull 34:5110–5114, 1986.
81. H Sato, Y Kurosaki, K Edo, M Mizugaki. Chem Pharm Bull 36:469–474, 1988.

82. K Edo, K Saito, M Kato, Y Akiyama, H Sato, M Mizugaki. Chem Pharm Bull 36: 4603–4607, 1988.
83. K Edo, H Sato, K Saito, Y Akiyama, M Kato, M Mizugaki, Y Koide, N Ishida. J Antibiot 34:535–540, 1986.
84. H Haga, S Hoshino, H Okada, N Hazemoto, Y Kati, Y Suzaki. Chem Pharm Bull 38:252–254, 1990.
85. K Edo, H Sato, M Kato, M Mizugaki, M Uchiyama. Chem Pharm Bull 33:3042–3045, 1985.
86. Y Kurosaki, H Sato, M Mizugaki. J Biolumin Chemilumin 3:13–19, 1989.

6

Application of Chemiluminescence in Inorganic Analysis

Xinrong Zhang
Tsinghua University, Beijing, P.R. China

Ana M. García-Campaña
University of Granada, Granada, Spain

Willy R. G. Baeyens
Ghent University, Ghent, Belgium

1. INTRODUCTION

In the past years, chemiluminescence (CL) analysis of inorganic compounds has been extensively developed in both gas and liquid phases. These methods typically rely on the oxidation or reduction of a chemically reactive agent and the subsequent emission of a photon from an electronically excited-state intermediate.

The primary attraction of CL detection of inorganic compounds is the excellent sensitivity that can be obtained over a wide dynamic range using simple instruments. A detection limit of 40 pg/mL gold in ore samples reported by us is a striking example to indicate the advantage of CL detection [1].

The vast majority of work done in the CL field has been devoted to solution-based flow analysis formats [2, 3]. The classical CL reagents for inorganic analysis include mainly luminol and lucigenin, among others. A number of new CL systems have also been explored in recent years for the determination of inorganic ions. Three types of CL reactions are currently being applied for analytical purposes: (1) based on the CL reaction between a reductant and an oxidant, e.g., luminol and I_2; (2) based on the catalytic behavior metal ions, e.g., Cr(III), Co(II), and Cu(II), . . . on the luminol-H_2O_2 CL reaction; (3) the determination of inorganic ions based on inhibiting effects (due to the limited number of available CL reactions).

Although CL analysis belongs to the powerful techniques for the determination of inorganic compounds due to high simplicity and rapidity, a limitation to the widespread application of this technique often lies in the poor selectivity of the CL detection as such, which is commonly based on the use of established CL reagents. It is essential to separate the analyte species from other ions or sample matrix components to improve selectivity of CL-based analysis. This can be achieved by online coupling liquid chromatography (LC) [4, 5] or capillary electrophoresis (CE) [6, 7] to a CL detector, although incompatibility of separation conditions and postcolumn reaction conditions may present unexpected problems.

A substantial number of papers have been published between the '60s and the '90s on the determination of inorganic analytes by CL-based techniques. The application of established methods to the analysis of inorganic compounds involves the areas of environmental, geographical, and biological sciences. Although many efforts have been undertaken in the past years, there still remains a challenge to apply CL-based techniques to routine analysis of inorganic elements, as the complex matrix of a real sample may cause unexpected effects on CL emission.

The aim of this chapter is to briefly introduce the methodology and application of CL-based techniques to the analysis of inorganic analytes. Since the number of relevant CL papers on inorganic analysis is large, only keynote references will be given. As many papers published in recent years originate from Chinese scientists, important selected sources from Chinese journals (in Chinese) had to be included.

2. POSITION OF INORGANIC CL ANALYSIS AMONG CURRENT TECHNIQUES

Despite the relative large number of sensitive analytical techniques for element determinations, CL still occupies a position as one of the more sensitive techniques for trace analysis of inorganic compounds. So far, only a few of techniques have been compared with CL detection. Table 1 illustrates the detection limits of inorganic analytes using different techniques including CL, atomic absorption spectrometry (AAS), inductively coupled plasma optical emission spectrometry (ICP-OES), and ICP–mass spectrometry (ICP-MS) for some typical metal ions. Although the techniques listed in Table 1 are considered the most sensitive techniques available today, there still exists no technique that exceeds the detection limits offered by CL for the determination of the elements, e.g., Au, Co, Cr, etc. Worth noticing is that the energy required for CL emission is produced by a chemical reaction; hence an excitation source and a spectral resolving system become unnecessary, leading to relatively simple equipment and low costs of instrumentation. A comparison of the prices of commercially available instruments as indicated in Figure 1 shows this advantage of the CL technique.

However, some intrinsic limitations and drawbacks of the CL technique should be taken into consideration. As listed in Table 1, for most analytes preference is given to only a few CL systems such as the luminol-H_2O_2 system. In many cases only poor selectivity is obtained when determining a typical element.

Table 1 Comparison of Detection Limits Offered by Important Analytical Techniques for Selected Inorganic Compounds (ng/mL)

Element	Au	Co	Cr	Cu	Mn
CL[a] (luminol system)	0.04	0.0074	0.0015	0.009	0.08
AAS[b]	20	2	2	4	0.8
ICP-OES[c]	0.6	0.21	0.15	0.18	0.05
ICP-MS[d]	0.06	0.01	0.01	0.02	0.03

[a] See Table 2.
[b] Data obtained from W. Robinson, *Atomic Spectroscopy*, New York: Marcel Dekker, 1990.
[c] Data obtained from the manual of JY-Ultima ICP-OES (John Yvon Emission, Longjumeau Cedex, France).
[d] Data obtained from the manual of Perkin Elmer ICP-MS (Perkin Elmer Corporation, Foster City, CA).

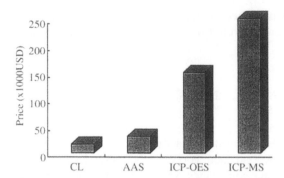

Figure 1 Comparison of the prices of commercially available instruments used for element analysis

For instance, Cr(III) ions may coexist with Co(II) and Cu(II) ions in a complex sample. The latter two ions may produce CL emission under similar conditions as for Cr(III). Fortunately, the formation rate of the Cr-EDTA complex is relatively slower, making possible the selective determination of this metal ion in waste water [8], urine, blood, and hair [9]. Owing to the small number of CL reagents explored in recent years, the elements covered by CL techniques are still rather limited. Up to now only a few elements have been found to produce direct CL emission when reacted with CL reagents. Most of the publications so far involve indirect methods for the detection of elements.

3.   PROGRESS IN ANALYTICAL CL METHODOLOGY

3.1   CL Measurements Based on Direct Oxidation, Catalysis, and Inhibition Effects

Inorganic compounds are mainly determined based on their catalytic effect on CL reactions. Direct oxidation reactions are also available. Determination of ions based on the inhibition of the catalytic reaction was explored to extend the analytical applications. More than 20 metals ions and some nonmetal compounds were determined with very high sensitivity, though poor selectivity, employing such methods. Without doubt, the most useful, versatile, and efficient CL system currently available for the determination of inorganic ions still remains the luminol reaction although the system has poor selectivity for some typical metal ions. To some extent, lucigenin was also extensively studied for the analysis of inorganic compounds. As they may not directly produce a CL reaction, some inorganic ions could be determined by coupling an enzyme reaction with a luminol-based CL reaction. For example, phosphate ion was determined using the reaction of phosphate-ion-dependent pyruvate oxidase, the hydrogen peroxide produced being detected by luminol CL [10].

In the oxalate ester area, *bis*-(2,4,6-trichlorophenyl)oxalate (TCPO) belongs to one of the most important reagents for CL detection of, e.g., amino-substituted polycyclic aromatic hydrocarbons and H_2O_2. Only a few reports refer to inorganic compounds by using this CL system, however. For instance, Quass and Klockow [11] developed a method for the determination of Fe(II) and H_2O_2 in atmospheric liquids by the CL reaction of oxygen with TCPO in the presence of the Fe(II) ion. The method gives detection limits below 100 nM Fe(II). Sato and Tanaka [12] reported a flow injection CL method for the determination of Al(III), Zn(II), Cd(II), and In(III). The method is based on the formation of fluorescent compounds of the metal ions with 8-hydroxyquinoline that produce CL emission in the reaction of TCPO and hydrogen peroxide. The detection limits for the metal ions were in the 20–70-ng/ml range.

An electrogenerated chemiluminescence (ECL) method was recently developed by several authors [13, 14] for the production of an (unstable) reagent for CL inorganic analysis. Zheng and Zhang [13] developed a novel electrolytic flow-through cell for the production of BrO^- from KBr. The method was applied to the determination of sulfide based on the strong enhancing effect upon the CL emission intensity in the reaction between BrO^- and luminol. A similar method was developed by Qin et al. [14] for determination of the ammonium ion based on chlorine production from the CL^- ion. To improve the sensitivity and selectivity of CL detection, Zhang et al. [15] developed a stripping CL method for preconcentration and determination of metal ions. The method consists of a concentration (preelectrolysis) step, in which the analyte metal ion is reduced at a controlled potential while being deposited as the metal on a solid electrode placed in an ECL flow cell, a subsequent media exchange step, and finally the stripping-CL step, in which the CL emission on the electrode surface is detected in situ when the CL reagent solution is delivered into the cell to react with the metal ion, just stripped by oxidation. Copper ion as a representative analyte was investigated using constant potential stripping and CL detection of a luminol-copper(II)-cyanide system. The low detection limit of 0.02 ng/mL could be achieved in a deposition time of 2 min. A similar design was also used for the development of a CL flow-through sensor for the copper ion [16].

Most of the described liquid CL systems are employed at ambient temperatures although CL emission may also be generated at very low temperatures. For instance, a low-temperature interaction of U(IV) and XeO_3 in frozen aqueous H_2SO_4 solutions accompanied by CL emission was studied by Lotnik et al. [17]. It was shown that the peak of luminescence at 195–200 K is related to CL of the excited uranyl ion (UO_2^{2+}) formed in the oxidation of U(V), an intermediate product of the interaction of U(IV) and XeO_3. This reaction, however, has not yet been applied for analytical purposes.

Major efforts have been undertaken in recent years for developing new CL systems. These studies extend the application of CL analysis to inorganic species. It should be noted, however, that there still exists no new CL system that exceeds the sensitivity offered by luminol for current metal ions. Table 2 lists some typical examples of inorganic analyses using different CL systems.

3.2 CL Measurements Based on Reduction Reactions

Several authors observed CL emission based on reduction reactions. Lu et al. [59] developed a method by applying a Jones reductor for producing unstable reductants. A column (100×3 mm i.d.) filled with Zn-Hg particles was inserted into the flow stream of a flow injection system. CL was measured using a home-made CL analyzer. Although the Jones reductor was more effective for the species studied in 0.5–5 mol/L H_2SO_4 solution, the authors found that a lower acid concentration improved the CL emission. The optimal pH was 6.5 for V(II), 2.5 for Mo(III), 3.5 for U(III), 3.0 for W(III), 3.0 for Cr(II), 2.5 for Ti(III), and 2.5 for Fe(II). The methods allowed determination of the above-mentioned species at μg/mL to ng/mL levels. It was assumed that the CL reactions were related to the production of superoxide radicals by dissolved oxygen in the solutions. The proposed methods could be successfully applied to the determination of V [60], Mo [61], and U [62] in water or steel samples.

It is well known that lucigenin produces CL emission when reacted with organic reducers such as ascorbic acid. Lu et al. [63–66] extended the reaction to inorganic ions. In the series of their reports they explored the reactions between lucigenin and reducers such as Cr(II), Fe(II), Mo(III), Ti(III), U(III), V(II), and W(III), among others, and found that CL emission can be produced by the reaction between the reductant and lucigenin. With this method, the ions mentioned above could be sensitively determined by using an online flow injection system applying a Jones reductor. It should be noticed that the detection limits for most ions studied were lower than those obtained using luminol reactions. Table 3 lists a comparison of the results.

3.3 CL Measurements Based on Coupling Reactions

Although many inorganic compounds cannot directly produce CL emission when reacting with the current CL reagents, they may be determined by coupling a CL reaction with another reaction, the latter producing a product playing a key role in the CL emission. A typical example is the determination of molybdenum(VI) ions based on the accelerating effect of the ions on the KBO_3-KI reaction [67].

Table 2 Analysis of Inorganic Species Using CL Systems

Analyte	CL system	Detection limit (ng/mL)	Ref.
Ag(I)	Luminol-$K_2S_2O_8$	10	18
	Lucigenin-H_2O_2	100	19
	Gallic acid-H_2O_2	500	20
As(V)	Luminol	10	21
As(III)	Luminol-Cr(VI)-H_2O_2	1	22
Au(III)	Luminol-H_2O_2	0.04	1, 23
	Tween-40-H_2O_2	0.2	24
	Coumarin-KOH-H_2O_2	0.6	25
Co(II)	Luminol-H_2O_2	0.0074	26
	Lucigenin-H_2O_2	0.02	27
	Gallic acid-H_2O_2	0.01	28
	Sulfosalicylic acid-H_2O_2	0.001	29
Cr(III)	Luminol-H_2O_2	0.0015	30
	Rhodamine-H_2O_2	28	31
Cu(II)	Luminol-KCN	0.009	32
	Luminol-H_2O_2	1	33
	4-Hydroxy-coumarin-H_2O_2-CTMAB	0.3	34
	o-Phenanthroline-H_2O_2-Sn(IV)	0.16	35
Fe(III)	Luminol-H_2O_2	0.04	36
Fe(II)	Luminol-O_2	0.1	37
	Gallocyanin-H_2O_2	0.07	38
Hg(II)	Luminol-Cu(II)-KCN	2	39
Ir(III)	Luminol-KIO_4	1.0	40
Mn(II)	Luminol-H_2O_2	0.08	41
	Gallic acid-H_2O_2	400	20
	MnO_4^--Na_2CO_3-KOH	0.1	42
Ni(II)	Luminol-H_2O_2	0.01	43
	Alizarine violet-CTMAB	0.1	44
	Anthracene green-H_2O_2	0.01	45
Os(VIII)	Luminol-H_2O_2	0.2	46
	1,10-Phenanthroline-H_2O_2	0.0004	47
Pb(II)	Luminol-H_2O_2	1000	33
	Gallic acid-H_2O_2	1000	20
Ru(III)	Luminol-H_2O_2	0.3	48
	Luminol-KIO_4	3	48
	Luminol-$K_2S_2O_8$	60	48
Rh(III)	Luminol-KIO_4	40	49
Si(IV)	Luminol	0.1	50
	Luminol-heteropoly acid	10	51
Cl_2	Luminol	0.2	52
	Luminol-H_2O_2	4	53
Br_2	Luminol	6	54
I_2	Luminol	10^{-9}(M)	55
NO_2	Luminol	0.46	56
CN^-	Luminol	1.2	57
	Luminol-Cu(II)	2	58

CTMAB, cethyltrimethyl amonium bromide.

Table 3 CL Determination of Inorganic Ions Based on Reduction Reaction

Analyte	Reduced form	CL reagent	Linear range (g/mL)	Ref.
Cr(VI)	Cr(II)	Luminol	1.0–10	59
		Lucigenin	0.1–10	63
Fe(III)	Fe(II)	Luminol	0.001–2.0	59
		Lucigenin	0.00001–10	64
Mo(VI)	Mo(III)	Luminol	0.001–0.1	61
		Lucigenin	0.0001–1.0	65
Ti(IV)	Ti(III)	Luminol	0.1–1.0	59
		Lucigenin	0.001–6.0	63
U(VI)	U(III)	Luminol	0.001–2.0	59
		Lucigenin	0.001–10	66
V(V)	V(II)	Luminol	0.0001–0.01	60
		Lucigenin	0.001–90	63
W(VI)	W(III)	Luminol	0.01–0.04	59
		Lucigenin	0.001–10	63
Sn(IV)	Sn(II)	Luminol	0.1–1.0	59

The CL signal arises from the reaction of luminol with I_2 produced by the latter reaction. Other reactions are also available including KIO_3-KI [68], H_2O_2-KI [69], and H_2O_2-$Na_2S_2O_3$ reactions [70] catalyzed by Mo(VI), among others. Similar reactions could also be used for the determination of trace vanadium(V) ions [71, 72]. The developed methods were applied to the determination of trace elements in natural waters and other matrices with detection limits below the ng/mL levels for traces of molybdenum and vanadium.

Another example for the determination of an element based on CL coupling reactions is the arsenic determination by coupling $K_2Cr_2O_7$-AsH_3 to the luminol-H_2O_2-Cr(III) reaction [22]. Arsine was produced by hydride generation and oxidized by $K_2Cr_2O_7$ in acidic medium. The Cr(III) ions produced in the first reaction could be reacted with luminol and H_2O_2 to generate CL emission. A similar method was used for the determination of SO_2 in air [73] and Na_2S in water [74].

An indirect method for the determination of lead by coupling reactions was developed based on the replacement of Fe(II) by Pb(II) from the Fe(II)-EDTA complex. The subsequent CL reaction was based on the Fe(II)-luminol-O_2 system. The method was used to determine lead in polluted water samples [75]. Such methods may be extended to other ions with proper complex constants as compared to the Fe(II)-EDTA complex, after HPLC separation. Analysis of elements based on indirect reactions is summarized in Table 4.

Table 4 Analysis of Elements Using Coupling Reactions

Element	Coupling reactions	Detection limit	Ref.
As(III), (V)	$Cr_2O_7^{2-} + AsH_3 \rightarrow Cr^{3+} + AsO_4^{3-}$	1 ng/mL As	22
	Luminol + $H_2O_2 \rightarrow$ CL light		
	$MoO_4^{2-} + I^- \rightarrow I_2$	40 pg/mL As	76
	Luminol + $I_2 \rightarrow$ CL light		
Cr(VI)	$CrO_4^{2-} + Fe(CN)_6^{2-} \rightarrow Cr^{3+} + Fe(CN)_6^{3-}$	20 pg/mL Cr	77
	Luminol + $Fe(CN)_6^{3-} \rightarrow$ CL light		
Ge(IV)	$MoO_4^{2-} + I^- \rightarrow I_2 + MoO_3^{2-}$	0.3 ng/mL Ge	78
	Luminol + $I_2 \rightarrow$ CL light		
Hg(II)	$Hg^{2+} + CN^- \rightarrow Hg(CN)_4^{2-}$	2.9 ng/mL Hg	79
	Luminol + $Cu^{2+} + CN^- \rightarrow$ CL light		
Mo(VI)	$BrO_3^- + I^- \rightarrow I_2 + Br^-$	3 ng/mL Mo	67
	Luminol + $I_2 \rightarrow$ CL light		
	$IO_3^- + I^- \rightarrow I_2 + Br^-$	3 ng/mL Mo	68
	Luminol + $I_2 \rightarrow$ CL light		
	$S_2O_3^{2-} + H_2O_2 \rightarrow SO_4^{2-} + H_2O$	0.12 ng/mL Mo	70
	Luminol + $H_2O_2 \rightarrow$ CL light		
Pb(II)	$Pb^{2+} + Fe\text{-EDTA} \rightarrow Pb\text{-EDTA} + Fe^{2+}$	—	75
	Luminol + $Fe^{2+} + O_2 \rightarrow$ CL light		
PO_4^{3-}	$MoO_4^{2-} + I^- \rightarrow PO_4^{3-} I_2 + MoO_3^{2-}$	4 ng/mL PO_4^{3-}	80
	Luminol + $I_2 \rightarrow$ CL light		
Rh(III)	$BrO_3^- + Mn^{2+} \rightarrow MnO_4^- + Br_2$	5 pg/mL Rh	81
	Luminol + $MnO_4^- \rightarrow$ CL light		
V(V)	$ClO_3^- + I^- \rightarrow I_2 + Cl^-$	2.8 ng/mL V	71
	Luminol + $I_2 \rightarrow$ CL light		
	$S_2O_3^{2-} + H_2O_2 \rightarrow SO_4^{2-} + H_2O$	0.43 ng/mL	72
	Luminol + $H_2O_2 \rightarrow$ CL light		
S^{2-}	$Fe^{3+} + S^{2-} \rightarrow Fe^{2+} + S$	0.4 nmol/L	82
	Luminol + $Fe^{2+} + O_2 \rightarrow$ CL light		
	$Cr_2O_7^{2-} + S^{2-} + H^+ \rightarrow Cr^{3+} + S$	0.25 ng/mL	74
	Luminol + $H_2O_2 \rightarrow$ CL light		

3.4 CL Measurements Based on Time-Resolved Techniques

Time-resolved CL analysis is based on the measurement of the difference of dynamic rates of CL reactions as proposed by Zhang et al. in 1989 [83]. It aims at improvement of the selectivity for analysis of metal ions in real samples, as it is known that generally such analyses suffer from poor selectivity since various metal ions may catalyze the same CL reaction—e.g., luminol/H_2O_2—under simi-

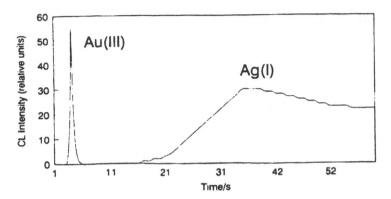

Figure 2 Typical dynamic profile of CL emission of Au(III) and Ag(I). (From Ref. 84, with permission.)

lar conditions. A typical example is the determination of silver in an environmental sample containing cobalt [84] or gold. Although it is well known that the latter two ions may produce strong CL emission similar to Ag(I) by catalyzing the reaction of luminol with persulfate in alkaline medium, a well-resolved Ag peak of CL emission was observed from the peaks of Co(II) or Au(III) due to a slower dynamic reaction of Ag with persulfate. Based on the kinetic distinction of Au(III) and Os(IV) in the Tween 80–KOH CL system, a time-resolved CL analysis for the simultaneous determination of traces of Au(III) and Os(IV) was proposed by Han et al. [85] offering detection limits of about 1 ng/mL for Au(III) and 10 ng/mL for Os(IV). The method was applied with satisfactory results to the determination of Au(III) and Os(IV) in metallurgical materials samples of noble metals [85]. Figure 2 shows the typical dynamic profile of CL emission of Au(III) and Ag(I).

3.5 CL Measurements on Solid Surfaces

In recent years, a series of luminol-based CL methods were developed for the determination of trace elements on a solid surface.

An early study referred to the determination of traces of gold in ores by absorbing this element on the surface of foamed plastic [23]. The sample containing the gold traces was first dissolved by a HCl-HNO₃ mixed solution and a piece of foamed plastic was then placed therein. After vibration for 30 min. on a vibrator, the solid material was removed, washed with tap water, and placed in a CL cell. A 2-mL volume of 1.0 mmol/L EDTA was added to the cell and the lid was closed; then 2 mL of 0.01 mmol/L luminol solution was injected into

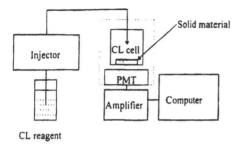

Figure 3 Schematic diagram of a typical instrument for measuring CL signals on a solid surface.

the cell and the CL signal was measured by using a laboratory-constructed CL photometer. With this procedure, traces of gold could be selectively determined in the range of 0.1–10 µg with an RSD of less than 10%.

Determination of silver in ores was also carried out on filter paper as developed by Zhang et al. [86]. Ag(I) was first separated by using a ring oven technique and determined by luminol-based CL measurement directly on the filter paper. The method permits selective determination of silver in the range of 0.5–50 ng with an RSD of 8.4%. Similar methods were also developed for the CL-based determination of Co(II) [87], Ni(II) [88], and Cr(VI) [89] on filter paper or on ion-exchange resins. The studies showed that solid-surface CL analysis (SSCL) has advantages over the suggested solid-surface fluorescence analysis (SSF) and solid-surface room temperature phosphorescence analysis (SSRTP) because the latter methods suffer from a high background emission from the solid substrate resulting from scattering of excitation light [90]. Figure 3 shows the schematic diagram of a typical instrument for measuring CL signals on a solid surface.

3.6 Hyphenated Techniques: Coupling HPLC/CE to CL Detection

High-performance liquid chromatography (HPLC) and, more recently, capillary electrophoresis (CE) [6] have been coupled to CL detection. These hyphenated techniques greatly improve the sensitivity and selectivity of analysis. For instance, Cr(III) and Cr(VI)-species were well separated on a Dionex AS4A anion exchange column containing a small portion of cation exchange groups. Very high sensitivity was achieved using a CL postcolumn reaction detector based on the catalytic oxidation of luminol. Linear calibrations were obtained over the range 0.1–500 ng/mL with detection limits of 0.05 and 0.1 ng/mL for Cr(III) and Cr(VI), respectively [91]. Similar work was also carried out by Gammelgaard et al. [92], who developed a method for Cr-species separation by using two Dionex ion-exchange guard columns in series. Cr(VI) was reduced by potassium sulfite, whereupon both species were detected by use of the luminol–hydrogen peroxide CL system. The detection limit was 0.12 ng/mL for chromium(III) and 0.09 ng/mL for chromium(VI). Although it was possible to speciate Cr(III) and Cr(VI) using a FIA system [8], Cr-species separation using the HPLC technique offers the advantage of high selectivity especially in waste water samples. Figure 4 shows a typical chromatogram of Cr species, separated and measured by coupling HPLC with CL detection.

Some inorganic anions were also separated and detected by coupling HPLC to a CL detector. For instance, Sakai et al. [93] developed a method for the determination of inorganic anions in water samples by ion-exchange chromatography with CL detection based on the neutralization reaction of nitric acid and potassium hydroxide (Fig. 5). The faint CL from the neutralization reaction was enhanced by addition of Fe(III) to the acid. The enhanced CL emission was suppressed by adding inorganic anions such as chloride, bromide, nitrite, nitrate, and sulfate to the base, which were separated by anion-exchange chromatography. For each anion, linear calibration ranges extending from 100 ng/mL to 100 μg/mL were obtained.

Fujiwara et al. [94] found that, when present as a heteropolyacid complex, molybdenum(VI), germanium(IV), and silicon(IV) produced CL emission from the oxidation of luminol, and similar CL oxidation of luminol was observed for arsenic(V) and phosphorus(V) but with the addition of the metavanadate ion to the acid solution of molybdate. A hyphenated method was therefore proposed for the sensitive determination of arsenate, germanate, phosphate, and silicate, after separation by ion chromatography. The minimum detectable concentrations of arsenic(V), germanium(IV), phosphate, and silicon(IV) were 10, 50, 1, and 10

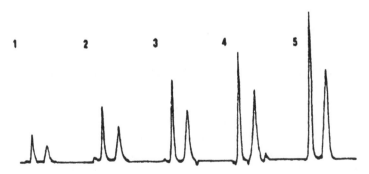

Figure 4   Chromatogram of Cr(VI) and Cr(III) in the range 1–5 μg/L (correlation coefficients >0.999 based on peak height measurements). Cr(VI) appears at 2.8 min, Cr(III) appears at 4.6 min. (From Ref. 92, with permission.)

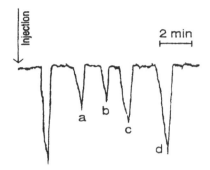

Figure 5 Chromatogram of a synthetic mixture of chloride (a), bromide (b), nitrate (c), and sulfate (d) separated on an anion-exchange column. (From Ref. 93, with permission.)

ng/mL, respectively. The method was applied to analyses of seaweed reference material, rice wine, and water samples. A gas-phase CL detector was also employed for selective detection of liquid-phase sulfur compounds (Fig. 6). This detector operates by converting sulfur-containing compounds in the liquid phase into sulfur monoxide. The photo emission resulting from the $SO + O_3$ reaction is monitored by a photomultiplier tube (PMT). Application of the detector to reversed-phase HPLC analysis of standards of sulfur-containing pesticides, proteins, and blood thiols, and to the FIA of acid-soluble thiols in rat plasma is presented [95].

Several authors have developed highly sensitive CL detection coupled to CE for ion analysis. Huang et al. [96] used luminol as a component of the separation electrolyte that prevented loss of the light signal. Detection limits of 20 zmol, 2 amol, 80 amol, 740 amol, and 100 fmol for Co(II), Cu(II), Ni(II), Fe(III), and

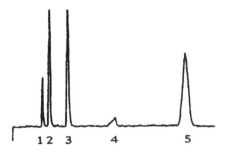

Figure 6 HPLC separation of a 4.0-nmol mixture of blood thiols detected by CL. (1) *N*-acetyl-*L*-cysteine; (2) reduced glutathion; (3) cysteine; (4) methionine; (5) oxidized glutathion. (From Ref. 95, with permission.)

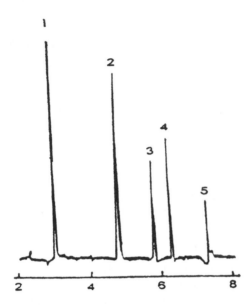

Figure 7 Typical electropherogram of the separation of metal ions using the CE-CL technique. (1) Co(II); (2) Cu(II); (3) Ni(II); (4) Fe(III); (5) Mn(II). (From Ref. 97, with permission.)

Mn(II), respectively, were reached. The CE-CL detector was used to separate five metal ions, Co(II), Cu(II), Ni(II), Fe(III), and Mn(II), within 8 min, with an average theoretical plate number of 4.6×10^5. A rapid separation and quantification of nitrite from nitrate ions in biological fluids using CE-CL was reported by Trushina et al. [97]. Nitrites and nitrates can be efficiently reduced to NO at 37°C using vanadium chloride (100%) or $HgCl_2$ (80%). However, these CE-derived conditions cannot be simply extrapolated to CL measurements. Vanadium(III) yields high backgrounds in the PMT, which diminishes the sensitivity of CL measurement to that outside of physiological ranges. The present method has been applied to measure nitrites and nitrates in biological fluids. Figure 7 shows a typical electropherogram from the separation of metal ions using the CE-CL technique.

4. CONCLUSIONS

The CL-based technique has been successfully applied to the determination of inorganic compounds, e.g., Cu(II), Co(II), Cr(III), and H_2O_2, among many others, with lowest detection limits among the current techniques available.

CL-based techniques require only simple instrumentation, which offers promising opportunities for analysts who are active in the trace analysis area but who do not have sophisticated instruments available in their laboratories.

It should be noted, however, that among the various analytical methods currently available for the determination of inorganic compounds, CL-based techniques are still not yet being considered as important techniques because only limited CL systems are available, so far.

Hence it is a great challenge for the analyst to develop new CL analytical systems and apply them to real samples and complicated matrices, though it would also be important to extend the applications of classical CL systems such as the luminol-based system as described in this chapter.

REFERENCES

1. JR Lu, XR Zhang. Fenxi Huaxue (Chinese J Anal Chem) 15:1120–1122, 1987.
2. WRG Baeyens, SG Schulman, AC Calokerinos, Y Zhao, AM García-Campaña, K Nakashima, D De Keukeleire. J Pharm Biomed Anal 17:941–953, 1998.
3. LP Palilis, AC Calokerinos, WRG Baeyens, Y Zhao, K Imai. Biomed Chromatogr 11:85–86, 1997.
4. K Nakashima. Bunseki 7:518–524, 1996.
5. WRG Baeyens, B Lin Ling, UATh Brinkman, SG Schulman. J Biolum Chemilum 4:484–499, 1989.
6. AM García-Campaña, WRG Baeyens, Y Zhao. Anal Chem 69:83A–88A, 1997.
7. WRG Baeyens, B Lin Ling, K Imai, AC Calokerinos, SG Schulman. J Microcol Sep 6:195–206, 1994.
8. R Escobar, Q Lin, A Guiraum, FF de la Rosa. Int J Environ Anal Chem 61:169–175, 1995.
9. R Escobar, MS García-Dominguez, A Guiraum, FF de la Rosa. Fresenius J Anal Chem 361:509–511, 1998.
10. H Nakamura, K Ikebukuro, S Mcniven, I Karube, H Yamamoto, K Hayashi, M Suzuki, I Kubo. Biosens Bioelectron 12:959–966, 1997.
11. U Quass, D Klockow. Int J Environ Anal Chem 60:361–375, 1995.
12. K Sato, S Tanaka. Microchem J 53:93–98, 1996.
13. XW Zheng, ZJ Zhang. Gaodeng Xuexiao Huaxue Xuebao (Chem J Chinese Univ) 20:209–213, 1999.
14. W Qin, ZJ Zhang, BX Li, YY Peng. Talanta 48:225–229, 1999.
15. CX Zhang, SC Zhang, ZJ Zhang. Analyst 123:1383–1386, 1998.
16. W Qin, ZJ Zhang, HJ Liu. Anal Chem 70:3579–3584, 1998.
17. SV Lotnik, LA Khamidullina, VP Kazakov. Kinet Catal 36:179–182, 1995.
18. XR Zhang, JR Lu, K Yang, ZJ Zhang. Fenxi Shiyanshi (Chinese J Anal Lab) 5:47–50, 1986.
19. AK Babko, AV Terletskaya, LI Dubovenko. Zh Anal Kim 23:932–934, 1968.
20. S Steig, TA Nieman, Anal Chem 49:1322–1325, 1997.

21. H Sakai, T Fujiwara, T Kumamaru. Bull Chem Soc Jpn 67:2317–2319, 1996.
22. J Ouyang, JR Lu, LD Zhang. Fenxi Shiyanshi 9:25–27, 1990.
23. JR Lu, XR Zhang, WZ Fan, ZJ Zhang. Anal Chim Acta 262:225–229, 1992.
24. F Zhang, ZM Liu. Fenxi Huaxue 16:485–489, 1988.
25. F Zhang, ZM Liu. Fenxi Huaxue 16:1119–1121, 1988.
26. SH Jia, ZJ Zhang. Huaxue Xuebao (Acta Chim Sinica) 42:1257–1261, 1984.
27. LA Montano, JD Ingle. Anal Chem 51:919–926, 1979.
28. F Zhang, YH Mei, H Chen. Fenxi Huaxue 16:5–8, 1988.
29. MF Ling, MG Lu, NB Tao, H Cui, XH Lu, F Yin. Fenxi Huaxue 14:941–943, 1986.
30. ZJ Zhang, JR Lu. Huaxue Tongbao 5:25–29, 1984.
31. F Zhang, YL Chen. Fenxi Shiyanshi 5:1–3, 1986.
32. ZJ Zhang and WB Dong. Kexue Tongbao 29:447–480, 1984.
33. S Pantel, H Weisz. Anal Chim Acta 74:275–280, 1975.
34. GH Xie, ZA Yu. Fenxi Huaxue 21:1052–1054, 1993.
35. H Li, ZA Yu. Gaodeng Xuexiao Huaxue Xuebao 14:1076–1078, 1993.
36. IE Kalinichenko, OM Grishchenko. Ukr Khim Zh 36:610–613, 1970.
37. WR Seitz, DM Hercules. Anal Chem 44:2143–2149, 1972.
38. SM Zhong, H Chen, WM Zhang. Fenxi Ceshi Tongbao 8:44–47, 1989.
39. AT Pilipenko, AV Terletskaya, TA Bogoslovskaya, NM Lukovskaya. Zh Anal Khim 38:807–810, 1983.
40. NM Lukovskaya, TA Bogoslovskaya, Ukr Khim Zh 48:842–844, 1982.
41. IE Kalinichenko. Ukr Khim Zh 7:755–757, 1969.
42. F Zhang, QX Lin. Talanta 40:1557–1561, 1993.
43. WR Seitz. In: MJ Cormier, DM Hercules, J Lee, eds. Chemiluminescence and Bioluminescence. New York: Plenum, 1973, p 427.
44. GZ Fang, L Liu. Fenxi Huaxue 24:743–743, 1996.
45. H Li, ZA Yu. Fenxi Huaxue, 21:1052–1054, 1993.
46. NM Lukovskaya, LV Markova, NF Evtushenko. Zh Anal Khim 29:767–772, 1974.
47. T Hara. Bull Chem Soc Jpn 60:2031–2035, 1987.
48. NM Lukovskaya, AV Terletskaya. Zh Anal Khim 31:751–756, 1976.
49. JL Burguera. Talanta 27:309–314, 1980.
50. H Sakai, T Fujiwara, T Kumamaru. Anal Chim Acta 302:173–177, 1995.
51. T Fujiwara, K Kurahashi, T Kumamaru, H Sakai. Appl Organomet Chem 10:675–681, 1996.
52. DF Marino, JD Ingle. Anal Chem 53:455–458, 1981.
53. DF Marino, JD Ingle. Anal Chim Acta 123:247–253, 1981.
54. Z Geng, JR Lu. Shaanxi Shida Xuebao 17:31–35, 1989.
55. WM Hardy, WR Seitz, DM Hercules. Talanta 24:297–302, 1977.
56. GE Collins, SL Ross-Pehrsson. Anal Chem 67:2224–2230, 1995.
57. JR Lu, XR Zhang, WD Han, ZJ Zhang. Fenxi Huaxue 20:575–577, 1992.
58. JZ Lu, W Qin, ZJ Zhang, ML Feng, YJ Wang. Anal Chim Acta 304:369–373, 1995.
59. JR Lu, XR Zhang, ML Feng, JQ Zhang, ZJ Zhang. Chem Res Chinese Univ 8:360–364, 1992.
60. JR Lu, XR Zhang, ZJ Zhu, ZJ Zhang, MQ Li. Gaodeng Xuexiao Huaxue Xuebao 12:1595–1598, 1991.
61. JR Lu, XR Zhang, ML Feng, ZJ Zhang. Fenxi Huaxue 21:1000–1003, 1993.

62. JR Lu, XR Zhang, ML Feng, ZJ Zhang. Fenxi Shiyanshi 13:3–7, 1994.
63. JR Lu, ZJ Zhu. Shaanxi Shida Xuebao 22:57–60, 1994.
64. JR Lu, ZJ Zhu. Shaanxi Shida Xuebao 22:39–42, 1994.
65. ZJ Zhu, JR Lu. Fenxi Huaxue 22:909–912, 1994.
66. ZJ Zhu, JR Lu. Shaanxi Shida Xuebao 23:59–62, 1995.
67. XR Zhang, JR Lu, JX Zhu, ZJ Zhang. Fenxi Huaxue 16:806–808, 1988.
68. XR Zhang, JR Lu, YP Xue, ZJ Zhang. Shaanxi Shida Xuebao 2:66–70, 1987.
69. ZJ Zhang, JR Lu, XR Zhang, YM Liang. Gaodeng Xuexiao Huaxue Xuebao 9:519–521, 1988.
70. WB Ma, W Zhang, ZJ Zhang. Fenxi Huaxue 16:818–820, 1988.
71. ZJ Zhang, XR Zhang, JR Lu, CS Pan. Yejin Fenxi 7:110–112, 1987.
72. ZJ Zhang, WB Ma, P Wang. Huaxue Xuebao 46:543–547, 1988.
73. JR Lu, XR Zhang, JK Tie. Shaanxi Shida Xuebao 17:42–47, 1989.
74. XR Zhang, JR Lu, XH Lu. Shaanxi Shida Xuebao 17:113–116, 1989.
75. WH Li, ZL Wang, JZ Li, ZJ Zhang, XQ Zhang. Fenxi Huaxue 26:219–221, 1998.
76. Z Geng, JC Wu, BJ Hui, ZJ Zhang. Fenxi Huaxue 20:423–425, 1992.
77. JR Lu, XR Zhang, ZJ Zhang. Fenxi Huaxue 20:61–63, 1992.
78. Z Geng, JG Zhang. Fenxi Shiyanshi 9:6–8, 1990.
79. ZZ Gao, FC Cai, L Wang, ZJ Zhang. Shaanxi Shida Xuebao 17:108–112, 1989.
80. Z Geng, JW Da. Shaanxi Shida Xuebao 16:47–50, 1988.
81. Z Geng, JC Wu, BJ Huang. Fenxi Shiyanshi 13:15–17, 1994.
82. JR Lu, XR Zhang, J Zhang. Fenxi Huaxue 17:542–544, 1989.
83. XR Zhang, JR Lu, ZJ Zhang. Huaxue Xuebao 47:481–484, 1989.
84. J Ouyang. Time-resolved chemiluminescence determination of Ag(I). Proceedings of the 4th Asian Chemical Congress, Beijing 2:144, 1991.
85. HY Han, QY Luo, XM Yu. Rare Met 16:195–200, 1997.
86. XR Zhang, JR Lu, ML Feng, ZJ Zhang. Fenxi Huaxue 21:575–577, 1993.
87. ZJ Zhang, WP Yang, JR Lu. Gaodeng Xuexiao Huaxue Xuebao 15:1146–1148, 1994.
88. WP Yang, ZJ Zhang. Fenxi Huaxue 22:27–30, 1994.
89. WP Yang, ZJ Zhang. Fenxi Huaxue 22:71–74, 1994.
90. ZJ Zhang, WP Yang, XR Zhang, JR Lu. Fenxi Huaxue 22:23–26, 1994.
91. HG Beere, P Jones. Anal Chim Acta 293:237–243, 1994.
92. B Gammelgaard, YP Liao, O Jons. Anal Chim Acta 354:107–113, 1997.
93. H Sakai, T Fujiwara, T Kumamaru. Anal Chim Acta 331:239–244, 1996.
94. T Fujiwara, K Kurashashi, T Kumamaru, H Sakai. Appl Organomet Chem 10:675–681, 1996.
95. TB Ryerson, AJ Dunham, RM Barkley, RE Sievers. Anal Chem 66:2841–2851, 1994.
96. B Huang, JJ Li, L Zhang, JK Cheng. Anal Chem 68:2366–2369, 1996.
97. EV Trushina, RP Oda, JP Landers, CT McMurray. Electrophoresis 18:1890–1898, 1997.

7

Mechanism and Applications of Peroxyoxalate Chemiluminescence

Malin Stigbrand, Tobias Jonsson, Einar Pontén, and Knut Irgum
Umeå University, Umeå, Sweden

Richard Bos
Deakin University, Geelong, Victoria, Australia

1. INTRODUCTION

Peroxyoxalate chemiluminescence (POCL) was first reported in 1963 by Chandross [1], who observed emission of light from a mixture consisting of oxalyl

141

chloride, hydrogen peroxide, and the fluorescent compound 9,10-diphenylanthracene dissolved in 1,4-dioxane. The light emitted was found to be spectrally identical to the fluorescence emission spectrum of 9,10-diphenylanthracene. POCL is thus an *indirect* or *sensitized* type of chemiluminescence, as the intermediates or reaction products from the primary reaction do not emit a significant quantity of light. Instead, the in situ–generated intermediate transfers its energy to an energy-accepting fluorophore (denoted F in the following scheme), which becomes electronically excited (denoted F^*) and subsequently emits light.

The steps in the light-generating process can be described as follows:

$$\text{Oxalic acid derivative} + H_2O_2 \rightarrow \text{intermediates} \tag{1}$$

$$\text{Intermediates} + F \rightarrow F^* \tag{2}$$

$$F^* \rightarrow F + h\nu \tag{3}$$

The intermediates generated in the POCL reaction are capable of exciting fluorophores that emit light from the near-ultraviolet to the near-infrared region; it is this defining characteristic that establishes the usefulness of this reaction to analytical chemistry.

Much of the early development of POCL was performed by researchers at American Cyanamid [2–4]. The primary focus of their research was the investigation of reagents and reaction formulations that could emit light efficiently. This led to the successful development of practical, spark-free light sources that serve as emergency lighting. A high quantum yield in combination with a long duration of the emission are regarded as optimal properties in these devices. The first study dealing with different oxalic acid derivatives was published in 1967 by Rauhut and co-workers [3]. They found that diaryloxalate esters with strongly electron-withdrawing substituents provided the highest quantum yields, whereas diaryloxalate esters with electron-donating or weak electron-withdrawing groups were found to result in either ineffective or poor production of chemiluminescence. The most efficient diaryloxalate ester reported was *bis*(2,4-dinitrophenyl) oxalate (DNPO), and although this reagent finds occasional use analytically [5–8], it has been surpassed in popularity by 2,4,6-trichlorophenyl oxalate (TCPO), which is not quite as efficient but has the virtue of being more conveniently handled.

An early investigation [9] of oxamide derivatives as alternate sources of excitation in POCL did not initially yield great success in the context of the research work performed by American Cyanamid. In common with the earlier work on diaryloxalate ester derivatives, there appeared to be a general relationship between the electron-withdrawing ability of the substituent(s) and POCL efficiency. In later works, they developed a new class of oxamide reagents, which resulted in a substantial improvement in CL performance of oxamide derivatives [10]. These oxamides featured the strongly electron-withdrawing trifluoromethylsulfonyl moiety attached to the oxamide nitrogen. One of the compounds described in the cited work, *N,N'-bis*(2,4,5-trichlorophenyl)-*N,N'-bis*(trifluoro-

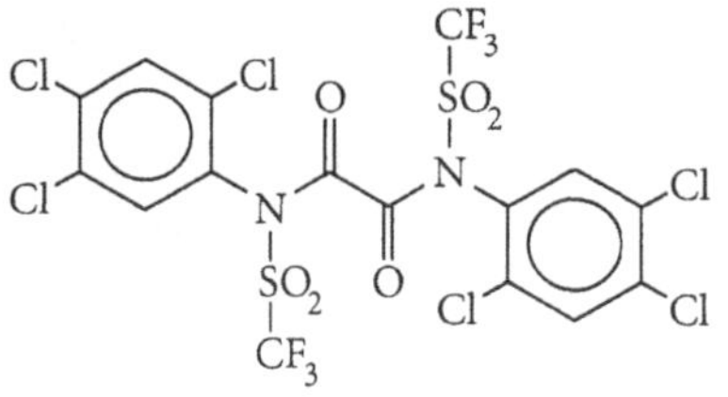

Figure 1 Structure of *N,N'-bis*(2,4,5-trichlorophenyl)-*N,N'-bis*(trifluoromethylsulfonyl) oxamide.

methylsulfonyl) oxamide (see Fig. 1) was reported (and still remains) to be the most efficient reagent presently known, with a claimed quantum yield of 34%.

The development of reagents for aqueous POCL received attention early. The impetus for this research was, again, to provide spark-free light sources for emergency use in both civilian and military applications. The first report published on aqueous POCL [11] attempted to adapt diaryloxalate ester reagents that had been found useful in nonpolar solvents, by adding functionalities that would assist in their water solubility. This approach gave rise to one reportedly useful compound, *bis*-{2,4-dichloro-6-[(2-dimethylaminoethyl)methylsulfamoyl]phenyl} oxalate dihydrochloride (see Fig. 2) that was claimed by the authors to be the first example of water-soluble POCL, producing a short-lived emission.

Further progress [12] in this field focused on developing the aforementioned trifluoromethylsulfonyl substituted oxamides [10]. Of the numerous molecules synthesized and tested, 4,4'-{oxalyl *bis*[(trifluoromethylsulfonyl)imino]-ethylene}*bis*(4-methylmorpholinium trifluoromethansulfonate) (METQ) and 2,2'-oxalyl-*bis*[(trifluoromethanesulfonyl)imino]ethylene-bis(*N*-methylpyridinium) trifluoromethanesulfonate (PETQ) (see Fig. 3) were found to be the most

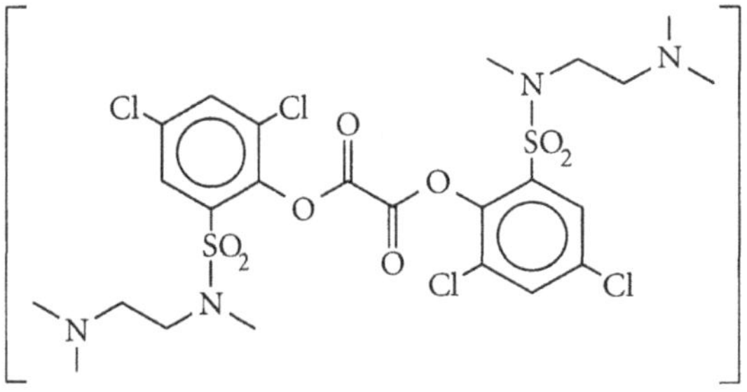

Figure 2 Structure of *bis*-{2,4-dichloro-6-[(2-dimethylaminoethyl)methylsulfamoyl]-phenyl oxalate dihydrochloride.

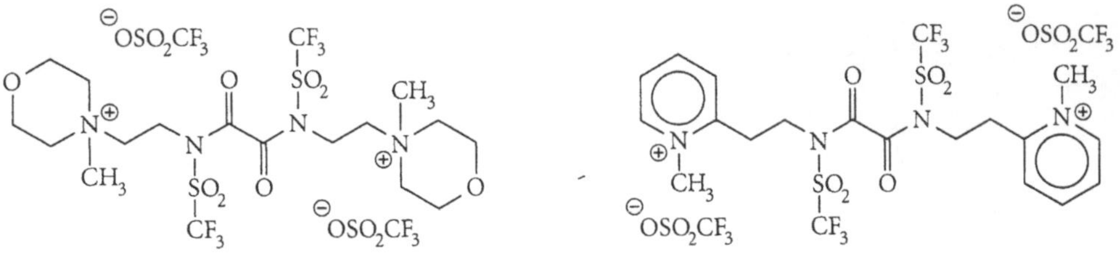

METQ   PETQ

Figure 3 Structures of 4,4′-{oxalyl *bis*[(trifluoromethylsulfonyl)imino]-ethylene}*bis* (4-methylmorpholinium trifluoromethansulfonate) (METQ) and 2,2′-oxalyl-*bis*[(trifluoromethanesulfonyl)imino]ethylene-*bis*(*N*-methylpyridinium) trifluoromethanesulfonate (PETQ).

efficient. Of these two molecules, only METQ [13–17] and a homolog 3,3′-oxa lyl-*bis*[(trifluoromethylsulfonyl)imino]trimethylene-*bis*(*N*-methylmorpholinium) trifluoromethanesulfonate (MPTQ) [18] have been employed analytically. Although the work was performed in a largely aqueous environment, the reagents were kept in acetonitrile prior to the reaction, to prevent preliminary hydrolysis. It has been reported that the half-life of METQ is of the order of 8 min under purely aqueous conditions [13].

Van Zoonen et al. [19,20] employed an alternative approach, in an attempt to overcome the limited aqueous solubility of diaryloxalate ester–type POCL reagents. In this work, granular TCPO was mixed with controlled pore glass and packed in a flow cell, forming a solid-state TCPO reactor. When this was used in conjunction with a flow system, some of the TCPO dissolved in the carrier solution. Numerous difficulties were encountered with this approach, namely, limited reactor lifetime (approximately 8 h) and low CL emission obtained as the carrier became more aqueous (a 90% reduction of CL intensity occurred when the aqueous content of the carrier stream comprised 50% water, as compared to pure acetonitrile). The samples also required dilution with acetonitrile to increase the solubility of TCPO in the sample plug.

The complex and sometimes competing issues of solubility, stability, and reactivity of reagents designed for purely aqueous POCL were addressed by Barnett et al. [21,22], who furthered the development of the trifluormethylsulfonyl-substituted oxamide class of reagents. They prepared disulfonic acid functionalised oxamides (see Fig. 4), which were found to possess a considerably improved degree of stability at both ambient and low temperature when compared to METQ, in addition to showing some promise as potential reagents in analytical applications.

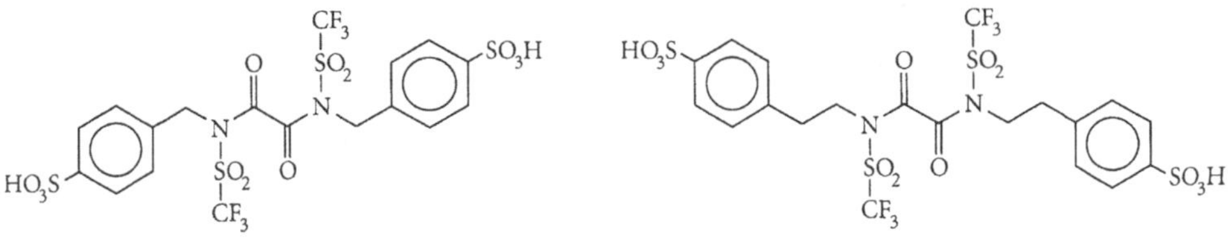

Figure 4 Structures of disulfonic acid functionalized oxamides.

As the POCL reaction began to find use in analytical chemistry, more researchers became interested in developing new reagents to address the specific requirements of this application [23–25]. In analytical flow systems, it is desirable for the light production to occur over a much shorter period of time, i.e., seconds instead of hours, and it is therefore usually inappropriate to employ reagents that have been developed with a long duration as the primary design criterion. Quantum yield is of course important, but equally critical for success in an analytical application is the intensity of the initial burst of light. In Table 1, some of the reagents used for analytical applications are listed.

The POCL reaction scheme can be used to detect several classes of analytes, depending on the limiting compound in the reaction, i.e.,

1. Hydrogen peroxide [26–33] and analytes converted to hydrogen peroxide via either an enzymatic [34–52] or a photochemical postcolumn reaction [53–55]
2. Fluorophores and analytes derivatized with a fluorophoric label [56–105]
3. Analytes that efficiently quench or enhance the CL reaction [106,107].

In addition to these three groups, there are some other examples, e.g., amines [108–110], phenols [111–113], metal ions [114–117], oxalic acid [118,

Table 1 The Most Commonly Used Oxalic Reagents (R-COCO-R) in Analytical Applications

Side chain, R-	Abbreviation	Ref.
2,4,6-Trichlorophenyl	TCPO	76, 78
2,4-Dinitrophenyl	DNPO	45, 78, 99, 183
2,4,5-Trichloro-6-carbopentoxyphenyl	CPPO	47
2-(3,6,9-Trioxadecyloxycarbonyl)-4-nitrophenyl	TDPO	91, 92, 109, 110, 160
2-Nitrophenyl	2-NPO	43, 79
Pentachlorophenyl	PCPO	111
Imidazole	ODI	29, 50, 151

119], and urea [120], which have been quantified with the POCL reaction. The high sensitivity attainable, combined with the wide scope of the reaction through the choice of the limiting or catalytic/quenching component, has resulted in a considerable number of analytical applications of peroxyoxalate chemiluminescence. The relative absence of interference is also a contributing factor to the wide variety of applications associated this chemistry [6–8].

Numerous fluorophores have been tested for their suitability as energy acceptors in the POCL reaction. Detection limits are generally in the low femtomole range and improvements of 10–100 times are possible compared to conventional fluorescence [75, 94, 121]. Another parameter that shows the contrast between POCL and fluorescence detection is the selectivity, where POCL, owing to its inability to excite all fluorescent compounds, is more selective than fluorescence and can provide simpler chromatograms, as has been demonstrated by Sigvardson and Birks [100]. Shale oil extracts were separated and detected with either CL or fluorescence detection. The chromatogram obtained with fluorescence detection was dominated by overlapping peaks, whereas when POCL chemistry was used, only well-resolved amino-polyaromatic hydrocarbons (amino-PAHs) were detected. This property can also be utilized in derivatization, wherein the 5-dimethylaminonaphthalene-1-sulfonyl (dansyl) group is the fluorescent moiety most commonly introduced. Dansylating derivatization can be directed toward many different kinds of analytes since this fluorophore has been synthesized with several functional groups, e.g., chloride [61], hydrazine [83], hydroxyl [122], ethanol amine [123], and aziridine [124].

It should also be noted that optimization of POCL detection is a relatively complex process due to the number of variable parameters in the reaction. Most papers dealing with practical applications of the POCL reaction also feature some degree of optimization studies. In addition, a number of studies have focused specifically on the optimization process [125–134], and Hadd and Birks [135] have recently summarized the most important aspects in a comprehensive overview.

2. REACTION MECHANISM

The mechanism of the POCL reaction is a complex multistep process and it has proved to be difficult to elucidate. Side reactions as well as light-generating reactions are fast and overlapping in time, and many of the intermediates are unstable. Because of this complexity, the complete reaction mechanism has still not been fully resolved, despite numerous investigations since its discovery.

Rauhut et al. [2, 3] conducted a number of mechanistic experiments aimed at elucidating the reaction mechanism of the POCL reaction. It was noted that fluorophores were not consumed in the reaction, and that the CL emission spec-

trum is essentially identical to the fluorescence emission spectrum of the fluorophore compound, demonstrating that the deexcitation of the fluorophore occurs
via the first singlet excited state. A 1:1 stoichiometry between oxalyl chloride
and hydrogen peroxide was observed, and the gaseous products of the reaction
were CO_2 and CO [2]. However, the ratio between these two gases varied substantially and difficulty was encountered in obtaining reproducible data. The changes
in the infrared absorption of carbonyl, acid chloride, and ester groups were studied during the course of the reaction. In the case of both oxalyl chloride [2]
and diaryloxalate esters [3], the absorbances from acid chloride and ester groups
decreased faster than the total carbonyl absorption, leading to the suggestion that
at least one intermediate contains a carbonyl group(s). Furthermore, the effect
of adding certain compounds was studied. Water increased the initial intensity
and the reaction rates, without affecting the quantum yield. Addition of ethanol
caused a side reaction in which no CL was produced, and the radical inhibitor
2,6-di-*tert*-butyl-4-methylphenol decreased the quantum yield, but did not affect
the reaction rates to a great extent. Some differences in behavior between oxalyl
chloride and the oxalate esters were noted. The reaction between oxalyl chloride
and hydrogen peroxide proceeded under acidic conditions [2], whereas the oxalate esters generally required neutral or basic conditions [3]. The quantum yield
increased with increasing hydrogen peroxide concentration in the oxalyl chloride
reaction, whereas a plateau was reached in the ester reaction. This led Rauhut et
al. to the conclusion that intermediates formed in the oxalyl chloride reaction
were short-lived, whereas in the ester reaction they were long-lived. These differences do not rule out a common intermediate; if there is one, it can be formed
through different reaction paths. Furthermore, competing side reactions, especially in the presence of certain additives, can affect the reaction paths significantly.

Recently, Orosz et al. [136] reviewed and critically reevaluated some of
the known mechanistic studies. Detailed mathematical expressions for rate constants were presented, and these are used to derive relationships, which can then
be used as guidelines in the optimization procedure of the POCL response. A
model based on the ''time-window concept,'' which assumes that only a fraction
of the exponential light emission curve is captured and integrated by the detector,
was presented. Existing data were used to simulate the detector response for
different reagent concentrations and flow rates.

2.1 Reactive Intermediates

Early research suggested that the ''key intermediate'' in the POCL reaction was
1,2-dioxetandione (structure II in Fig. 5) [2, 3], which is formed following the
multistage nucleophilic attack by the hydroperoxide ion on one of the carbonyl
carbons in the oxalate compound. This series of reactions is proposed to occur

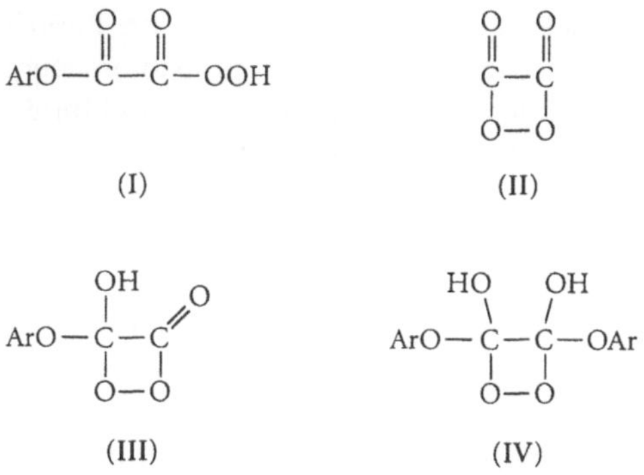

Figure 5   Proposed intermediates in the POCL reaction.

as follows: deprotonation of the newly formed hydroperoxyoxalate·ester group on the other carbonyl carbon is followed by a nucleophilic attack, which in turn triggers an intramolecular cyclization. This would explain why the nature of the aryl groups has a major influence on the efficiency in the POCL reaction. 1,2-Dioxetandione was once claimed to have been positively identified using mass spectrometry [137], but in a later study [138], it was shown that the peak corresponding to the dioxetandione (m/z 88; $C_2O_4^+$) was formed in the ion source itself.

Catherall et al. [139, 140] continued to study the mechanism and claimed that there was no experimental evidence to support the existence of a common intermediate. Instead they suggested another "key intermediate," one that still has a side chain (R) attached to it (structure III in Fig. 5). These workers also claimed that based upon their research, the key intermediate exhibited a lifetime of approximately 5×10^{-7} s at ambient temperatures.

Rauhut and Semsel [141] and Steinfatt [142] have suggested an unusual mechanism whereby the proposed intermediate 1,2-dioxetandione forms a structural (isoelectronic) isomer as outlined in route B in Figure 6. Route A represents a dark reaction. The isomerized molecule is then thought to decompose into singlet oxygen and carbon trioxide [143].

A kinetic investigation of photoinduced peroxyoxalate chemiluminescence led to the postulation of a new series of high-energy intermediates thought to be capable of inducing the chemiexcitation of fluorophore compounds [143]. Photoinitiation of the POCL reaction was performed by irradiation with a laser pulse to generate hydrogen peroxide in situ from oxygen in the presence of a hydrogen donor, 2-propanol. The kinetic behavior of the ensuing reaction(s) between TCPO, hydrogen peroxide, and a fluorophore was monitored. Complex reaction

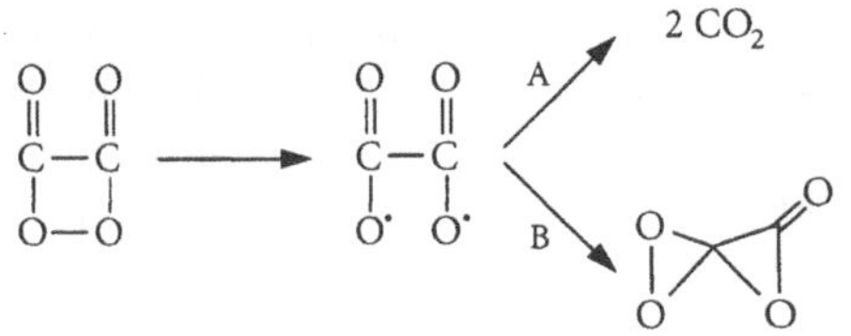

Figure 6 Proposed formation of an alternative C_2O_4 structure in the excitation mechanism for peroxyoxalate chemiluminescence.

schemes were proposed, giving rise to a number of reactive intermediates whose existence was considered by the authors to be speculative.

When the POCL reaction is initiated using diaryloxalate ester reagents in the absence of a fluorophore molecule, a weak emission may be observed. This emission, subsequently referred to as background emission, was found by an early study [100] to be a fundamental characteristic of the POCL reaction. Mann and Grayeski [144] studied the background emission from three different diaryloxalate esters and one oxamide compound. The results indicated that there were two emission bands, one with a maximum centered at 450 nm with a weaker band at a longer wavelength, which appeared to be reagent dependent. The two emission bands were attributed to two separate key intermediates of the POCL reaction, formed and decomposed at different rates. Recently, Barnett et al. [145] demonstrated that the background emission associated with oxalate esters was absent in the case of certain oxamides, but if the oxamides were mixed with chlorinated phenols, emission spectra exhibiting the same spectral characteristics as those of the corresponding *bis*(chlorinated phenyl) oxalate esters were seen. The intensity of the background emission arising from diaryloxalate ester reagents was also found to be significantly enhanced upon further addition of the corresponding phenol to the reaction. Accordingly, the background emission observed with diaryloxalate ester reagents was postulated to originate from phenoxyl radical intermediates, produced in situ as a result of the oxidizing conditions, and not from key intermediates in the POCL reaction.

The formation of stable intermediate(s) during the storage of POCL reagents under cold, acidic conditions has been observed [146]. Although the exact nature of the intermediate(s) is not yet understood, it is known that following prolonged cold storage ($\sim -20°C$) of either TCPO or *bis*(4-nitro-2-(3,6,9-trioxadecyloxycarbonyl)phenyl) oxalate (TDPO) with hydrogen peroxide in the presence of a strong acid such as trifluoroacetic acid, a stable compound(s) of unknown structure was formed. When either of these aged solutions was reacted with a fluorophore under suitable conditions, chemiluminescence was observed.

The POCL mechanism has also been studied by NMR spectroscopy by Chokshi and co-workers [147], who used *bis*(2,6-difluorophenyl) oxalate as reagent and measured the ^{19}F signals. Only one intermediate was found at the concentration level detectable with NMR, and the authors attributed it to a hydroperoxyoxalate ester (structure I in Fig. 5). They proposed that this was the key intermediate in the POCL reaction. In a later paper, Stevani et al. [148] were able to synthesize an analogous structure, 4-chlorophenyl-*O,O*-hydrogen monoperoxyoxalate. This compound did not transfer its excess energy to fluorophores without the addition of a base catalyst to deprotonate the molecule. This result indicated that this type of molecule was not a ''key intermediate'' but instead an important precursor that was capable of generating CL following deprotonation by a base. In a similar study, Hohman et al. [149] synthesized two new oxalate esters, triisopropylsilylperoxy 2,6-difluorophenyl oxalate and triisobutylsilylperoxy 2,6-difluorophenyl oxalate. These oxalate derivatives were found to be stable precursors of the half-ester peroxy acid and were intended for mechanistic studies, in pursuit of the ''key intermediate.'' Following the addition of fluoride ion in the presence of a fluorescer, a bright, short-lived chemiluminescence was observed.

In the case of the experiments performed by Hohman and co-workers [149], the fluoride anion would readily displace the silicon-leaving group. The peroxide anion could then further react via an intramolecular nucleophilic attack, resulting in cyclization to form the reactive intermediate responsible for the chemiluminescence that was observed. A recent kinetic study by Stevani and Baader [150] of the reaction of 4-chlorophenyl-*O,O*-hydrogen monoperoxyoxalate with various oxygen and nitrogen bases suggested that the intermediate formed must be 1,2-dioxetandione.

2.2   General-Base and Nucleophilic Catalysis

Over the years, there has also been considerable confusion regarding the interaction mechanisms of compounds that appear to catalyze the POCL reaction. The complexity of the reaction and its apparent dependency upon a significant number of parameters has resulted in slow progress in the understanding of the role of these catalysts. Recent work in this area [151–157] has considerably extended our knowledge of the catalyst reaction mechanisms in the POCL reaction, widening the possibilities for development of new and more carefully designed catalysts.

In the first systematic study of the reaction between several different diaryloxalates, hydrogen peroxide, and fluorophores [3], it was recognised that the chemiluminescence reaction was highly sensitive to base catalysis by potassium hydroxide or benzyltrimethylammonium hydroxide, and that acidic conditions markedly diminished the light production. The addition of bases was noted to both accelerate the reactions and, when less reactive oxalates were used, increase the quantum efficiency [4]. A reaction rate increase of about 100 times was observed when 0.1 molar percent of 2,6-lutidine was present in the reaction with DNPO [158]. Almost two decades later, a comprehensive report on POCL kinetics [139] stated that reproducible kinetics could not be obtained in the reaction between *bis*-(pentachlorophenyl) oxalate (PCPO) and hydrogen peroxide, unless the base catalyst sodium salicylate was present. It was also concluded that the reaction was general-base-catalyzed, and that the attack of the hydroperoxide anion was the step limiting the rate of the light production. Moreover, the quantum yield was found to decrease with increasing base concentrations, and the rate of light decay declined when the base concentration exceeded the oxalate concentration. The authors ascribed this to the formation of a complex between the diaryloxalate and the base catalyst, and noted that the quantum yield decrease was larger than what could be explained by a 1:1 stoichiometry in the assumed complex. The studies above supported the impression of general-base catalysis, but it should be emphasised that none of the bases employed are known to be capable of engaging in modes of reaction initiation other than simple deprotonation.

Following the successful introduction [159, 160] of imidazole as a buffer component in a postcolumn POCL reaction detection in liquid chromatography, the catalytic ability of several bases was compared in static systems [125, 161]. The reaction intensity versus time profiles were recorded under semiaqueous conditions with catalyst buffers containing either imidazole, pyrazine, pyridine, aniline, diethylamine, mono-, di-, and triethylamine, *tris*(hydroxymethyl)aminomethane (TRIS), or salts of phthalate, phosphate, and tetramethylammonium ions. The use of imidazole was found to generate light intensities approximately 10 times higher than any other catalyst buffer. The kinetics of the light production was also highly dependent upon the concentration of imidazole [125]. Under dry conditions in aprotic media, 4-dimethylaminopyridine has been found to be a superior catalyst compared to sodium salicylate and tetrabutylammonium perchlorate [162]. As a result of these observations, the authors suggested that the pH of the medium was of prime importance to the reaction rate.

Following the disclosure of the outstanding catalytic ability of imidazole compared to other bases, the catalysis of the POCL reaction by imidazole was studied in more detail [163], and it was concluded that the POCL reaction mechanism included the concurrent catalysis by two imidazole molecules, by what was described as "general-base" and "nucleophilic" pathways, respectively. The mechanism for this was suggested to be a "base catalysis of imidazole catalysis by imidazole itself" as previously reported for imidazole-catalyzed reaction of esters [164, 165]. Despite this, it was not until the introduction [151] of 1,1'-oxalyldiimidazole (ODI) as a chemiluminescence reagent, and the postulation of its intermediate appearance in the imidazole-catalysed POCL reaction, that the true catalytic action of imidazole was revealed. The identity of this transient intermediate formed in the nucleophilic replacement reaction between TCPO and imidazole was later confirmed by UV spectroscopy to be ODI [153, 154]. Moreover, the ^{13}C NMR results [154] of the proceeding reaction could also be explained by ODI appearing as an intermediate. From these studies it was thus evident that the POCL reaction was subject to nucleophilic catalysis by imidazole, which is also apparent by an observed autocatalytic breakdown of ODI caused by the presence of imidazole [152]. The use of pure reagents and meticulously dry solvents was thus the key to obtaining these results. The subsequent reaction of ODI with hydrogen peroxide has recently been confirmed [155] to be the prime reaction pathway in imidazole-catalyzed POCL.

The nucleophilic catalytic reaction route was found to be second-order with respect to imidazole with moderately reactive diaryloxalates [153–155] whereas only one imidazole molecule was involved in the reaction with more electronegatively substituted diaryloxalates [154]. Analogous to the imidazole-catalyzed hydrolysis of esters [164, 165], this was rationalized [153, 154] by general-base catalysis of one imidazole molecule on another imidazole molecule attacking the diaryloxalate, and by a difference in the leaving-group ability of the substituted phenol on these types of reagents [157]. The rate-limiting step of ODI formation appeared to depend upon the types of leaving group on the reagent [157]. With TCPO [154, 156] the first replacement was slowest, whereas with 4-nitrophenyl oxalate and DNPO [153, 154] the second replacement was the rate-limiting step of the reaction. Recently, substituted imidazoles and other azoles have been shown to catalyze the POCL reaction of TCPO via the formation of intermediates [166], although the intensity of the chemiluminescence was lower than that observed with imidazole.

In attempting to summarize these extensive investigations, it is clear that the POCL reaction is subject to both nucleophilic catalysis and general-base catalysis by concurrent mechanisms, and that the nucleophilic catalysis can itself be influenced by general-base catalysis. From the POCL reaction mechanism (see Fig. 7), the general-base catalysis can readily be accounted for. The nucleophilic attack of hydrogen peroxide on the diaryloxalate substrate is subject to catalysis by a base that can assist in the removal of a proton from the peroxide. The hydroperoxide anion is a stronger nucleophile than hydrogen peroxide and thus reacts with the diaryloxalate and forms the arylperoxyacid at a far faster rate. Likewise, the suspected and postulated cyclization of the arylperoxyacid is accelerated by proton acceptors. Both the intermolecular and intramolecular peroxide attacks proceed at very slow rates, or not at all, with less reactive oxalates. The nucleophilic catalysis path involves the stepwise in situ formation of different oxalic acid derivatives. This may be either a monosubstituted or disubstituted reagent, depending on the relative reactivity of the leaving group from the oxalate and

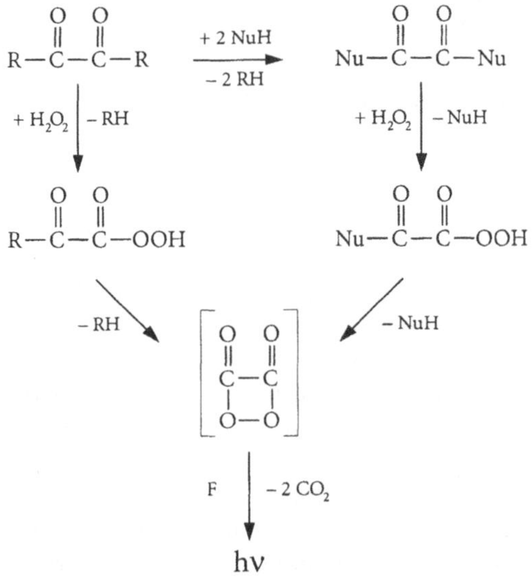

Figure 7 Schematic diagram of the POCL reaction with general-base catalysis according to the left route and nucleophilic (Nu) catalysis following the right route.

the nucleophilic catalyst at a certain pH. The in situ–generated intermediate then reacts with hydrogen peroxide in the same manner as the original reagent would have done, but at a faster rate.

General-base catalysis can, as the name suggests, be accomplished by any adequately strong base, whereas very special demands are placed upon compounds acting as nucleophilic catalysts. The efficiency of these catalysts depends on three factors: basicity, nucleophilicity, and leaving-group ability [166]. Each of these characteristics is in turn the combined result of several attributes.

At low base concentrations, very complex reaction kinetics have been observed by several authors [143, 156, 167, 168]. Emission profiles containing dual maxima has been reported with triethylamine [167] and sodium salicylate [168], which are known to act as general-base catalysts, and also with the nucleophilic catalyst imidazole [156, 168]. From these observations, reaction mechanisms containing two or more speculative intermediates suggestively capable of exciting the fluorophore, plus some metastable "storage intermediate," have been proposed [143, 154, 167, 168]. Another, and equally likely, explanation is that at low catalyst concentrations there is simply not enough catalyst present to maintain constant reaction conditions throughout the light-generating reaction. The variable reaction conditions cause changes in the reaction kinetics and thus induce an apparent pulse in the intensity-time profile.

2.3 Chemically Initiated Electron Exchange Luminescence (CIEEL)

The POCL reaction is thought to follow the CIEEL mechanism proposed by Schuster [169], later modified by McCapra et al. to fit the POCL reaction [170]. According to this mechanism, the intermediate forms a charge complex with the fluorophore, which donates one electron to the intermediate. The radical anion of 1,2-dioxetandione is believed to decompose into carbon dioxide and carbon dioxide radical anion, and the latter transfers its extra electron back to the fluorophore at a higher energy level, resulting in an excited fluorophore. The energy of the intermediates has been determined to be 105 kcal/mol, which corresponds to an excitation wavelength of 270 nm [171].

The logarithm of the excitation efficiency is a linear function of the oxidation potential of the fluorophore, as experimentally verified in several studies [67, 171]. Among the most efficient energy acceptors are polyaromatic hydrocarbons (PAHs) [100] and amino-PAHs [102]. Kang and Kim [172] studied the effect of substitution on PAHs. The results showed that CL intensity was increased with substitution by phenyl, phenylethenyl, and amino groups. In contrast, cyano and carboxyl substituents lowered the CL efficiency. The higher efficiency of amino-PAHs compared to other PAHs is not only explained by a lowering of the oxidation potential; it is also believed to be due to differences in solvation energies between the compounds and the radical ions formed in the charge-transfer complex.

In conclusion, the intermediate(s) responsible for excitation and the process of the energy transfer step(s) in POCL appear to be extremely difficult to both observe and characterize owing to their complex and inherently unstable nature. The final definitive elucidation of the mechanism and the positive identification of the key intermediates of this enigmatic reaction will continue to provide a significant challenge to researchers.

3. ANALYTICAL APPLICATIONS

3.1 Determination of Hydrogen Peroxide

Hydrogen peroxide plays an important role in many processes in the atmosphere and in natural aqueous systems. It affects numerous redox reactions, which in turn influence the stability and transport of other chemical substances, e.g., pollutants. In the atmosphere, hydrogen peroxide is believed to be involved in several important oxidation reactions, e.g., conversion of sulfur dioxide to sulfuric acid and nitrous acid to nitric acid [173, 174]. In the atmosphere, the concentrations of hydrogen peroxide can be as high as 0.65 parts per million volume (ppmv). Measured concentration levels range from 5 to 500 nM in seawater [175], 0 to 66 nM in groundwater [176], and up to as much as 80 μM in rainwater [175]. There is one significant problem when working with detection of low levels of hydrogen peroxide, and that is its presence as a trace component in purified water. Several methods are available to reduce this background level of hydrogen peroxide. Lazrus et al. [177] used catalase, which decomposes hydrogen peroxide to water and oxygen. Dasgupta and Hwang [178] reported the use of granular manganese dioxide packed in a column and incorporated into a flow system. This latter method was used with some modifications [29] to enable protection of hydrogen peroxide "free" water without incorporating a manganese dioxide reactor in the system. Helium-degassed distilled water was circulated through a column filled with manganese dioxide. This water contained approximately 5–10 nM hydrogen peroxide, which corresponded to a 75% decrease in blank signals compared to untreated Milli-Q water. Low concentration standards were prepared by diluting stock solutions of hydrogen peroxide with this water, and to prevent light-induced formation of hydrogen peroxide, these standards were kept in dark, high-density polyethene bottles. It should, however, be emphasized that it is very difficult to prepare reliable standards with a concentration lower than 25 nM.

The POCL reaction has been used for trace determinations of hydrogen peroxide, most commonly in environmental and clinical analysis [26–52]. The latter applications often include various enzyme systems [34–52], where a number of substrates can be indirectly determined by measuring the hydrogen peroxide that is produced as a by-product in the enzymatic reaction.

3.1.1 Direct Determination of Hydrogen Peroxide

Klockow and Jacob have used POCL chemistry to study the concentration levels of hydrogen peroxide in different kinds of precipitates and air. In their first study [26], determinations of hydrogen peroxide in rainwater were carried out. The instrumental configuration consisted of a dual-line flow injection (FIA) manifold, where one aqueous carrier delivered the analyte and the other contained TCPO and perylene dissolved in acetone. Extensive studies of potential interferents, including metal ions, nitrite, sulfite, formaldehyde, hydrogen sulfide, organic hydroperoxide, and ozone, were made. The results demonstrated that none of these had any influence when they were added in 100-fold excess over hydrogen peroxide. At higher concentration levels, nitrite, sulfite, and formaldehyde caused signal depression, whereas hydrogen sulfide and organic hydroperoxides gave a positive signal. Furthermore, they concluded that ozone present at normal atmospheric levels (30 pptv) corresponds to a signal equivalent to 0.7 pM hydrogen peroxide, i.e., several orders of magnitude lower than the detection limit of the method (10 nM). The rainwater collected in Dortmund, Germany had a concentration of hydrogen peroxide ranging from 2 to 2200 ppb during daytime and 0 to 440 ppb at night. The same instrumental setup was utilized to determine gaseous hydrogen peroxide [27]. A sample was collected offline in a cryogenic trap, subsequently melted, and analyzed using the FIA system. A sampling time of approximately 2 h was required; i.e., the time resolution was relatively low. On the other hand, impressive detection limits were obtained, between 0.2 and 6 pptv depending on the relative humidity. The air in Dortmund had a daily mean concentration of hydrogen peroxide in the range 5–60 pptv and maximum levels in the range 10–220 pptv. Later, the same group measured hydrogen peroxide in Antarctic air, snow, and firn cores [28]. The air in Antarctica contained higher levels of hydrogen peroxide, 100–1050 ppt, and diurnal variations were not observed.

An online direct method for determination of gaseous hydrogen peroxide in ambient air was presented by Stigbrand et al. [29]. The analyte was collected continuously in a diffusion scrubber, in which gaseous hydrogen peroxide was captured by diffusion into a scrubber liquid. Sample plugs of aqueous hydrogen peroxide were subsequently injected into a single-line FIA system using stabilized ODI and an immobilized fluorophore. Interferences from naturally occurring organic peroxides were investigated and revealed a selective response for hydrogen peroxide, consistent with early research on POCL that investigated the effects of different organic peroxides. The detection limit for gaseous hydrogen peroxide was estimated to be 23 pptv; the linear range was limited by the operational range of the mass flow controllers rather than the chemistry, with a linear response obtained between 0.6 and 3.4 ppbv. The sample throughput was high, 120 samples per hour, resulting in a good time resolution. Van Zoonen et al. [30] determined the amount of hydrogen peroxide in rainwater using a solid-state TCPO reactor in combination with immobilized fluorophores in a single-line FIA system. Nakashima et al. [31] developed a liquid chromatographic method to determine hydrogen peroxide in cola drinks, ascorbic acid, and commercially available organic peroxides.

Recently the POCL reaction was used to study the role of hydrogen peroxide in antitumor activity [32, 33]. Addition of sodium 5,6-benzylide L-ascorbate or ascorbic acid increased the production of hydrogen peroxide, which in turn induced cell death.

3.1.2   Detection of Hydrogen Peroxide Generated in a Postcolumn Reactor

Oxidases are capable of converting specific substrates and molecular oxygen into hydrogen peroxide or water, and other products [179]. By quantifying the content of hydrogen peroxide, the substrate concentration can thus be indirectly determined. Immobilized enzyme reactors (IMERs) offer several advantages over the use of dissolved enzymes, most important being the reduction in cost and operational work. Owing to the high enzyme load in the IMERs and favorable equilibrium constants for the enzymatic reactions, 100% conversion of the substrate is commonly attained [179]. The enzymes can be either class- or compound-selective, and depending upon the selectivity, different instrumental configuration may be employed. A FIA system is sufficient if the enzyme is compound-selective, whereas with a class-selective enzyme a separation step is often included. This separation can be achieved with a reversed-phase LC system having a low percentage of organic modifier in the mobile phase or some kind of ion exchange separation. Higher amounts of organic solvents are in most cases incompatible with the IMER [180] owing to a loss of enzymatic activity.

Many different kinds of detection principles have been used to determine the hydrogen peroxide formed and chemiluminescent reactions are an attractive approach since the selectivity in the CL reaction combined with the highly selective enzyme results in an extremely selective and sensitive detection. The advantage of POCL compared to other CL reactions capable of determining hydrogen peroxide is its optimal pH close to 7, the same as most enzymatic reactions. On the other hand, most POCL reactions require organic solvents, which are usually incompatible with enzymes. Generally, a completely aqueous carrier/eluent transports the sample plug through the IMER, which then merges with a second carrier containing a fluorophore and the POCL reagent dissolved in a water-miscible organic solvent. In Table 2, applications dealing with POCL detection of enzymatically generated hydrogen peroxide are listed.

One of the most frequently studied enzymatic reaction systems involves the use of glucose oxidase:

$$\text{Glucose} + O_2 + H_2O \rightarrow \text{gluconic acid} + H_2O_2 \qquad (4)$$

This reaction was employed by Williams et al. [34] in the first application of the POCL reaction dealing with the determination of hydrogen peroxide. Several studies of the same enzymatic reaction have been performed since then [35–39], probably because glucose oxidase is relatively inexpensive, highly stable, and well characterized. Rigin used several different enzymatic reactions coupled to the POCL chemistry; cholesterol oxidase [40], L-amino acid oxidase [41], and aldehyde oxidase [42] were all immobilized on glass supports and evaluated in a FIA system. Jansen et al. [43] also studied the stereoselective L-amino acid oxidase. Eight different L-amino acids were separated in a reversed-phase liquid chromatographic system and detected by the use of a flow cell containing immobilized 3-aminofluoranthene. Beer and urine samples were run, as well as standards. Recently, the less common amino acid isomer, the D-form, was determined in human plasma samples using FIA-IMER-POCL [44].

Table 2 Summary of Analytes that Have Been Converted to Hydrogen Peroxide and Subsequently Quantified with the POCL Reaction

Analyte	Enzyme	Ref.
Glucose	Glucose oxidase	34–39
Cholesterol	Cholesterol oxidase	40
L-Amino acids	L-Amino acid oxidase	41, 43
Formaldehyde, formic acid	Aldehyde oxidase	42
D-Amino acids	D-Amino acid oxidase	44
Polyamines	Putrescine oxidase	45, 46
	Polyamine oxidase	
Uric acid	Uricase	37, 47
Choline, acetylcholine	Choline oxidase	48–50
	Acetylcholine esterase	
Choline-containing phospholipids	Phospholipase D	51
	Choline oxidase	
L-Glutamic acid	Glutamate oxidase	52

Kamei et al. [45] separated spermine, spermidine, putrescine, and cadaverine in an ion-pair reversed-phase LC system and detected the hydrogen peroxide formed in the reaction catalyzed by the enzymes putrescine oxidase and polyamine oxidase with POCL. The same analytes were determined in a later study [46], together with the acetyl derivatives. The sensitive determination of uric acid, selectively converted to hydrogen peroxide by uricase, has been investigated by several authors [37, 47].

The determination of choline and acetylcholine using an IMER coupled with POCL has been performed [48–50]. Acetylcholine esterase and choline oxidase were coimmobilized in a single IMER and the analytes separated either with an ion-pair reversed-phase LC system [48] or on a cation exchanger column [49,50]. Van Zoonen et al. [49] combined these enzymes with a solid-state TCPO reactor and immobilized 3-aminofluoranthene, and one of the problems encountered was the limited water solubility of TCPO and its by-product trichlorophenol (TCP). The coupling of IMERs to ODI chemiluminescence was therefore investigated, with special attention being paid to compatibility with aqueous solutions due to a higher solubility of ODI and imidazole [50]. Two different enzyme systems were investigated: glucose oxidase and acetylcholine esterase/choline oxidase. First, different compositions (acetonitrile/water) of the carrier delivering the analyte were investigated. Surprisingly, the sensitivity was found to be essentially independent of the solvent composition. Even purely aqueous buffers can be used without loss in sensitivity, which must be regarded as a major advantage with ODI. An IMER containing glucose oxidase was incorporated in the carrier delivering the analyte and highly sensitive determinations of glucose were accomplished. The detection limit for glucose was 3 nM and the response was linear over the range 30 nM–10 μM. With a second IMER containing acetylcholine esterase and choline oxidase, a cation exchange separation step was included to separate the analytes. In this setup, the detection limits were higher (50 nM), mainly due to band broadening occurring in the analytical column. The response was linear from the detection limit up to 10 μM. To test the applicability of the systems, urine samples were analyzed for glucose, acetylcholine, and choline. The concentrations found in the urine were 580 ± 20 μM β-glucose and 45 ± 1 μM choline, while no acetylcholine was detected. Chromatograms obtained for the choline/acetylcholine system are presented in Figure 8.

Choline-containing phospholipids have been determined in human serum using an IMER consisting of coimmobilized phospholipase D and choline oxidase [51]. Recently, immobilized glutamate oxidase was used to determine L-glutamic acid in culture media [52].

Hydrogen peroxide can also be formed postcolumn by a photochemical reaction. The Birks group has generated hydrogen peroxide from good hydrogen atom donors such as aliphatic alcohols and molecular oxygen, the reaction being catalyzed by quinones. This has been utilized in two different modes. In the first, quinones were determined [53, 54] and because of their catalytic nature, up to 100 molecules of hydrogen peroxide can be formed for each quinone molecule. Consequently, detection limits in the range of tenths of femtomoles were obtained. In the other mode, the hydrogen atom donor is determined instead and a wide range of compounds lacking chromophoric groups (e.g., isopropanol, octanol, glucose, and ascorbic acid) were determined in the low-nanogram range [55].

3.2 Detection of Fluorescent Compounds

Since a relatively small number of analytes of interest have native fluorescent properties, derivatization reactions are frequently employed to enable this detection technique to be extended to a broader range of compounds. This is an excellent means of increasing the detectability for a whole range of molecules, but it is important to realize that there are certain limitations. First, it is difficult to obtain quantitative yields at low analyte concentrations. This implies that in some cases, the obtainable detection limit are not limited by the detector sensitivity, but instead by low yields in the derivatization reaction. Furthermore, to shift the equilibrium toward the product side at low analyte concentrations, as much as 10^4 times excess of fluorescent label may be necessary. Low concentrations of impurities in the label can be present at levels greater than the analytes of interest and as a result, numerous interfering peaks in the chromatograms may be observed. These problems are discussed in detail in Ref. 181.

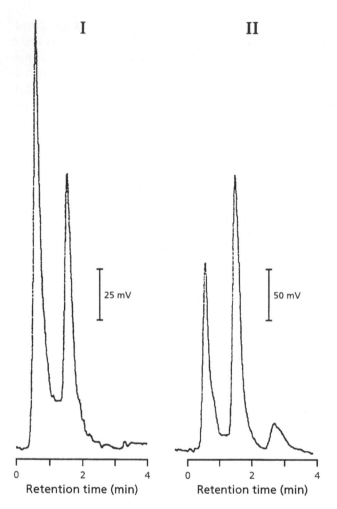

Figure 8 Chromatograms of diluted urine (I) and diluted urine spiked with 2.1 μM choline and 2.3 μM acetylcholine (II). Elution order: Hydrogen peroxide (0.7 min), choline (1.6 min), and acetyl choline (2.9 min). See Ref. 50 for further details.

As can be seen in Table 3, a wide range of analytes derivatized with different labels have been detected using the POCL reaction. Most of these applications have employed flow injection or liquid chromatographic techniques. An area of growing interest is the combination of capillary electrophoresis with chemiluminescence. Several strategies have been used to detect analytes with fluorescent

Table 3 Labeled Analytes Detected by the POCL Reaction

Analyte	Label	Ref.
Amino acids	5-Dimethylaminonaphthalene-1-sulfonyl chloride	60–64
Carboxylic acids	N-(Bromoacetyl)-N'-[5-(dimethylamino)naphthalene-1-sulfonyl]piperazine	65
Carboxylic acids	3-Aminoperylene	67
Carboxylic acids	7-(Diethylamino)-3-[4-((iodoacetyl)amino)phenyl]-4-methylcoumarin	68
Carboxylic acids	9-Anthracenemethanol	69
Carboxylic acids	6,7-Dimethoxy-1-methyl-2(1H)-quinoxalinone-3-propionylcarboxylic acid hydrazine	70
Carboxylic acids	Lumarine-4	71
Aliphatic amines	5-Dimethylaminonaphthalene-1-sulfonyl chloride	72, 75
Aliphatic amines	4-Chloro-7-nitrobenzo-1,2,5-oxadiazole o-Phthaldialdehyde	72
Ebiratide	4-(N,N-dimethylaminosulfonyl)-7-fluoro-2,1,3-benzoxadiazole	73
Primary amines	Naphthalene-2,3-dialdehyde Anthracene-2,3-dialdehyde	74
Amines	Lumarine-1	75
Amphetamine-related compounds	4-Fluoro-7-nitrobenzoxadiazole Naphthalene-2,3-dicarboxaldehyde	76, 77
Catecholamines	Fluorescamine	78
Nitrosamines	5-Dimethylaminonaphthalene-1-sulfonyl chloride	79, 80
Amantadine	CY5	81
Oxosteroids, oxo bile acids	5-Dimethylaminonaphthalene-1-sulfonyl hydrazine	82, 83
Medroxyprogesterone acetate	4-(N,N'-Dimethylaminosulfonyl)-7-hydrazino-2,1,3 benzoxadiazole	85
Estradiol	5-Dimethylaminonaphthalene-1-sulfonyl chloride Lissamine rhodamine B sulfonyl chloride	87
Carbonyl compounds	5-Dimethylaminonaphthalene-1-sulfonyl hydrazine	90
Unsaturated disaccharides	5-Dimethylaminonaphthalene-1-sulfonyl hydrazine	91, 92
Aldehydes, ketones	3-Aminofluoranthene	93
Alkyl-, nitro- and chlorophenols	5-Dimethylaminonaphthalene-1-sulfonyl chloride	94

properties (native or labeled) with the promising fusion of capillary electrophoresis and POCL [56–59].

3.2.1 Derivatization of Amino Acids

The first application describing POCL detection of fluorophores was published in 1977 and dealt with the detection of dansylated amino acids [60] separated by thin-layer chromatography. The amino acids were labeled by a reaction between dansyl chloride and the primary amine group. A few years later, Kobayashi and Imai [61] detected the same analytes with a similar detection system, but using HPLC. Gradient elution of amino acids has been successful [62–64], although significant baseline drift occurs. Miyaguchi et al. [64] extended the determination of amino acids to include N-terminal sequencing of peptides. Bradykinin was dansylated and subjected to acid hydrolysis, and the released dansylated amino acid was identified in an LC system with gradient elution.

3.2.2 Derivatization of Carboxylic Acids

Kwakman et al. [65] described the synthesis of a new dansyl derivative for carboxylic acids. The label, N- (bromoacetyl)-N'-[5-(dimethylamino)naphthalene-1-sulfonyl]-piperazine, reacted with both aliphatic and aromatic carboxylic acids in less than 30 min. Excess reagent was converted to a relatively polar compound and subsequently separated from the derivatives on a silica cartridge. A separation of carboxylic acid enantiomers was performed after labeling with either of three chiral labels and the applicability of the method was demonstrated by determinations of racemic ibuprofen in rat plasma and human urine [66]. Other examples of labels used to derivatize carboxylic acids are 3-aminoperylene [67], various coumarin compounds [68], 9-anthracenemethanol [69], 6,7-dimethoxy-1-methyl-2(1H)-quinoxalinone-3-propionylcarboxylic acid hydrazide (quinoxalinone) [70], and a quinolizinocoumarin derivative termed Lumarin 4 [71].

3.2.3 Derivatization of Amines

Amines are another important group of analytes. Mellbin and Smith [72] compared three different fluorescent reagents, dansyl chloride, 4-chloro-7-nitrobenzo-1,2,5-oxadiazole, and o-phthaldialdehyde, for derivatization of alkylamines. The dansyl tag was found to be the most effective. Hamachi et al. [73] described the application of an HPLC-POCL method for determination of a fluorescent derivative of the synthetic peptide ebiratide. Another comparative study was done by Kwakman et al. [74], where naphthalene-2,3-dialdehyde and anthracene-2,3-dialdehyde were evaluated as precolumn labeling agents for primary amines. The anthracene-2,3-dialdehyde derivatives were not stable, especially in the presence of hydrogen peroxide, and the POCL detection of these derivatives was therefore not suitable. Lumarine 1 (a 7-aminocoumarin derivative) [75], 4-fluoro-7-nitro-benzoxadiazole (NBD-F) [76, 77], and fluorescamine [78] have also been used to derivatize amines. Nitrosamines have been determined after a denitrosation step and labeling with dansyl chloride [79,80]. Recently, Ellingson and Karnes derivatized amantadine hydrochloride with a range of different far-red dyes. Among the tested ones, a dye termed CY5 proved optimal [81].

3.2.4 Derivatization of Steroids

Oxosteroids can be labeled with a fluorescent tag containing a hydrazine group. Two major products are formed, the anti- and syn-isomer of the hydrazone derivative [82]. The dansyl isomers are normally separated during reversed-phase chromatographic conditions, but Imai et al. [83] discovered that if tetrahydrofuran was used in the eluent as the organic modifier, the two isomers eluted in a single peak. This nonresolving eluent in combination with removal of excess reagent improved detection limits significantly compared to a previous study [84]. Uzu et al. [85] used another hydrazine derivative, 4-(N, N-dimethylaminosulfonyl)-7-hydrazino-2,1,3-benzoxadiazole, to determine medroxyprogesterone acetate in serum. Hydroxysteroids have been converted to ketosteroids by an enzymatic reaction using immobilized hydroxysteroid dehydrogenase, and in a second step derivatized with dansyl hydrazine [86]. Direct labeling using dansyl chloride has also been practiced, with the derivatives being separated by both reversed-phase [87] and normal-phase chromatography [88]. Appelblad et al. [89] separated 10 dansylated ketosteroids and detected them using the ODI POCL reaction. A chromatogram is depicted in Figure 9.

3.2.5 Derivatization of Miscellaneous Compounds

Dansyl hydrazine has been used to label other carbonyls as well. Porous glass particles have, e.g., been impregnated with this labeling agent to derivatize carbonyl compounds in ambient air [90]. Glycosaminoglucans have been digested by specific enzymes to the corresponding unsaturated disaccharides and subsequently derivatized with dansyl hydrazine [91, 92]. The method was applied to rat peritoneal mast cells, from which glycosaminoglucans were isolated. One of the most efficient fluorophores in the POCL reaction, 3-aminofluoranthene, has been used to derivatize aldehydes and ketones [93]. Kwakman et al. [94] derivatized various phenols with dansyl chloride. Excess label was removed by allowing the organic phase to pass through an amino-bonded solid-phase extraction column. After the phenols had been separated, a photochemical reaction was allowed to take place, where the derivatives fractured into dansyl hydroxide or dansyl methoxide, and phenol. This conversion was done to prevent the efficient intramolecular quenching by phenols carrying electronegative substituents. Finally,

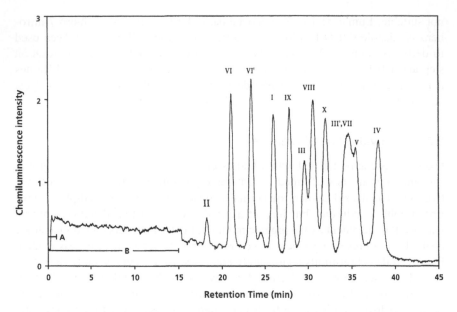

Figure 9   Chromatogram showing a derivatized mixture containing 1 μM of 10 steroids: 3β-hydroxy-5-pregnen-20-one (I), 3α,21-dihydroxy-5β-pregnane-20-one (II), 4-pregnene-3,20-dione (III), 5α-pregnane-3,20-dione (IV), 5β-pregnane-3,20-dione (V), 20α-hydroxy-4-pregnen-3-one (VI), 3α-hydroxy-5α-pregnan-20-one (VII), 3β-hydroxy-5α-pregnan-20-one (VIII), 3α-hydroxy-5β-pregnan-20-one (IX), 3β-hydroxy-5β-pregnan-20-one (X).

the method was applied to analysis of Rhine river water samples for their content of phenols.

Recently the continuous-addition-of-reagent (CAR) technique [182] was applied for the determination of fluorophores by POCL chemistry [95–99]. The applicability of this technique was demonstrated by the determination of natively fluorescent acepromazine in horse plasma [95], the alkaloid harmaline in plasma [96], and other dansylated alkaloids [97]. A separation step has also been included and applied to postcolumn detection of PAHs [98] and dansylated β-carboline alkaloids [99].

Among the natively fluorescent compounds determined by the POCL reaction are PAHs in different matrices such as coal tar [100] and biomass emissions [101], and amino-PAHs in shale oil, coal oil, and coal gasifier tar [102]. Nitro-PAHs have no fluorescent properties, but have been reduced online (either pre- or postcolumn) to the corresponding amino-PAHs [103]. The two fluorescent drugs dipyridamole and benzydamine have been determined in rat plasma by a method developed by Nishitani et al. [104]. Urinary porphyrines have been successfully detected by the POCL chemistry in a FIA setup, in which common urinary metabolites (e.g., riboflavin and bilirubin) did not interfere [105].

3.3 Miscellaneous Applications

It was observed early that halide ions quenched the POCL reaction efficiently [183] and this behavior has been exploited analytically. The mechanism is not fully understood, but probably involves a radiationless decomposition of the charge-transfer complex; i.e., less excited fluorophores are formed [184]. It has also been observed that the quenching efficiency of a certain compound is independent of the nature and the concentration of the fluorophore and of the concentration of hydrogen peroxide, but dependent upon the nature of the oxalate [106, 183]. Van Zoonen et al. [106] used a single-line FIA setup containing a solid-state TCPO reactor and immobilized 3-aminofluoranthene. The carrier consisted of hydrogen peroxide and TRIS buffer dissolved in acetonitrile/water. A wide range of analytes was screened, and nitrite, sulfite, some anilines, and organic sulfur compounds were found to quench the CL reaction. The concentrations at the detection limits were in the low-nanogram range, and as in most indirect modes of detection, the linear range was relatively small (two to three orders of magnitude). In a more recent study [107], a separation step was included, and the solid-state TCPO reactor removed and replaced by a postcolumn solution containing 2-nitrophenyloxalate. This change of reagent increased the peak heights approximately 10-fold.

DeVasto and Grayeski [108] used the same indirect approach to determine amines in a static system. Aliphatic and aromatic amines could be determined, but in this case the linear range was even smaller (one order of magnitude). In addition, nonlinear calibration curves were obtained for the aromatic amines. Amines have been determined in a direct approach as well [109, 110]. The analytes were separated on a C_8 column with an eluent containing hydrogen peroxide, but no buffer. A postcolumn flow containing TDPO and fluorophore merged with the eluent and light evolved as amines having catalytic properties eluted. The best detection limits were obtained for aliphatic amines and imidazoles (at the best 1 nM); some aromatic and heterocyclic amines were not detectable at all.

Two different approaches have been used to determine phenols without derivatization. In the first, the corresponding oxalate esters were synthesized in the traditional way (i.e., using oxalyl chloride and triethylamine) [111, 112]. Pentachlorophenol, 1-naphthol, bromofenoxim, bromoxynil, and p-cyanophenol were treated this way, after which the POCL resulting from their reaction was measured in a static system. The second approach exploits the oxidation reaction between imidazole and hydroxyl compounds at an alkaline pH, where hydrogen peroxide is formed [113]. Polyphenols, e.g., pyrogallol, pyrocatechol, and dopamine, were most sensitively detected, but monophenols and sugars could also be determined.

Even metal ions have been determined by the POCL reaction. Steijger et al. [114] observed in batch experiments an enhancement of the total light emitted whenever metal ions were present. Copper gave the highest increase, approximately an order of magnitude, but other metal ions such as Cr^{3+}, Pb^{2+}, Fe^{3+}, Mn^{2+}, Zn^{2+}, and Al^{3+} also showed increases in signal intensity. The mechanism for this enhancement is not completely understood; the authors suggested that it was due to stabilization of the oxalate ester-imidazole complex by the metal ion. In a later study, the influence of Cu^{2+} was further investigated [115]. Quass and Klockow [116] developed a method for the determination of hydrogen peroxide and Fe^{2+} in atmospheric water. To increase the selectivity, catalase was added to the reaction mixture when Fe^{2+} was determined, and 1,10-phenanthroline, which forms a complex with Fe^{2+}, was added when hydrogen peroxide was determined. In a FIA system, detection limits were 40 nM for hydrogen peroxide and 28 nM for Fe^{2+}. Sato and Tanaka [117] used a different approach to determine metal ions. The CL emission due to the formation of a fluorescent complex between metal ions and 8-hydroxyquinoline was measured. A separation of Zn^{2+}, Al^{3+}, and In^{3+}, all as 8-hydroxyquinoline complexes, was successfully detected by means of the POCL reaction.

A mixture consisting of oxalic acid, carbodiimide, fluorescer, and hydrogen peroxide is known to produce strong visible light [185]. Albrecht et al. used this reaction to determine oxalate in urine [118] and serum [119].

Capomacchia et al. [120] utilized the background emission present when DNPO and hydrogen peroxide are mixed to detect ouabain and urea. In the presence of these analytes, an intensity enhancement was observed and detection limits were in the picomole range.

4. CONCLUSIONS AND FUTURE TRENDS

In analytical chemistry there is an ever-increasing demand for rapid, sensitive, low-cost, and selective detection methods. When POCL has been employed as a detection method in combination with separation techniques, it has been shown to meet many of these requirements. Since 1977, when the first application dealing with detection of fluorophores was published [60], numerous articles have appeared in the literature [6–8]. However, significant problems are still encountered with derivatization reactions, as outlined earlier. Consequently, improvements in the efficiency of labeling reactions will ultimately lead to significant improvements in the detection of these analytes by the POCL reaction. A promising trend is to apply this sensitive chemistry in other techniques, e.g., in supercritical fluid chromatography [186] and capillary electrophoresis [56–59]. An alternative approach to derivatization reactions involves the enzymatic conversion of substrates to hydrogen peroxide. The number of applications appearing in the literature now appears to be limited by the number and purity of commercially available enzymes; this field is expected to continue to be an area of significant growth.

Although POCL is regarded as a highly sensitive technique, a considerable fraction of the energy may be lost in dark-side reactions in aqueous solutions. A better understanding of the mechanism and clarification of the identity of reactive intermediates are important to minimize side reactions and to exploit the full potential of the POCL reaction. The discovery of ODI as an intermediate in imidazole-mediated POCL has led to a better understanding of the initial reaction steps. Many of the mechanistic studies undertaken earlier have been hampered by the complex role of this catalyst. Now the light-generating reactions occurring at a later stage in the reaction chain can be studied without being masked by the ODI-forming steps. A better understanding of the POCL reaction is therefore to be expected in the future.

The screening of new catalysts is important to guide the synthesis of new amide-based oxalic acid derivatives. Most reagents presently employed in this chemistry are based on different substituted phenyl esters, which indeed provide good quantum yields. However, their limited solubility in aqueous solutions and their inherent background emission are serious drawbacks when used in analytical applications. ODI is by no means the optimal reagent, but it offers certain advantages compared to the commonly used oxalate esters. First, ODI provides quantum yields comparable to the combination of TCPO and imidazole, but with faster kinetics. Second, ODI has a high solubility in acetonitrile and it does not precipitate in water, which simplifies the coupling of completely aqueous separations or FIA sample streams with POCL detection. Finally, if combined with immobilized fluorophores, a single-line flow system is sufficient for sensitive determinations of hydrogen peroxide, since no additional catalyst is required. The expansion of POCL chemistry into purely aqueous chemical environments, such as those required in immunoassay and in situ bioassays, may ultimately be possible with further refinements to low-background-emission, water-soluble oxamide POCL reagents.

REFERENCES

1. EA Chandross. Tetrahedron Lett 12:761–765, 1963.
2. MM Rauhut, BG Roberts, AM Semsel. J Am Chem Soc 88:3604–361, 1966.
3. MM Rauhut, LJ Bollyky, BG Roberts, M Loy, RH Whitman, AV Iannotta, AM Semsel, RA Clarke. J Am Chem Soc 89:6515–6522, 1967.
4. MM Rauhut. Acc Chem Res 2:80–87, 1969.

5. GJ de Jong, PJM Kwakman. J Chromatogr 492:319–343, 1989.
6. K Robards, PJ Worsfold. Anal Chim Acta 266:147–173, 1992.
7. PJM Kwakman, UATh Brinkman. Anal Chim Acta 266:175–192, 1992.
8. AR Bowie, MG Sanders, PJ Worsfold. J Biolumin Chemilumin 11:61–90, 1996.
9. DR Maulding, RA Clarke, BG Roberts, MM Rauhut. J Org Chem 33:250–254, 1968.
10. S-S Tseng, AG Mohan, LC Haines, LS Vizcarra, MM Rauhut. J Org Chem 44: 4113–4116, 1979.
11. AG Mohan, RG Dulina, AA Doering. Final Report to N.A.S.A., Contract NAS5-22303, American Cyanamid Company, Chemical Research Division, Bound Brook, NJ, 1976.
12. AG Mohan, S-S Tseng, MM Rauhut, FJ Arthen, RG Dulina, VM Kamhi, DE McKay, RJ Manfre, DJ Oldfedt, LS Vizcarra. Final Report to N.A.S.A., Contract N0014-77-C-0634, American Cyanamid Company, Chemical Research Division, Bound Brook, NJ, 1982.
13. ML Grayeski, EJ Woolf, PJ Helly. Anal Chim Acta 183:207–215, 1986.
14. EJ Woolf, ML Grayeski. In: D Eastwood, LJ Cline Love, eds. Progress in Analytical Luminescence. Philadelphia: American Society for Testing and Materials, 1988, pp 67–74.
15. EJ Woolf, ML Grayeski. J Lumin 39:19–27, 1987.
16. R DeLavalle, ML Grayeski. Anal Biochem 197:340–346, 1991.
17. N Dan, M-L Lau, ML Grayeski. Anal Chem 63:1766–1771, 1991.
18. K Nakashima, S Kawaguchi, RS Givens, S Akiyama-Anal Sci 6:833–836, 1990.
19. P van Zoonen, DA Kamminga, C Gooijer, NH Velthorst, RW Frei. Anal Chim Acta 167:249–256, 1985.
20. P van Zoonen, DA Kamminga, C Gooijer, NH Velthorst, RW Frei, G Gübitz. Anal Chim Acta 174:151–161, 1985.
21. NW Barnett, R Bos, SW Lewis, RA Russell. Analyst 123:1239–1245, 1998.
22. NW Barnett, R Bos, RN Evans, RA Russell. Anal Chim Acta 403:145–154, 2000.
23. K Honda, K Miyaguchi, K Imai. Anal Chim Acta 177:103–110, 1985.
24. K Imai, H Nawa, M Tanaka, H Ogata. Analyst 111:209–211, 1986.
25. K Nakashima, K Maki, S Akiyama, WH Wang, Y Tsukamoto, K Imai. Analyst 114:1413–1416, 1989.
26. D Klockow, P Jacob. Chemistry of Multiphase Atmospheric Systems Berlin: Springer, 1986, pp 117–130.
27. P Jacob, TM Tavares, D Klockow. Fresenius Anal Chem 325:359–364, 1986.
28. P Jacob, D Klockow. Fresenius Anal Chem 346:429–434, 1993.
29. M Stigbrand, A Karlsson, K Irgum. Anal Chem 68:3945–3950, 1996.
30. P van Zoonen, I de Herder, C Gooijer, NH Velthorst, RW Frei. Anal Lett 191: 1949–1961, 1986.
31. K Nakashima, M Wada, N Kuroda, S Akiyama, K Imai. J Liq Chromatogr 17: 2111–2126, 1994.
32. H Sakagami, M Hosaka, H Arakawa, M Maeda, K Satoh, Y Ida, K Asano, T Hisamitsu, M Takimoto, H Ota, M Inagaki, K Sasuga, S Sho, T Tanaka, N Utsumi, T Oi, M Kochi. Anticancer Res 18:2519–2524, 1998.
33. M Tajima, M Toguchi, Y Kanda, S Kunii, M Hosaka, H Arakawa, M Maeda, K Satoh, K Asano, M Kochi, H Sakagami. Anticancer Res 18:1697–1702, 1998.

34. DC Williams III, GF Huff, WR Seitz. Anal Chem 48:1003–1006, 1976.
35. K Nakashima, K Maki, S Kawaguchi, S Akiyama, Y Tsukamoto, K Imai. Anal Sci 7:709–713, 1991.
36. N Kiba, H Koemado, M Furusawa. Anal Sci 11:605–609, 1995.
37. K Nakashima, N Hayashida, S Kawaguchi, S Akiyama, Y Tsukamota, K Imai. Anal Sci 7:715–718, 1991.
38. VI Rigin. J Anal Chem USSR 34:619–623, 1979.
39. MS Abdel-Latif, GG Guilbault. Anal Chem 60:2671–2674, 1988.
40. VI Rigin. J Anal Chem USSR 33:1265–1270, 1978.
41. VI Rigin. J Anal Chem USSR 38:1328–1330, 1983.
42. VI Rigin. J Anal Chem USSR 36:1111–1115, 1981.
43. H Jansen, UATh Brinkman, RW Frei. J Chromatogr 440:217–223, 1988.
44. M Wada, N Kuroda, S Akiyama, K Nakashima. Anal Sci 13:945–950, 1997.
45. A Kamei, S Ohkubo, S Saito, S Takagi. Anal Chem 61:1921–1924, 1989.
46. M Wada, N Kuroda, T Ikenaga, S Akiyama, K Nakashima. Anal Sci 12:807–810, 1996.
47. G Scott, WR Seitz, J Ambrose. Anal Chim Acta 115:221–228, 1980.
48. K Honda, K Miyaguchi, H Nishino, H Tanaka, T Yao, K Imai. Anal Biochem 153: 50–53, 1986.
49. P van Zoonen, C Gooijer, NH Velthorst, RW Frei, JH Wolf, J Gerrits, FJ Flentge. J Pharm Biomed Anal 5:485–492, 1987.
50. M Emteborg (b. Stigbrand), K Irgum, C Gooijer, UATh Brinkman. Anal Chim Acta 357:111–118, 1997.
51. M Wada, K Nakashima, N Kuroda, S Akiyama, K Imai. J Chromatogr 678:129–136, 1996.
52. T Hasebe, E Hasegawa, T Kawashima. Anal Sci 12:881–885, 1996.
53. JR Poulsen, JW Birks, G Gübitz, P van Zoonen, C Gooijer, NH Velthorst, RW Frei. J Chromatogr 360:371–383, 1986.
54. JR Poulsen, JW Birks. Anal Chem 62:1242–1251, 1990.
55. I Aichinger, G Gübitz, JW Birks. J Chromatogr 523:163–172, 1990.
56. N Wu, CW Huie. J Chromatogr 634:309–315, 1993.
57. T Hara, S Kayama, H Nishida, R Nakajima. Anal Sci 10:223–225, 1994.
58. K Tsukagoshi, A Tanaka, R Nakajima, T Hara. Anal Sci 12:525–528, 1996.
59. LL Shultz, S Shippy, TA Nieman, JV Sweedler. J Microcol Sep 10:329–337, 1998.
60. TG Curtis, WR Seitz. J Chromatogr 134:343–350, 1977.
61. S Kobayashi, K Imai. Anal Chem 52:424–427, 1980.
62. K Miyaguchi, K Honda, K Imai. J Chromatogr 303:173–176, 1984.
63. N Hanaoka, H Tanaka. J Chromatogr 606:129–132, 1992.
64. K Miyaguchi, K Honda, T Toyo'oka, K Imai. J Chromatogr 352:255–260, 1986.
65. PJM Kwakman, H-P van Schaik, UATh Brinkman, GJ de Jong. Analyst 116:1385–1391, 1991.
66. T Toyo'oka, M Ishibashi, T Terao. J Chromatogr 627:75–86, 1992.
67. K Honda, K Miyaguchi, K Imai. Anal Chim Acta 177:111–120, 1985.
68. ML Grayeski, JK DeVasto. Anal Chem 59:1203–1206, 1987.
69. SW Lewis, PJ Worsfold, A Lynes, EH McKerrell. Anal Chim Acta 266:257–264, 1992.

70. BW Sandmann, ML Grayeski. J Chromatogr 653:123–130, 1994.
71. M Tod, M Prevot, J Chalom, R Farinotti, G Mahuzier. J Chromatogr 542:295–306, 1991.
72. G Mellbin, BEF Smith. J Chromatogr 312:203–210, 1984.
73. Y Hamachi, K Nakashima, S Akiyama. J Liq Chrom Rel Technol 20:2377–2387, 1997.
74. PJM Kwakman, H Koelewijn, I Kool, UATh Brinkman, GJ de Jong. J Chromatogr 511:155–166, 1990.
75. M Tod, M Prevot, M Poulou, R Farinotti, J Chalom, G Mahuzier. Anal Chim Acta 223:309–317, 1989.
76. K Hayakawa, K Hasegawa, N Imaizumi, OS Wong, M Miyazaki. J Chromatogr 464:343–352, 1989.
77. K Nakashima, K Suetsugo, K Yoshida, S Akiyama, S Uzo, K Imai. Biomed Chromatogr 6:149–154, 1992.
78. S Kobayashi, J Sekino, K Honda, K Imai. Anal Biochem 112:99–104, 1981.
79. C Fu, H Xu. Analyst 120:1147–1151, 1995.
80. C Fu, H Xu, Z Wang. J Chromatogr 634:221–227, 1993.
81. A Ellingson, HT Karnes. Biomed Chromatogr 12:8–12, 1998.
82. R Weinberger, T Koziol, G Millington. Chromatographia 19:452–456, 1984.
83. K Imai, S Higashidate, A Nishitani, Y Tsukamota, M Ishibashi, J Shoda, T Osuga. Anal Chim Acta 227:21–27, 1989.
84. T Koziol, ML Grayeski, R Weinberger. J Chromatogr 317:355–366, 1984.
85. S Uzu, K Imai, K Nakashima, S Akiyama. J Pharm Biomed Anal 10:979–984, 1992.
86. S Higashidate, K Hibi, M Senda, S Kanda, K Imai. J Chromatogr 515:577–584, 1990.
87. PJM Kwakman, DA Kamminga, UATh Brinkman, GJ de Jong. J Pharm Biomed Anal 9:753–759, 1991.
88. O Nozaki, Y Ohba, K Imai. Anal Chim Acta 205:255–260, 1988.
89. P Appelblad, T Jonsson, T Bäckström, K Irgum. Anal Chem 70:5002–5009, 1998.
90. L Nondek, RE Milofsky, JW Birks. Chromatographia 32:33–39, 1991.
91. H Akiyama, T Toida, T Imanari. Anal Sci 7:807–809, 1991.
92. H Akiyama, S Shidawara, A Mada, H Toyoda, T Toida, T Imanari. J Chromatogr 579:203–207, 1992.
93. B Mann, ML Grayeski. J Chromatogr 386:149–158, 1987.
94. PJM Kwakman, DA Kamminga, UATh Brinkman. J Chromatogr 553:345–356, 1991.
95. J Cepas, M Silva, D Peréz-Bendito. Anal Chem 66:4079–4084, 1994.
96. J Cepas, M Silva, D Pérez-Bendito. Anal Chim Acta 314:87–94, 1995.
97. J Cepas, M Silva, D Pérez-Bendito. Analyst 121:49–54, 1996.
98. J Cepas, M Silva, D Pérez-Bendito. Anal Chem 67:4376–4379, 1995.
99. J Cepas, M Silva, D Pérez-Bendito. J Chromatogr 749:73–80, 1996.
100. KW Sigvardson, JW Birks. Anal Chem 55:432–435, 1983.
101. AN Gachanja, PJ Worsfold. Anal Proc 29:61–63, 1992.
102. KW Sigvardson, JM Kennish, JW Birks. Anal Chem 56:1096–1102, 1984.
103. KW Sigvardson, JW Birks. J Chromatogr 316:507–518, 1984.

104. A Nishitani, Y Tsukamoto, S Kanda, K Imai. Anal Chim Acta 251:247–253, 1991.
105. M Lin, W Huie. Anal Chim Acta 339:131–138, 1997.
106. P van Zoonen, DA Kamminga, C Gooijer, NH Velthorst, RW Frei. Anal Chem 58:1245–1248, 1986.
107. P van Zoonen, H Bock, C Gooijer, NH Velthorst, RW Frei. Anal Chim Acta 200: 131–141, 1987.
108. JK DeVasto, ML Grayeski. Analyst 116:443–447, 1991.
109. M Katayama, H Takeuchi, H Taniguchi. Anal Chim Acta 287:83–88, 1994.
110. M Katayama, H Taniguchi, Y Matsuda, S Akihama, I Hara, H Sato, S Kaneko, Y Kuroda, S Nozawa. Anal Chim Acta 303:333–340, 1995.
111. J Chimeno, R von Wandruszka. Anal Lett 22:2059–2064, 1989.
112. MA Abubaker, R von Wandruszka. Anal Lett 24:93–102, 1991.
113. O Nozaki, T Iwaeda, Y Kato. J Biolum Chemilum 10:339–344, 1995.
114. OM Steijger, PHM Rodenburg, H Lingeman, UATh Brinkman, JJM Holthuis. Anal Chim Acta 266:233–241, 1992.
115. OM Steijger, HCM den Nieuwenboer, H Lingeman, UATh Brinkman, JJM Holthuis, AK Smilde. Anal Chim Acta 320:99–105, 1996.
116. U Quass, D Klockow. Int J Environ Anal Chem 60:361–375, 1995.
117. K Sato, S Tanaka. Mikrochem J 53:93–98, 1996.
118. S Albrecht, H Brandl, W-D Böhm, R Beckert, H Kroschwitz, V Neumeister. Anal Chim Acta 255:413–416, 1991.
119. S Albrecht, H Hornak, H Brandl, T Freidt, WD Böhm, K Weis, A Reinschke. Fresenius J Anal Chem 343:175–176, 1992.
120. AC Copomacchia, RN Jennings, SM Hemingway, P D'Souza, W Prapaitrakul, A Gingle. Anal Chim Acta 196:305–310, 1987.
121. S Uzu, K Imai, K Nakashima, S Akiyama. Analyst 116:1353–1357, 1991.
122. PJM Kwakman, UATh Brinkman, RW Frei, GJ de Jong, FJ Spruit, NGFM Lammers, JHM van den Berg. Chromatographia 24:395–399, 1987.
123. PJ Ryan, TW Honeyman. J Chromatogr 312:461–466, 1984.
124. EP Lankmayr, KW Budna, K Müller, F Nachtmann, F Rainer. J Chromatogr 222: 249–255, 1981.
125. N Hanaoka, RS Givens, RL Schowen, T Kuwana. Anal Chem 60:2193–2197, 1988.
126. R Gohda, K Kimoto, T Santa, T Fukushima, H Homma, K Imai. Anal Sci 12:713–719, 1996.
127. N Hanaoka, H Tanaka, A Nakamoto, M Takada. Anal Chem 63:2680–2685, 1991.
128. GJ de Jong, N Lammers, FJ Spruit, RW Frei, UATh Brinkman. J Chromatogr 353: 249–257, 1986.
129. W Baeyens, J Bruggeman, B Lin. Chromatographia 27:191–193, 1989.
130. W Baeyens, J Bruggeman, C Dewaele, B Lin, K Imai. J Biolum Chemilum 5:13–23, 1990.
131. PD Bryan, AC Capomacchia. J Pharm Biomed Anal 9:855–860, 1991.
132. N Hanaoka. J Chromatogr 503:155–165, 1990.
133. RJ Weinberger. J Chromatogr 314:155–165, 1984.
134. GJ de Jong, N Lammers, FJ Spruit, UATh Brinkman, RW Frei. Chromatographia 18:129–133, 1984.
135. AG Hadd, JW Birks. Selective Detectors. New York: Wiley, 1995, pp 209–240.

136. G Orosz, RS Givens, RL Schowen. Crit Rev Anal Chem 26:1–27, 1996.
137. HF Cordes, HP Richter, CA Heller. J Am Chem Soc 91:7209, 1969.
138. JJ DeCorpo, A Baronavski, MV McDowell, FE Saalfeld. J Am Chem Soc 94:2879–2880, 1972.
139. CLR Catherall, TF Palmer, RB Cundall. J Chem Soc Faraday Trans 2 80:823–834, 1984.
140. CLR Catherall, TF Palmer, RB Cundall. J Chem Soc Faraday Trans 2 80:837–849, 1984.
141. MM Rauhut, AM Semsel. Final Report to the Office of Naval Research, Contract Nr. N00014-73-C-0343, 1974. American Cyanamid Co. Chemical Research Division, Bound Brook, NJ.
142. MFD Steinfatt. Bull Soc Chim Belg 94:85–86, 1985.
143. RE Milofsky, JW Birks. J Am Chem Soc 113:9715–9723, 1991.
144. B Mann, ML Grayeski. Anal Chem 62:1532–1536, 1990.
145. NW Barnett, R Bos, SW Lewis, RA Russel. Anal Comm 34:17–20, 1997.
146. P Prados, T Santa, H Homma, K Imai. Anal Sci 11:575–580, 1995.
147. HP Chokshi, M Barbush, RG Carlson, RS Givens, T Kuwana, RL Schowen. Biomed Chromatogr 4:96–99, 1990.
148. CV Stevani, LP de Arruda Campos, WJ Baader. J Chem Soc Perkin Trans 2:1645–1648, 1996.
149. JR Hohman, RS Givens, RG Carlson, G Orosz. Tetrahedron Lett 37:8273–8276, 1996.
150. CV Stevani, WJ Baader. J Phys Org Chem 10:593–599, 1997.
151. M Stigbrand, E Pontén, K Irgum. Anal Chem 66:1766–1770, 1994.
152. M Emteborg (b. Stigbrand), E Pontén, K Irgum. Anal Chem 69:2109–2114, 1997.
153. H Neuvonen. J Chem Soc Perkin Trans 2:951–954, 1995.
154. AG Hadd, JW Birks. J Org Chem 61:2657–2663, 1996.
155. AG Hadd, AL Robinson, KL Rowlen, JW Birks. J Org Chem 63:3023–3031, 1998.
156. CV Stevani, DF Lima, VG Toscano, WJ Baader. J Chem Soc Perkin Trans 2:989–995, 1996.
157. H Neuvonen. J Biolum Chemilum 12:241–248, 1997.
158. EH White, PD Wildes, J Wieko, H Doshan, CC Wei. J Am Chem Soc 95:7050–7058, 1973.
159. K Miyaguchi, K Honda, K Imai. J Chromatogr 136:501–505, 1984.
160. K Imai, Y Matsunaga, Y Tsukamoto, A Nishitani. J Chromatogr 400:169–176, 1987.
161. K Imai, A Nishitani, Y Tsukamoto, W-H Wang, S Kanda, K Hayakawa, M Miyazaki. Biomed Chromatogr 4:100–104, 1990.
162. G Orosz, K Torkos, J Borossay. Acta Chim Hung 128:911–917, 1991.
163. M Orlovic, RL Schowen, RL Givens, F Alvarez, B Matuszewski, N Parekh. J Org Chem 54:3606–3610, 1989.
164. JF Kirsch, WP Jencks. J Am Chem Soc 86:833–837, 1964.
165. JF Kirsch, WP Jencks. J Am Chem Soc 86:837–846, 1964.
166. T Jonsson, M Emteborg (b. Stigbrand), K Irgum. Anal Chim Acta 361:205–215, 1998.

167. FJ Alvarez, JP Parekh, B Matusewski, RS Givens, T Higuchi, RL Schowen. J Am Chem Soc 108:6435–6437, 1986.
168. JH Lee, SY Lee, K-J Kim. Anal Chim Acta 329:117–126, 1996.
169. GB Schuster. Acc Chem Res 12:366–373, 1979.
170. F McCapra, K Perring, RJ Heart, RA Hann. Tetrahedron Lett 22:5087–5090, 1981.
171. P Lechtken, NJ Turro. Mol Photochem 6:95–99, 1974.
172. SC Kang, K-J Kim. Bull Korea Chem Soc 11:224–227, 1990.
173. JG Calvert, A Lazrus, GL Kok, BG Heikes, JG Walega, J Lind, CA Cantrell. Nature 317:27–35, 1985.
174. Y-N Lind, JA Lind. J Geophys Res 91:2793–2800, 1986.
175. D Price, PJ Worsfold, RFC Mantoura. Anal Chim Acta 298:121–128, 1994.
176. TR Holm, GK George, MJ Barcelona. Anal Chem 59:582–586, 1987.
177. AL Lazrus, LK Gregory, JA Lind, SN Gitlin, BG Heikes, RE Shetter. Anal Chem 58:594–597, 1986.
178. PK Dasgupta, H Hwang. Anal Chem 57:1009–1012, 1985.
179. L Gorton, G Marko-Varga. In: S Lam, G Malikin, eds. Analytical Applications of Immobilized Enzyme Reactors. London: Blackie Academic & Professional, Chapman & Hall, 1994, pp 1–21.
180. LD Bowers, PR Johnson. Biophys Biochim Acta 661:100–105, 1981.
181. P Kwakman. Thesis, Free University Amsterdam, 1991.
182. A Velasco, M Silva, D Peréz-Bendito. Anal Chem 64:2359–2365, 1992.
183. K Honda, J Sekino, K Imai. Anal Chem 55:940–943, 1983.
184. C Gooijer, NH Velthorst. Biomed Chromatogr 4:92–95, 1990.
185. MM Rauhut, RA Sheehan, RA Clarke, AM Semsel. Photochem Photobiol 4:1097–1110, 1965.
186. BW Sandmann, ML Grayeski. Chromatographia 38:163–167, 1994.

8

Kinetics in Chemiluminescence Analysis

Dolores Pérez-Bendito and Manuel Silva
University of Córdoba, Córdoba, Spain

1. INTRODUCTION

Chemiluminescence (CL) is the emission of the electromagnetic (ultraviolet, visible, or near infrared) radiation by molecules or atoms resulting from a transition from an electronically excited state to a lower state (usually the ground state) in which the excited state is produced in a chemical reaction. The CL phenomenon is relatively uncommon because, in most chemical reactions, excited molecules lose their excitation energy via nonradiactive pathways (e.g., heat). This type of luminescence provides interesting analytical applications thanks to its high sensitivity [1]. This entails considering its kinetic aspects in theoretical and applied terms.

For a reaction to produce detectable CL emission, it must fulfill the following conditions: (1) it should be exothermic so that sufficient energy for an electronically excited state to be formed (at least 180 kJ/mol for emission in the visible region) can be provided; (2) there should be a suitable reaction pathway for the excited state to be formed; and (3) a radiactive pathway (either direct or via energy transfer to a fluorophore) for the excited state to lose its excess energy should exist.

CL is observed in the liquid, gas, and solid phases. In the last decade, there has been growing interest in CL as a detection technique for quantitative analysis, particularly in the liquid (aqueous) phase [2, 3], which will be solely dealt with in this chapter, because of the excellent sensitivity and wide dynamic ranges that can be achieved by using relatively simple and inexpensive instrumentation.

This chapter focuses on analytical CL methodologies, with emphasis on the kinetic connotations of typical approaches such as the stopped-flow, the continuous-addition-of-reagent (a new kinetic methodology) and the pulse perturbation technique developed for oscillating reactions, among others. Recent contributions to kinetic simultaneous determinations of organic substances using CL detection (kinetometric approaches included) are also preferentially considered here.

2. KINETIC ASPECTS OF CHEMILUMINESCENCE

One key aspect of CL techniques is the transient signal that results from the underlying spectroscopic, chemical, and physical kinetics (see Fig. 1). This is primarily the result of the spectroscopic phenomenon (i.e., the emission of light by a molecule in an excited electronic state on return to its ground state) being intrinsically kinetic in nature. Because excited CL states are produced by a chemical reaction, chemical kinetics is also involved in the process. In fact, the intensity of the CL signal, I_{CL}, is related to the reaction rate, v, via the CL quantum yield, ϕ. The way the ingredients of a CL reaction are mixed in an aqueous medium also influences the CL signal, particularly its initial portion. As a result, the physical kinetics inherent in the fluid dynamics must also be considered.

The typical profile of a CL transient signal (a plot of CL intensity vs. time) is a kinetic response curve that corresponds to a first-order sequence of two consecutive steps, namely: (1) generation of the light-emitting product by mixing of the chemical ingredients (the substrate and oxidant), and (2) formation of the end product (Fig. 2). The rate at which each step takes place depends on the formation

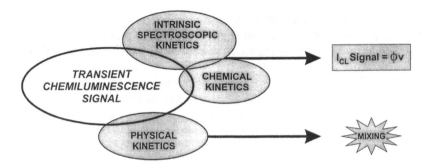

Figure 1 Scheme of the spectroscopic, chemical, and physical kinetics involved in the CL signal (I_{CL}, intensity of the CL signal; v, reaction rate; ϕ, CL quantum yield).

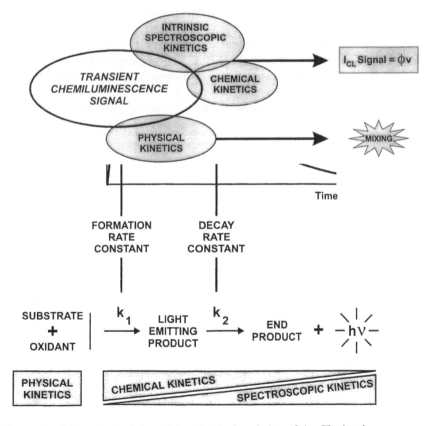

Figure 2 Schematic depiction of the kinetic foundation of the CL signal.

and decay rate constants, k_1 and k_2, which correspond to the rising and falling portion, respectively, of the transient signal. Physical kinetics plays a role in the prior mixing process, whereas chemical and spectroscopic kinetics are involved in both steps.

The relationship between CL intensity and time is expressed by a kinetic equation including the reaction rate constants and the substrate concentration. Such is the case with the specific equation for the CL of the luminol reaction, which is one of the most widely studied in this context:

$$I_{CL} = k(\text{substrate}) \frac{k_1 k_2}{k_1 - k_2} (e^{-k_2 t} - e^{-k_1 t}) \tag{1}$$

The CL phenomenon possesses three distinct analytical connotations, namely: (1) the chemical system, where the analyte may be directly or indirectly involved (for example, in immunoassay); (2) the way the process is implemented (i.e., whether a conventional or an innovative methodology is used); and (3) the measured parameters.

The formation of a light emitter (i.e., an excited product) can be accomplished by mixing the substrate and oxidant in the presence or absence of a catalyst or cofactor. The ingredients can also be electrochemically generated in situ, using so-called "electrogenerated CL" (ECL).

One must distinguish between two different alternatives: direct and indirect CL. In direct CL, the excited product returns to its ground state to give the starting product and emitted light. In indirect CL, also referred to as "energy-transfer process," the excited product interacts with a fluorophore, which must be a fluorescent molecule, to form the product; simultaneously, the fluorophore is promoted to its excited state, from which it subsequently returns to its ground state with light emission.

In direct CL methods, the target analyte can be the substrate, the oxidant, or the catalyst. In some cases, the analyte can also be an inhibitor that decreases the intensity of the CL signal. This has helped expand the scope of direct methodologies, which was formerly restricted to the few available CL reactions. In indirect CL methods, the analyte is usually the fluorophore. These methods have a broader scope than their direct counterparts as a result of the wide variety of— mainly organic—substances that can act as fluorophore, either as such or following derivatization into fluorescent molecules.

The most widely used chemical systems in direct CL methods are based on classical substrates such as luminol, lucigenin, lophine, and pyrogallol, which are oxidized with hydrogen peroxide or dissolved oxygen [4, 5]. In these CL reactions, free and complexed metal ions (Cu, Co, Fe, Mn), and peroxidases, have prominent catalytic effects; also, some inorganic anions and organic ligands for metal catalysts have substantial inhibitory effects. Most of the target analytes used in this context are either the catalyst (a metal ion) or inhibitor (an organic compound); others are the oxidant or even the substrate of the CL reaction. In this context, it is worth mentioning immunoassay, which uses CL detection as an advantageous alternative to radioimmunoassay. A number of CL labels have been proposed for this purpose [6].

The most commonplace substrates in energy-transfer analytical CL methods are aryl oxalates such as *bis*(2,4,6-trichlorophenyl) oxalate (TCPO) and *bis*(2,4-dinitrophenyl) oxalate (DNPO), which are oxidized with hydrogen peroxide [7, 8]. In this process, which is known as the peroxyoxalate-CL (PO-CL) reaction, the fluorophore analyte is a native or derivatized fluorescent organic substance such as a polynuclear aromatic hydrocarbon, dansylamino acid, carboxylic acid, phenothiazine, or catecholamines, for example. The mechanism of the reaction between aryl oxalates and hydrogen peroxide is believed to generate dioxetane-1,2-dione, which may itself decompose to yield an excited-state species. Its interaction with a suitable fluorophore results in energy transfer to the fluorophore, and the subsequent emission can be exploited to develop analytical CL-based determinations.

Although the ECL phenomenon is associated with many compounds, only four major chemical systems have so far been used for analytical purposes [9, 10], i.e., (1) the ECL of polyaromatic hydrocarbons in aqueous and nonaqueous media; (2) methods based on the luminol reaction in an alkaline solution where the luminol can be electrochemically produced in the presence of the other ingredients of the CL reaction; (3) methods based on the ECL reactions of ruthenium(II) *tris*(2,2'-bipyridine) complex, which is used as an ECL label for other non-ECL compounds such as tertiary amines or for the quantitation of persulfates and oxalate (this is the most interesting type of chemical system of the four); and (4) systems based on analytical properties of cathodic luminescence at an oxide-coated aluminum electrode.

Before dealing with specific analytical CL methodologies, it is worth commenting on the parameters typically used to relate the transient signal to the analyte concentration in the sample. Such parameters can be of the classical type (e.g., peak height and peak area, which are kinetic in nature). It should be noted that the intrinsically kinetic nature of the CL signal requires using special systems to acquire and process data. In this context, CL chemometrics has opened up interesting avenues for development, some of which are discussed below.

In addition to conventional measured parameters (peak height or area under the CL response signal), one can use typically kinetic parameters such as CL formation and decay rates, both of which are directly related to the analyte concentration. These parameters can be easily determined from the straight segments of the rising and falling portions of the response curve, using a computer to acquire and process data. These alternative kinetic parameters result in improved selectivity and precision in CL analyses, as shown in Sec. 3.2.

3. ANALYTICAL METHODOLOGIES

This section deals briefly with classical methods based on conventional mixing of the sample and reagents such as the batch mode and low-pressure flow mixing methods, as well as the use of CL detection in continuous separation techniques such as liquid chromatography and capillary electrophoresis for comparison with the unconventional mixing mode.

Analytical CL methodologies share a number of advantages, including: (1) low detection limits (in the nanogram- or even subnanogram-per-milliliter region); (2) wide dynamic ranges (up to six orders of magnitude); (3) high signal-to-noise ratios resulting from the absence of a light source and the consequent absence of noise; (4) absence of Rayleigh and Raman scattering; (5) instrumental simplicity and affordability; and (6) absence of toxic effects from the usual CL reagents.

On the other hand, the most severe constraint of CL analyses is their relatively low selectivity. One major goal of CL methodologies is thus to improve selectivity, which can be accomplished in three main ways: (1) by coupling the CL reaction to a previous, highly selective biochemical process such as an immunochemical and/or enzymatic reaction; (2) by using a prior continuous separation technique such as liquid chromatography or capillary electrophoresis; or (3) by mathematical discrimination of the combined CL signals. This last approach is discussed in Sec. 4.

3.1 Conventional Mixing

Batch CL analyses are conducted by using commercially available "chemiluminometers." The last ingredient of the CL reaction is injected, by means of a syringe, into a reaction vessel accommodated in a light-tight housing placed in front of a mirror. In many cases, a filter or monochromator is inserted between the sample light source and detector. The detector is usually a classical high-voltage photomultiplier tube (PMT) or, increasingly frequently, a low-power single photodiode or a linear two-dimensional photodiode array (referred to as a charge coupled device, CCD), which provides a CL spectrum [11, 12]. Chemiluminometers also include a signal processor, which can be connected to an analog-to-digital converter or a computer to obtain an oscilloscope-type fast recording or printout, respectively. The main drawbacks of the batch CL mode are that mixing is not efficient enough and the process is difficult to automate. On the other hand, its most salient advantage is simplicity.

Flow injection methodologies are highly suitable for implementing CL analyses using low-pressure continuous mixing. There are many reported applications of this type including immobilized reactants [13] or enzymes [14]. One recent example is the flow injection manifold used for the determination of polynuclear aromatic hydrocarbons [15], which act as fluorophores in an energy-transfer CL process. The sample is inserted into an acetonitrile stream that is merged with another resulting from the continuous mixing of aryl oxalate (DNPO) and hydrogen peroxide solutions. The merging point is near a planar coiled quartz flow cell, where the reaction takes place, which is located in front of a low-power photodiode. The detection limits thus achieved range from 0.6 to 70 ng/ml and the system is quite useful for screening purposes.

The flow injection/CL coupling provides a number of interesting advantages including: (1) automated mixing; (2) easy online coupling to automated sample treatment systems; (3) ready adaptation to the requirements of CL reactions thanks to the high flexibility of the flow injection technique; (4) low cost; (5) easy assembling from available parts; and (6) commercial availability. However, this combined approach also has some drawbacks, such as: (1) medium to low mixing efficiency owing to the low pressure in the merging streams; (2) high dead volumes resulting from the mixing point being located outside the flow cell; and (3) no selectivity improvement relative to the conventional batch mode.

The lack of selectivity can be circumvented by coupling a postcolumn flow system to a liquid chromatograph. This has promoted the development of a number of efficient liquid chromatography-CL approaches [16, 17]. Eluted analytes are mixed with streams of the substrate and oxidant (in the presence or absence of a catalyst or inhibitor) and the mixed stream is driven to a planar coiled flow cell [18] or sandwich membrane cell [19] in an assembly similar to those of flow injection–CL systems. Many of these postcolumn flow systems are based on an energy-transfer CL process [20]. In others, the analytes are mixtures of metal ions and the luminol–hydrogen peroxide system is used to generate the luminescence [21].

In addition to its high sensitivity, automated mixing capability, and easy assembly, liquid chromatography–CL monitoring has the advantage of the high selectivity resulting from efficient discrimination on the chromatographic column relative to flow injection configurations. On the other hand, the medium-to-low mixing efficiency and high dead volumes of this coupled approach add up to compatibility problems of CL reactions with chromatographic mobile phases and to a high cost of the equipment required.

Capillary electrophoresis–CL detection [22] is an effective marriage of convenience where the main drawbacks of each partner are offset by the advantages of the other. In fact, the high discriminating capacity of capillary electrophoresis more than compensates for the low selectivity of CL reactions. Conversely, the high sensitivity of such reactions compensates for the low sensitivity of its partner. Despite its promising prospects, this coupling is not yet mature enough for use in routine analyses. In fact, the interfaces involved (mixing points, the CL cell) are rather complicated to implement and result in scarcely robust approaches. Also, commercially available equipment is scant. By contrast, a large number of capillary electrophoresis–CL systems have been reported [23], some of which include ECL detection [24]. By way of example, two recent applications, both of which use a metal-catalyzed luminol system, have been proposed. One is the resolution of a mixture of amino acids in an aspirated sample where the analytes act as CL inhibitors by complexing copper(II) ion dissolved in the buffer carrier, which is mixed in a capillary with a stream of luminol and hydrogen peroxide near the CL detection cell [25]. The other is the resolution of a mixture of metal ions in an aspirated sample [26]. A weak acid buffer carrier containing luminol and a ligand is used to form soluble complexes with the analytes to avoid precipitation. A stream of alkaline hydrogen peroxide is merged with the carrier in the capillary, near the CL detection cell.

3.2 Stopped-Flow Technique

The above-described drawbacks inherent in the conventional mixing mode are circumvented by using appropriate CL assemblies based on the high-pressure stopped-flow, the continuous-addition-of-reagent, or the analyte pulse perturbation technique. The first two have largely been used in kinetic methods involving fast reactions as they are particularly well suited to the monitoring of transient signals such as those produced by a CL reaction. These approaches meet the two principal requirements of CL analyzers, namely: (1) rapid, highly efficient mixing of sample and reagents; and (2) rapid detection of the light emitted after the CL reaction has started.

The combination of the high-pressure stopped-flow technique and a CL reaction is a new approach called "stopped-flow CL spectrometry" (SF-CLS) [27] that allows one to obtain entire CL profiles, which is impossible with continuous and flow injection systems owing to the fugacity of the CL signal. Figure 3 depicts a modular instrument for implementing high-pressure stopped-flow CL analyses. The sample and reagents are mixed at a high pressure from two propelling syringes actuated by manual or automatic pneumatic pulses in a flow cell or mixing chamber that also acts as the observation cell [28]. The cell is accommodated in the detector cell compartment of a commercially available spectrofluorimeter. When the flow is abruptly stopped by actuating the stopping syringe, the variation of the CL signal as a function of time is recorded and the data thus obtained are processed by a computer to deliver the analytical results [27]. Both the flow cell and the detector cell compartment are thermostated. The spectrofluorimeter's light source is maintained off throughout the process.

This approach has several advantages with regard to the above-described conventional mixing mode including rapid, highly efficient mixing of sample and reagents; immediate detection of the CL signal; the ability to record the whole CL intensity-versus-time profile, even for extremely fast reactions; and the ability to use new measured parameters such as formation and decay rates. These param-

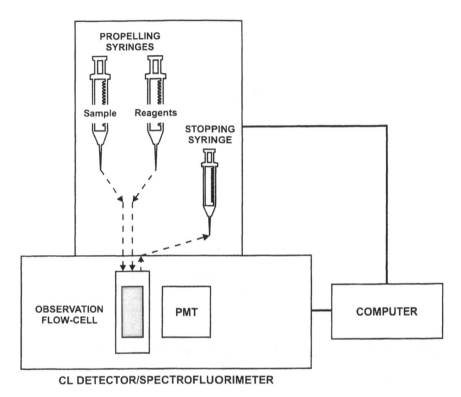

Figure 3   Instrumental setup used to implement SF-CLS by using a modular stopped-flow device for mixing the sample and reagents.

eters are obtained from the slopes of the straight portions shown in Figure 4, both of which are proportional to the analyte concentration. The use of these kinetic parameters provides some advantages over conventional peak height and peak area measurements, specifically [27]:

1. They contain a greater amount of useful information for optimizing variables and performing mechanistic studies.
2. They provide a more accurate method for the simultaneous determination of formation and decay rate constants using curve-fitting software.
3. The use of formation and decay rates as measured parameters results in improved precision, selectivity, and throughout, which facilitates application to routine analyses.

By using the modular stopped-flow/CL approach and the above-mentioned kinetic parameters, analytes such as hydrogen peroxide [27], hypochlorite [29],

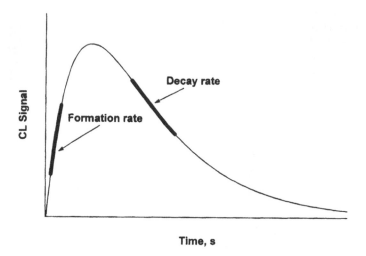

Figure 4 CL profile provided by a modular stopped-flow device showing the portions in which formation and decay rates are measured.

periodate [30], and manganese(II) [31], which take part in the luminol reaction, can be advantageously determined in terms of sensitivity, throughput, and precision (see Table 1).

Luminol in a carbonate buffer solution at pH 10 is used to fill one of the drive syringes of the stopped-flow system, and the analyte in the sample to fill the other. The maximum emission wavelength is 425 nm and the operating temperature 20–25°C. Worth special note is the stopped-flow determination of manganese(II) with luminol, which was carried out in the absence of hydrogen peroxide, dissolved oxygen acting as the oxidant since the reactants were mixed at a high pressure so the activity of dissolved oxygen was very high and comparable to that of an oxidant such as hydrogen peroxide. Under these conditions, high concentrations of sodium chloride enhance the CL emission and provide a very sensitive method for the determination of manganese [31]. This enhancing effect can be ascribed to catalytic cleavage of a potential $Mn(II)-O_2$-luminol-activated complex resulting in a chloride-manganese interaction in the presence of dissolved oxygen via the following reaction [31]:

$$MnCl_4^{2-} + O_2 \leftrightarrow MnCl_4^- + H_2O^-$$

This reaction is consistent with: (a) the observed enhanced reactivity of chloride ion toward manganese relative to other halide ions (e.g., bromide ion); and (b) the presence of H_2O^- species, which is indispensable for the subsequent formation

Table 1 Analytical Figures of Merit and Applications of Low-Pressure Stopped-Flow CL Spectrometry in Routine Analyses

Analyte	System	Linear dynamic range, M	Detection limit, M	Sampling frequency, h^{-1}	RSD, %	Applications	Ref.
Direct determinations							
Hydrogen peroxide	Luminol-Co(II)	10^{-8}–5×10^{-5}	25 pmol	60	1.7		27
Hypochlorite	Luminol	7.5–5000 ng/mL		80	1.0	Tap water	29
Periodate	Luminol-Mn(II)-TEA	10^{-6}–2.5×10^{-4}	1.2×10^{-7}	100	0.7		30
Manganese(II)	Luminol-O_2-TEA	2×10^{-7}–10^{-5}	1.8×10^{-8}	100	0.9	Biological materials	31
Indirect determinations							
Penicillins	Luminol-I_2	10^{-6}–1.2×10^{-4}		120	2.0	Pharmaceuticals	32
Tartrate	Luminol-IO_4^--Mn(II)-TEA	5×10^{-6}–5×10^{-5}	1.5×10^{-6}	—	1.8	Pharmaceuticals	30

TEA, triethanolamine; RSD, relative standard deviation.

of the CL system. The presence of manganese(III) in the reaction medium is commonplace in manganese-catalyzed reactions.

Based on the analytical figures of merit of the methods in Table 1, the best precision and selectivity are accomplished by using the decay rate rather than the formation rate or conventional CL-measured parameters such as the peak height or area under the CL curve. Table 2 gives the selectivity factor, expressed as decay-rate and peak-height tolerated concentration ratio, for the CL determination of hydrogen peroxide using SF-CLS. As can be seen, the selectivity factor was quite favorable in most instances.

The luminol reaction has also been used for the CL determination of organic substances such as penicillins [32] and tartrate ion [30] in pharmaceutical preparations by their inhibitory effect on the luminol-iodine and luminol-periodate-manganese(II)-TEA system, respectively. As can be seen from Table 1, the results were quite satisfactory. In the indirect determination of penicillins by their inhibitory effect on the luminol-iodine system, the stopped-flow technique improves the accuracy and precision of the analytical information obtained, and also the sample throughput [32]. Thus, in only 2–3 s one can obtain the whole CL signal-versus-time profile and calculate the three measured parameters: formation and

Table 2 Selectivity of Various Measurement Methods Used in the Determination of 20 ng/mL Hydrogen Peroxide

Species tested	Tolerated concentration, ng/mL		Selectivity factor, V_D/h
	Decay rate, V_D	Peak height, h	
Fe(III)	60	20	3.0
Cu(II)	80	20	4.0
Cr(III)	30	10	3.0
Mn(II)	10	10	1.0
Ni(II)	75	30	2.5
Hg(II)	2000	2000	1.0
Ce(IV)	100	50	2.0
MnO_4^-	100	30	3.3
$[Fe(CN)_6]^{3-}$	3000	1000	3.0
ClO^-	120	60	2.0
S^{2-}	1000	500	2.0
SO_3^{2-}	1000	500	2.0
NO_2^-	2000	2000	1.0

Source: Adapted from Ref. 27.

decay rates, and peak height. From these, penicillins such as penicillin G, ampicillin, amoxycillin, and carbenicillin can be determined.

Many of these CL reactions are very fast, so they require the use of the stopped-flow technique. Analytical results can be obtained within seconds, so the sampling frequency is quite high. Figure 5 illustrates the short time needed for the maximum peak in the CL curve to be reached in the previous examples. Such a time is 500 and 700 ms in the hydrogen peroxide and hypochlorite determinations, respectively, and 200 ms in the periodate one. The fastest CL reactions are those of the luminol-dissolved oxygen-Mn(II) and luminol-iodine systems for the determination of manganese and penicillins, respectively. The entire transient CL signal can be acquired within about 300 ms and the maximum peak height is reached at ca. 50 ms.

Other high-pressure stopped-flow/CL systems are based on special mixing modes, the main technical difference of which is that the CL observation cell

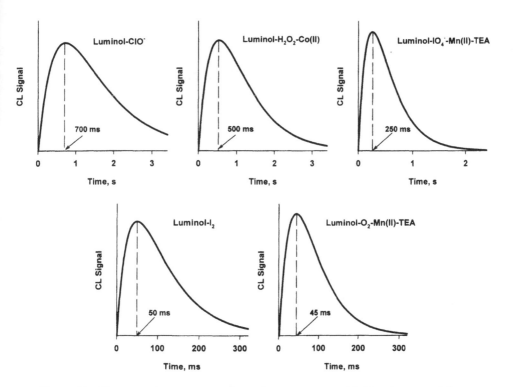

Figure 5 CL-versus-time responses for various systems provided by the stopped-flow technique.

and mixing point are not integrated; in fact, detection takes place at a chamber located before the observation flow cell [33, 34], as shown in Figure 6. Figure 6a depicts a system similar to that of Figure 3 but based on conventional propelling—three instead of two—and stopping syringes; however, the mixer and flow cell are separate [33]. In the second scheme (Fig. 6b), the mixer and flow cell are also separate, but three liquid chromatographic pumps—instead of the typical propelling syringes—are used and the stopping syringe is replaced with a valve for stopping the flow before the sample and reagent are mixed [34]. These assemblies also provide the whole CL profile and are used mainly as fitted to a liquid chromatograph in the determination of organic substances using peroxyoxalate systems.

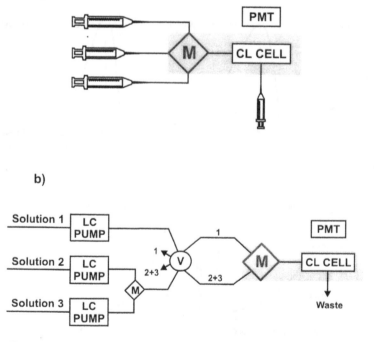

Figure 6 Schematic diagrams of atypical high-pressure SF-CL systems based on special mixing modes: (a) system based on conventional propelling and stopping syringe; and (b) system based on liquid chromatographic (LC) pumps and stopping the flow before the mixer (M). V, six-ports valve. (Adapted from Refs. 33 and 34.)

3.3   Continuous-Addition-of-Reagent Technique

The continuous-addition-of-reagent (CAR) technique is one other unconventional mixing mode used with CL reactions. This approach, developed in the authors' laboratory [35], is based on the continuous addition, at low pressure and constant rate, of one reaction ingredient (usually the reagent) over a vessel containing the analyte and the other reaction ingredients. This technique is specially suited to fast reactions and offers a major alternative to high-pressure SF-CLS. In fact, it is more flexible for the kinetic study of these reactions, possesses a higher analytical potential, and uses a more affordable instrumentation [35–37]. As shown below, these features endow the CAR technique with a high analytical usefulness in CL analysis, known as "continuous-addition-of-reagent CL spectroscopy" (CAR-CLS).

In general, if an irreversible reaction, $A + R \rightarrow P$ (where A is the analyte, R the reagent, and P the products), is developed by using the CAR technique (i.e.,if a solution of the reagent is added at a low pressure and constant rate u to a volume V_0 of a solution containing the analyte in a reaction vessel) the overall reaction rate of the process will depend on two factors: one chemical (the reaction rate proper) and the other physical (the rate of dilution of the species present in the reaction vessel). If the reagent concentration in the addition unit, $[R]_0$, is assumed to be much higher than that of the analyte in the reaction vessel, $[A]_0$, and that the reagent uptake always will be much smaller than the amount of reagent added, then the integrated rate equation for this process is given by [35]:

$$ln \frac{S_\infty - S_t}{S_\infty} = -k[R]_0 t + \frac{k[R]_0 V_0}{u} ln \frac{V_0 + ut}{V_0} - ln \frac{V_0 + ut}{V_0} \tag{2}$$

where k is the second-order rate constant, and S_t and S_∞ are the signals at time t and ∞ (total reaction development), respectively, provided the reaction is monitored via the product, using spectrophotometric detection (absorbance measurements), which is the usual choice with this technique.

The kinetic profile (signal-vs.-time plot) obtained for a chemical reaction developed by using the CAR technique is peculiar; and as can be seen in Figure 7a, it consists of three distinct portions, namely: (1) an initial, concave segment where the analytical signal is directly proportional to t^2, this dependence being the basis for the so-called "initial-reaction method" for the determination of the analyte; (2) a wide, linear, intermediate portion in which changes in the analyte signal with time (maximum reaction rate) depend on various factors according to

$$\left[\frac{dS}{dt} \right]_{max} = K[A]_0 \sqrt{\frac{k[R]_0 u}{V_0}} \tag{3}$$

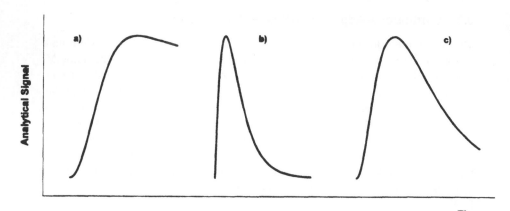

Figure 7   Typical signal-versus-time profiles obtained from: (a) a chemical reaction carried out by the CAR technique with photometric detection; (b) a CL reaction developed by using the SF technique; and (c) a CL reaction performed by using the CAR technique to mix the sample and reagents.

where K is a constant that depends on the detection system used; this linear dependence between the maximum reaction rate and the analyte concentration is the key to kinetic analytical determinations with the CAR technique; (3) a final, convex portion that responds to the dilution of the products formed and is of little analytical interest.

In the case of a CL reaction, such as $A + R \rightarrow P + h\nu$, the response curve corresponds to two first-order consecutive reaction steps; taking into account the possible rate equations that can be formulated for each reaction step, the integrated equation can be formulated as [27]:

$$S_{CL} = Ck_2[R]_0 \frac{k_1}{k_1 - k_2} (e^{-k_2 t} - e^{-k_1 t}) \tag{4}$$

where S_{CL} is the CL signal at any time t, k_1 and k_2 are the rate constants corresponding to two opposite simultaneous first-order processes, and C is a constant related to instrumental features. Figure 7b shows the typical CL profile (signal vs. time) obtained in this case, the peak height, and the area under the CL curve, which are used as measured parameters for quantitative determinations.

When a CL reaction is developed by using the CAR technique, the shape of the resulting CL signal-versus-time plots follows a differential equation that is a combination of the integrated Eq. (2) and (4), and is very difficult to obtain. However, the kinetic curve exhibits the characteristic initial concave and wide linear portions that correspond to a reaction (see Fig. 7c). Therefore, the maximum reaction rate can also be used as measured parameter for kinetic determinations in CL analyses with CAR-CLS technique.

The instrumental setup used to implement the CAR technique in CL kinetic-based determinations is very simple and inexpensive (see Fig. 8). It consists of three basic units, such as: the addition unit, the detector housing, and the data acquisition system. The addition unit is composed of an autoburette, a cylindrical glass reaction vessel with a suitable capacity to achieve V_0 of ca. 1.0 mL, and a small magnetic stirrer to homogenize the reaction mixture. The most important feature of this addition unit is that the addition rate of the autoburette can be programmed over a wide range according to the requirements (half-life) of the CL reaction under study—it can be changed from sub 0.1 to above 10 mL/min. Two alternatives can be used to detect the generated CL signal: (1) a commercial spectrofluorimeter with its light source switched off the sample holder of which is replaced with the above-mentioned small magnetic stirrer and the reaction vessel; and (2) a homemade photomultiplier (PMT) detector housing (black box) with the corresponding high voltage supply and built-in voltage divider and current-to-voltage conversion circuits, among other components. In both configurations, the distance of the reaction vessel from the PMT is critical to acquire as much emitted CL as possible, which can also be increased by using

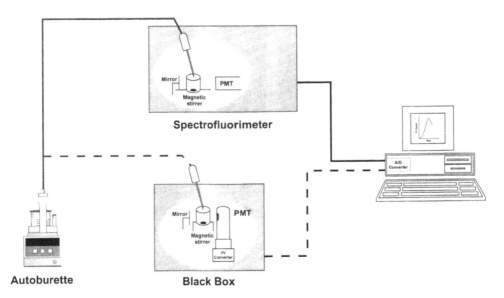

Figure 8 Instrumental setup used to implement the CAR technique in CL kinetic-based determinations according to the detection system used: a spectrofluorimeter or a black box including a PMT.

a mirror in front of the photomultiplier tube. Finally, the processing unit consists of a PC-compatible computer equipped with an analog-to-digital converter. A trigger allows data acquisition to be synchronized with reagent addition from the autoburette. This simple, inexpensive instrumentation—available in many analytical laboratories— is an effective tool for monitoring CL reactions with relatively short emission lifetimes because the CL signal is simultaneously monitored by the PMT during development of the reaction.

The first attempt at assessing the potential of the CAR technique for CL analysis involved determining copper by its catalytic effect on the reaction between luminol and hydrogen peroxide [38]. One important step in implementing the CAR-CLS technique is selecting the appropriate reagent for addition from the autoburette (luminol or hydrogen peroxide in this case). There is no general rule to make the decision, which is based on experimental results (signal-to-noise ratio, precision). One other important consideration in optimizing a CAR system is selecting the reagent concentration and its rate of addition from the autoburette. In fact, both variables are closely related via Eq. (3); the actual reagent concentration $[R]$ in the reaction vessel depends on both according to

$$[R] = \frac{ut[R]_0}{V_0} \tag{5}$$

In this case, it is generally preferable to use a higher concentration of reagent (if it is soluble enough) and add it at a lower rate than the opposite to ensure more reproducible mixing in the reaction vessel.

Under these conditions, a CL intensity-versus-time curve can be obtained to determine cupric ion at the ng/mL level using the luminol/hydrogen peroxide system. Luminol, at a 3.5×10^{-5} M concentration, was placed in the reaction vessel and the reaction was started by adding a 1.5 M solution of hydrogen peroxide at a rate of 1 mL/min from the autoburette, under stirring. The CL signal was detected at an emission wavelength of 425 nm, using a spectrofluorimeter. The CL response curve spanned a relatively long time interval (between 1 and 30 s), and the peak height was obtained within about 6 s, which is long relative to, for example, commercially available chemiluminometers and SF-CLS (in this latter, the time needed to obtain the peak height in the determination of manganese(II) with the luminol system is only about 50 ms [31]). This can be ascribed to the special way in which sample and reagent are mixed in the CAR technique, which provides a way for CAR-CLS to modify the rate of the CL reaction. In fact, from Eq. (3) it can be inferred that the half-lives of the reactions will be directly proportional to the concentration of reagent, $[R]_0$, and its addition rate. This facilitates data acquisition and improves the analytical performance of the ensuing analytical method. Thus, as can be seen in Table 3, the maximum reaction-rate method provides a wider dynamic linear range and better preci-

Table 3 Comparison of the Performance of the Reaction-Rate and Peak Height Methods in the CL Determination of Copper(II) Using Luminol and the CAR Technique

Figure of merit	Maximum reaction-rate method	Peak-height method
Linear dynamic range (ng/mL)	1–700	1–250
Detection limit (ng/mL)	0.30	0.25
Precision (RSD) (%)	2.48	3.66
Sampling frequency (h^{-1})	48	30

Source: Adapted from Ref. 38.

sion, throughput, and selectivity (e.g., selectivity factor of 2.0 for copper and manganese) than does the peak-height method (the classical choice in CL analysis) in the CL determination of copper.

The scope of CAR-CLS in analytical determinations has been expanded with one other type of CL reaction (luminol-based CL reactions are restricted to direct determinations of metal ions and some indirect ones). The so-called "energy transfer CL" is one interesting alternative, with a high analytical potential. As stated above, PO-CL systems based on the reaction between an oxalate ester and hydrogen peroxide in the presence of a suitable fluorophore (whether native or derivatized) and an alkaline catalyst are prominent examples of energy transfer CL. This technique has proved a powerful tool for the sensitive (and occasionally selective) determination of fluorophores; its implementation via the CAR technique is discussed in detail later.

One other form of energy transfer CL involves the production of very weak luminescence by an excited state formed in a reaction of the analyte; the resulting emission intensity can be greatly enhanced by adding a luminescent energy acceptor (sensitizer) to the reaction medium. This alternative—of a lower analytical potential than PO-CL reactions—is useful in some cases as it enables some CL analytical determinations that are inaccessible via PO-CL reactions. Such is the case with the CL determination of low-molecular-mass aliphatic tertiary amines using the CAR technique. In fact, these amines are not fluorescent, and also impossible to derivatize to fluorophore with fluorescent probes owing to the lack of an appropriate structure. However, if tertiary aliphatic amines are reacted with sodium hypochlorite in an alkaline medium containing fluorescein as sensitizer, the CL signal is significantly enhanced. Using this reaction, a simple CAR method was developed for the CL determination of these species [39]. A 0.9-M sodium hypochlorite solution was added from the autoburette at 10 mL/min to the reaction vessel containing the analyte and the other reaction ingredients (borate buffer and fluorescein as sensitizer). The CL signal was measured by using a laboratory-assembled detector housing (black box). The method thus developed allows the determination of trimethylamine over a wide linear range (10–2500 nmol) with good precision (relative standard deviation, RSD, 2.6%), and compares favorably with recent alternatives to the determination of this amine in fish tissue. Primary and secondary amines such as methyl- and ethylamine can be indirectly determined at the µmol level by their inhibitory effect on the CL determination of trimethylamine.

PO-CL detection has two major advantages over classical spectrofluorimetry, specifically: (1) improved selectivity resulting from the fact that not all fluorophores are equally efficient as energy acceptors and (2) enhanced detection arising from suppression of source instability and source scatter in fluorescence detection. However, the background CL observed in these reactions in the absence of fluorophore is a serious analytical drawback [40–42]. This background emission has been ascribed to the presence of reaction intermediates, mainly phenoxy radicals generated in the oxidation by hydrogen peroxide of phenolate anion produced during the formation of the excited intermediate [42]. Specifically, the TCPO/hydrogen peroxide system gives a broad emission band centered at ~450 nm, together with a weak emission band at 540 nm, both which are formed relatively slowly [41]. In addition, in excess hydrogen peroxide, only the intermediate peaks at ~450 nm seem to prevail. Accordingly, background emission can be minimized by (a) keeping the hydrogen peroxide in excess and (b) recording the CL intensity immediately after mixing to avoid the presence of these intermediates, which emit light at ~450 nm and are formed relatively slowly. The continuous-addition-of-reagent technique meets these requirements since the aryl oxalate is added from an autoburette over a reaction vessel containing excess hydrogen peroxide and the analyte, and the CL signal is recorded as soon as it is produced, using an integrated mixing/detection system. With kinetic methodology, this takes only about 1 s. Background emission is thus avoided and the signal-to-noise ratio is increased, which ultimately leads to a lower detection limit for the analyte.

Table 4 illustrates the use of the CAR technique to develop CL kinetic-based determinations for various analytes in different fields. As can be seen, the dynamic range, limit of detection, precision, and throughput (~80–100 samples/h) are all quite good. All determinations are based on the use of the TCPO/hydrogen peroxide system; by exception, that for β-carboline alkaloids uses TCPO and DNPO. A comparison of the analytical figures of merit for these alkaloids reveals that DNPO results in better sensitivity and lower detection limits. However, it also leads to poorer precision as a result of its extremely fast reactions with the analytes. Finally, psychotropic indole derivatives with a chemical structure derived from tryptamines have also been determined, at very low concentrations, by CAR-CLS albeit following derivatization with dansyl chloride.

Table 4 Analytical Figures of Merit of the CL Kinetic-Based Determination of Drugs and Hallucinogenic Alkaloids Using the CAR Technique

Analyte	Dynamic range (ng/mL)	Detection limit (ng/mL)	Precision (RSD, %)	Comments	Ref.
Phenothiazines (acepromazine, propio-promazine, promazine, trimepra-zine, methotrimeprazine, thiorida-zine, chlorpromazine)	7.5–64,000	2.2–66.5	1.0–3.7	Acepromazine was determined in horse plasma with good results; the method is suitable for pharmacoki-netic studies	43
Psychotropic indol derivatives (psilo-cin, bufetonine, N-methyl-, 5-meth-yl- and α-methyltryptamine)	0.6–140 pmol	138–230 pmol	1.5–2.2	Psilocin was quantified in specimens of the fungal species *Psylocibe semilanceata*	44
β-Carboline alkaloids (harmaline, har-malol, harmane, harmol, harmine)	0.3–150	0.25–90	2.9–3.7	Harmaline was determined in plasma samples; DNPO was used as aryl oxalate	45

DNPO, *bis*(2,4-dinitrophenyl)oxalate.

One of the most severe shortcomings of PO-CL reactions is their relatively low selectivity; this problem can be overcome in two ways: by physical discrimination (i.e., by sequentially separating the analytes prior to their determination), or by mathematical discrimination, the usefulness of which is restricted to only a few components. The former approach is the more interesting; in fact, PO-CL is widely used as very sensitive detection method for HPLC [46, 47]. The latter approach is a novel alternative in CL analysis; it is discussed in detail in Sec. 4.

However, an essential condition to ensure high sensitivity with PO-CL detection in HPLC is that the reactants (aryl oxalate and hydrogen peroxide) should be mixed with the column eluate at a point as close to the PMT as possible. In other words, a zero-dead-volume detector is required to avoid losses of emitted light and hence of sensitivity. Reported devices for PO-CL detection based on flow systems have non-zero-dead volumes, so some compromises must be made in their optimization. Based on the principles and technical aspects of the CAR technique, our group has developed a zero-dead-volume integrated derivatization CL detection system coupled to a liquid chromatograph [48]. This combined system circumvents the shortcomings of conventional postcolumn flow CL detection systems. As can be seen in Figure 9, one such system consists of a standard 1.0-cm spectrofluorimetric quartz cell in which the two reactant solutions (a) aryl oxalate in ethyl acetate and b) hydrogen peroxide, Tris buffer in an isopropyl alcohol:water medium, delivered by a peristaltic pump, and the eluate from the column are mixed. All three are simultaneously delivered and an additional chan-

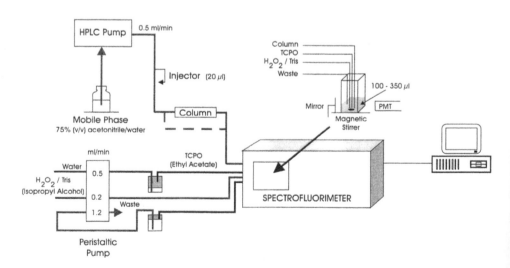

Figure 9 Schematic depiction of the zero-dead-volume CL detection system for liquid chromatography. (From Ref. 48.)

nel is used to keep the volume of the reaction mixture in the cell constant. The resulting CL response is monitored in real time by the PMT. Isopropyl alcohol is used as cosolvent to improve the solubility of the aryl oxalate solution (ethyl acetate) in water. This device allows more flexible manipulation of the reactant concentrations and flow rates relative to the mobile phase.

The analytical potential of this zero-dead-volume detection system was evaluated in the HPLC determination of polycyclic aromatic hydrocarbons (PAHs) [48]. The cell volume, or, specifically, the volume of the reaction mixture in the cell, was found to be the most influential variable on its performance. Thus, the minimum volume used to ensure reproducible results was 100 µL, and a reaction volume of 350 µL was recommended as a compromise between sensitivity and band broadening (peak width only ~1.5 min) in the determination. Various PAHs were thus determined over a wide concentration range (0.2–6000 ng), with good precision (RSD 3.5–5.7%). The ensuing method is more sensitive than the gas chromatography/mass spectrometry tandem and supercritical fluid chromatography with high-resolution laser-induced fluorescence detection in PAH determinations. This integrated derivatization-CL detection system has also been used for the HPLC determination of hallucinogenic alkaloids at the picomole to nanomole level using DNPO as aryl oxalate reagent in this case [49]. The method allows the determination of these alkaloids (the identification and quantification of some were reported for the first time) in *Heliconiini* butterfly specimens with good results.

In summary, this zero-dead-volume integrated CL detection approach provides two major advantages, namely:

1. Increased sensitivity by virtue of the high hydrogen peroxide-to-aryl oxalate ratios used, which facilitate suppression of background emission and thus raise the signal-to-noise ratios.
2. a zero-dead time between mixing of the reactants and reaction with the fluorophore/analyte, which is unfeasible in a typical flow system. This results in increased efficiency of the PO-CL reaction and avoids the usual instability of some aryl oxalate solutions (e.g., those of DNPO) in the presence of hydrogen peroxide.

3.4 Analyte Pulse Perturbation (with oscillating reactions)

The last unconventional approach considered in this chapter is low-pressure "analyte pulse perturbation-CL spectroscopy" (APP-CLS). This approach is highly dynamic as it relies on the combination of an oscillating reaction, which is a particular case of far-from-equilibrium dynamic systems, and a CL reaction.

The APP technique, recently introduced [50, 51], uses a continuous stirring tank reactor (CSTR) and relies on the sequential perturbation of an oscillating reaction by successive addition of analytes (or standards) after regular oscillations are restored. The perturbation results in a change in the oscillation period or amplitude the magnitude of which is proportional to the analyte concentration. Provided the optimum experimental conditions are maintained, the system remains in an oscillating state, acting as a continuous indicator system, for at least 8 h. This provides a rapid, simple method for performing many determinations on the same oscillation system. This operating mode offers obvious advantages over classical discrete systems and endows the APP technique with a high practical potential.

The potential of the APP technique with CL detection has been assessed by coupling the following reactions [52]: the Epstein-Orban chemical oscillator reaction, the ingredients of which are hydrogen peroxide, potassium thiocyanate, and copper sulfate in the presence of sodium hydroxide, and the CL reaction involving the oxidation of luminol by hydrogen peroxide, catalyzed by copper(II) in alkaline medium. In this case, the CL reaction acts as detector or indicator system of the far-from-equilibrium dynamic system. In fact, the CL oscillating response may be due to interaction between the yellow hydroxyl radical–cuprous ion complex involved in the positive-negative feedback loop of the oscillating reaction and luminol to give the light-emitting form of luminol according to the following reaction:

$$HO_2 - Cu(I)_{yellow} + LumH^- \rightarrow Cu^{2+} + 2OH^- + Lum^{*-} \tag{6}$$

When this oscillating system is perturbed by a pulse of an analyte such as vitamin B_6, it undergoes a change in its amplitude or period (amplitude for this vitamin) that is proportional to the concentration and can be used to construct a calibration plot.

Figure 10 shows the instrumental setup used to implement the APP-CLS approach. It consists of: (a) a CSTR that is a thermostated 10-mL glass reaction vessel accommodated in a commercially available spectrofluorimeter (a Hitachi F2000 model in this case); (b) a four-channel peristaltic pump with three channels used to dispense the reagent solutions and the fourth to keep the volume of the reaction mixture in the CSTR constant; the three reagent solutions are as follows: (1) 0.15 M hydrogen peroxide; (2) 0.15 M sodium thiocyanate, 0.15 M sodium hydroxide, and 1.95×10^{-3} M luminol; and (3) 6.0×10^{-4} M copper(II) sulfate; (c) an autoburette from which pulses (μL) of the analyte solution are added; and (d) a compatible computer for recording and processing the CL oscillating signal.

Figure 11 shows typical CL oscillating responses of this system as perturbed by vitamin B_6 pulses, which decrease the oscillation amplitude. Arrowheads indicate the times at which analyte pulses were introduced. Zone A corresponds to the oscillating steady state; zone B to the response of the oscillating system to vitamin B_6 perturbations; and zone C to the recovery following each perturbation (second response cycle), which was the measured parameter. This

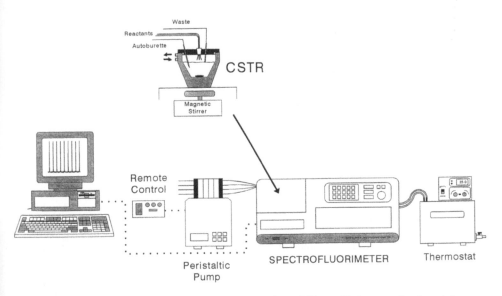

Figure 10 Experimental setup for implementation of CL oscillating reaction-based determinations. CSTR, continuous stirring tank reactor. (From Ref. 52.)

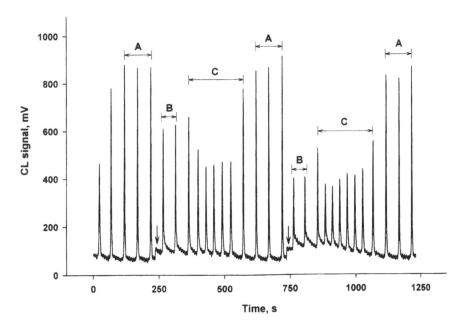

Figure 11 Typical profiles for the H_2O_2-KSCN-$CuSO_4$-NaOH-luminol CL oscillating reaction in the presence and absence of a vitamin B_6 perturbation. Arrows indicate the times at which oscillations were perturbed. (From Ref. 52.)

system allows determination of this vitamin at the µmol level (0.5–20) with good precision (RSD, 3%) and acceptable throughput (9 samples/h). After evaluating potential interfering effects from structurally related species on the performance of the method, this was satisfactorily validated by determining the vitamin in real samples such as pharmaceutical preparations.

4. RECENT APPROACHES TO MULTICOMPONENT CHEMILUMINESCENCE-BASED DETERMINATIONS

As stated in previous sections, the use of kinetic CL-based determinations has grown considerably in recent years as a result of their high sensitivity and expeditiousness, and also of their instrumental simplicity. In this context, SF-CLS and CAR-CLS have been two important supports for development as they provide kinetic information of a higher quality than does the low-pressure continuous-flow alternative. Despite these developments, the low selectivity achieved in some CL-based determinations restricts their scope. Although this problem can be solved by using coupled CL reactions as detection systems, with chromatographic techniques, such as liquid chromatography (and, recently, capillary electrophoresis), a more inexpensive alternative exists that solves the analytical problem with major practical advantages. This section deals with the potential of kinetic methods of analysis for multicomponent determinations based on CL-time primary data.

Multicomponent CL-based determinations are based on different principles depending on the particular system used to mix the sample and reagents to generate the CL signal, and also on the half-life of the CL reaction involved. Thus, when the CL reaction is carried out in a low-pressure flow system (i.e., by using the so-called CL-FIA), the manifold configuration is the actual key to accomplishing discrimination among the analytes in the sample. In this case, an instrumental approach, rather than differences in reaction rate among the analytes, enables the multicomponent determination, even though some kinetic aspects are involved owing to the transient nature of the CL signal. In other cases, a kinetic discrimination is the actual basis for the multicomponent CL-based determination, which can be approached in two ways depending on the relative half-lives of the CL reactions for each analyte in the mixture. Thus, when the measured CL intensity for each analyte peaks at a very different time, the sequential determination is feasible and the approach is designated time-resolved CL. When the kinetic behavior of the mixture components is very similar, then the resulting complex dynamic system is generally modeled with computational aids based on mathematical algorithms. This field of kinetic methods of analysis, which combines chemometric and kinetic aspects, has recently been termed *kinetometrics* [53].

4.1 Instrumental Discrimination

Although the subject of CL in FIA is treated elsewhere in this book, it might be of interest to briefly comment on the most salient features of the resolution of mixtures using these flow systems. Thus, the simultaneous flow injection determination of acetylcholine and choline [54], and that of lysine and glucose [55], are based on measurements of the hydrogen peroxide produced from the analytes in enzyme reactions, which is detected by CL using the luminol-cobalt(II) system. The enzyme reactors (used to generate the active species in the CL reaction) and the delay coils (employed to resolve CL signals in time for their sequential detection) before the detector allow the mixture to be successfully resolved (Fig. 12a). A similar principle has also been used for the speciation of metal ions such as ferric-ferrous ions [56], and chromium(III) and chromium(VI) [57] (see Fig. 12b). Because only iron(II) and chromium(III) are catalytically active on the luminol-dissolved oxygen and luminol-H_2O_2 reaction, respectively, the resolution of the

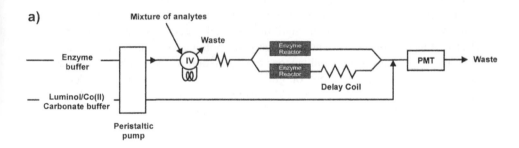

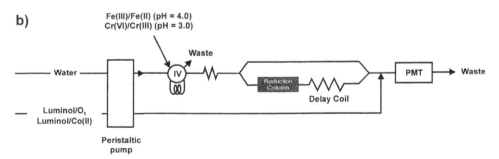

Figure 12   General flow injection manifold used for the simultaneous determination of (a) organic species involved in enzymatic reactions; and (b) inorganic ions, using a reduction column both with CL detection. IV, injection valve.

mixtures involves two steps: first, the iron(II) and chromium(III) are directly determined; and second, the sample is passed through a reduction column (copper-plated zinc granules and copper-coated zinc granules, respectively) to convert iron(III) to iron(II) and chromium(VI) to chromium(III). CL is again measured and the concentrations of oxidized species are determined by difference. The methods provide good analytical performance; e.g., iron can be determined at the nM level with a high sample throughput (60 runs/h); they were validated with real samples such as cobalt in waste water from a chromium-plating factory.

Sequential injection analysis (SIA) is a further development of flow injection methods [58] with a high potential for simultaneous determinations. An SIA system with CL detection typically consists of a double-piston pump, a multiposition valve, a holding coil, enzyme reactors (for enzymatic determinations), and a PMT detector [59]. Basically, two operations are performed: (1) the sample and reagents are aspirated in a given sequence into the holding coil; and (2) the plug of partially layered reagents and sample is dispensed to the detector. Between the first aspiration of sample and reagents into the holding coil and the final dispensing of the well-mixed plug to the detector, there may be several intermediate stages to develop additional suitable reactions—enzymatic reactions, stopped-flow, etc.—and back into the holding coil. This approach has mainly been used for the simultaneous determination of species based on enzymatic reactions such as those of glucose-lactic acid-penicillin [59], glucose-penicillin [60], and H_2O_2-glucose [61], using the CL luminol system with good analytical results and minimal reagent consumption.

4.2   Time-Resolved Chemiluminescence

This approach uses a kinetic sequential principle to carry out multicomponent CL-based determinations. In fact, when the half-lives of the CL reactions involved in the determination of the analytes in mixture are appreciably different, the CL intensity-versus-time curve exhibits two peaks that are separate in time (in the case of a binary mixture); this allows both analytes to be directly determined from their corresponding calibration plots. In general, commercially available "chemiluminometers" have been used in these determinations, so the CL reaction was initially started by addition of one or two reaction ingredients. Thus, in the analysis of binary mixtures of cysteine and gluthatione, appropriate time-resolved response curves were obtained provided that equal volumes of peroxidase and luminol were mixed and saturated with oxygen and that copper(II) and aminothiol solutions were simultaneously injected [62, 63].

Table 5 shows the most salient features of reported methods for multicomponent CL-based determinations using time-resolved CL spectroscopy. As can be seen, the CL luminol system has been widely used for this purpose because

Table 5 Multicomponent CL-Based Determinations Using Time-Resolved CL Spectroscopy

Mixtures	CL chemical system	Dynamic range (μM)	Precision (RSD, %)	Comments	Ref.
Cysteine/glutathione	Luminol-peroxidase-Cu(II)	1–50/3–50	2.5/2.8	Solutions were saturated with oxygen	62, 63
Cobalt(II)/copper(II)	Luminol-H_2O_2-cysteine	0.5–10/2–20	2.5/3.0	Metal ion solutions were saturated with oxygen	64
Gold(III)/silver(I)	Luminol-$K_2S_2O_8$-ethylenediamine	0.05–5/0.1–10	–/–	A YHF-1 liquid CL analyzer was used	65
Ascorbic acid/tartaric acid	*Tris*(bipyridyl)ruthenium(II)-Ce(IV)	0.05–1.7/1.4–140	4.8/1.7	The method was applied to wine samples	66
Oxalic acid/tartaric acid	*Tris*(bipyridyl)ruthenium(II)-Ce(IV)	0.14–14/2.9–150	2.4/4.8	The method was applied to synthetic urine	67

it provides appropriate time-resolved response curves (i.e., CL curves in which the two peaks are well apart in many cases). Thus, in the simultaneous determination of cobalt(II) and copper(II) with the luminol-hydrogen peroxide-cysteine system, light emission appearing immediately in the presence of cobalt(II) and after ~7 min in that of copper(II) [64]. Although this situation is less favorable with other CL systems such as the *tris*(bipyridyl)ruthenium(II)-Ce(IV) one, where the CL intensity for ascorbic and tartaric acids peaks at 2 and 40 s, respectively, the multicomponent determination is also feasible [66]. On the other hand, the linear dynamic ranges for the analytes in Table 5 are in the µM region and precision is quite good. In some cases, a short half-life in the first peak results in poor reproducibility; such is the case with the above-mentioned determination of ascorbic acid in mixtures with tartaric acid [66], where the RSD was 4.8% versus 1.7% for tartaric acid.

4.3   Kinetometric Approaches

One other way of improving selectivity in kinetic CL-based determinations is by mathematical discrimination using kinetometric approaches. In general, chemical systems involved in multicomponent CL-based determinations have drawbacks that make the use of statistical methods inadvisable. Thus, the differential equations that describe these chemical systems are unknown; also, synergistic effects are often present. Because of this marked nonlinearity, only powerful chemometric tools provide adequate accuracy to resolve these mixtures. Among them, least-squares matrix methodology [68] and computational neural networks (CNNs) [69] have been used for this purpose. On the other hand, when the CL system behaves linearly, various approaches allow one or more components in mixtures to be determined from kinetic differences. Such is the case with the resolution of mixtures of oxalate and proline with electrogenerated *tris*(bipyridine)-ruthenium(III) [70], in which a differential reaction rate method based on least-squares regression of the first decay data obtained by the stopped-flow technique was used to resolve the analytes. The method should be useful for the simultaneous determination of unresolved analytes of pharmaceutical and environmental interest.

Kinetic analysis usually employs concentration as the independent variable in equations that express the relationships between the parameter being measured and initial concentrations of the components. Such is the case with simultaneous determinations based on the use of the classical least-squares method but not for nonlinear multicomponent analyses. However, the problem is simplified if the measured parameter is used as the independent variable; also, this method resolves for the concentration of the components of interest being measured as a function of a measurable quantity. This model, which can be used to fit data that are far from linear, has been used for the resolution of mixtures of protocatechuic and caffeic acids, which act as enhancers in the CL reaction of luminol induced by hexacyanoferrate(III) [68]. The phenolic acids can be quantified both separately and together—in the presence of synergistic effects—using the stopped-flow technique and monitoring the intensity at 0.3 s and the initial rate. A two-factor factorial design was used to obtain the experimental response and a first-order equation was employed to fit it as closely as possible. The method allows the determination of the analytes in mixtures in concentration ratios ([protocatechuic acid]:[caffeic acid]) from ca. 1:3 to 2:1 with good sensitivity (at the nM level) and recoveries between 95 and 112%.

CNNs are among the most exciting recent developments in computational science [71] and have grown enormously in popularity in different scientific fields including analytical chemistry [72–74]. However, CNNs have scarcely been used in connection with kinetic methods of analysis; our group has employed this kinetometric approach for the estimation of kinetic analytical parameters [75] and also in multicomponent kinetic analyses with photometric detection [76–78]. Based on the good results achieved with these methods, and taking into account that multilayer feed-forward neural networks based on different versions of standard back-propagation learning algorithm have been used by several authors as highly powerful tools to study uniform approximation of an unknown continuous function that can be derived up to a p-th order (C^p-function), we developed a new methodology for multicomponent CL-based determinations.

The simultaneous determination of trimeprazine and methotrimeprazine in mixtures using the classical peroxyoxalate system based on the reaction between TCPO and hydrogen peroxide was used to validate the new methodology. The reaction was implemented by using the CAR technique, which increased nonlinearity in the chemical system studied by virtue of its second-order kinetic nature. In addition, both drugs exhibited a similar kinetic behavior and synergistic effects on each other, as can be inferred from the individual and combined (real and theoretical) CL-versus-time response curves.

One important kinetic problem in this simultaneous determination is selecting an appropriate time domain of the CL curve as the source of inputs to CNN. In this case, the optimal interval of the CL signal-versus-time plot for data preprocessing (selection of CNN inputs) is the initial concave portion of the plot and part of the linear portion (see Fig. 13). When the linear portion is wider, the additional information collected does not contribute to increased discrimination. Using this time domain, the filtration step (data preprocessing) was carried out by using principal component analysis (PCA), a widely employed technique for reducing the dimensions of multivariate data while preserving most of the variance. By using a heuristic method, nine significant principal components were selected as input to CNNs, the final architecture being 9:5s:2/ (see Fig. 13).

Under these conditions, and to test the generalizing capacity of the selected network for the simultaneous determination of phenothiazine derivatives subject

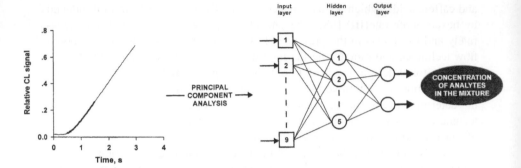

Figure 13　Schematic depiction of the foundation and architecture of the CNN used for multicomponent CL kinetic-based determinations.

to synergistic effects, various synthetic mixtures were assayed. The magnitude of the relative errors in the concentration of each component in the mixture is apparent from Table 6, which shows the results for various synthetic binary mixtures containing variable amounts of trimeprazine and methotrimeprazine. Mixtures in ratios from 10:1 to 1:10 were satisfactorily resolved. The relative errors

Table 6　Analysis of Various Trimeprazine/Methotrimeprazine Mixtures Using the Optimal Network Design

Concentration taken (µg/mL)		Trimeprazine		Methotrimeprazine	
Trimeprazine	Methotrimeprazine	Found (µg/mL)	Relative error (%)	Found (µg/mL)	Relative error (%)
3.00	3.00	2.99	−0.33	2.99	−0.33
2.00	4.00	1.99	−0.50	3.99	−0.25
2.00	5.00	2.01	0.50	5.01	0.20
1.00	3.00	0.98	−2.00	2.99	−0.33
2.00	8.00	1.92	−4.00	8.10	1.25
1.00	10.0	1.12	12.0	10.0	0.00
9.00	1.00	8.76	−2.66	1.10	10.0
10.0	2.00	10.01	0.10	2.03	1.50
8.00	2.00	8.00	0.00	2.00	0.00
6.00	2.00	5.97	−0.50	1.99	−0.50
10.0	4.00	9.99	−0.10	4.01	0.25
10.0	5.00	10.05	0.50	5.04	0.80

Source: Ref. 69.

made were less than 5% (except at both ends of the range, where they rose to ca. 10%) and hence quite acceptable for such a wide range of concentration ratios.

ACKNOWLEDGMENT

The authors gratefully acknowledge financial support from Spain's Dirección General Interministerial de Ciencia y Tecnología (DGICyT) in the framework of Project PB96-0984.

REFERENCES

1. K Robards, PJ Worsfold. Anal Chim Acta 266:147–173, 1992.
2. SR Crouch, TF Cullen, A Scheeline, ES Kirkor. Anal Chem 70:53R–106R, 1998 (see also previous reviews).
3. LJ Kricka, PE Stanley. J Biolumin Chemilumin 13:157–184, 1998.
4. ML Grayeski. Anal Chem 59:1243A–1254A, 1987.
5. A Townshed. Analyst 115:495–500, 1990.
6. HAH Rongen, RMW Hoetelmans, A Bult, WP van Bennekom. J Pharm Biomed Anal 12:443–462, 1994.
7. B Yan, SW Lewis, PJ Worsfold, JS Lancaster, A Gachanja. Anal. Chim Acta 250: 145–155, 1991.
8. OM Steijger, PHM Rodenburg, H. Lingeman, UATh Brickman, JJM Holthuis. Anal Chim Acta 266:233–241, 1992.
9. AW Knight, GM Greenway. Analyst 119:879–890, 1994.
10. AW Knight, GM Greenway. Analyst 121:101R–106R, 1996.
11. PE Stanley. J Biolumin Chemilumin 12:61–78, 1997.
12. A Gachanja, PJ Worsfold. Spectrosc Eur 6:19–25, 1994.
13. W Qin, Z Zhang, H Liu. Anal Chim Acta 357:127–132, 1997.
14. N Kiba, A Itagaki, S Fukumura, K Saegusa, M Furusawa. Anal Chim Acta 354: 205–210, 1997.
15. KN Andrew, MG Sanders, S Forbes, PJ Worsfold. Anal Chim Acta 346:113–120, 1997.
16. J Ishida, T Yakabe, H Notta, M Yamaguchi. Anal Chim Acta 346:175–181, 1997.
17. H Li, Y Ci, L Huang. Anal Sci 13:821–824, 1997.
18. M Mäkinen, V Pironen, A Hopia, J Chromatogr A 734:221–229, 1996.
19. JL Pérez-Pavón, E Rodriguez-Gonzalo, GD Christian, J Ruzicka. Anal Chem 64: 923–929, 1992.
20. M Emteborg, K Irgum, C Gooijer, UA Th Brinkman. Anal Chim Acta 357:111–118, 1997.
21. Special issues of the International Symposia on Quantitative Luminescence Spectrometry in Biomedical Sciences. Anal Chim Acta 290:1–248, 1994 and Anal Chim Acta 303:1–148, 1995.

22. AM García-Campaña, WRG Baeyens, Z Zhao. Anal Chem News Features February 1:85A–88A, 1997.
23. L Dadoo, AG Seto, LA Colon, RN Zare. Anal Chem 66:303–306, 1994.
24. GA Forbes, TA Nieman, JV Sweedler. Anal Chim Acta 347:289–293, 1997.
25. SY Liao, CW Whang. J Chromatogr A 736:247–254, 1996.
26. B Huang, J Li, L Zhang, J Cheng. Anal Chem 68:2366–2369, 1996.
27. D González-Robledo, M Silva, D Pérez-Bendito. Anal Chim Acta 217:239–247, 1989.
28. A Loriguillo, M Silva, D Pérez-Bendito. Anal Chim Acta 199:29–40, 1987.
29. D González-Robledo, M Silva, D Pérez-Bendito. Anal Chim Acta 228:123–128, 1990.
30. A Gaikwad, M Silva, D Pérez-Bendito. Analyst 119:1819–1824, 1994.
31. A Gaikwad, M Silva, D Pérez-Bendito. Anal Chim Acta 302:275–282, 1995.
32. S Ventura, M Silva, D Pérez-Bendito. Anal Chim Acta 266:301–307, 1992.
33. PD Bryan, AC Capomacchia. J Pharm Biomed Anal 9:855–860, 1991.
34. N Hanoaka, H Tanaka, A Nakamoto, M Takada. Anal Chem 63:2680–2685, 1991.
35. A Velasco, M Silva, D Pérez-Bendito. Anal Chem 64:2359–2365, 1992.
36. M Silva. Analyst 118:681–688, 1993.
37. D Pérez-Bendito, A Gómez-Hens, M Silva. J Pharm Biomed Anal 14:917–930, 1996.
38. J Cepas, M Silva, D Pérez-Bendito. Anal Chim Acta 285:301–308, 1994.
39. M Cobo, M Silva, D Pérez-Bendito. Anal Chim Acta 356:51–59, 1997.
40. K Sigvardson, JW Birks. Anal Chem 55:423–435, 1983.
41. B Mann, ML Grayeski. Anal Chem 62:1532–1536, 1990.
42. NW Barnett, R Bos, SW Lewis, RA Russell. Anal Comm 34:17–20, 1997.
43. J Cepas, M Silva, D Pérez-Bendito. Anal Chem 66:4079–4084, 1994.
44. J Cepas, M Silva, D Pérez-Bendito. Analyst 121:49–54, 1996.
45. J Cepas, M Silva, D Pérez-Bendito. Anal Chim Acta 314:87–94, 1995.
46. SW Lewis, PJ Worsfold. Anal Proc 29:10–11, 1992.
47. PJM Kwakman, UATh Brinkman. Anal Chim Acta 266:175–192, 1992.
48. J Cepas, M Silva, D Pérez-Bendito. Anal Chem 67:4376–4379, 1995.
49. J Cepas, M Silva, D Pérez-Bendito. J Chromatogr 749:73–80, 1996.
50. R Jiménez-Prieto, M Silva, D Pérez-Bendito. Anal Chem 67:729–734, 1995.
51. R Jiménez-Prieto, M Silva, D Pérez-Bendito. Analyst 123:1R–8R, 1998.
52. R Jiménez-Prieto, M Silva, D Pérez-Bendito. Talanta 44:1463–1472, 1997.
53. D Pérez-Bendito, M Silva. Trends Anal Chem 15:232–240, 1996.
54. T Hasebe, J Nagao, T Kawashima. Anal Sci 13:93–98, 1997.
55. AM Almuaibed, A Townshend. Anal Chim Acta 338:149–154, 1997.
56. HS Zhang, XC Yang, LP Wu. Fenxi Huaxue 24:220–223, 1996.
57. HS Zhang, XC Yang, LP Wu. Fenxi Huaxue 23:1148–1150, 1995.
58. J Ruzicka, GD Marshall. Anal Chim Acta 237:329–343, 1990.
59. RW Min, J Nielsen, J Villadsen. Anal Chim Acta 312:149–156, 1995.
60. RW Min, J Nielsen, J Villadsen. Anal Chim Acta 320:199–205, 1996.
61. DJ Tucker, B Toivola, CH Pollema, J Ruzicka, GD Christian. Analyst 119:975–979, 1994.
62. T Kamidate, T Tani, H Watanabe. Anal Sci 14:725–729, 1998.

63. T Kamidate, T Tani, H Watanabe. Chem Lett 2049–2052, 1993.
64. T Kamidate, A Ishikawa, T Segawa, H Watanabe. Chem Lett 1123–1126, 1994.
65. H Zhang, J Lu, Z Zhang. Huaxue Xuebao 47:481–483, 1989.
66. Z He, XY Li, H Meng, SF Lu, GW Song, LJ Yuan, Y Zeng. Anal Lett 31:1553–1561, 1998.
67. ZK He, H Gao, LJ Yuan, QY Luo, YE Zeng. Analyst 122:1343–1345, 1997.
68. A Navas-Díaz, JA González-García. Anal Chem 66:988–993, 1994.
69. C Hervás, S Ventura, M Silva, D Pérez-Bendito. J Chem Inf Comput Sci 38:1119–1124, 1998.
70. LL Shultz, TA Nieman. J Biolumin Chemilumin 13:85–90, 1998.
71. J Zupan, J Gasteiger. Neural Networks for Chemists. An Introduction. Weinheim: VCH, 1993.
72. K Hornik, M Stinchcombe, H White. Neural Networks 2:359–366, 1989.
73. H Schulz. GIT Fachz Lab 39:1009–1010, 1995.
74. J Zupan, M Novic, I Ruisanchez. Chemom Intell Lab Syst 38:1–23, 1997.
75. S Ventura, M Silva, D Pérez-Bendito, C Hervás. Anal Chem 67:1130–1139, 1995.
76. S Ventura, M Silva, D Pérez-Bendito, C Hervás. Anal Chem 67:4458–4461, 1995.
77. S Ventura, M Silva, D Pérez-Bendito, C Hervás. J Chem Inf Comput Sci 37:287–291, 1997.
78. S Ventura, M Silva, D Pérez-Bendito, C Hervás. J Chem Inf Comput Sci 37:517–521, 1997.

9

Electrogenerated Chemiluminescence

Andrew W. Knight
University of Manchester Institute of Science and Technology,
Manchester, England

1. INTRODUCTION

Electrogenerated chemiluminescence (ECL) is the process whereby a chemiluminescence emission is produced directly, or indirectly, as a result of electrochemical reactions. It is also commonly known as electrochemiluminescence and electroluminescence. In general, electrically generated reactants diffuse from one or more electrodes, and undergo high-energy electron transfer reactions either with one another or with chemicals in the bulk solution. This process yields excited-state molecules, which produce a chemiluminescent emission in the vicinity of the electrode surface.

Much of the study of ECL reactions has centered on two areas: electron transfer reactions between certain transition metal complexes, and radical ion-annihilation reactions between polyaromatic hydrocarbons. ECL also encompasses the electrochemical generation of conventional chemiluminescence (CL) reactions, such as the electrochemical oxidation of luminol. Cathodic luminescence from oxide-covered valve metal electrodes is also termed ECL in the literature, and has found applications in analytical chemistry. Hence this type of ECL will also be covered here.

1.1 Brief Historical Background

Some of the first observations of luminescence accompanying electrolysis were reported by Bancroft in 1914, when halides were electrolyzed at mercury and other anodes [1]. Thirteen years later luminescence was observed by Dufford et al. when Grignard compounds in anhydrous ether were electrolyzed, at the anode or cathode, by applying 500 to 1500 volts [2], and subsequently by Harvey for luminol in alkaline solution at the anode, by applying 2.8 volts [3].

However, ECL was not then studied in detail until 1963 [4, 5]. At this time ECL from solutions of aromatic hydrocarbons was first recorded, and mechanisms involving electron transfer between electrically generated radical anions and cations were proposed. Between the mid-1960s and late 1980s there was considerable interest in the phenomenon of ECL. More than 60 publications in the literature focused almost solely on the mechanism of ECL reactions, identifying the excited states, rate constants, quantum efficiencies, and the effects of temperature and various solvents and electrolytes. These were determined by a variety of spectroscopic and electron spin resonance methods, and by observing the effect of magnetic fields on ECL emissions. Developments during this period have been thoroughly reviewed [6–9]. Over 100 additional papers reported on the use of ECL in the study of novel photochemical and electrochemical properties of a wide range of compounds, complexes, and clusters. These included polyaromatic hydrocarbons and their derivatives; organometallics and other transition metal complexes including those of ruthenium, osmium, platinum, palladium, chromium, iridium, and rare earth elements; and molybdenum and tungsten clusters [10–12].

The potential of ECL in analytical chemistry has only more recently been investigated, but has rapidly gained recognition as both a sensitive and selective method of detection. Most reported applications have utilized the *tris*(2,2′-bipyridyl) ruthenium(II) [Ru(bpy)$_3^{2+}$] ECL reaction, or else the electrochemical initiation of more conventional CL reactions, but many other potentially useful systems have been investigated. The applications of ECL in analytical chemistry have recently been the subject of comprehensive reviews [12–16].

1.2 The Potential of ECL in Analytical Science

While retaining the selectivity and sensitivity inherent to conventional CL analysis, ECL has several potential advantages as an analytical technique that are currently being realized.

1.2.1 Advantages from the Electroinitiation of CL Reactions

Since the CL reaction is being produced by an electrochemical stimulus, greater control is gained over the initiation, rate, and course of the CL reaction. Indeed this control can be to the extent that the CL reaction may be "switched on and off," allowing for synchronous detection, effective background correction, and ready automation with computer control. By careful selection of the electrode material, surface treatment, and applied potential, an additional degree of electrochemical selectivity can also be introduced.

1.2.2 Advantages from the Electromanipulation of the CL Reaction

Some CL reagents and intermediates can be electrochemically regenerated allowing them to take part in a CL reaction with analyte molecules once again. This often allows lower concentrations of expensive reagents to be used. In some cases, principally Ru(bpy)$_3^{2+}$, such reagents can be immobilized on the electrode surface, removing the need to continually add the reagent to the system and creating an ECL sensor. Some analytes, such as polyaromatic hydrocarbons, or species labeled with a $Ru(bpy)_3^{2+}$ derivative, can be electrochemically regenerated to their active form, thus allowing each analyte molecule to produce many photons per measurement cycle. In such cases extremely low limits of detection can be achieved. It may also be possible to electrochemically modify compounds considered inactive for CL to form new species that can take part in a CL reaction, and hence extend the range of analytical applications of a particular CL method.

1.2.3   Simplifying Existing CL Methodology

CL analysis methods using flowing streams (e.g., flow injection analysis or HPLC) can be simplified by reducing the number of reagents needed to be pumped separately to the site of mixing and light detection. This is possible because active reagents can be electrochemically produced from passive precursors in the carrier stream. For example, hydrogen peroxide for the luminol CL reaction may be electrochemically generated by reduction of molecular oxygen dissolved in the carrier medium. In many flow-through CL methods the speed and completeness of mixing of the redox reagents in view of a light transducer are crucial to the reproducibility and sensitivity of the technique. However in ECL, it is often the case that all of the reagents can be thoroughly and reproducibly mixed prior to their passage into the detection cell, whereupon the CL reaction is electroinitiated. Similarly in conventional CL methods the optimization of the physical distance between the point of mixing and the light transducer, primarily dependent on the kinetics of the CL reaction, is important. This is in order for the reaction mixture to be passing the light detector at the point of maximum CL emission. In ECL however, the light-producing reaction is largely confined to the immediate vicinity of the electrode surface, which can be shaped and positioned directly in front of the light transducer for maximum sensitivity. In some cases unstable CL reagents or intermediates can be electrochemically generated, and allowed to react as soon as they are formed at the electrode surface, or some point downstream in a flow-through system. In this way reproducibility is improved and the problem of working with analytical reagents that rapidly deteriorate is removed. Examples of this include the electrochemical generation of peroxides, $Ru(bpy)_3^{3+}$, and radical species of polyaromatic hydrocarbons.

1.2.4   Advantages from the Combination of a Spectroscopic and Electrochemical Technique

ECL gives the opportunity of gaining additional analytical information by monitoring the electrochemical activity of the analyte, by a range of electroanalytical techniques, alongside recording the light output. Similarly measurements of chemiluminescent light output can be used to elucidate and visualize electrochemical processes, and measure reaction kinetics. For example, one particular reaction that produces a CL emission can be selectively monitored from many other reactions occurring simultaneously at the electrode.

1.2.5 Limitations

Electrogenerated chemiluminescence does, however, have some limitations, which have not yet been fully overcome. The combination of electrochemical and CL techniques brings together two sets of species that can potentially interfere with ECL determinations, either constructively or destructively. These include species that preferentially react with the electrogenerated CL reagents and intermediates, in nonchemiluminescent reactions; or that quench excited species; or that are electrochemically active and interfere with, or swamp, electrochemical reactions. As a direct consequence of this, the majority of analytical ECL methods now being developed include a chromatographic separation step prior to ECL detection. Similarly optimization is complicated by the need to find conditions, such as pH or choice of electrolyte, to suit both the electrochemical and CL reactions, and as a result a compromise often has to be reached. Additionally, some factors such as the pH and optimum applied potential also interact and are partially dependent on each other. Optimization is complicated still further if the medium is to support an enzymic reaction or act as a mobile phase for a chromatographic separation. As a result multivariate optimization techniques are often required.

Most ECL reactions involve a complex series of electrochemical and CL reaction steps, such that the exact mechanism of all but the simplest ECL reactions has not been explicitly determined. This may be a problem in the search for new analytical applications, when certain compounds show marked ECL activity, while other closely related compounds are inactive, or when trying to produce a linear calibration for a particular analyte. Many applications have also been hindered by poor reproducibility in the ECL measurements, which can only be overcome by frequent refreshing of the electrode surface, the reasons for which are not well understood.

2. SURVEY OF ECL MECHANISMS AND ANALYTICAL APPLICATIONS

2.1 Organic Ion Annihilation ECL

Energetic electron transfer reactions between electrochemically generated, short-lived, radical cations and anions of polyaromatic hydrocarbons are often accompanied by the emission of light, due to the formation of excited species. Such ECL reactions are carried out in organic solvents such as dimethylformamide or acetonitrile, with typically a tetrabutylammonium salt as a supporting electrolyte. The general mechanism proposed for these reactions is as follows.

Radical formation:

$$A \rightarrow A^{+\cdot} + e^- \qquad\qquad \text{Electrochemical oxidation} \qquad (1)$$

$$B + e^- \rightarrow B^{-\cdot} \qquad\qquad \text{Electrochemical reduction} \qquad (2)$$

Two main CL pathways are then possible: If sufficient energy is available, an electron transfer reaction where the singlet excited state of A is accessible (3); otherwise an energy-deficient route whereby the energy from two triplet excited-state species are "pooled," to provide sufficient energy to form the singlet excited state, in what is termed a "triplet-triplet annihilation reaction," (4) and (5).

$$A^{+\cdot} + B^{-\cdot} \rightarrow {}^1A^* + B \qquad\qquad\qquad\qquad (3)$$

or

$$A^{+\cdot} + B^{-\cdot} \rightarrow {}^3A^* + B \qquad\qquad\qquad\qquad (4)$$

$${}^3A^* + {}^3A^* \rightarrow {}^1A^* + A \qquad\qquad\qquad\qquad (5)$$

Followed by light emission,

$${}^1A^* \rightarrow A + h\nu \qquad\qquad\qquad\qquad (6)$$

where A is a polyaromatic hydrocarbon or aromatic derivative, B is the same or another polyaromatic hydrocarbon, ${}^1A^*$ and ${}^3A^*$ are the singlet and triplet excited states of A, respectively, and $h\nu$ is a fluorescence emission [12].

The reactions can be carried out by producing the radical cations and anions at separate electrodes in close proximity to each other, or more elegantly using a single electrode and applying a square-wave potential alternating between the oxidation and reduction potentials of the species concerned. Either A or B can be the analyte species in the reaction, and in most cases after the CL reaction has occurred, barring destructive side reactions, both the original analyte and reagent species are regenerated. Common molecules used for these reactions include rubrene, 9,10-diphenylanthracene, *N,N,N',N'*-tetramethylphenylenediamine, and benzophenone.

Organic ion annihilation ECL (IAECL) reactions have not, however, found significant numbers of applications in analytical chemistry. This is principally due to the practical problems arising because these reactions take place only in organic solvents, from which water and dissolved oxygen have been rigorously excluded, to prevent quenching of the CL reaction. This invariably involves thoroughly drying solvents, recrystallizing electrolytes, and using enclosed apparatus. Additionally, analysis of the emission intensity from these short-lived radical species involves complex mathematical treatment of the electrode diffusion process, which takes into account species stability. Hence quantitative data are difficult to establish. Nevertheless during the 1980s Hill et al. developed IAECL as a detection technique for reverse-phase HPLC, using acetonitrile as the mobile phase, and tetrabutylammonium perchlorate and chloride as the supporting electrolyte [17]. Chromatograms were obtained for a variety of polyaromatic hydrocarbons and aromatic pesticide compounds including, DDT, DDE, and Barban, and calibrations performed in the concentration range 10–1000 ng/mL. The limitation of incompatibility with aqueous solutions has recently been partially overcome by Richards and Bard using water-soluble derivatives of polyaromatic hydrocarbons [18]. They observed that the anodic oxidation of sodium 9,10-diphenylanthracene-2-sulfonate and 1- and 2-thianthrene-carboxylic acid in the presence of tripropylamine as a coreactant produced ECL in aqueous solution. It was speculated that these compounds may be useful as ECL labels in immunoassay or DNA probe studies.

The use of IAECL for analytical applications has now almost entirely been surpassed by techniques based on certain transition metal complexes, from which ECL reactions can occur in aqueous solution.

2.2 Inorganic Electron Transfer ECL

Analytical applications of inorganic electron transfer ECL have almost exclusively focused on the reactions of *tris*(2,2′-bipyridine) ruthenium(II), shown in Figure 1, or its derivatives. This is because this compound undergoes fully reversible, one-electron, electrochemical redox reactions, at easily attainable potentials, forming stable reduced or oxidized species. These species can then take part in a host of ECL reactions, producing a phosphorescence emission from the excited state of $Ru(bpy)_3^{2+}$, which is regenerated in the reaction. Such ECL reactions can be performed in fully aqueous solutions, a range of organic solvents, or mixtures of the two. The reactions take place over a wide pH range, in the presence of oxygen and many other impurities and at room temperature. $Ru(bpy)_3^{2+}$ ECL reactions exhibit very high efficiencies, and hence are often used as standards

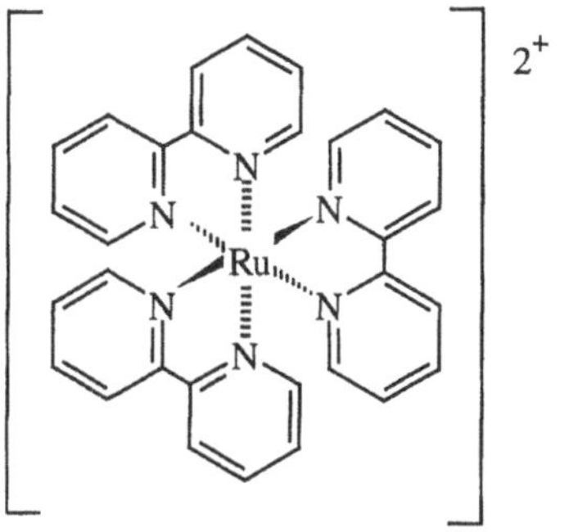

Figure 1   Tris(2,2′-bipyridine) ruthenium(II).

in determining the ECL efficiency of other compounds [19]. $Ru(bpy)_3^{2+}$ is also relatively nontoxic and not prohibitively expensive. Analytical applications involving $Ru(bpy)_3^{2+}$ ECL and CL have been the subject of several recent fundamental reviews [12, 14–16].

$Ru(bpy)_3^{2+}$ itself can undergo electron transfer reactions to produce ECL in an analogous fashion to polyaromatic hydrocarbons, thus;

$Ru(bpy)_3^{2+} \rightarrow Ru(bpy)_3^{3+} + e^-$	Electrochemical oxidation	(7)
$Ru(bpy)_3^{2+} + e^- \rightarrow Ru(bpy)_3^+$	Electrochemical reduction	(8)
$Ru(bpy)_3^{3+} + Ru(bpy)_3^+ \rightarrow Ru(bpy)_3^{2+}* + Ru(bpy)_3^{2+}$	Electron transfer	(9)
$Ru(bpy)_3^{2+}* \rightarrow Ru(bpy)_3^{2+} + h\nu$ ($\lambda_{max} = 620$ nm)	Chemiluminescence	(10)

The analytical usefulness of this reaction, stems mainly from that fact that the electrochemically generated $Ru(bpy)_3^{3+}$ species can be reduced by a large number of potential analyte compounds, or their electrochemical derivatives, via high-energy electron transfer reactions, to produce the $Ru(bpy)_3^{2+}*$ excited species, without the need for an electrochemical reduction step. The converse is also true. The reduction of peroxodisulfate ($S_2O_8^{2-}$) for example, in the presence of $Ru(bpy)_3^{2+}$, produces the $Ru(bpy)_3^{2+}*$ excited species and an ECL emission, from the reaction of $Ru(bpy)_3^+$ and $SO_4^{-\bullet}$ [20]. Although this latter system has been used for the determination of both $Ru(bpy)_3^{2+}$ [21] and $S_2O_8^{2-}$ [22], the vast majority of analytical applications use the co-oxidation route.

2.2.1   Determination of Oxalate and Other Organic Acid Salts

One of the first analytically useful ECL reactions of $Ru(bpy)_3^{2+}$ was documented by Chang et al. [23] for the simultaneous oxidation of $Ru(bpy)_3^{2+}$ and oxalate. The mechanism later proposed by Rubinstein and Bard [24] is as follows;

$Ru(bpy)_3^{2+} \rightarrow Ru(bpy)_3^{3+} + e^-$	Electrochemical oxidation	(7)
$Ru(bpy)_3^{3+} + C_2O_4^{2-} \rightarrow Ru(bpy)_3^{2+} + C_2O_4^{-\bullet}$		(11)
$C_2O_4^{-\bullet} \rightarrow CO_2^{-\bullet} + CO_2$		(12)

followed by either;

$Ru(bpy)_3^{3+} + CO_2^{-\bullet} \rightarrow Ru(bpy)_3^{2+}* + CO_2$	Electron transfer	(13)

or

$Ru(bpy)_3^{2+} + CO_2^{-\bullet} \rightarrow Ru(bpy)_3^+ + CO_2$	Electron transfer	(14)
$Ru(bpy)_3^{3+} + Ru(bpy)_3^+ \rightarrow Ru(bpy)_3^{2+}* + Ru(bpy)_3^{2+}$	Electron transfer	(9)

and finally,

$Ru(bpy)_3^{2+}* \rightarrow Ru(bpy)_3^{2+} + h\nu$	Chemiluminescence	(10)

Oxalate has been determined using this reaction in a diverse range of biological and industrial samples, including vegetable matter [25], urine, blood plasma [26], and alumina process liquors [27]. Other examples include ascorbic acid, which has been determined in soft drinks and fruit juices post HPLC separation by Chen and Sato [28], ECL in this case arising from the reaction of $Ru(bpy)_3^{3+}$ and a product of the electrochemical oxidation of ascorbic acid. Only a few organic acid salts are capable of being oxidized directly by $Ru(bpy)_3^{3+}$ in a single electron reaction to form a reducing radical anion. However, the reaction can be extended to related organic acid salts if a more powerful oxidizing agent than $Ru(bpy)_3^{3+}$ is used. For example, pyruvate has been determined by simultaneously oxidizing $Ru(bpy)_3^{2+}$ and cerium(III) nitrate in the presence of pyruvate in dilute sulfuric acid [29]. A detection limit of 3×10^{-7} M was achieved in synthetic samples. Ce(III) is oxidized to Ce(IV), which, unlike $Ru(bpy)_3^{3+}$, is able to oxidize the pyruvate to initiate an ECL reaction analogous to that for oxalate. The reaction proceeds as follows;

$$Ru(bpy)_3^{2+} \rightarrow Ru(bpy)_3^{3+} + e^- \qquad \text{Electrochemical oxidation} \qquad (7)$$

$$Ce^{3+} \rightarrow Ce^{4+} + e^- \qquad \text{Electrochemical oxidation} \qquad (15)$$

$$Ce^{4+} + CH_3COCO_2^- \rightarrow Ce^{3+} + CH_3COCO_2^{-\bullet} \qquad (16)$$

$$CH_3COCO_2^{-\bullet} \rightarrow CH_3CO^{\bullet} + CO_2 \qquad (17)$$

$$Ru(bpy)_3^{3+} + CH_3CO^{-\bullet} + H_2O$$

$$\rightarrow Ru(bpy)_3^{2+*} + CH_3CO_2H + H^+ \qquad \text{Electron transfer} \qquad (18)$$

$$Ru(bpy)_3^{2+*} \rightarrow Ru(bpy)_3^{2+} + h\nu \qquad \text{Chemiluminescence} \qquad (10)$$

2.2.2 Determination of Amines

The greatest exploitation of $Ru(bpy)_3^{2+}$ ECL has been in the determination of analytes with an amine functional group, with the highest sensitivities achieved for tertiary and secondary aliphatic amines. A broad range of potential analytes contain amine functionality including many pharmaceuticals, alkaloids, amino acids, and other biologically important molecules. Trialkylamines and closely related compounds are difficult to detect by other methods since they do not absorb well in the ultraviolet/visible region of the spectrum, and are extremely difficult to derivatize. Many amines can be determined by $Ru(bpy)_3^{2+}$ ECL since, like oxalate, they can be readily oxidized to form species that subsequently react to produce radical products, which in turn are capable of reducing $Ru(bpy)_3^{3+}$ back to $Ru(bpy)_3^{2+}$ in an excited state. The mechanism of this reaction has been investigated by many workers [16, 30] and is believed to be as indicated in Figure 2.

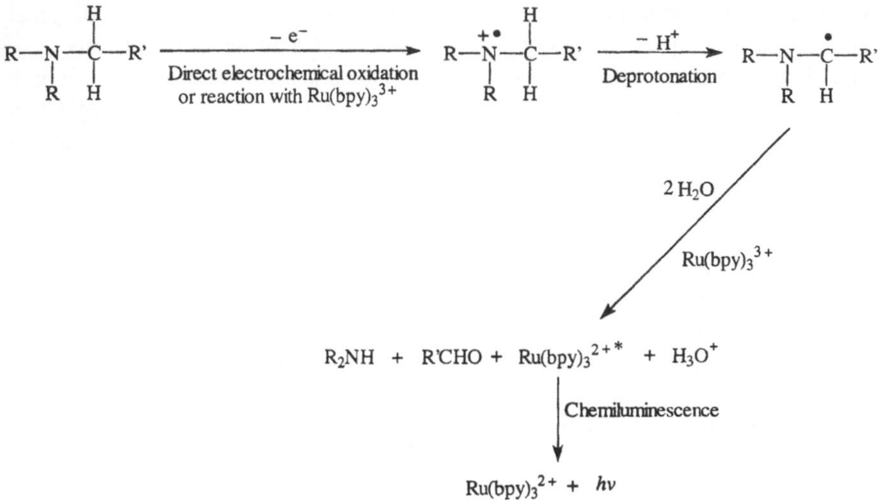

Figure 2   General reaction mechanism for the ECL reaction of $Ru(bpy)_3^{2+}$ with a tertiary amine.

Since the critical step in the reaction is the oxidation and subsequent deprotonation of the amine, the efficiency of ECL reactions of amines is generally very pH dependent. For example, the variation of ECL intensity with pH for tripropylamine in phosphate buffer is shown in Figure 3. The pH must be sufficiently high to promote the deprotonation reaction of the amine; however, since the maximum ECL signal is generally obtained below the pKa of the amine, it is likely that the acidity of the amine radical cation, and not the basicity of the amine, is most important in determining the pH dependence of ECL [30]. The optimum pH varies from compound to compound but is generally in the range pH 4–9. Higher pHs are generally to be avoided since the CL reaction of $Ru(bpy)_3^{3+}$ with hydroxide ions produces a significant background signal [16].

The greatest area of applications of this type of ECL has been in the analysis of pharmaceutical compounds with amine functionality. The reader is directed toward the previously mentioned review articles and Table 1 for further details [12, 14–16]. Many methods have also been successfully applied to real samples in the form of body fluids or pharmaceutical preparations, although sample pretreatment such as deproteinization, centrifugation, and neutralization followed by a chromatographic step to remove interfering species is often required. Limits of detection are typically in the range 10^{-9}–10^{-12} M. Figure 4 shows examples of some classes of pharmaceutical compounds that have been determined by $Ru(bpy)_3^{2+}$ ECL.

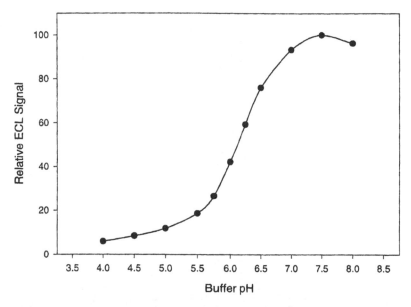

Figure 3 Variation of ECL intensity with pH for the ECL reaction of tripropylamine with $Ru(bpy)_3^{2+}$. (From Ref. 31.)

It is not the case, however, that all amine compounds take part in ECL reactions with $Ru(bpy)_3^{2+}$. Many amines produce intense ECL, while other closely related compounds produce virtually no ECL emission. While there are no strict rules governing ECL activity in amines, several workers have observed general trends that appear to explain some of these variations [31]. In general, ECL efficiencies increase in the order primary $<$ secondary $<$ tertiary amines, with tertiary amines having the lowest detection limits. Primary aliphatic amines have been determined by ECL, but most sensitively after prior derivatization with divinyl sulfone, $(CH_2CH)_2SO_2$, to form a cyclic tertiary amine [32]. It is also known that the nature of α-carbon substituents on the amine affects ECL activity. In general, electron-withdrawing substituents, such as carbonyl, halogen, or hydroxyl groups, tend to cause a reduction of ECL activity, probably by making the lone pair of electrons on the nitrogen less available for reaction, or by destabilizing the radical intermediate. Electron-donating groups such as alkyl chains have the opposite effect. Aromatic amines and those with a double bond that can conjugate the radical intermediate consistently give a very low ECL response. This may be due to resonance stabilization of the radical intermediate, rendering it less active toward $Ru(bpy)_3^{3+}$. Molecules that are hindered in attaining planar geometry after oxidation, such as quinine, show ECL intensity orders of magni-

Table 1 Survey Demonstrating the Range of Analytes Detectable by ECL, with Examples of Sample Matrices Used and Typical Limits of Detection

Class of compound	Selected examples	Samples analyzed	Typical LOD[a]
Ru(bpy)$_3$$^{2+}$ ECL reactions			
Ru(bpy)$_3$$^{2+}$ itself	Ru(bpy)$_3$$^{2+}$	Synthetic	10^{-13} mol L^{-1}
Organic acids	Oxalate	Synthetic	10^{-6} mol L^{-1}
		Urine	10^{-6} mol L^{-1}
		Blood plasma	10^{-6} mol L^{-1}
	Pyruvate	Synthetic	27 ppb
	Ascorbic acid	Fruit juice	2 pmol
Amines			
Trialkylamines	Tripropylamine	Synthetic	0.28 pmol
Alkaloids	Nicotine	Synthetic	0.4 pmol
	Atropine	Synthetic	1.5 pmol
	Codeine	Synthetic	10^{-11} mol L^{-1}
		Industrial process streams	5×10^{-9} mol L^{-1}
Pharmaceuticals	Thiazides	Pharmaceutical capsules	0.06 pmol (hydroflumethiazide), 2.1 pmol (cyclothiazide)
	Antihistamines	Pharmaceutical tablets, syrup/urine	9 pmol (brompheniramine)
	Tricyclic antidepressants	Synthetic	3×10^{-7} mol L^{-1} (amitriptyline)
	Local anesthetics	Pharmaceutical preparation	7×10^{-8} mol L^{-1} (lignocaine)
	β blockers	Synthetic	35 nmol L^{-1} (oxprenolol)
	Antibiotics	Urine/blood plasma	0.05 μmol L^{-1} (erythromycin)
		Pharmaceutical preparations	10 ppb (clindamycin)
Amino acids	D-, L-Amino acids	Synthetic	100 fmol–22 pmol (arginine, valine, leucine, proline)
	Tryptophan	Blood plasma	0.2 pmol
	Dansylated amino acids	Synthetic	0.1 μmol L^{-1} (glutamine)
	Primary amino acids	Synthetic	0.06 pmol (alanine)–1 pmol (serine) (derivatized with divinyl sulfone)
Other biochemicals	NADH	Synthetic	10^{-6} mol L^{-1}
	Glucose (enzymatically via NADH)	Synthetic	10^{-6} mol L^{-1}
	Thyrotropin (TSH)	Serum	<mIU L^{-1}
	DNA PCR products	Synthetic	10–200 amol
	HIV-1 viral RNA	Plasma	6×10^{2} RNA copies per 0.1 mL

Category	Analyte	Sample	Detection limit
Miscellaneous	Ethanol (enzymatically via NADH)	Synthetic	10 μmol L^{-1}
	Benzene (by enhancement of $Ru(bpy)_3^{2+}$/oxalate ECL)	Synthetic	μg L^{-1} level
	Hydrazine (gas phase)	Synthetic	0.035 μg L^{-1}
	Virulent bacteria	Food and environmental samples	1000 cells mL^{-1} (*Escherichia coli* O157, *Salmonella*, *Bacillus anthrax* spores)
Radical ion–annihilation ECL			
	Polyaromatic hydrocarbons	Synthetic	10^{-5}–10^{-7} mol L^{-1} (anthracene, rubrene, pyrene)
	Barban/DDT/DDE	Synthetic	10 ng mL^{-1}
Luminol ECL			
Luminol itself			
Luminol-labeled species	Luminol	Synthetic	1×10^{-10} mol L^{-1}
	Ibuprofen	Synthetic	0.15 pmol
	Histamine	Synthetic	1.5 pmol
Peroxides	Hydrogen peroxide	Synthetic	66 pmol
	Methyl linoleate hydroperoxide	Synthetic	0.3 nmol
Metal ion catalysts	Cobalt (II)	Synthetic	10^{-8} mol L^{-1}
	Copper (II)	Synthetic	10^{-7} mol L^{-1}
Biochemicals	Glucose (enzymatically via hydrogen peroxide)	Fruit juice/human serum	8 μg mL^{-1}
Miscellaneous	Sulfite (via inhibition of luminol ECL)	Synthetic	1 μmol L^{-1}
	Atrazine (enzymatic immunoassay)	Synthetic	<0.1 ppb
Cathodic luminescence			
Metal ions	Thallium (I)	Synthetic	10^{-10} mol L^{-1}
	Copper (II)	Synthetic	5×10^{-9} mol L^{-1}
	Terbium (III)	Synthetic	10^{-13} mol L^{-1}
Organic compounds	Salicylic acid	Synthetic	3×10^{-8} mol L^{-1}
	Diphenylanthracene	Synthetic	10^{-8} mol L^{-1}
Miscellaneous ECL reactions	Antioxidants	Synthetic	1 μmol^{-1} (α-tocopherol, β-carotene)
	Indole/tryptophan	Synthetic	1×10^{-7} mol L^{-1}
	Biodegradation products of explosives	Synthetic	<1 ppm

[a] As quoted in the original publication. The list has been compiled from Ref. 12, 15, and 16, but is not exhaustive. The reader is directed to these review articles for other examples and further experimental details of those analytes shown here.

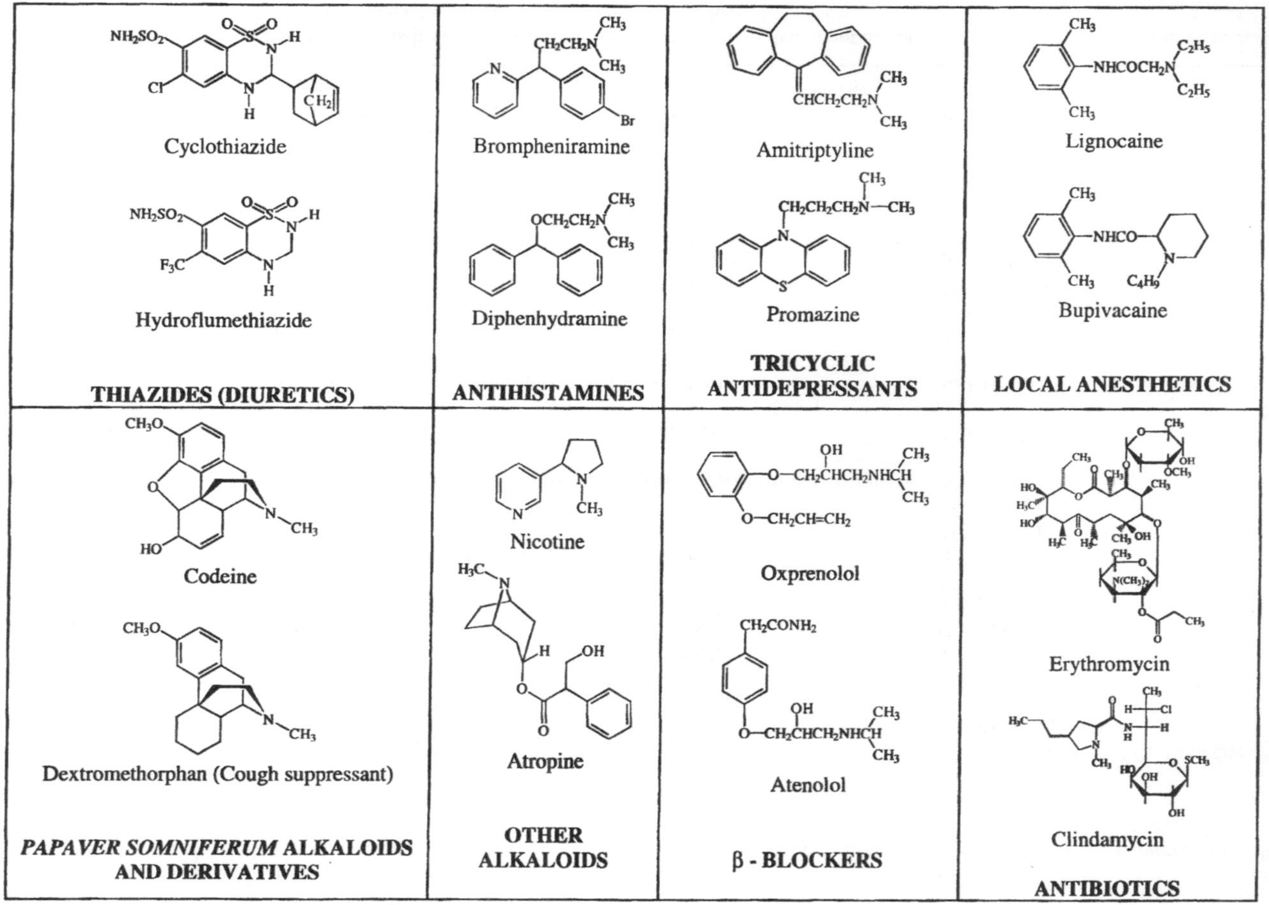

Figure 4 Examples of classes of pharmaceutical compounds with amine functionality, determined by Ru(bpy)$_3^{2+}$ ECL.

tude less than other amines that are able to tend toward trigonal planar geometry, resulting in effective delocalization of the charge. Finally amines that have other functional groups that are more readily oxidized than the nitrogen atom may take part in alternative, ultimately nonchemiluminescent reactions.

2.2.3 Determination of Amino Acids

The determination of amino acids by ECL is seen as having many potential applications, including protein sequencing, and has been investigated extensively by Bobbitt et al. [33, 34] among others [35]. Most amino acids have no inherent chromophore or fluorophore, and hence usually have to be derivatized prior to conventional spectroscopic determination. However, amino acids encompass a diverse group of compounds, with a correspondingly diverse range of ECL activities. Proline, a secondary aliphatic amine, has been determined with the highest sensitivity by $Ru(bpy)_3^{2+}$ ECL, with a detection limit of 100 fmol M [34]. Many other amino acids have been determined in the range 10^{-6}–10^{-9} M. Again, in general, amino acids with electron-withdrawing α-carbon substituents and aromatic amines give lower ECL responses. Amino acids containing indole or imadazole functional groups, however, generally give higher ECL responses. For example, Uchikura and Kirisawa [36] determined D- and L-tryptophan enantiomers in plasma by HPLC-ECL, with a limit of detection of 0.2×10^{-12} M. Interestingly, their method is selective for tryptophan under acidic conditions, approximately pH 3, where interference from other amino acids and amines is diminished.

Enhancements in the sensitivity with which amino acids containing a primary amine group can be determined have been achieved by derivatization. Chen and Sato [37] reported derivatization with divinyl-sulfone-reduced limits of detection by several orders of magnitude, while Lee and Nieman [38] reported derivatization with dansyl-chloride-reduced limits of detection by a factor of three.

2.2.4 Other $Ru(bpy)_3^{2+}$ Analytical Applications

NADH, containing a tertiary amine functional group, has been readily determined by $Ru(bpy)_3^{2+}$ ECL. However the oxidized form, NAD^+, containing an aromatic secondary amine group produces virtually no ECL signal. This had led to a variety of indirect enzymic methods of analysis, where the activity of the enzyme results in the conversions between NAD^+ and NADH. These are discussed in Sec. 8.

$Ru(bpy)_3^{2+}$ itself can be determined with great sensitivity in an excess of an amine to subpicomolar levels [39]. This has led to the development of electrochemiluminescent labels based on $Ru(bpy)_3^{2+}$ derivatives that have found successful applications in ECL immunoassay and DNA probe analysis. These are discussed in Sec. 9.

Recently ECL reactions of $Ru(bpy)_3^{2+}$ with other reducing agents have been documented, such as various β-diketone and some methylene compounds that have cyano and carbonyl groups [40]. Hence with further research, analytical applications should arise for many classes of compounds other than amines that can act as reductants, or electrochemical precursors of reductants, capable of reacting with $Ru(bpy)_3^{3+}$ to produce ECL.

2.2.5   Methods of Generation of $Ru(bpy)_3^{3+}$

The various approaches to the generation of the active ECL reagent $Ru(bpy)_3^{3+}$ have been reviewed by both Lee [14] and Gerardi et al. [16]. Methods of generation include purely chemical, photochemical, external electrochemical, and in situ electrochemical approaches.

Chemical oxidants such as lead dioxide and cerium(IV) are commonly used to produce $Ru(bpy)_3^{3+}$ from $Ru(bpy)_3^{2+}$, and such methods are rapid and relatively simple. $Ru(bpy)_3^{3+}$ can be prepared in bulk, offline, using a batch procedure, or online, using flow methods, the latter either by combining a stream of $Ru(bpy)_3^{2+}$ with that of an oxidant at a mixing T, or by passing $Ru(bpy)_3^{2+}$ over an immobilized oxidant, such as lead dioxide on silica gel. However, the majority of applications use electrochemical means to produce $Ru(bpy)_3^{3+}$, owing to problems of maintaining a stable and reproducible supply of reagent.

External electrochemical generation of $Ru(bpy)_3^{3+}$ involves the bulk electrolysis of a reservoir of acidified $Ru(bpy)_3^{2+}$, at about 1.1–1.3 V (vs. Ag/AgCl), over a period of typically 30–45 min, during which time the solution will change color from orange to green. Working electrodes are fabricated from materials with a large surface area, such as platinum gauze or glassy carbon sponge, to maximize conversion. The counterelectrode is usually kept separate from the bulk solution to prevent reduction and electrodeposition of ruthenium at this electrode. Owing to the instability of the reagent, solutions of $Ru(bpy)_3^{3+}$ are often maintained at around 0°C. This method of reagent production is time consuming, and the reagent, once formed, has a limited lifetime. Hence Uchikura developed a flow-through, in-line, electrochemical reactor for continuous $Ru(bpy)_3^{3+}$ generation [16]. This device employs a working electrode consisting in part of a porous plug of tightly packed glassy carbon particles though which acidified $Ru(bpy)_3^{2+}$ is pumped. More commonly in situ generation is used, producing both $Ru(bpy)_3^{3+}$ and the CL reaction with the analyte, together at the electrode surface. $Ru(bpy)_3^{2+}$ may be in the bulk solution or immobilized at the electrode surface.

These different approaches to reagent generation have recently been compared by Lee and Nieman, for analytical determinations of a range of analytes using flowing streams [41]. The external generation mode gave the most intense emissions and highest sensitivity; however, working curves had poor linearity and ECL intensities were heavily dependent on experimental variables such as flow rate. Using in situ immobilized reagents produced the widest linear dynamic ranges and consumed the least amount of reagent, but gave the lowest sensitivity. In situ generation from solution-phase reagents proved the most satisfactory, in terms of convenience, rapidity, and reproducibility, and it is from this method that many of the advantages of ECL, discussed in Sec. 1.2, arise.

2.3 Electrochemical Generation of Conventional CL

Many conventional CL reactions can be initiated electrochemically. However, the most studied and exploited reaction has been that of luminol due to its versatility in analytical determinations. The mechanism of luminol ECL is thought to be similar to that of its chemiluminescence [42] and has been investigated in detail by Haapakka and Kankare [43]. The generally accepted mechanism is shown in Figure 5. In alkaline solution the luminol anion undergoes a single electron electro-oxidation to form a diazaquinone, which is further oxidized by peroxide or superoxide to give 3-aminophthalate in an excited state, which emits

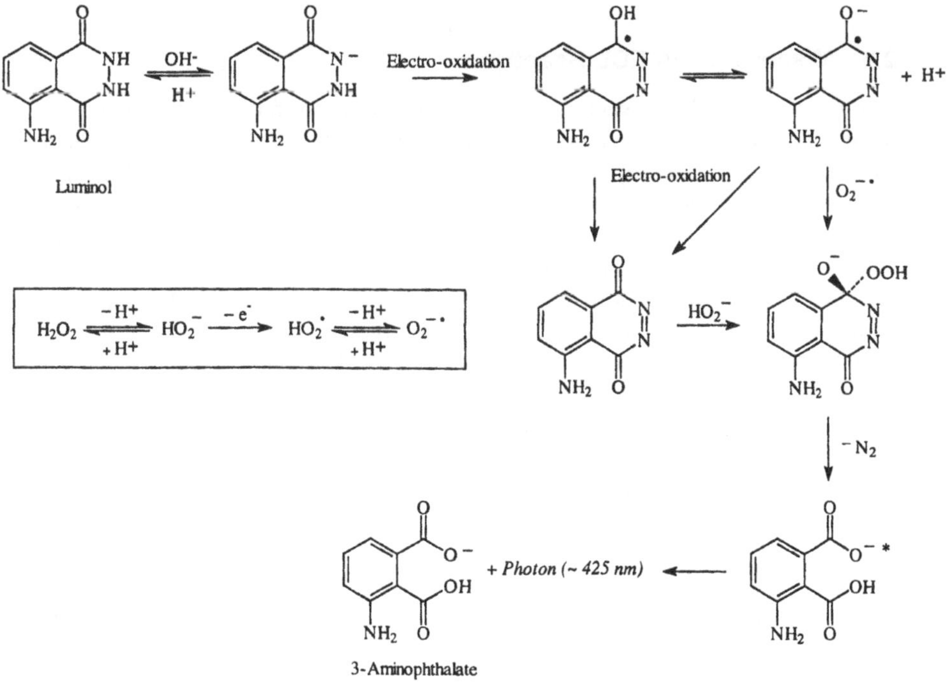

Figure 5 Proposed mechanism for the ECL reaction of luminol. (Reprinted from Ref. 15, with permission from, and copyright, Elsevier Science.)

light at 425 nm. Luminol ECL has been used for a diverse range of analytical applications [12, 15]. These have included; the determination of luminol or species such as phenylalanine, ibuprofen, and hisidine labeled with luminol derivatives; hydrogen and other peroxides; and biochemical analytes that are substrates for enzymic reactions that produce hydrogen peroxide, such as glucose with glucose oxidase. Luminol ECL can also be used to indirectly determine species that either catalyze the reaction, such as the transition metal ions cobalt(II), copper(II), and nickel(II), or species that inhibit the reaction, such as sulfite. Analytical ECL applications of other conventional CL reactions have been briefly investigated. These include; lucigenin for trace metals, *bis*(2,4,6-trichlorophenyl) peroxyoxalate (TCPO)/fluorescer for trace O_2, and acridinium ester–labeled compounds [12].

Methods of analysis based on electrochemical generation of conventional CL reported in the literature have declined in recent years since it is now generally recognized that where a conventional CL method exists, the added complications in methodology and instrumentation needed to produce an ECL system outweigh the potential advantages. Also in most cases, as with luminol, the CL reagents cannot be electrochemically regenerated.

2.4 Miscellaneous ECL Reactions

A variety of other ECL reactions are known that do not fall into the categories mentioned thus far, and some of these have found analytical applications. Such CL reactions can generally only be readily produced by electrochemical means, and often exhibit only very low ECL intensities, resulting in relatively high limits of detection. Examples include the determination by direct electrolysis of indole and tryptophan in the presence of hydrogen peroxide, and saccharides and alcohols that have neighboring hydroxyl groups; the determination of 2,4- and 3,4-diaminotoluene, biodegradation products of nitrotoluene explosives, which form weakly electrochemiluminescent complexes with gold(I) and copper(II) ions; the determination of antioxidants that inhibit the ultraweak anthracene sensitized ECL from the anodic oxidation of sodium citrate, methanol, and O_2; and the determination of yttrium(III) and silver(I) ions that catalyze the ECL reaction of certain phthalazine derivatives [12, 15].

2.5 Cathodic Luminescence

The mechanism of cathodic luminescence is distinctly different from other ECL systems. Light is emitted from oxide-covered, so-called valve metal, electrodes, namely aluminium and tantalum, during the reduction of peroxodisulfate, hydrogen peroxide, or oxygen, in aqueous solution, at relatively low potentials (<10 V). The mechanism involving persulfate, for example, is as follows. A conductance band electron is transferred to the peroxodisulfate ion ($S_2O_8^{2-}$) resulting in the formation of a sulfate radical ($SO_4^{-\cdot}$). If its standard reduction potential matches the valence band edge of the semiconductor, it captures an electron from the valence band thus injecting a hole in this band. The recombination of an electron from the conduction band with the valence band hole produces light emission termed band-gap electroluminescence. However, cathodic luminescence having energy less than the semiconductor band-gap energy is often observed caused by recombination via surface states, i.e., energy levels localized at the electrode surface. Many species from inorganic ions to organic compounds can form attachments with the hydroxylated oxide surface of the electrode, and can hence act as surface states to enhance this sub-band-gap ECL, which occurs at 500–700 nm [44].

Haapakka and Kankare have studied this phenomenon and used it to determine various analytes that are active at the electrode surface [44–46]. Some metal ions have been shown to catalyze ECL at oxide-covered aluminum electrodes during the reduction of hydrogen peroxide in particular. These include mercury(I), mercury(II), copper(II), silver(I), and thallium(I), the latter determined to a detection limit of $<10^{-10}$M. The emission is enhanced by organic compounds that are themselves fluorescent or that form fluorescent chelates with the aluminum ion. Both salicylic acid and micelle solubilized polyaromatic hydrocarbons have been determined in this way to a limit of detection in the order of 10^{-8}M.

3. ECL INSTRUMENTATION

Unlike conventional CL, no readily available commercial instrumentation exists for developing and utilizing methods of ECL analysis, and this is still a major drawback of the technique. One relatively simple option is to convert existing laboratory instrumentation. For example, by enclosing electrochemical apparatus in a light-tight box and incorporating a light detector, or by placing an electrochemical cell within a spectrophotometer in which the excitation light source has been disabled. However, in most cases researchers have opted to build their own instrumentation. ECL detectors need to be light-tight, yet still allow for easy sample introduction and removal; incorporate working, counter, and reference electrodes and a suitable phototransducer with a geometry such that light is not generated on an electrode surface obscured to the light detector; and be robust yet relatively easy to dismantle for electrode cleaning.

3.1 Configuration and Nature of the Electrodes

Three main electrode configurations have been used in ECL work: rotating ring-disk (RRD), dual and single electrodes. With the RRD electrode, the disk may, for example, be set at a potential to produce the oxidized species and the ring set to produce a reduced species. The rotation of the disk sweeps the oxidized species out to the ring to react with the reduced species on the face of the ring [47]. This system is useful in the study of ECL reaction kinetics by varying the rotation rate; however, it does not readily lend itself to flow-though systems, which are used for most analytical applications. Dual-electrode configurations use direct-current potentials such that each species needed for the ECL reaction is produced continually at separate electrodes. For stable species, one electrode can be placed downstream of the other. Thus ECL is produced when species formed at the first electrode are transported downstream to react with species formed at the second electrode. In the case of relatively unstable intermediates, the electrodes are placed in close proximity to each other, and species formed at each electrode diffuse and react together in the small interelectrode gap. Two configurations that have been used are interdigitated electrodes and two plate electrodes in thin-layer geometry placed approximately 100 μm apart. In the case of the latter, at least one electrode is transparent. Tin oxide on glass has been used, although lack of robustness is a problem [48]. The simplest, most widely adopted solution is to use a single electrode, and where more than one species needs to be electrochemically produced for the ECL reaction, an alternating potential is applied switching between the redox potentials of the particular species desired.

Various materials have been used for the working electrode where the ECL reaction takes place. These include platinum, gold, glassy carbon, and carbon paste. No one particular electrode material is suitable for a specific ECL reaction and similar electrodes have shown marked differences in sensitivity in the hands of different workers. In each case the condition of the electrode surface has a marked effect on the ECL signal, and the intensity and reproducibility of ECL measurements often fall as the surface becomes fouled during continuous use. Relatively little is known about these fouling processes and various approaches are used to minimize the effect, including repolishing, chemical treatment, and redox cycling [34, 49]. After regeneration of the electrode surface, electrodes are generally stable for approximately 40 h of continuous use.

3.2 ECL Flow Cells

Although the earliest ECL work was carried out in simple batch cells, the need for a rapid throughput of samples, and reproducible sampling and mixing with reagents, has led to the almost universal adoption of ECL flow cells. Most are developments of CL laminar flow cells, where the test solution flows in a thin layer, sandwiched between a glass observation window and the working electrode surface, the volume and shape of the cell being defined by an inert spacer. An example devised by Jackson and Bobbitt [34] is shown in Figure 6. Generally

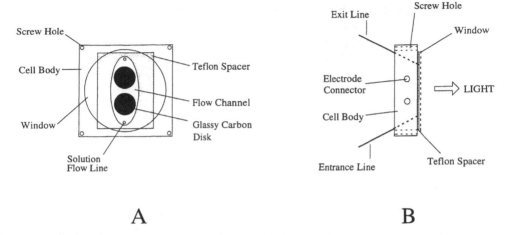

Figure 6 Schematic diagram of an ECL flow cell developed by Jackson and Bobbitt. (A) Front view; (B) Side view. (Reprinted from Ref. 34, with permission from, and copyright, Elsevier Science.)

the counterelectrode is placed downstream of the working electrode, such that any species that may form at the counterelectrode cannot interfere with the ECL reaction. The counterelectrode can be within the main body of the cell or, more commonly, made of a short section of stainless steel tubing forming part of the outflow line from the cell. Standard reference electrodes are used, either incorporated within the body of the cell or positioned close to the output line of the cell. However, pseudoreference electrodes, such as plain silver or platinum metal, are often used. Such electrodes have been shown to provide a stable reference, have the advantages of being simple in construction and easy to refresh, produce minimum sample contamination, and are compatible with both organic and aqueous electrolyte systems.

Standard commercial potentiostats are usually used, although since some ECL methods require rapid switching of potentials and the application of complex waveforms, a flexible computer-controlled potentiostat is preferable. To achieve useful limits of detection with ECL a photomultiplier tube (PMT) is most commonly used as the light detector, although photodiodes have also been successfully employed [50]. Photodiode structures have the advantage that they can become an intrinsic part of the flow cell itself, have peak sensitivity at the red end of the visible spectrum where the ruthenium metal complexes emit, and are small, robust, cheap, and run from a low-voltage source. Wavelength discrimination is rarely used and all light emitted is recorded in the analytical signal. Hence monochromators are not usually a component of ECL devices.

A novel flow cell for the analysis of gaseous analytes has recently been developed by Collins and Rose-Pehrsson [51]. The device incorporates an electrochemical cell containing a $Ru(bpy)_3^{2+}$ solution contained beneath a Teflon diffusion membrane, over which air is continually sampled. $Ru(bpy)_3^{3+}$ is generated in situ at a platinum gauze working electrode. The gaseous analyte flowing through the cell passes through the membrane to react with the $Ru(bpy)_3^{3+}$, generating an ECL emission in view of a PMT. The cell was used for determination of gas-phase hydrazine and derivatives to low ppb levels. Later designs have used a gold-coated cellulose membrane for the working electrode separating the $Ru(bpy)_3^{2+}$ solution on one side from the gas flow on the other [52].

3.3 ECL Probes

Recently various ECL probes have been developed. Early designs had a small reservoir of reagents, incorporating the electrodes, facing a fiberoptic bundle to carry the light to a PMT. In a novel design by Kuhn et al. [53] a gold coated optical fiber was polished flat such that the gold formed a micro-ring electrode surrounding the optical fiber. In each case the probes had to be used within a light-tight enclosure. Preston and Nieman [54], however, developed a probe that does not require a dark box, and is similar in construction and operation to a pH probe. The probe contains working, counter, and reference electrodes, an optical fiber, and baffled channels to admit the test solution, while at the same time shielding the optical transducer from ambient light. This device has been successfully used with both $Ru(bpy)_3^{2+}$ ECL and luminol ECL coupled with an immobilized oxidase enzyme.

4. REAGENT IMMOBILIZATION AND ECL SENSORS

The ability to immobilize ECL reagents on the electrode surface and thus produce sensors is seen as being very advantageous, especially if the activity of the reagent can be electrochemically regenerated following the ECL reaction. This is because the need to continually deliver reagent to the reaction cell is removed, simplifying both the instrumentation and methodology, and dramatically reducing reagent consumption. In ECL this has been achieved with limited success for $Ru(bpy)_3^{2+}$. The most favored approach has been to use the perfluorinated, sulfonated cation exchange polymer Nafion, to which $Ru(bpy)_3^{2+}$ electrostatically binds. Nafion has been widely used for modifying electrodes owing to its high chemical, mechanical, and thermal stability, and selectivity for large cations such as $Ru(bpy)_3^{2+}$, while excluding potential anionic interferences such as OH^-. Electrodes can be simply prepared by brief immersion into a solution of a few percent Nafion dissolved in alcohol and evaporating to dryness, followed by immersion for approximately 30 min in an acidic solution of 5 mM $Ru(bpy)_3^{2+}$. Electrodes prepared in this manner take on a deep-orange color since the polymer film takes up significant amounts of the cation. Rubinstein and Bard first investigated $Ru(bpy)_3^{2+}$ immobilized in this way [55], and noted that charge transfer through the polymer film partly results from diffusion of the electroactive species [14]. This approach has then been subsequently exploited by Downey and Nieman for the determination of oxalate, alkylamines, and NADH [56].

However, there are significant limitations. Sensors made in this may have been universally observed to have limited long-term stability, in the worst cases losing up to 85% sensitivity when stored overnight in buffer, and total loss of activity if allowed to dry out between measurements [56]. The decrease is speculated to be due to diffusion of $Ru(bpy)_3^{2+}$ into hydrophobic regions of the Nafion film restricting charge transport, rather than desorption of $Ru(bpy)_3^{2+}$.

Another approach, developed by Egashira et al. [57], used a carbon paste electrode. Graphite powder was blended with 2% w/w bis(2,2'-bipyridine)-(4,4'-dinonadecyl-2,2'-bipyridine) ruthenium(II), mineral oil, and a solvent. After thorough mixing the solvent was evaporated and the resulting paste packed into a glass tube. The electrode, as part of a fiberoptic ECL sensor, was used for the determination of oxalate ions, toward which selectivity was enhanced by the hydrophobic environment produced by the long alkyl chains of the ruthenium complex. However, the ECL response rapidly decayed under continuous operation for only 10 min, attributed to the degradation of the complex, and the electrode material had to be removed from the tube, blended, and repacked before being reused. Electropolymerization of monomers based on $Ru(bpy)_3^{2+}$ has also been carried out, and the resulting polymers show ECL activity. For example, Abruña and Bard demonstrated a chemiluminescent polymer based on $tris$(4-vinyl-4'-methyl-2,2'-bipyridyl) ruthenium(II) [58]. However, once again the analytical potential of such systems is limited by instability, the ECL emission persisting for only 20 min of continuous operation.

5. ECL IN FLOW INJECTION ANALYSIS

Flow injection analysis (FIA) has been widely adopted in ECL, although primarily for analytes in simple matrices such as synthetic mixtures or pharmaceutical preparations that have few potential interfering species [15, 16]. Flow cells incorporating a working electrode, as described in Sec. 3.2, are employed for methods using in situ generation of ECL reagents, and more conventional CL flow cells are used for methods that employ external electrochemical generation of ECL reagents such as $Ru(bpy)_3^{3+}$. Figure 7 shows an example of the FIA calibration and analysis of the diuretic pharmaceutical, hydrochlorothiazide, in tablet form, using external electrochemical generation of $Ru(bpy)_3^{3+}$ [59]. For in situ genera-

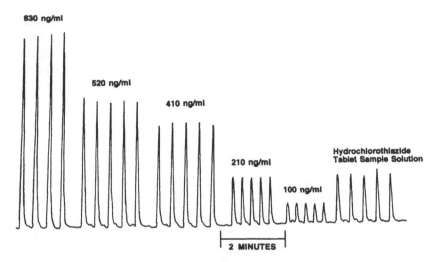

Figure 7   Flow injection peaks of hydrochorothiazide standards and samples, showing reproducibility. (Reprinted from Ref. 59, with permission from, and copyright, Elsevier Science.)

tion of ECL, the electrode is usually permanently charged and the injected sample produces light as it passes over the electrode, which is recorded as a familiar FIA peak response.

Consideration should be given to the flow rate of the sample through the detection cell. Shultz and co-workers have demonstrated the wide variability in reaction kinetics between ECL reactions, and hence the influence of flow rate on ECL intensity [60]. For example, the rate constants (k) of the $Ru(bpy)_3^{2+}$ ECL reactions of oxalate, tripropylamine, and proline were calculated to be 1.482, 0.071, and 0.011/s, respectively. Maximum ECL emission was obtained at low linear velocities for slow reactions ranging up to high linear velocities for fast reactions. That is, the flow rate and flow cell volume should be optimized such that the light-emitting species produced is still resident within the flow cell, in view of the light detector, when emission occurs.

6. ECL IN LIQUID CHROMATOGRAPHY

Increasingly ECL detection is being coupled with chromatography, to allow the determination of analytes in more complex matrices, such as blood plasma or foodstuffs, and samples containing more than one ECL active compound of interest [15, 16]. However, such methods are complex to optimize, as ideally a medium should be found that suits not only the CL and electrochemical reactions, but also the chromatographic separation. Two basic strategies are evident that have been contrasted by Lee for Ru(bpy)$_3^{2+}$ ECL [14]. These are the postcolumn mixing method where the ECL reagents are added after the separation usually at a mixing T, or the addition of the ECL reagents directly to the mobile phase. The postcolumn mixing method is useful where the pH or other solution conditions for ECL are different from those of the separation, as both ECL reagents and mobile-phase modifiers can be added at this stage. However, additional instrumentation such as an extra pump may be needed, and this method causes

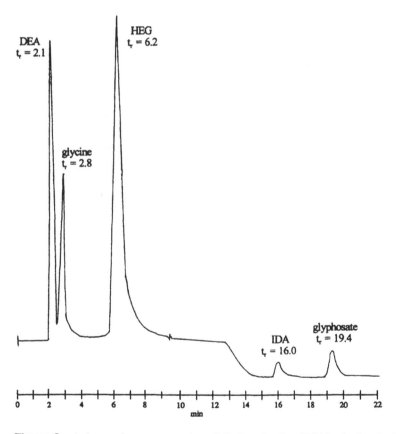

Figure 8 Anion exchange separation of diethanolamine (DEA), glycine, hydroxyethyl glycine (HEG), iminodiacetic acid (IDA), and glyphosate followed by ECL detection. Mobile phase consisting of 0.01 mM Ru(bpy)$_3^{2+}$ in 10% acetonitrile, 90% 0.01 M phosphate buffer at pH 9.8. (Reprinted from Ref. 61, with permission from, and copyright, Elsevier Science.)

further sample dilution and band broadening. Nevertheless this is currently the most popular method for LC-ECL and is the simpler to optimize. The addition of ECL reagents to the mobile phase is a simpler method in principle, provided the separation and ECL reaction can occur in the same mobile phase, the ECL reagents are not strongly bound by the column, and the efficiency of the separation and retention times are not adversely affected. The flow cells previously described are commonly used for post-LC-ECL detection, although cell volumes may need to be reduced.

Ru(bpy)$_3^{2+}$ ECL has been the most successfully coupled to chromatography as the reactions occur in a wide range of organic and aqueous solutions, and over a wide pH range. Indeed Ru(bpy)$_3^{2+}$ ECL is often enhanced by the presence of an organic modifier such as acetonitrile or methanol. Figure 8 shows an example of an HPLC separation with ECL detection of a mixture of glyphosate and related compounds by Ridlen et al. [61]. A novel indirect detection method for liquid chromatography based on the ECL of luminol has been reported by Wang and Yeung [62]. The mobile phase used contained luminol, hydrogen peroxide, and cobalt(II), which produces a steady ECL emission on passing through a postcolumn ECL flow cell. When an analyte is eluted, it displaces components of the ECL reaction thus decreasing the ECL emission and producing a negative peak. In principle a vast range of non-ECL-active analytes can be determined by this method, crucially those for which conventional ultraviolet absorbance is insensitive.

7. ECL IN CAPILLARY ELECTROPHORESIS

Over the past decade capillary electrophoresis (CE) has proved a rapid, efficient, and versatile method of separating analytes in extremely small sample volumes, with particular suitability for biochemical analysis. During this period several CE methods with chemiluminescence detection have been documented based on acridinium esters, luminol, peroxyoxalate, and firefly luciferase using a range of different apparatus. These have been reviewed by García Campaña et al. [63]. More recently ECL detection has been coupled with CE. The flow cells previously described for FIA and LC analysis are, however, not ideally suitable for coupling to CE systems since their internal volumes would lead to excessive band broadening of the nanoliter volume analyte peaks. Several slightly different approaches to the development of a CE-ECL system have been reported for use with both Ru(bpy)$_3^{2+}$ [64–67] and luminol [68] ECL. An example of such a system is shown in Figure 9. In general, the working electrode is either a fine platinum wire or carbon fiber, which is placed very close to the end of the CE capillary. The end of the capillary and working electrode are placed within a reservoir holding a few milliliters of buffer forming the ECL cell, which is housed

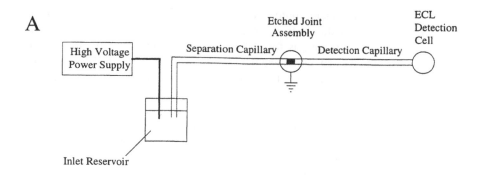

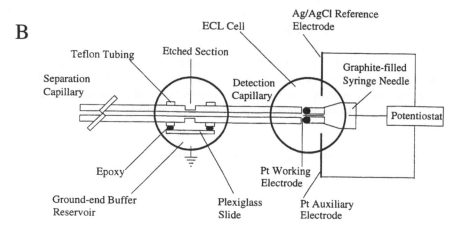

Figure 9 Schematic diagram for capillary electrophoresis with postcapillary ECL detection. (A) Overview of the apparatus. (B) Enlarged view of the etched joint and ECL detection cell. The PMT (not shown) is positioned directly over the Pt working electrode, and the entire apparatus placed inside a light-tight housing. (From Ref. 64.)

within a light-tight enclosure. Other conventional reference and counterelectrodes are also placed within the reservoir. There is obviously a need to isolate the high voltage used to drive the CE separation from the capillary output where the low-voltage ECL detection occurs. To do this it is necessary to ground the capillary near the outlet end. This is achieved by making a section of the capillary porous, by etching or cracking, and encapsulating it in a buffer-filled reservoir that is connected to ground. The high-voltage circuit is then between the sample injection end of the capillary and this ground buffer reservoir. Bulk flow of the liquid then carries the separated analytes onto the "ECL cell."

There are three main approaches to the introduction of the ECL reagents for CE. First is the addition of all necessary reagents to the eluent. This is the preferred method; however, as for HPLC, solution conditions have to be optimized to suit both the separation and the ECL reaction. Second, the ECL reagents may be mixed online with the CE stream at a mixing junction. However, this is not ideal, again owing to band broadening of the analyte peaks. Finally, the simplest solution is to add the necessary reagents to the ECL cell reservoir. The small volume of analytes yields a low ECL signal such that sensitive detectors (i.e., cooled PMTs) or photon-counting methods are often used. The geometry of the electrode arrangement is also far from ideal for efficient light collection, so parabolic mirrors are often employed for maximum light collection. Fiber optic bundles placed either side of the electrode have also been used, which carry the light to the phototransducer external to the cell. In addition, the buildup of analytes, reagents, and eluent in the ECL detection cell causes a significant background signal. The consequence of these factors is that only relatively high limits of detection are currently achievable of $10^{-5}-10^{-7}$ M. However, since the sample volumes are generally so small, these can equate to mass detection limits of attomoles. Nevertheless with ECL moving increasingly toward biochemical applications, the design and construction of more robust and efficient CE-ECL detection cells should allow greater exploitation of CE in ECL analysis.

8. ENZYME-COUPLED ECL FOR BIOSENSING

Enzyme-coupled ECL enables the selective determination of many clinical analytes that are not in themselves directly electrochemiluminescent, but that can act as substrates for a variety of enzymic reactions. There are two general strategies for ECL: the use of dehydrogenase enzymes, which convert NAD^+ to NADH, and oxidase enzymes, which produce hydrogen peroxide.

As can be seen in Figure 10, NADH contains a tertiary amine functional group, and has been determined by $Ru(bpy)_3^{2+}$ ECL to µM levels [14, 69]. However, in NAD^+ this group has been converted to an aromatic secondary amine, which gives virtually no ECL response. (Similarly it has been noted that NADPH and $NADP^+$ show the same contrasting ECL activity.) This difference has been exploited for the determination of analytes such as glucose and ethanol, for which oxidation by the appropriate dehydrogenase enzyme is accompanied by the conversion of added NAD^+ to NADH, which is subsequently detected by $Ru(bpy)_3^{2+}$ ECL. The ECL reaction of NADH and $Ru(bpy)_3^{3+}$ also regenerates NAD^+ and $Ru(bpy)_3^{2+}$. Similarly analytes that take part in enzymic reactions in which NADH is oxidized can also be detected, the ECL intensity being inversely proportional to the concentration of analyte. Bicarbonate has been determined in this way [70].

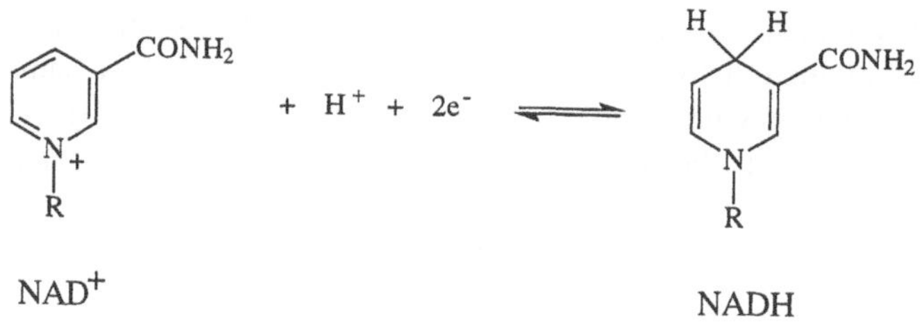

Figure 10 Structures of NAD$^+$ and NADH showing the part of the molecule where reversible reduction occurs, changing the ECL reactivity.

Hydrogen peroxide produced as a result of reactions of oxidase enzymes with analyte substrates can be sensitively determined, both directly by luminol ECL and indirectly by Ru(bpy)$_3^{2+}$ ECL. For the latter, hydrogen peroxide is detected on the basis of its ability to diminish the ECL reaction between Ru(bpy)$_3^{2+}$ and added oxalate, by reacting with, and depleting the concentration of, oxalate. Thus ECL intensity is inversely proportional to the concentration of analyte. This principle has been used, for example, to determine cholesterol [70].

In general, enzyme-coupled ECL produces detection limits in the 10^{-6}–10^{-8} M range, which are within or below the clinical range for typical analytes. Calibrations are often nonlinear since enzyme kinetics, ECL reactivity, and mass transfer effects, when immobilized reagents are used, all affect the sensitivity of the method. Ru(bpy)$_3^{2+}$ ECL is compatible with the optimum pH for many enzyme reactions, whereas luminol ECL occurs at a much higher pH. However, in biologically derived samples, many species interfere with Ru(bpy)$_3^{2+}$ ECL reactions, either constructively, such as oxalate, NADH, and other amines, or destructively, such as vitamin C, uric acid, and paracetamol. Hence sample pretreatment, the use of chromatographic separation, or dialysis through a membrane may be required.

In many FIA methods an inline immobilized enzyme reactor may be used. Alternatively the enzyme may be immobilized at the electrode surface. Wilson et al. [71] covalently attached glucose oxidase to an aminosilanized indium tin oxide (ITO)-coated glass wafer, which was used as the working electrode in a fountain flow cell. Light was recorded at the back of the transparent electrode. The cell was used for the luminol ECL determination of glucose to a limit of detection of 0.4 mM. Martin and Nieman [72] have used polymer entrapment to coimmobilize both an enzyme and Ru(bpy)$_3^{2+}$ to form a range of biosensors. Coimmobilization of both these components in the same film led to a much reduced enzyme activity; thus various multilayer configurations were tried. Ru(bpy)$_3^{2+}$ was absorbed into Nafion, while the enzyme was immobilized using a water-soluble Eastman AQ polymer. Stacking the two layers on top of each other gave the greatest sensitivity; however, the arrangement lacked stability and robustness. Preferred orientations were using adjacent layers such that the samples flowed first over the enzyme polymer and then over $Ru(bpy)_3^{2+}$-polymer-coated electrode, or placing the $Ru(bpy)_3^{2+}$-polymer-coated electrode and enzyme-polymer on opposite sides of a thin flow channel.

9. ECL IMMUNOASSAY AND DNA PROBES

The use of ECL detection for immunoassays has several advantages. No radioactive isotopes are used, thus reducing the problems of storage, handling, and disposal of samples and reagents. The labels are very stable with shelf lives in excess of 1 year at room temperature. Calibrations can be performed over a wide linear dynamic range, extending over up to six orders of magnitude in some cases. In addition, ECL labels based on $Ru(bpy)_3^{2+}$ can be very sensitively detected using an excess of an amine coreactant, and such labels are small allowing multiple labeling of an analyte without affecting its immunoreactivity [14–16].

The potential of ECL detection for immunoassays has led to the development of automated commercial instrumentation, first by IGEN and subsequently by Perkin-Elmer and Boehringer Mannheim GmbH. The latest commercial instruments use the following immunomagnetic method. Small magnetic beads are supplied coated with streptavidin to which biotin molecules are attached, in turn bound to a selected antibody, antigen, or DNA probe. These beads are combined with the analyte that binds to the immobilized group on the bead. Next $Ru(bpy)_3^{2+}$-labeled antibodies or antigens, specific for the analyte, are added that bind to the analyte attached to the bead. In a flow-through system the beads are magnetically captured onto the working electrode, and the residual sample matrix and labels swept clear. ECL is then produced from the $Ru(bpy)_3^{2+}$-based labels by applying a potential to the working electrode in the presence of an excess of tripropylamine. Rapid assays have been developed for a range of applications, including the diagnosis of thyroid diseases, tumors, infectious diseases, anemia, cardiological function, pregnancy, and sex function [73]. Figure 11 shows the basic principle of this technology.

Yu and Stopa have recently used an ECL immunoassay for the detection of virulent pathogenic bacteria in a range of environmental, food, and other biological samples, using $Ru(bpy)_3^{2+}$-labeled antibodies [74]. Traditional methods for bacterial detection and identification are laborious and time consuming, typically taking 24–48 h to complete. The ECL method, however, was shown to detect *Bacillus anthrax* spores, *Escherichia coli* O157:H7, and *Salmonella typhimurium* to a limit of detection of 1000 cells/mL, in under 1 h, including sample preparation.

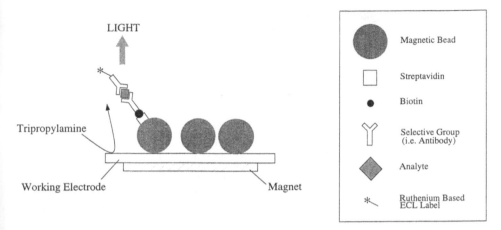

Figure 11   The basic principle of ECL immunoassay using streptavidin-coated magnetic beads, and labels based on Ru(bpy)$_3$$^{2+}$.

A few other ECL immunoassay and DNA probe methods have been designed. An enzymatic immunoassay based on luminol ECL has been developed by Wilson et al. for the detection of pesticides in drinking water to less than 0.1 ppb [75]. Transparent, aminosilanized, ITO-coated glass electrodes were derivatized with aminodextran covalently modified with atrazine caprioic acid. Antibodies to atrazine were labeled with glucose oxidase enzyme and used in an indirect competitive binding assay to detect atrazine in solution. Labeled antibodies that bound to the immobilized atrazine at the electrode surface were detected by an ECL reaction between oxidized luminol and H$_2$O$_2$ produced by the enzyme with the addition of glucose. Kankare et al. demonstrated that oxide-bound terbium(III) can be excited at an aluminum electrode surface in aqueous solution to yield narrow-band terbium emissions (548 nm) in the presence of persulfate [76]. When chelated with 2,6-*bis*[*N,N-bis*(carboxymethyl)aminomethyl]-4-benzoylphenol, terbium cathodic luminescence had a significantly longer lifetime, such that time-resolved ECL was possible, the intensity measured a short time after electroexcitation, once short-lived background ECL had diminished. This terbium chelate has subsequently been used as an immunoassay label for the determination of human pancreatic phospholipase A$_2$ to 10 ng/ml. Xu and Bard developed a biosensor based on single-stranded DNA immobilized on an electrode covered with alkanebisphosphonate, which could recognize a complimentary strand of DNA [77]. Hybridization of unlabeled DNA could be determined using ECL produced during the oxidation of Ru(1,10-phenanthroline)$_3$$^{2+}$ intercalated with

DNA and tripropylamine. Alternatively, target DNA immobilized on the film could be determined by hybridization with a $Ru(bpy)_3^{2+}$-labeled complementary DNA strand.

10. ECL IN MINIATURIZED ANALYTICAL SYSTEMS

Recently the miniaturization of analytical systems has received considerable interest, with the aim of realizing the whole analytical process of sampling, separation, and detection, incorporated on a single chip of perhaps a few square centimeters in area [78]. Such chips usually consist of multilayered structures of micromachined or etched glass, silicon, or polymer, with narrow interlocking channels for the passage and manipulation of reagents. These are moved either by external, conventional pumps or by electrokinetic means on the chip itself. ECL has potential as a detection system in such systems, since the components needed for ECL generation and detection are easily miniaturized. However, to date only a few devices using ECL have been reported [39, 79–81].

These miniature devices can be mounted close to a photodetector, but with a more elegant arrangement using photodiodes, the photodetector can become an intrinsic component of the device itself. This can be achieved either by bonding the photodiode to the exterior surface of a transparent window of the cell [80], or by direct fabrication in chips made from silicon substrates [81]. Plain platinum or silver metal pseudoreference electrodes that are readily miniaturized can replace more complex reference electrode constructions. Electrode formats used include fine wires, metal foils, or metal coatings typically 1 μm thick. With use of transparent ITO electrodes [80] or interdigitated electrodes [81], ECL can be detected by an underlying photodetector while the top surface of the electrode is in contact with the test solution. In addition, since ECL and light emission is only stimulated in the direct vicinity of the electrode, they can be positioned such that only a small volume of the flow channel is probed at any one time. Hence it is possible to analyze extremely small volumes of solution. For example, Arora et al. used a multilayer polymer flow cell chip to detect $Ru(bpy)_3^{2+}$ with a limit of detection of 5×10^{-13} M in an effective cell volume of 100 nL, corresponding to just 30,000 molecules [39].

11. ECL IN ELECTRODE CHARACTERIZATION AND THE VISUALIZATION OF ELECTROCHEMICAL PROCESSES

Recently ECL has been utilized as a diagnostic tool to visualize electrochemical processes, to characterize electrode surfaces in terms of their uniformity or microarchitecture, and to reveal the coverage and heterogeneity of layers of compounds attached to the electrode surface. Much electrochemical data relies on the assumption that the observed current is representative of the electrode reaction at all parts of the electrode surface, which is uniformly active and accessible. ECL is being used to nondestructively probe electrodes where current density is not evenly distributed across the electrode. This effect may be due to absorbed species, oxide or other pacifying films, differential reactivities at various crystal faces at the electrode surface, analyte depletion downstream in a flow cell, diffusion effects, or the use of composite electrodes [82]. Equipment and procedures for the ECL visualization of faradic processes and electrode surface heterogeneity have been described by Kukoba and Rozhitskii [83].

The extent of coverage of electrodeposited films can be imaged by ECL to a resolution of about 0.5 μm, a limit set by light microscopy and the lifetime of the excited species as they diffuse away from the electrode surface [83–85]. ECL is generally inhibited and absent from areas of the electrode covered by the film. Photographs give qualitative information of electrode activity, yet quantitative data can be obtained by tracking a detector, such as a PMT, across the electrode surface, or by using image capture on a CCD camera. Many ECL reactions have been used for imaging, depending on the electrode material or absorbate under study. Since luminol and $Ru(bpy)_3^{2+}$ are negatively and positively charged, respectively, they can be used to visualize different types of oxidation sites on an electrode. For example, Hopper and Kuhr demonstrated that surface oxide on carbon fiber microelectrodes inhibited luminol ECL while it facilitated $Ru(bpy)_3^{2+}$ ECL [85]. In organic solvents IAECL has been used, with rubrene or diphenylanthracene as the fluorescer [83].

ECL has also been produced from organized monolayers of $Ru(bpy)_3^{2+}$, or its derivatives such as surfactants formed by using ligands with long-chain hydrocarbons, on a variety of electrode surfaces. Measurements can be used to determine absorption isotherms. Thiol derivatives have a particular affinity for gold electrodes and self-assemble into monolayers [14]. Although fairly intense ECL has been observed from absorbed monolayers of ECL active species, their current lack of robustness limits their analytical potential in terms of producing a sensor. However, by coupling with imaging techniques it should be possible to optically detect the arrival and interaction of small numbers of reactant species at the monolayer/solution interface [14].

12. CONCLUSIONS

In conclusion, ECL should continue to have a promising future in analytical science. The versatility of $Ru(bpy)_3^{2+}$ ECL will continue to be exploited for an increasingly diverse range of analytical determinations, especially in biomedical applications and immunoassay. Inevitably, for all but the most simple sample matrices, ECL will be coupled with a separation step such as HPLC or CE to remove interfering species; however, ECL is wholly compatible with these techniques. There are also good prospects for ECL to be used as a detection technique for miniaturized analytical systems. The production of a really robust and reproducible ECL sensor incorporating immobilized ECL reagents holds great potential but is yet to be fully realized, with further research required in this area. Finally, ECL should be a useful technique to nondestructively visualize electrochemical processes and electrode surfaces, and it is hoped that this will be further exploited in the future.

REFERENCES

1.   WD Bancroft, HB Weiser. J Phys Chem 18:762–781, 1914.
2.   RT Dufford, D Nightingale, LW Gaddum. J Am Chem Soc 49:1858–1864, 1927.
3.   N Harvey. J Phys Chem 33:1456–1459, 1929.
4.   T Kuwana, B Epstein, ET Seo. J Phys Chem 67:2243–2244, 1963.
5.   DM Hercules. Science 145:808–809, 1964.
6.   LR Faulkner, AJ Bard. Electroanal Chem 10:1–95, 1977.
7.   LR Faulkner. Methods Enzymol 57:494–526, 1978.
8.   S-M Park, DA Tryk. Rev Chem Intermed 4:43–79, 1981.
9.   JG Velasco. Bull Electrochem 10:29–38, 1994.
10. HA Sharifian, S-M Park. Photochem Photobiol 36:83–90, 1982.
11. A Vogler, H Kunkely. In: Am Chem Soc Symp Ser, Vol 333, High-Energy Processes in Organometalic Chemistry, 1987, pp 155–168.
12. AW Knight, GM Greenway. Analyst 119:879–890, 1994.
13. NN Rozhitskii. J Anal Chem USSR (Eng Transl) 47:1288–1310, 1992.
14. W-Y Lee. Mikrochim Acta 127:19–39, 1997.
15. AW Knight. Trends Anal Chem 18:47–62, 1999.
16. RD Gerardi, NW Barnett, SW Lewis. Anal Chim Acta 378:1–41, 1999.
17. E Hill, E Humphreys, DJ Malcolme-Lawes. J Chromatogr 370:427–437, 1986.
18. TC Richards, AJ Bard. Anal Chem 67:3140–3147, 1995.
19. WL Wallace, AJ Bard. J Phys Chem 83:1350–1357, 1979.
20. HS White, AJ Bard. J Am Chem Soc 104:6891–6895, 1982.
21. D Ege, WG Becker, AJ Bard. Anal Chem 56:2413–2417, 1984.
22. K Yamashita, S Yamazaki-Nishida, Y Harima, A Segawa. Anal Chem 63:872–876, 1991.
23. M-M Chang, T Saji, AJ Bard. J Am Chem Soc 99:5399–5403, 1977.
24. I Rubinstein, AJ Bard. J Am Chem Soc 103:512–516, 1981.
25. N Egashira, H Kumasako, Y Kurauchi, K Ohga. Anal Sci 8:713–714, 1992.
26. DR Skotty, TA Nieman. J Chromatogr B 665:27–36, 1995.
27. NW Barnett, TA Bowser, RA Russell. Anal Proc 32:57–59, 1995.
28. X Chen, M Sato. Anal Sci 11:749–754, 1995.
29. AW Knight, GM Greenway. Analyst 120:2543–2547, 1995.

30. JK Leland, MJ Powell. J Electrochem Soc 137:3127–3131, 1990.

31. AW Knight, GM Greenway. Analyst 121:101R–106R, 1996.

32. K Uchikura, M Kirisawa, A Sugii. Anal Sci 9:121–123, 1993.

33. SN Brune, DR Bobbitt. Anal Chem 64:166–170, 1992.

34. WA Jackson, DR Bobbitt. Anal Chim Acta 285:309–320, 1994.

35. L He, KA Cox, ND Danielson. Anal Lett 23:195–210, 1990.

36. K Uchikura, M Kirisawa. Anal Sci 7:971–973, 1991.

37. X Chen, M Sato. J Flow Inject Anal 12:216–222, 1995.

38. W-Y Lee, TA Nieman. J Chromatogr A 659:111–118, 1994.

39. A Arora, AJ de Mello, A Manz. Anal Commun 34:393–395, 1997.

40. K Saito, S Murakami, S Yamazaki, A Muromatsu, S Hirano, T Takahashi, K Yokota, T Nojiri. Anal Chim Acta 378:43–46, 1999.

41. W Lee, TA Nieman. Anal Chem 67:1789–1796, 1995.

42. G Merenyi, J Lind, TE Eriksen. J Biolumin Chemilumin 5:53–56, 1990.

43. KE Haapakka, JJ Kankare. Anal Chim Acta 138:263–275, 1982.

44. K Haapakka, J Kankare, O Puhakka. Anal Chim Acta 207:195–210, 1988.

45. K Haapakka, J Kankare, S Kulmala. Anal Chim Acta 171:259–267, 1985.

46. K Haapakka, J Kankare, K Lipiäinen. Anal Chim Acta 215:341–345, 1988.

47. KE Haapakka, JJ Kankare. Anal Chim Acta 138:253–262, 1982.

48. H Schaper, H Köstlin, E Schnedler. J Electrochem Soc 129:1289–1294, 1982.

49. NJ Kearney, CE Hall, RA Jewsbury, SG Timmis. Anal Commun 33:269–270, 1996.

50. AW Knight, GM Greenway, ED Chesmore. Anal Proc 32:125–127, 1995.

51. GE Collins, SL Rose-Pehrsson. Sens Actuat B 34:317–322, 1996.

52. GE Collins. Sens Actuat B 35:202–206, 1996.

53. LS Kuhn, A Weber, SG Weber. Anal Chem 62:1631–1636, 1990.

54. JP Preston, TA Nieman. Anal Chem 68:966–970, 1996.

55. I Rubinstein, AJ Bard. J Am Chem Soc 103:5007–5013, 1981.

56. TM Downey, TA Nieman. Anal Chem 64:261–268, 1992.

57. N Egashira, N Kondoh, Y Kurauchi, K Ohga. Denki Kagaku 60:1148–1151, 1992.

58. HD Abruña, AJ Bard. J Am Chem Soc 104:2641–2642, 1982.

59. JA Holeman, ND Danielson. Anal Chim Acta 277:55–60, 1993.

60. LL Shultz, JS Stoyanoff, TA Nieman. Anal Chem 68:349–354, 1996.

61. JS Ridlen, GJ Klopf, TA Nieman. Anal Chim Acta 341:195–204, 1997.

62. Y Wang, ES Yeung. Anal Chim Acta 266:295–300, 1992.

63. AM García Campaña, WRG Baeyens, Y Zhao. Anal Chem 69:83A–88A, 1997.

64. JA Dickson, MM Ferris, RE Milofsky. J High Res Chromatogr 20:643–646, 1997.

65. GA Forbes, TA Nieman, JV Sweedler. Anal Chim Acta 347:289–293, 1997.

66. K Tsukagoshi, K Miyamoto, E Saiko, R Nakajima, T Hara, K Fujinaga. Anal Sci 13:639–642, 1997.

67. DR Bobbitt, WA Jackson. US Patent No. 5 614 073, March 25, 1997.

68. SD Gilman, CE Silverman, AG Ewing. J Microcol Sep 6:97–106, 1994.

69. AF Martin, TA Nieman. Anal Chim Acta 281:475–481, 1993.

70. F Jameison, RI Sanchez, L Dong, JK Leland, D Yost, MT Martin. Anal Chem 68: 1298–1302, 1996.

71. R Wilson, J Kremeskötter, DJ Schiffrin, JS Wilkinson. Biosens Bioelectron 11:805–810, 1996.

72. AF Martin, TA Nieman. Biosens Bioelectron 12:479–489, 1997.
73. Boehringer Mannheim GmbH, Mannheim. Germany. Company literature.
74. H Yu, PJ Stopa. In: Am Chem Soc Symp Ser, Vol 646, Environmental Immuno-chemical Methods, 1996, pp 297–306.
75. R Wilson, MH Barker, DJ Schiffrin, R Abuknesha. Biosens Bioelectron 12:277–286, 1997.
76. J Kankare, K Haapakka, S Kulmala, V Näntö, J Eskola, H Takalo. Anal Chim Acta 266:205–212, 1992.
77. X-H Xu, AJ Bard. J Am Chem Soc 117:2627–2631, 1995.
78. SJ Haswell. Analyst 122:R1–R10, 1997.
79. J Kremeskötter, R Wilson, DJ Schiffrin, BJ Luff, JS Wilkinson. Measurement Sci Technol 6:1325–1328, 1995.
80. Y-T Hsueh, RL Smith, MA Northrup. Sens Actuat B Chem 33:110–114, 1996.
81. GC Fiaccabrino, NF deRooij, M Koudelka-Hep. Anal Chim Acta 359:263–267, 1998.
82. JE Vitt, RC Engstom. Anal Chem 69:1070–1076, 1997.
83. AV Kukoba, NN Rozhitskii. Russian Electrochem 29:345–351, 1993.
84. RC Engstrom, PL Nohr, JE Vitt. Colloids Surfaces A: Physicochem Engin Asp 93:221–227, 1994.
85. P Hopper, WG Kuhr. Anal Chem 66:1996–2004, 1994.

10

Applications of Bioluminescence in Analytical Chemistry

Stefano Girotti, Elida Nora Ferri, Luca Bolelli, Gloria Sermasi, and Fabiana Fini
University of Bologna, Bologna, Italy

1. AN INTRODUCTION TO BIOLUMINESCENCE

The emission of light after chemical excitation is called chemiluminescence (CL). If it occurs in biological systems it is known as bioluminescence (BL) or, described in a more detailed manner, BL is a case of enzyme-catalyzed CL (Fig. 1) [1, 2].

For most people, BL is represented by the flash of the firefly or the "phosphorescence" that frequently occurs on agitating the surface of ocean water. Chemical excitation, luminescent reactions occurs in almost all zoological kingdoms (bacteria, dinoflagelates, crustacea, worms, clams, insects, and fishes) except higher vertebrates: BL is not found in any organisms higher than fish. In most cases this phenomenon occurs within specialized cells called photocytes [3–5]. As shown in Table 1, BL occurs in many terrestrial forms but is most common in the sea, particularly in the deep ocean, where the majority of species are luminescent [6].

BL has independently evolved many times; some 30 different independent systems are still existent [7]. Thus the responsible genes are unrelated in the various organisms, the enzymes show no homology to each other, and the substrates are chemically unrelated. There is, however, one common thread tying different systems at molecular level. All involve exergonic reactions of molecular

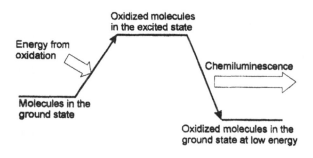

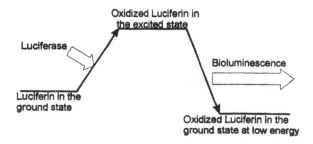

Figure 1 Scheme of chemiluminescent and bioluminescent light emission.

Table 1 Different Kinds of Bioluminescent Organisms

Type and examples	Bioluminescence max range, in vivo λ_B (nm)	Essential factors
Bacteria		
Photobacterium, Vibrio fischeri and *Vibrio harveyi*	478–505	Luciferine (an aldehyde), luciferase FMNH$_2$ and O$_2$
Protozoans		
Dinoflagellates: *Gonyaulux, Pyrocystis*	470	Luciferine (a biliar pigment), luciferase and O$_2$
Coelenterates		
Hydrozoan jellyfish (*Aequorea*)	508	Photoproteins, Ca^{2+}, a green fluorescent protein
Anthozoans: *Renilla, Ptilosarcus*	509	Luciferase, green fluorescent protein, O$_2$
Annelids		
Polychaetes: *Chaetopterus*	465	Photoprotein, Fe^{2+}, H$_2$O$_2$, O$_2$
Oligochetes (earthworms)	500	Luciferine (an aldehyde), luciferase, H$_2$O$_2$
Molluscs		
Limpet: *Latia*	535	Luciferine, luciferase, purple protein, O$_2$
Bivalve: *Pholas*	490	Luciferine (a protein) luciferase, (or Fe^{2+}), O$_2$
Squids: *Encleoteuthis, Chiroteutis*	416–540	Luciferine (coelenterazine), luciferase, O$_2$
Crustacea (*Cypridina*)	465	Luciferine (*Cypridina luciferin*), luciferase, O$_2$
Shrimps and decapods		
Oplophorus, Sergestes	462	Luciferine (coelenterazine), luciferase, O$_2$
Millipedes: *Liminodesmus*	496	Photoproteins, ATP, Mg^{2+}, O$_2$
Insects:		
Firefly (Lampiridae): *Photinus, Photuris* and others	496	Luciferine, luciferase, ATP, Mg^{2+}, O$_2$
Emicordates		
Balanoglossus		Luciferine, luciferase, (Peroxidase), H$_2$O$_2$
Fish		
Batracoides: *Porichthys*	459	Luciferine (*Cypridina luciferin*), luciferase, O$_2$

oxygen with different substrates (luciferins) catalyzed by enzymes (luciferases), resulting in photons of light ($\cong 50$ kcal). Luciferase and luciferin are generic terms referring to an enzyme that catalyzes the oxidation of a substrate, such as luciferin, and to a reduced compound that can be oxidized in an appropriate environment to produce an electronically excited singlet state. Light is produced on its return to the ground state. All involve a luciferase-bound peroxy-luciferin intermediate, the breakdown of which provides energy for excitation. Under in vitro conditions the quantum yield of such reactions can be as high as 0.9. In Table 2 the properties of some bioluminescent systems are listed. The maximum wavelength of the light emitted is often in the range 460–560 nm; then the color ranges from the red of worm through the deep blue characteristic of most marine creatures. Several factors affect the color of BL [6]. In the simplest case, the emission matches the fluorescence of an excited luciferase-bound product of the reaction. The luciferase structure can itself alter the color, as in the firefly, where single amino acid substitutions result in significant shifts in the emission spectrum. In bacteria and coelenterates, the chromophores of accessory proteins associated with luciferases may serve as alternate emitters, such as the yellow fluorescent protein (YFP) in bacteria and the green fluorescent protein (GFP) in coelenterates, now used as reporter genes and cellular markers [8]. The cell biology and regulation of BL differ among groups. While bacteria and some other systems emit light continuously, in many the luminescence occurs as flashes, typically of 0.1–1 s duration.

BL research has increasingly drawn scientists' renewed attention during the last years. One of the main reasons for this concern is due to the development of gene technology and the applications of its many new methods to study BL at the molecular level. Progress in the fundamental knowledge of BL has led to numerous gene-reporting techniques, thanks to which new basic knowledge is

Table 2 Properties of Some Bioluminescent Systems

Organism	Protein involved (Mass-Da, subunits)
Bacteria	Luciferase (76,000; α,β)
	Lumazine protein (21,200)
Dinoflagellates	Luciferase (130,200)
	Luciferin-binding protein
Anthozoa	Luciferase (35,000)
	Green-fluorescent protein (52,000; α_2)
	Luciferin-binding protein
Firefly	Luciferase (60,000)
Crustaceans: *Cypridina*	Luciferase (68,000; α_6)

now acquired in many areas of biology and medicine. As a consequence, there is a growing interest and demand of research institutions other than biomedical to utilize highly sensitive bioluminescent techniques for analytical purposes.

This review deals mainly with BL analytical applications in the last 10–15 past years, but some previous fundamental works are also listed. In Table 3 some fundamentals references of general interest and the findings of recent symposia on this topic are collected. In the journal *Luminescence, the Journal of Biological and Chemical Luminescence* (previously *Journal of Bioluminescence and Chemiluminescence*) are also reported surveys of the recent literature on selected topics (like ATP or GFP applications), instruments, and kits commercially available.

2. ANALYTICAL APPLICATIONS OF THE MAIN BIOLUMINESCENT SYSTEMS

2.1 Firefly Luciferase

The most popular system in mechanistic and model studies as well as in analytical applications (clinical, food, environmental) appears to be that of firefly luciferin and luciferin-type-related model luminescence [3, 5, 23, 57]. The luciferase from *Photinus pyralis*, *Photinus* luciferin 4-monooxygenase (ATP-hydrolyzing), EC 1.13, 12.7, is a hydrophobic enzyme that catalyzes the air oxidation of luciferin in the presence of ATP and magnesium ions to yield light emission:

$$\text{ATP} + \text{reduced luciferin} + O_2 \xrightarrow{\text{Luciferase, Mg}^{2+}} \text{AMP} + \text{oxyluciferine} + \text{h}\nu$$

The mechanism is more complex than reported above, and starting with the pioneer studies of DeLuca and McElroy [58], it has been the object of deep investigations. A more detailed scheme is reported in Figure 2. ATP is consumed as a substrate and photons at a wavelength of 562 nm are emitted. The quantum yield of this reaction is 0.9 einstein $\times$ mol/L of luciferin. Considering the stoichiometry of the reaction for one ATP molecule consumed, approximately one photon is emitted. This property, together with the high nucleoside specificity of the enzyme, makes this reaction an ideal analytical system for assaying ATP presence, ATP production or consumption in dependence of enzymatic activity, and for quantification of substrates linked to the ATP metabolism. ATP is the most important and central coupling agent between exergonic and endergonic processes and it is ubiquitous in living organisms where it functions as an allosteric effector, as a group-carrier coenzyme, and as a substrate. Because of the essentiality of ATP and of the related enzymes and substrates in metabolism, accurate, sensitive,

Table 3 Recently Reviewed Analytical Applications of Bioluminescent Systems

Topic	Ref.
General	
Chemiluminescence and bioluminescence	9
Bioluminescence and chemiluminescence	10
Analytical luminescence: its potential in the clinical laboratory	11
ATP determination with firefly luciferase	3
Chemiluminescent and bioluminescent methods in analytical chemistry	12
Luminometry	13
Biological diversity, chemical mechanism, and the evolutionary origins of bioluminescent systems	14
Evolutionary origins of bacterial bioluminescence	15
Bioluminescence and chemiluminescence, Part B	1
Fluorescence and bioluminescence measurement of cytoplasmic free calcium	16
Chemiluminescence: principles and applications in biology and medicine, several chapters on bioluminescence	7
Clinical and biochemical applications of luciferase and luciferins	17
Bioluminescence and chemiluminescence-based fiberoptic sensors	18
Bioluminescence/chemiluminescence-based sensors	19
Genetics of bacterial bioluminescence	4
Chemistries and colors of bioluminescent reactions—a review	6
Chemiluminescence and bioluminescence	2
Luminescent techniques applied to bioanalysis	20
Bioluminescence	5
Immunoassay, nucleic acid, and reporter gene assays	
Immunoassay	21
Bioluminescent immunoassay and nucleic acid assay	22

Table 3 Continued

Topic	Ref.
Bioluminescence and chemiluminescence, 1994 part 2	44
Bioluminescence and chemiluminescence, 1994 part 3	45
Bioluminescence and chemiluminescence, 1995 part 1	46
Bioluminescence and chemiluminescence, 1995 part 2	47
Bioluminescence and chemiluminescence, 1995 part 3	48
Green fluorescent protein	49
Bioluminescence and chemiluminescence, 1996	50
Bioluminescence and chemiluminescence, 1997 part 1	51
Bioluminescence and chemiluminescence, 1997 part 2	52
Commercial available luminometers, imaging devices, and reagents, survey update 5	53
Commercial available luminometers, fluorometers, imaging devices, and reagents, survey update 6	54
Teaching	
Creatures that glow: a book about bioluminescent animals	55
Animals that glow	56
Web sites on bioluminescence	
The bioluminescence web page	http://lifesci.ucsb.edu/~biolum/
Scripps Institution of Oceanography	http://siobiolum.ucsd.edu/Biolum_intro.html
Bioluminescence: a proctor project by R Abaza	http://www.biology.lsa.umich.edu/~www/bio311/projects/ronney/biochem.shtml
Bioluminescence and biological fluorescence	http://www.herper.com/Bioluminescence.html
Bioluminescence studies research and resources by John E. Wampler (*Renilla* green fluorescent protein and other organisms)	http://bmbiris.bmb.uga.edu/wampler/biolum/index.html#web
Video	
David Attenborough, "Talking to Strangers," a program in the Trials of Life series	
The last 15 min of this video talks about bioluminescence	

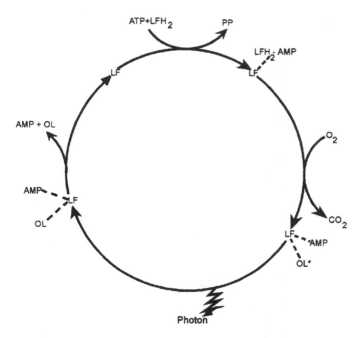

Figure 2 A more detailed scheme of ATP firefly luciferase reaction. LF = luciferase; OL = oxiluciferine; LFH$_2$ = reduced luciferase; PP = pirophosphate; OL* = oxiluciferine in the excited state.

and simple methods are required for its determination. These methods are applicable to medicine, biology, environmental studies, agriculture, and industry, as well as in the laboratory, pointing out the great versatility of the system. Some of the numerous applications of ATP bioluminescent assay are listed in Table 4.

One problem with the ATP assay in aqueous media is that the enzyme requires hydrophobic media; the reaction rate of luciferase-catalyzed reactions is variously affected by the presence of detergents [117, 118]. The presence of cationic liposomes improves sensitivity by a factor of five times compared to that in water alone [119].

Another characteristic to take into account is that ATP is an endogenous component of the cells, both somatic and bacterial. Therefore, an extraction step must to be included in the assay protocol; it is very simple and quick to perform. Several extraction methods have been reported, both physical and chemical, such as heating and the use of surfactants, trichloroacetic acid, and organic solvents [89, 120, 121]. The chemical methods are generally preferred; the addition of a surfactant can be effective in most cases. The use of mild or strong extraction

Table 4 Analytical Applications of BL ATP assay

Analyte		Ref.
General	Biotin	59
	Microorganisms trapped on filter	60
	Genetic	61
Medical application	Duchenne muscular dystrophy	62
	Mapping arterial wall	63
Immunoassay	General	25, 64–66
	PCR	67
Bacteria	General	68–71
	Flow	72
	Cosmetic	73
	Milk	74–77
	Vegetables and water	78
	Beer	79, 80
HACCP	General	81, 82
	Hygiene in food industry	83
	Poultry industry	84
	Beef and pork carcasses	85
	Milk	86
	Wine	87
	Beer	88
Drugs		89–93
Blood		94–97
Cancer		98–104
Cells	General	105, 106
	Energetic state of the cells	107–109
Environment	General	110
	Air	111, 112
	Marble	113
	Sludge	114
	Pesticide	115
	Wastewater	116

agents also allows differentiating somatic and bacterial cells, disrupted by mild and strong agents, respectively. In any case the effects of detergents on luminescence must be taken in account.

Beside that of *Photinus pyralis*, other luciferases have been described in the literature such as those from *Luciola cruciata, Luciola lateralis* [122], and *Luciola mingrelica*. Among them luciferase from the latter one has been employed for analytical purposes [123].

2.1.1 Biomedical Applications

In the study of biological alterations or during the clinical treatment of several pathologies it can be useful to evaluate the cells' content of ATP, which is often directly related to the alterations under study. ATP determination can be used, for example, in the diagnosis of transformed cells: in general, cancer cells have low ATP levels and their ATP values fall with the length of culturing, but in polyploidal cells the ATP level increases with the polyploidy. Measurements in single cells have proved to be particularly informative in resolving the sequence of events occurring in cells exposed to pathological insults. A great cell-to-cell variability was found in the time course of the ATP decline in response to metabolic poisons. For example, ATP changes were monitored in single hepatocytes, at a concentration of 1 mM or lower, allowing the monitoring of chemical hypoxia [105].

Analogously ATP BL can be used for cytotoxicity assays, i.e., to determine cell response to antibiotics [91], to antineoplastic agents, even in association with ionizing radiation [100], and to the solvents commonly used for solvation of xenobiotic agents [106]. The assay, compared to the classical Ames test [104], showed a very good correlation between the BL assay and plates method. The strong potential of ATP BL to become a clinical assay for chemosensitivity testing therefore appeared clear. It proved to be a highly quantitative assay, which measures cellular response of the cultured cells evaluating cellular ATP instead of counting colonies. The assay was found to be reproducible and reliable and to have a clinical applicability rate of more than 90%, making it well suited for clinical application [98, 99].

ATP is an ideal indicator of cell viability. Blood or blood cell concentrates prepared for transfusion are stored for periods of a few days to several weeks in the blood bank. Viability checking of the blood cells is necessary to avoid posttransfusional reactions [94]. This quality control of the conserved red blood cells and platelets can easily be performed by measuring the ATP concentration as an expression of their integrity. By the same measurement it was possible to confirm the diagnosis and monitor the treatment effects in various cases of platelet disease [97]. The possibility of determining cells' viability can be exploited to examine more free cells or tissue, as in the spermatozoa viability test, based on the correlation between ATP content and mobility.

Blood platelets release ATP, ADP, serotonin, and other compounds and this plays an important role in thrombosis and hemostasy. To study the mechanism of the release, in vitro release of ATP was followed using the firefly luciferase luminescence method [124].

Firefly luciferase, together with luciferases from other organisms, can be used as the labeling enzyme in immunoassays and nucleic acid assays [25]. Recently, a highly sensitive BL ELISA using firefly luciferase was applied to thyreotropine detection [125] and a BL homogeneous enzyme-binding assay for biotin detection was developed [59]. These are among the few applications reported on this topic. Actually, the enzyme proved to have some problems when employed in this application, being not very stable, and expensive; large batch-to-batch variations were also recorded. To develop such immunoassays, several strategies must therefore be applied, making them not competitive compared to other labeling systems [21, 22]. In the past some luciferins, when derivatized with 6-N-acetyl-L-phenylalanine or o-phosphate, were used in immunoassays [126–128]. In the latter case the reaction with alkaline phosphatase, acting as the labeling enzyme, allowed release of the free luciferin that reacts with subsequent light emission.

Luminometric assay of ATP can also be applied to measurement of other substrates such as ADP, AMP, and ATP-specific enzymes. Other examples are reported in Table 4.

2.1.2   Rapid Microbiology

The obvious presence of ATP inside bacterial cells allows setting up new microbiological assays, quicker, easier, more sensitive but at the same time offering the same reliability as classical microbiological plate count. The main difference between the classical plate count method and ATP BL is that BL is able to detect all the cells (bacteria, fungi, yeast, somatic, etc.), both alive and stressed, i.e., not capable of reproducing in that medium, while the plate method measures only the cells that are alive and able to reproduce. For this reason ATP can be used as indicator of biological contamination in different fields [68]. The amount of ATP per cell is essentially proportional to the intracellular volume; consequently most bacterial cells contain around 2×10^{-18} mol ATP/cell (approx. 10^{-15} g/cell). However, differences in ATP content were found in different bacterial species and in yeast, for which the amount of ATP is 100–1000 times higher than for bacteria. When a cell dies from natural causes, intracellular enzymes rapidly degrade ATP. If cells are killed in a way that also inactives the enzymes, some ATP may appear even outside cells [129, 130].

The sensitivity obtained with ATP BL is around 1000–10,000 colony-forming units (CFU) per milliliter. Better detection was obtained by determining adenylate kinase instead of ATP: 10–1000 cells/100 µL [71].

Since the U.S. Department of Agriculture, Food Safety and Inspection Service (USDA-FSIS) published the pathogen reduction hazard analysis and critical control point (HACCP), this system has been frequently applied in the quality management and assurance field to increase safety of foods. This is achieved by monitoring and controlling all the steps of foods processing, and especially checking the surfaces in contact with food. Generally, microbiological methods to verify that the processing plant, equipment, and products have been properly sanitized are time consuming and expensive. There has consequently been considerable interest in rapid microbiological enumeration methods not only in agriculture and food industries but in a wide list of fields: environmental quality control, monitoring systems at sludge and sewage water plants, pharmaceutical and cosmetics industries, vitality and quality of biomasses involved in fermentation processes, bacterial contamination in spaces otherwise sterile (surgical rooms), microorganisms' contribution to alterations affecting artistic works. ATP BL is one of these alternative methods and has already been widely applied (Table 4). Several commercial kits are available and also directly usable in the field with portable instruments. When ATP is found on a surface, this indicates it is not perfectly clean; the contamination may come from microorganisms, food residues, or human contact. The latter two of these sources can both provide an ideal environment for bacterial growth and neither of them can be detected by conventional microbiological methods. The advantage of ATP measurement in the evaluation of cleanliness is its ability to detect these product residues and organic debris in addition to living microorganisms. At the same time, the fact that the method does not differentiate between ATP from different sources can be a limitation, unless additional selective steps are performed. When ATP BL is employed in monitoring sanitation, it is important to consider that cleaning agents and sanitizers can interfere with ATP detection [81].

Monitoring of ATP levels also proved useful to quantify air contamination in the work environment [111]. Comparison with other analytical methods shows the usefulness of the BL assay because of its rapidity, in spite of being aspecific [131].

ATP measured by luciferin-luciferase BL assay was used to examine the effect of toxic substances on whole microbial communities in activated sludge mixed liquid samples [114]. It was used to detect whether wastewater had an effect on the biodegradation capability of the resident population of microorganisms. Actually ATP BL represents an important rapid toxicity test that utilizes waste treatment natural microorganisms to determine the toxicity of wastes discharged to the sewer [132].

Another interesting application of the ATP assay is related to alterations affecting artistic stonework, which can be due to various factors. Air pollution is recognized as the major cause, but microbiological attack also plays an important role. These alterations have been observed on several Italian monuments such as Scalone dei Giganti in Venice, the facade of Certosa di Pavia, the Duomo di Siena, and the Duomo di Orvieto, always on Carrara marble. For more than 50 years studies have been carried out on the causes of these alterations and only recently has their biological origin been taken into consideration. Bioluminescent assays were used to confirm the presence of microorganisms [113].

2.1.3   Firefly Luciferase as Reporter Gene

Genetically modified microorganisms hold great promise for many applications including analytical uses in the environmental, industrial, and agricultural fields [24, 25, 41]. A key component of this technique is a reliable detection method, specific and extremely sensitive, to study the fate of the modified organism and/ or engineered DNA in environmental samples or field tests. The firefly luciferase gene has been proposed as an ideal marker for genetically modified organisms to be released into the environment, having the requested characteristics. In addition, the bioluminescent assays are extremely fast and easy, relatively inexpensive, there is no phenotype in the absence of luciferin, and *luc*-tagged cells can be detected by a variety of complementary methods: this provides flexibility in choosing the appropriate method for specific applications [133]. The methods for detection of the *luc* marker gene can rely on cultivation of the tagged organisms and detection of whole cells, detection of cell extracts, or detection of the gene by polymerase chain reaction (PCR), each one having advantages and disadvantages [133]. In particular, the luciferase gene marker is ideally detected by PCR amplification since it is not naturally present in the microbial population.

An entirely different class of luciferase enzyme has been isolated from the firefly *Photinus pyralis* and related species, comparing the properties of the recombinant ones to the crystalline native enzyme [134]. The firefly luciferase gene *luc* was shown to act as a specific marker for monitoring of genetically modified bacteria, yeast, and plant leaves [135, 136]. The eukaryotic enzyme was approximately 10 times more sensitive than the corresponding bacterial one. Specific strains of very common bacteria, for example *Escherichia coli*, were often used as the recipient strain for recombinant plasmids. Using a multicopy-number plasmid as vector for *luc* gene often enhances sensitivity in the detection of *luc*-tagged cells.

An interesting example of sensitive and convenient biosensing of environmental pollutants was developed by fusing the firefly luciferase gene to the TOL plasmid of *Pseudomonas putida*. This plasmid encodes a series of enzymes for degradation of benzene and its derivatives. The gene fusion resulting plasmid was used to transform *E. coli*, and then applied to the environmental biosensing of benzene derivatives: the expression of luciferase was induced in the presence of aromatic compounds and at low detection limits (5 µM for m-xylene) [137]. In a second step the transformed *E. coli*, bearing firefly luciferase gene fused to TOL plasmid, was immobilized at one end of a fiberoptic demonstrating the possibility to obtain a luminescent remote biomonitoring device for protection against environmental deterioration [138].

Since the genetic code is universal among all living organisms, the coding region of a bacterial enzyme can be successfully expressed as a reporter gene not only in microorganisms, but also in animal cells and plants. The *luc* gene was successfully utilized as reporter gene to study transcriptional regulation in bacteria, yeasts, *Dictyostelium*, plants, viruses, cultured animal cells, and transgenic animals [139]. Recently the BL assay was applied to visualize the kinetics of tumor-cell clearance in living animals, also allowing evaluation of chemotherapeutic and immunotherapeutic treatments [140].

Other applications dealt with the development of a luciferin ester substrate to measure the luciferase activity in living cells [141], the detection of toxic compounds such as sodium azide, fluoroacetic acid, and antibiotics [142], the development of a biosensor for the determination of bioavailable mercury [143], the use of eukaryotic luciferases as bacterial markers with different colors of luminescence [144], the determination of complement-mediated killing of bacteria [145], and the development of a bioassay for the determination of HIV type 1 virus and HIV-1 Tat protein activity, valuable also for analysis of HIV-inhibitory agents [146].

2.2 Bacterial Luminescence

As mentioned above, most of the luminescent organisms are sea-living organisms, ranging in complexity from microscopic bacteria and plankton to many species of fishes. Luminous bacteria are the most abundant and widely distributed of the light-emitting organisms and are found in marine, freshwater, and terrestrial environments [14, 15, 147]. These bacteria are all gram-negative motile rods and can function as facultative anaerobes. Although the phenomenon is so widespread, it is possible to affirm that to date less than 1% of the known luminous species have been studied in great detail and the main part of this knowledge concerns the marine bacteria of three genera: *Photobacterium*, *Vibrio*, and *Photorhabdus*. Most studies have centered mainly on *Photobacterium* (*Vibrio**) *fischeri* and *Vibrio harveyi* [4].

The bioluminescent enzyme system from marine bacteria consists of a NAD(P)H:FMN oxidoreductase and a luciferase that emits light at 490 nm in the presence of FMN, NAD(P)H, a long-chain aliphatic aldehyde, and molecular oxygen, according to reactions (1) and (2) [148, 149]. The luminescent reaction in bacteria involves the oxidation of reduced riboflavin phosphate and a long-chain fatty aldehyde with the emission of blue-green light. The whole reaction mechanism is very complex and is still under study. A general reaction scheme is:

$$\text{NAD(P)H} + \text{FMN} + \text{H}^+ \xrightarrow{\text{NAD(P)H:FMN oxidoreductase}} \text{NAD(P)}^+ + \text{FMNH}_2 \quad (1)$$

---

* *Vibrio fischeri* has recently been reclassified as *Photobacterium fischeri*.

$$\text{FMNH}_2 + \text{RCHO} + \text{O}_2 \xrightarrow{\text{Luciferase}} \text{FMN} + \text{RCOOH} + \text{H}_2\text{O} + \text{Light} \quad (2)$$

Bacterial luciferase is a heterodimeric enzyme of 77 kDa that can be included, in general, under the group of the pyridine-nucleotide-linked systems. Bacterial luciferase is highly specific for FMNH_2, but the enzyme also shows weak activity toward other flavins. Only aliphatic aldehydes with a chain length of eight or more carbon atoms are effective in the luminescent reaction. The total light production is proportional to the amount of each of the substrates (O_2, FMNH_2, RCHO) when they are present in limited quantities. Generally, the emission spectra are characterized by a broad emission with a maximum near 478–505 nm [14]. Various substances of biological interest and enzyme activities can be analyzed by coupling the luciferase and the oxidoreductase to a third reaction, which produces or consumes NADH or NADPH.

2.2.1 Applications of the Isolated Bacterial Luminescent System

The possibility of isolating the components of the two above-reported coupled reactions offered a new analytical way to determine NADH, FMN, aldehydes, or oxygen. Methods based on NAD(P)H determination have been available for some time and NAD(H)-, NADP(H)-, NAD(P)-dependent enzymes and their substrates were measured by using bioluminescent assays. The high redox potential of the couple NAD^+/NADH tended to limit the applications of dehydrogenases in coupled assay, as equilibrium does not favor NADH formation. Moreover, the various reagents are not all perfectly stable in all conditions. Examples of the enzymes and substrates determined by using the bacterial luciferase and the NAD(P)H:FMN oxidoreductase, also coupled to other enzymes, are listed in Table 5.

2.2.2 Applications of Whole Luminescent Bacteria

Toxic pollutants are found everywhere, in water, in the air, and in the soil, causing environmental damage. Therefore, it is increasingly important that their early detection but at the same time the use of animals for toxicity testing has come under critical surveillance. For ethical and economic reasons various techniques have been developed and proposed as potential alternatives, among them the luminescent bacteria toxicity test [167]. Luminescent bacteria emit light when they find themselves in an optimal environment. In vivo luminescence is a sensitive indicator of xenobiotic toxicity to microorganisms because it is directly coupled to respiration via the electron transport chain, and thus reflects the metabolic status of the cell. If noxious substances are present, the luminescence decreases. The higher the degree of toxicity, the less the amount of light emitted by the bacteria. Thus, the presence of toxic substances can be evaluated and several commercial kits and dedicated instruments are now available. The luminescent

Table 5 Applications of Isolated BL Bacterial Systems

Application	Ref.
Cancer cells measurement—glucose-6-phosphate dehydrogenase	150
Inhibitors of bacterial luciferase (cyanide)	151
Detection of one molecule of β-D-galactosidase produced from *Escherichia coli*	152
Bovine serum albumin increases initial light intensity and eliminates the adverse effects of short-chain alcohols on luciferase	153
Pyridine nucleotide contents extracted from cell monolayers	154
Estrogens and luteinizing hormone (LH)	155
Heroin and metabolites—detection limits: 89 ng/mL heroin and 2.0 ng/mL morphine	156
Glutathione semiquantitative assay in small volume of samples	157
Glycogen detection: detection limit 0.12 nmol	158
Regional determination of glucose in brain sections	159
Intermediates of lactose synthesis	160
Automated analysis of several cellular metabolites	161
BL enzyme fiber optic probe	162
Glucose and lactate in the human cornea	163
Several cellular metabolites (creatine, creatine phosphate, pyruvate, succinate, and lactate)	164
Quinones and phenols influence on alcohol dehydrogenase/NADH:FMN-oxidoreductase/luciferase system	165
Salt lake water quality monitoring	166

bacteria toxicity test, compared with other bioassays, proved that its average sensitivity is well within the same order of magnitude as the other tests in evaluating organic or inorganic pollutants [168–170]. It is, however, acknowledged that the "battery of test" approach, utilizing several different short-term biological tests, would be preferred in any monitoring scheme.

Luminescent bacteria also allow detection of the carcinogenic effect of genotoxics. A dark mutant of a *Photobacterium* or *Vibrio* strain that can revert back to luminescence at an increased rate in the presence of base-substitutes or frameshifts agents, DNA-damaging agents, DNA synthesis inhibitors, and DNA intercalating agents can be employed [171, 172].

Additional examples of the analytical applications of BL bacterial bioassays are listed in Table 6.

2.2.3 Bacterial *lux* Genes as Reporter Gene

Cloning and expression of the DNA coding for luciferases from different luminescent organisms have provided the basis for the rapid expansion in the knowledge of molecular biology of luminescence [195, 196] and in its use. It was found

Table 6 Applications of BL Bacteria Bioassays

Application	Ref.
Toxicity and genotoxicity tests	
Toxicity in estuarine sediments—use of Mutatox and Microtox to evaluate the acute toxicity and genotoxicity of organic sediments	173
Toxicity tests for the analysis of pore water sediment: a comparison of 4 tests	174
Dark mutant for genotoxicity and toxicity of river waters and sediment extracts, comparison with other bioassays	168
Water toxicity—use of *Photobacterium phosphoreum* on concentrated water extracts of rivers Tormes (Spain) and Po (Italy)	175, 176
BL and Ames assays for screening of contaminated sediment: a comparison	
Toxicity of sediments: comparison between *Photobacterium phosphoreum* and Microtox system	177
Toxic organic constituents in industrial waters	116, 178
Organotin compounds toxicity: comparison between BL bacteria assay and submitocondrial particle assay	179
Sewage sludge toxicity of Cu, Cd, Pb, Zn	180
Analysis of six metals by microplate and microluminometric technologies	181
Predictive relative toxicity of nine bivalent metal ions (Ca, Cd, Cu, Hg, Mg, Mn, Ni, Pb, and Zn)	182
Toxicity of HCl-treated and nontreated extracts from soil and water	183
Sulfur toxicity from acetonitrile extracts from sediments	184
Toxicity of products released by three alloys for orthopedic implants	185
Safety evaluation of medical devices	186
Reproducibility evaluation of toxicity test of phenylmercuric nitrate	187
Hepatotoxic (microcystin-LR-containing) cyanobacteria	188
Genotoxicity test of proflavine, aflatoxin B_1, benzo(a)pyrene and N-methyl-N-nitro-nitrosoguanidine	172
Cytotoxicity and BL bioassay for 709 medical devices and biomaterial extracts: a comparison	169
Three years' interlaboratory comparison studies of the BL bacteria test	189
Chronic toxicity evaluation of wastewater and treatment-plant effluents: a comparison between BL bacterial, invertebrate, and fish assays	170
General or basic studies	
Use of biology Gn combined with API 20e or BBL crystal ID plate to differentiate some seawater luminous bacteria	190
Growth and luminescence of four luminous bacteria are promoted by agents of microbial origin	191
Determination of glucose and toxic compounds by a flow system with immobilized bacteria	192
Environmental pollution detection: three bacterial bioluminescent systems are investigated to determine the toxic effects of nine substances (quinones and phenols)	193
Data elaboration	
Use of an algorithm to eliminate the inner-filter effect in a bioreactor	194
Comparison between natural and genetically marked luminescent bacteria	199

Table 7 BL Bioassays Using Transformed Bacteria

Application	Ref.
Comparison and test of natural, genetically marked luminescent bacteria, and nitrification respiration inhibition assays on sewage sludge	200
Construction and application of *lux*-based nitrate biosensor using a plasmid-borne transcriptional fusion between *Escherichia coli* nitrate reductase promoter and *Photorhabdus luminescens lux* operon	201
Development of a mercury biosensor using *lux* genes of *Vibrio fischeri* fused to the regulatory elements of the mercury detoxification genes of *Serratia marcescens*	202
Toxicity of chlorophenols by commercial BL strain and *lux*-marked biosensor	203
Use of *lux* biosensor to evaluate the paper mill sludge toxicity	204
Single cell determination of *lux* genes cloned *Pseudomonas syringae* pv. *Phaseolicola* by charge coupled-device-enhanced microscopy	205
Nitric-oxide-releasing compounds detection by bioluminescent *Escherichia coli*	206
Al^{3+}, Cr^{6+}, Hg^{2+}, and Li^+ toxic effects on BL *Escherichia coli*, library of 3000 *Escherichia coli* clones	207
Toxicity of six toxicants (Zn^{2+} > ethidium bromide > sodium pentachlorophenate > Cu^{2+} > 2,4-dichlorophenoxyacetic acid; sodium dodecyl sulfate = no) on a recombinant *Escherichia coli*	208
Toxicity of metals in soils amended with sewage sludge	209
Toluene dioxygenase-*lux* gene was fused into *Pseudomonas putida* to evaluate toluene and trichloroethylene effects	210
Toxicity of chlorobenzenes	211
Toxicity of *n*-alkanols	212
Control of bioremediation of benzene, toluene ethylbenzene, and xylene (BTEX)-contaminated sites	213
Genetically engineered bioluminescent surfactant-resistant bacteria are useful to detect toxicity of nonpolar narcotics	214
Determination of antimonite and arsenite at the subattomole level	215, 216
Rapid assay for *Escherichia* O157-H7 in yogurts and cheeses	217
Measurement of genotoxicity kinetics by BL *Salmonella typhimurium* test	218
SOS-Lux test for environmental genotoxins	219
Direct determination of *Salmonella* live-vaccine strains	220

that significant differences exist between the *lux* systems from luminescent strains of the same species. In general, the bacterial *lux* genes can be routinely transferred into *E. coli* by transformation using a variety of different plasmids. For high expression a promoter is usually required on the plasmid [195, 197].

The ability to introduce the *lux* phenotype into different bacterial species provides a convenient method for rapidly screening in a simple and sensitive way the presence of specific bacteria and for monitoring their growth and distribution in the environment [198]. Another application of transformed bacteria deals with specific susceptibility in toxicity tests: the presence of agents that disrupt or kill the bacteria destroys the metabolism, thus eliminating light emission. Some examples are listed in Table 7.

3. IMMOBILIZED BIOLUMINESCENT SYSTEMS

The extreme sensitivity and specificity of the light-producing systems may be improved by using immobilized enzymes on solid supports. The immobilized enzymes are often more stable than the soluble forms and they can be used for several analyses, reducing costs. The sensitivity of the assays is generally increased, owing to the creation of a microenvironment with locally high concentrations of the involved reagents. Min et al. [221] reported eight times higher BL activity of immobilized FMN:NAD(P)H oxidoreductase and luciferase than the free enzymes. They developed inexpensive and sensitive biosensors for measuring the substrates of NAD(P)H-dependent enzymes, immobilizing the two in vivo biotinylated enzymes on avidin-conjugated agarose beads.

Various supports are used for immobilizing proteins, among them polyacrylic hydrazide, controlled-porosity glass beads, amino alkylated glass rods, cellophane films, cellulose films, collagen, sepharose, epoxy methacrylate, acylazide-activated-collagen strips, and nylon [222]. Several procedures are available for coupling proteins with those supports. In general, the chemical methods give better yields of active immobilized enzymes than the physical ones, whose main disadvantage is the fragility of the binding, sensitive to temperature, pH, and ionic strength variations. The chemical methods use various substances, such as glutaraldehyde, to covalently bind the protein through amino acidic residues. The most important factor in the choice of the suitable method of immobilization is conservation of the enzymatic activity. New methods of immobilization minimizing enzyme inactivation are continuously being investigated. Recently, a new protocol for luciferase immobilization was reported, which does not require prior silanization and glutaraldehyde activation, thus saving preparation time [223]. The method is based on coimmobilization by adsorption of luciferase and poly-L-lysine on nonporous glass strips. Luciferase immobilized in this way has minimal variation in intersample activity and good stability.

Another important aspect of immobilized enzymes is that they can be incorporated into flow cells where they can be used for multiple assays. The continuous-flow format offers greater possibilities than a single-batch system because it leads to rapid and sensitive assays. Flow assays are characterized by extremely accelerated kinetics: a very high surface-area-to-volume ratio is obtained and the reactions do not have to rely on passive diffusion to bring reagents together. Many analytes can be detected at pmol levels, with good precision and a wide range of linearity. Moreover, the analysis time is reduced to a few seconds and small specimens are needed.

Biosensors are generally described as probe-type devices made up of a selective biological layer with very sharp molecular-recognition capacity and of a physicochemical transducer [224]. Enzymes, antibodies, and nucleic acid sequences are reported as the most common biological element used in biosensors, because of their strong affinity for distinctive target molecules. Whole bacteria were also used to develop microbial biosensors [225]. As a transducer, strong interest was shown in optical systems and, between them, in those based on luminescence: they do not require light sources and monochromators, and they display selectivity, sensitivity, and versatility. Coulet and Blum developed several bioluminescent-based sensors using commercially available preactivated polyamide membranes to immobilize luminescent enzymes. The tubes connected the enzymatic membrane to an 8-mm-diameter, 1-m-long glass fiber bundle through a screw cap [19, 226, 227]. The light generated by the enzymatic reaction in the presence of the target analytes was conducted to the photomultiplier tube (PMT) of a luminometer. This fiberoptic sensor proved to be suitable for any kind of luminescent enzymes, allowing detection of several analytes.

Table 8 lists analytes that were determined by using immobilized luminescent enzymes [20].

A typical example of these analytical systems is a manifold using bacterial luciferase for L-phenylalanine assay [228] developed with two separate nylon coils, as shown in Figure 3. The first one contained the specific L-phenylalanine dehydrogenase (L-PheDH) enzyme.

$$\text{L-Phenylalanine} + \text{NAD}^+ \xrightleftharpoons{\text{L-PheDH}} \text{L-phenylpyruvate}$$

$$+ \text{NADH} + \text{H}^+ + \text{NH}_4^+$$

The second one contained the bacterial bioluminescent enzymes. This system made it possible to reach a detection limit of 0.5 µmol/L.

Sensitive flow-injection analyses of aspartate, glutamate, 2-oxoglutarate, and oxaloacetate were developed using immobilized bacterial luciferase enzymes.

Table 8 Analytical Performance of the Immobilized Bioluminescent Enzymes

Analyte	Range (pmol)	Detection limit (μmol/l)	Imprecision	
			Content (pmol)	RSD (%)
NADH (B)[a]	1–2500	0.1	2.5	3.7
NADPH (B)	5–1500	1	10	5.3
α-Bile acid (B)	1–1000	0.5	2.7	5.9
3α-Bile acid (B)	1–1000	0.5	2.8	7.1
12α-Bile acid (B)	10–10000	1	5.2	6.3
Ethanol (B)	50–2500	1	100	8.3
Glycerol (B)	50–500	2	70	7.2
Acetaldehyde (B)	50–1000	2	100	6.8
L-Alanine (B)	5–500	0.5	75	7.4
Branched-chain amino acids (B)	20–2000	0.5	75	7.8
ATP (L)	0.05–100	0.02	1.5	8.0
ADP (L)	0.6–100	0.1	15	8.2
ATP (M, r-LM)	1–500	0.3	10	9.3
ATP (r-LN)	0.3–100	0.06	10	5.2
L-Lactic acid (B)	1–500	0.1	7	5.2
D-Lactic acid (B)	10–500	5	50	7.8
L-Glutamic acid (B)	50–1000	10	500	3.7
L-Phenylalanine (B)	1–100	0.5	50	4.3
Magnesium (L)	0.05–6.7	50	0.67	4.5
Magnesium (r-LN, M)	0.01–6.7	10	0.99	2
Lactate dehydrogenase (B)	3–2000[a]	1[a]	151[a]	8.8
Creatine kinase (L)	0.1–100[a]	0.1[a]	17[a]	8.2

[a] B, bioluminescent bacterial system on nylon; L, bioluminescent firefly system on nylon; M, bioluminescent firefly system on methacrylate beads; r-LM, recombinant firefly luciferase on methacrylate beads; r-LN, recombinant firefly luciferase on nylon.
[a] μmol/min L.

Precision was generally excellent, and sensitivities were 100–1000-fold higher than with spectrophotometric methods. The immobilized-enzymes preparations were stable for several months and each reactor could be used for 600–800 analyses [229].

A flow assay was reported for determination of inorganic pyrophosphate: a pyrophosphatase was coimmobilized with luciferase on Sepharose beads with continuous flow of saturating concentrations of substrates. The instrument allowed automation with a throughput of approximately one sample every 4 min.

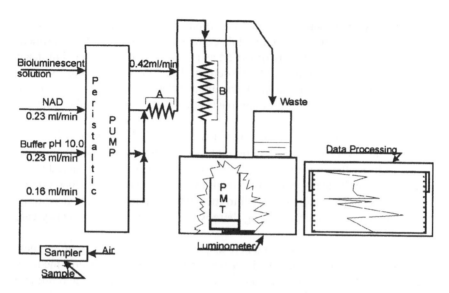

Figure 3 Example of a typical luminescent flow sensing device [manifold for pesticide chemiluminescent flow assay with one (A) or two (B) columns using immobilized enzymes]. A = immobilized dehydrogenase (Phe-DH) 1.0 m coil; B = immobilized bioluminescent enzymes 0.5 m coil.

Less than 1 pmol pyrophosphate was determined in a volume of 20 μL with a coefficient of variation approximately equal to 4% [230].

A continuous-flow system was developed for the assay of magnesium (II) in serum, drugs, and beverages, using firefly luciferase (*luc*) or recombinant luciferase (r-*luc*) from *E. coli*, immobilized both enzymes on nylon coil and on epoxy methacrylate [231]. The volume of the analyzed specimens was 10 μL with a sample assay rate of 20 samples per hour and no carryover in the system. The detection limit of the assay was 0.05 mmol/L for LUC-nylon coil and 0.01 mmol/L for LUC-Eupergit column and r-*luc* nylon coil. No interference from ions present in the samples (Ca^{2+}, Mn^{2+}, Fe^{2+}, Cu^{2+}, Zn^{2+}, Co^{2+}) was detected. The nylon-immobilized enzymes have a relatively high stability (1–4 months), despite low recovery in terms of activity with respect to the soluble forms; Eupergit C gave better sensitivity and activity recovery, but lower enzyme stability (3 days–1 month).

Other analytes recently determined by means of immobilized bioluminescent enzymes are europium(III) [232], D- and L-lactate in beer [233], and D-sorbitol [234]. A novel approach of the sensing layer design was reported for this latter assay. In biosensors based on multienzymatic sequential reactions, the different enzymes are usually coimmobilized on the same support. Compartmentalization of the sensing layer is proposed as an alternative, enabling significant improvement in the response of the biosensor. The enhancement was assumed to result from hyperconcentration of the intermediate product, common to the two sequential reactions. Such an approach proved applicable to other systems [235].

Flow sensors were also used coupled with microdialysis probe [236], providing several advantages: the reaction took place in few minutes allowing continuous analysis; the sensitivity was in the order of pmoles; the microdialysis probe allowed biological specimens to be drawn without proteins or macromolecules. This technique can be extended to the analysis of analytes that need to be detected continuously such as during therapy monitoring, or in emergency care units.

The variety of assays reported here displays the great versatility of luminescent detection systems. As already described, in all luminescent systems the main advantages are the high sensitivity and specificity, which reduce to the minimum the sample treatment, and the ease of use of the reagents and the luminometer. Immobilized systems greatly reduce the cost per assay; on the other hand, their preparation requires expertise, especially in the surface activation step on nylon tubes.

4. OTHER BIOLUMINESCENT SYSTEMS

As previously mentioned, luminescent reactions occur in almost all zoological kingdoms. The literature is therefore full of papers reporting bioluminescent systems other than those just described [49, 237].

Cloning of the cDNAs that code for several luciferases from a bioluminescent click beetle, *Pyrophorus plagiophtalatus*, dates back to 1989 [238]. The clones code for luciferases of four types, distinguished by the colors of BL they catalyze. Owing to these different colors, these clones may be useful when multiple reporter genes are needed. An application was reported in which the different wavelengths allow both a target and control signal to be incorporated into each cell, providing a means of differentiating between specific effects of a genetic sensing system and other nonspecific interfering influences [239].

The oral bacterium *Streptococcus mutans*, containing the luciferase gene of *Pyrophorus plagiophtalatus*, was used as rapid assay to estimate the effects of various antimicrobial treatments [93].

Pholasin is the protein-bound luciferin from the bivalve mollusc *Pholas dactylus*. This substrate reacts with its luciferase and molecular oxygen to produce light. The photoprotein, commercially available, undergoes an oxidative reaction to oxypholasin in the presence of superoxide anion, peroxidases, or other oxidants. Actually, it acts as an ultrasensitive detector of reactive oxygen species, such as those derived from activated leukocytes. Therefore, pholasin is a useful tool to quantify inflammation and infections, in particular through recognition of granulocyte conditions: quiescent, activated, prone to degranulation, degranulated, and up- or down-regulated in response to receptor stimulants [240].

Cypridina luciferin analogs are widely used for several analytical applications (determination of substrates, enzymes, active oxygen species such as superoxide), but they are mainly related to CL [241, 242].

Other bioluminescent microorganisms such as dinoflagellates have some applications, but generally only for teaching purposes [243].

Some additional luminescent systems of some importance are further described in detail.

4.1 *Aequorea victoria*

Aequorea victoria is a hydromedusan jellyfish found mainly at Friday Harbor, Washington [244], which emits a greenish light (λ_{max} 508 nm) from the margins of its umbrella. The bioluminescent system consists of two proteins, a calcium-binding photoprotein named aequorin and a green fluorescent protein (GFP). As a consequence of the binding of calcium ions, the aequorin reacts, in vitro, yielding blue light, a blue fluorescent protein (BFP), and carbon dioxide. In the jellyfish, where aequorin is considered to be closely associated with GFP, the excited state of BFP undergoes radiationless energy transfer to GFP, which than emits green light relaxing to its ground state (Fig. 4.)

4.1.1 Aequorin

Aequorin is the most widely studied of the Ca^{2+}- binding photoproteins described to date. It is composed of an apoprotein (apo-aequorin; 189-amino-acid residues), molecular oxygen, and a luciferin named coelenterazine. The protein contains three EF-hand Ca^{2+}-binding sites, and when Ca^{2+} occupies these sites, aequorin is converted into an oxygenase (luciferase), which oxidizes coelenterazine. The latter is converted to a highly unstable dioxetanone intermediate, which releases carbon dioxide to form the excited phenolate anion of coelenteramide. Coelenteramide remains noncovalently bound to the protein, emitting blue light when relaxing to the ground state [244].

As light emission from aequorin is dependent on Ca^{2+}, the protein has been widely employed for determination of this ion. In particular, the protein was used in the past in the highly sensitive measurement of intracellular calcium concentration in several kind of cells [16]. More recently immobilized aequorin was used to develop an optical biosensor for measurement of calcium ions in complex

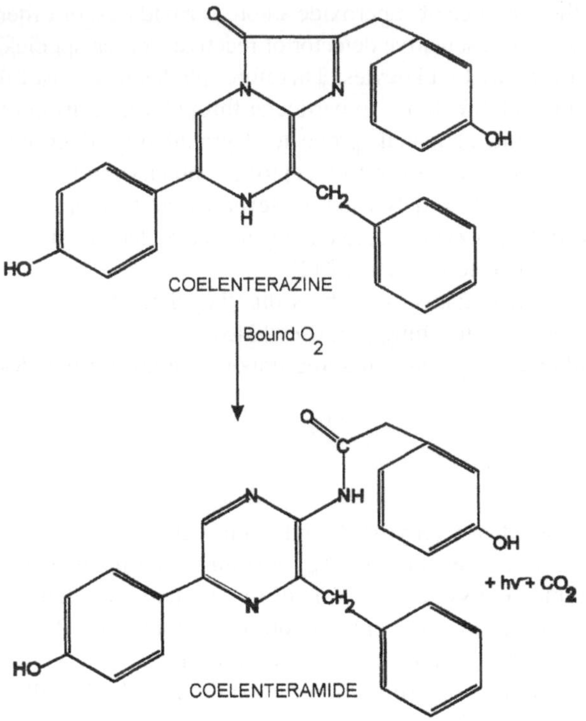

Figure 4 Reaction scheme for light emission of aequorin.

matrices such as human serum and milk samples. A porous sol-gel glass environment was used for the immobilization, having negligible effects on the protein stability and calcium-ion-binding activity [245].

Cloning of the gene encoding apo-aequorin synthesis allowed recombinant expression of the photoprotein [246], which led to genetic transformation of bacteria, yeast, plant, and animal cells. Several applications have arisen with the use of recombinant aequorin. One of the most interesting was the development of methods to assay Ca^{2+} in a discrete subcellular domain [247]. A further application was the use of recombinant aequorin as a quantitative label in different analytical assays. Several characteristics make aequorin an ideal label: it can be readily conjugated, it generates a high signal-to-noise ratio with a broad range of sensitivity, and the signal is generated rapidly [244]. Lizano et al. pointed out the importance of aequorin as a label in competitive binding assays for biotin [248]. The interaction of the vitamin biotin with the binding protein avidin, which is characterized by strong affinity, is currently employed to enhance the sensitivity of competitive binding assay in a variety of samples. It was demonstrated that the Ca^{2+}-triggered luminescent reaction of biotinylated aequorin can be inhibited by the presence of avidin; the reduction of bioluminescent intensity is correlated to the biotin concentration in a sample [249]. The reported assay had a detection limit of 10^{-14} M with high reproducibility. The same technique was applied to the detection of prostate-specific antigen (PSA) mRNA [250]. PSA mRNA from a single cell, in the presence of one million cells that do not express PSA, was detected with a signal-to-background ratio of 2.5 and typical CVs of 6%. Recently this aequorin BL was also used for detecting biomolecules in picoliter vials [251], which were fabricated on glass substrates using a laser ablation technique. The assay in such small-volume vials can find potential applications in a variety of fields, such as microanalysis and single-cell analysis, where the amount of the sample is limited.

The aequorin label, commercially available, proved to be useful in clinical immunoassay [252] and for the detection of nucleic acids. Aequorin was used to develop a rapid diagnostic assay for qualitative evaluation of cytomegalovirus presence in clinical samples [253]. The assay could be performed in less than 2 h and is amenable to automation and to processing hundreds of samples per hour. In a recent study aequorin-based bioluminescent technology was employed for direct quantification of cytokine mRNA by reverse transcriptase PCR techniques (RT-PCR) and for the investigation of induction of human cytokine expression [254]. The bioluminescent assay proved to be 30- to 60-fold more sensitive than radioimaging in detecting human IL-2 and CD3-delta amplicons and allowed processing of hundreds of RT-PCR samples per day.

4.1.2 Green Fluorescent Protein

Green fluorescent protein (GFP) is made up of 238 amino acid residues in a single polypeptide chain. The fluorophore is formed by the cyclization and oxidation of three amino acids-Ser-Tyr-Gly [255]. Wild-type GFP absorbs blue light (peaks at 395 nm and 475 nm) and emits green light with a maximum at 509 nm. Several characteristics, together with the molecular cloning of GFP cDNA [256], have led to widespread use of GFP in a variety of fields, including environmental monitoring, cell biology, and molecular biology. Despite being rather thermosensitive, GFP is very resistant to denaturation and does not require any cofactor or substrate. Mutational analysis of GFP has generated a variety of GFP, which differ from the native protein, such as increase in brightness and in stability. As a consequence, use of GFP has even increased.

One of the main applications of GFP is as reporter gene. GFP has been expressed in a variety of organisms, including animals [257], plants [258], bacteria [259], and viruses [260], for monitoring gene expression. The attractiveness unique to GFP as a reporter allows nondestructive in vivo fluorescence visualization. GFP can also be employed to investigate organelle structure and can function as a protein tag, as it tolerates N- and C-terminal fusion to a variety of proteins, many of which retain their native functions.

GFP has also been proposed as a successor to the Ames and SOS chromotest. Billinton et al. [8] obtained a reporter system, employed as genotoxicity biosensor, that uses eukaryotic cells (the baker yeast *Saccharomyces cerevisiae*) instead of bacteria. The strain produces green fluorescent protein, codon optimized for yeast, when DNA damage has occurred. It was demonstrated that the reporter does not falsely respond to chemicals that delay mitosis, and responds appropriately to the genetic regulation of DNA repair.

In a recent work [261] a recombinant herpesvirus, PrV, expressing a genetically modified version of the cDNA encoding *gfp* was constructed. GFP proved to be useful for monitoring the virus infection in cell culture and in rat brain tissue and can be used to follow the spreading route of the virus in the nervous system, which corresponds to the synaptic linkage of the neurons. The generated viral strain, with the virulence appropriately reduced, is also potentially applicable as a vehicle for expressing GFP in permissive cells, including use for transneural tract tracing.

Regarding environmental applications, GFP appears to be ideally suited for in situ detection of specific bacteria in environmental samples. GFP as a marker has the advantage of being stable during starvation conditions, which is a common state for bacteria in natural environments. Moreover, the fluorescence intensity of GFP is so strong that even bacteria with a single copy of *gfp* can be detected [262]. A *gfp*-transformed *Mycobacterium smegmatis* strain was employed to investigate the impact of biofilms on pathogen persistence in potable water [263]. Biofilms are ubiquitous in drinking-water-distribution plants. The attachment and retention of the *gfp*-transformed strain was monitored in laminar flow cells exposed to different concentrations of chlorination. Using a transformed organism allowed tracking its response to disturbances in its environment in real time.

4.2 Obelin

Obelin is a Ca^{2+}-activated bioluminescent photoprotein that has been isolated from the marine polyp *Obelia longissima*. Binding of calcium ions determines a luminescent emission. The protein consists of 195 amino acid residues [264] and is composed of apoobelin, coelenterazine, and oxygen. As aequorin, it contains three EF-hand Ca^{2+}-binding sites and the luminescent reaction may be the result of coelenterazine oxidation by way of an intramolecular reaction that produces coelenteramide, CO_2, and blue light. As for aequorin, the luminescent reaction of obelin is sensitive to calcium and the protein was used in the past as an intracellular Ca^{2+} indicator. More recently, the cloning of cDNA for apoobelin led to the use of recombinant obelin as a label in different analytical systems.

Recombinant obelin was derivatized with biotin and employed in avidin-biotin immunoassays for detection of biotinylated targets immobilized on microtiter wells. Frank and Vysotski [265] applied this technique to detect alpha-fetoprotein (AFP), a level of monitoring widely used for early cancer detection and prenatal diagnosis of some genetic diseases. The results obtained were found to be similar to those obtained for radioimmunoassay. Another method, based again on obelin BL, was successively reported for the assay of alpha-fetoproteins and human luteinizing hormone [266]. A bifunctional protein (proZZ-Obe) was constructed, which has both the luminescent activity of obelin and the ability to bind mammalian IgG. It was demonstrated that the chimeric protein proZZ-Obe can be used as a universal marker to determine almost any antigen in a sandwich-format immunoassay with the $F(ab)_2$ fragment of immunoglobulins.

Recently it was found that obelin mRNA can be a useful tool for evaluating the efficiency of cell-free translation and for screening of translation inhibitors.

4.3 *Renilla* Luciferase

Renilla luciferase is a monomeric protein contained in the bioluminescent sea pansy *Renilla reniformis*. The luciferase catalyzes the oxidation of coelenterazine to produce light emission at 482 nm. In vivo an energy transfer to green fluorescent protein occurs and light is emitted at 509 nm. Cloning of the *Renilla luciferase* gene [267] allowed its use as a reporter gene. Its suitability for this use is confirmed by the several works in the recent literature on this topic. In one of these papers the characteristics of *Renilla* luciferase were combined with those for GFP through the engineering of a new protein [268]. This bifunctional polypeptide may become a useful tool based on fluorescence for identification of transformed cells at the single-cell level. Simultaneously, it may allow quantifying of promoter activation in transformed tissues and transgenic organisms by measurement of luciferase activity. More recently *Renilla* luciferase gene was associated with the firefly one to develop dual-luciferase reporter systems. The assay was used to measure translation-coupling efficiency of recording mechanisms such as frame shifting or readthrough [269]. In another work a dual-luciferase assay was employed to simultaneously screen agonist activity at two G-protein-coupled receptors in a 96-well format, resulting in significant time and cost savings [270]. Recently the use as reporter of gene expression in living cells has been extended to the chloroplast genome [271].

5. CONCLUSIONS

The wide collection of analytical applications of the main bioluminescent systems here reported, even necessarily incomplete, is supposedly exhaustive enough to illustrate the great versatility of these biological reagents and to stir the curiosity of researchers who never used these techniques to solve their analytical problems.

The intensive research carried out to achieve good stability and easy handling of the BL reagents allowed development of many simple, rapid, highly sensitive and specific assays, many of which are available as commercial kits. More recently application of the newest techniques of molecular biology greatly accelerated the continuous improvement of BL assay performance.

Moreover, national and international laws and directives, such as International Standard Organization (ISO) documents and HACCP instructions, help to increase the use of BL techniques and kits as official method of analyses.

Today, the most advanced applications are in the field of rapid microbiology and molecular biology to perform a great variety of biochemical, biomedical, environmental, and food analyses.

REFERENCES

1. MA DeLuca, WD McElroy. Bioluminescence and Chemiluminescence, Part B. Methods in Enzymology. vol 133. New York: Academic Press, 1986.
2. LJ Kricka. Anal Chem 67:499–502, 1995.
3. FR Leach. J Appl Biochem 3:473–517, 1981.
4. EA Meighen. Annu Rev Genet 28:117–139, 1994.
5. T Wilson, JW Hastings. Annu Rev Cell Dev Biol 14:197–230, 1998.
6. JW Hastings. Gene 173:5–11, 1996.
7. AK Campbell. Chemiluminescence: Principles and Applications in Biology and Medicine. Chichester: Ellis Horwood, 1988.
8. N Billinton, MG Barker, CE Michel, AW Knight, WD Heyer, NJ Goddard, PR Fielden, RM Walmsley. Biosens Bioelectron 13:831–838, 1998.
9. S Rudolf, PN Michael. Anal Chem 46:188–202, 1974.
10. MA DeLuca. Methods Enzymol 57, 1978.
11. TP Whitehead, LJ Kricka, TJN Carter, GHG Thorpe. Clin Chem 25:1531–1546, 1979.
12. LK Kricka, HG Thorpe. Analyst 108:1274–1296, 1983.
13. K Wulff. In: HU Bergmeyer, ed. Methods of Enzymatic Analysis, 3rd ed, vol 1. Weinheim: Verlag Chemie, 1983, pp 340–368.
14. JW Hastings. J Mol Evol 19:309–321, 1983.
15. DJ O'Kane, DC Prasher. Mol Microbiol 6:443–449, 1992.
16. PH Cobbold, TJ Rink. J Biochem 248:313–328, 1987.
17. LK Kricka. Anal Biochem 175:14–21, 1988.
18. LJ Blum, PR Coulet. Biosensor Principles and Applications. New York: Marcel Dekker, 1991.
19. PR Coulet, LJ Blum. Trends Anal Chem 11:57–61, 1992.
20. S Girotti, EN Ferri, S Ghini, F Fini, M Musiani, G Carrea, A Roda, P Rauch. Chem Listy 91:477–482, 1997.

21. HC Rongen, RMW Hoetelmans, A Bult, WP Van Bennekom. J Pharmaceut Biomed Anal 12:433–464, 1994.
22. LJ Kricka. J Biolum Chemilum 13:189–193, 1998.
23. UK Winkler. Phys Stat Sol 130:K81–K84, 1991.
24. I Bronstein, J Fortin, PE Stanley, GSAB Stewart, LJ Kricka. Anal Biochem 219: 169–181, 1994.
25. RL Price, DJ Squirrell, MJ Murphy. J Clin Ligand Assay 21:349–357, 1998.
26. M Pazzagli, E Cadenas, LJ Kricka, A Roda, PE Stanley. J Biolum Chemilum 4: 1–646, 1989.
27. A Szalay, LJ Kricka, PE Stanley. Bioluminescence and Chemiluminescence: Status Report, 7th International Symposium on Bioluminescence. Chichester: Wiley, 1993.
28. AK Campbell, LJ Kricka, PE Stanley. Bioluminescence and Chemiluminescence: Fundamentals and Applied Aspects. New York: Wiley, 1994.
29. JW Hastings, LJ Kricka, PE Stanley. Bioluminescence and Chemiluminescence: Molecular Reporting with Photons. New York: Wiley, 1997.
30. PE Stanley, R Smither, WJ Simpson. A Practical Guide to Industrial Uses of ATP-Luminescence in Rapid Microbiology. Lingfield: Cara Technology, 1997.
31. A Roda, M Pazzagli, LJ Kricka, PE Stanley. Bioluminescence and Chemilumines-cence: Perspectives for the 21st Century. New York: Wiley, 1999.
32. LJ Kricka, PE Stanley. J Biolum Chemilum 6:45–67, 1991.
33. NN Ugarova. J Biolum Chemilum 6:139–145, 1991.
34. LJ Kricka, PE Stanley. J Biolum Chemilum 6:203–220, 1991.
35. LJ Kricka, PE Stanley. J Biolum Chemilum 6:289–296, 1991.
36. LJ Kricka, PE Stanley. J Biolum Chemilum 7:47–73, 1992.
37. LJ Kricka, PE Stanley. J Biolum Chemilum 7:133–142, 1992.
38. LJ Kricka, PE Stanley. J Biolum Chemilum 7:223–228, 1992.
39. LJ Kricka, PE Stanley. J Biolum Chemilum 7:263–298, 1992.
40. LJ Kricka, O Nozaki, PE Stanley. J Biolum Chemilum 8:169–182, 1992.
41. PJ Hill, GSAB Stewart, PE Stanley. J Biolum Chemilum 8:267–291, 1993.
42. LJ Kricka, PE Stanley. J Biolum Chemilum 9:87–105, 1994.
43. LJ Kricka, PE Stanley. J Biolum Chemilum 9:379–388, 1994.
44. LJ Kricka, PE Stanley. J Biolum Chemilum 10:133–146, 1995.
45. LJ Kricka, PE Stanley. J Biolum Chemilum 10:301–322, 1995.
46. LJ Kricka, PE Stanley. J Biolum Chemilum 10:361–370, 1995.
47. LJ Kricka, PE Stanley. J Biolum Chemilum 11:39–48, 1996.
48. LJ Kricka, PE Stanley. J Biolum Chemilum 11:107–122, 1996.
49. LJ Kricka, PE Stanley. J Biolum Chemilum 12:113–134, 1997.
50. LJ Kricka, PE Stanley. J Biolum Chemilum 12:261–270, 1996.
51. LJ Kricka, PE Stanley. J Biolum Chemilum 13:41–59, 1997.
52. LJ Kricka, PE Stanley. J Biolum Chemilum 13:157–184, 1997.
53. PE Stanley. J Biolum Chemilum 12:61–78, 1997.
54. PE Stanley. Luminescence 14:201–213, 1999.
55. M Berger. Creatures that Glow: A Book About Bioluminescent Animals. New York: Scholastic, 1996.
56. JJ Presnall. Animals that Glow. New York: F Watts, 1993.

57. J Lee. In: KC Smith, ed. The Science of Photobiology, 2nd ed. New York: Plenum, 1989, pp 391–417.
58. MA DeLuca, WD McElroy. Methods Enzymol 57:1–124, 1978.
59. H Arakawa, M Maeda, A Tsuji. Anal Lett 25:1055–1063, 1992.
60. S Tsasuya, M Shigeharu. Jpn Kokai Tokkyo Koho JP 11 56,393 [99 56,393], 1997.
61. I Feliciello, G Chinali. Anal Biochem 212:394–401, 1993.
62. A Drousiotou, O Ioannou, T Georgiou, E Mavrikiou, G Christopoulos, T Kyria-kides, M Voyasianos, A Argyriou, L Middleton. Genet Test 2:55–60, 1998.
63. M Levin, T Bjornheden, M Evaldsson, S Walenta, O Wiklund. Arterioscler Thromb Vasc Biol 19:950–958, 1999.
64. IK Goto, TS Murakami, M Maeda, S Shioda, A Arimura. Anal Chim Acta 382: 245–251, 1999.
65. NH Chiu, TK Christopoulos. Clin Chem 45:1954–9, 1999.
66. OH Abe, KI Kosaka, M Maeda. Anal Chim Acta 395:265–272, 1999.
67. H Arakawa, A Kokado, S Yoshizawa, M Maeda, A Tokita, Y Yamashiro. Anal Sci 15:943–949, 1999.
68. OH Hansen, DM Karl. Methods Enzymol 57:73–85, 1978.
69. DS Askgaard, A Gottschau. Biologicals 23:55–60, 1995.
70. AA Mafu, D Roy, L Savoie, J Goulet. Appl Environ Microbiol 57:1640–1643, 1991.
71. R Blasco, MJ Murphy, MF Sanders, DJ Squirrell. J Appl Microbiol 84:661–666, 1998.
72. JN Miller, MB Nawawi, C Burgess. Anal Chim Acta 266:339–343, 1992.
73. P Nielsen, E Van Dellen. J Assoc Off Anal Chem 72:708–711, 1989.
74. W Reybroek, E Schram. J Neth Milk Dairy 49:1–14, 1995.
75. SC Murphy, SM Kozlowsky, DK Bandler, KJ Boor. J Diary Sci 81:817–820, 1998.
76. G Wahlstrom, PE Saris. Appl Environ Microbiol 65:3742–3745, 1999.
77. S Girotti, EN Ferri. Quim Anal 16(Suppl 1):S111–S117, 1997.
78. T Miyamoto, Y Kuramitsu, A Ookuma, S Trevanich, K Honjoh, S Hatano. J Food Prot 61:1312–1313, 1998.
79. MJ Waites, K Ogden. J Inst Brew 93:30–32, 1987.
80. E Storgårds, A Haikara. Ferment 9:352–360, 1996.
81. TA Green, SM Russel, DL Fletcher. J Food Prod 62:86–90, 1999.
82. H Tanaka, T Shinji, K Sawada, Y Monji, S Seto, M Yajima, O Yagi. Water Res 31:1913–1918, 1997.
83. JA Poulis, M de Pijper. Int J Food Microbiol 20:109–116, 1993.
84. DA Bautista, JP Villancourt. Poultry Sci 73:1673–1678, 1994.
85. GR Siragusa, CN Cutter, WJ Dorsa, M Koohmaraie. J Food Prot 58:770–775, 1995.
86. C Bell, PA Stallard, SE Brown, JTE Standley. J Int Diary 4:629–640, 1994.
87. D Champiat, P Granier. La vigne avril:35–36, 1992.
88. E Storgårds, R Juvonen, L Vanne, A Haikara. Detection Methods in Process and Hygiene Control. European Brewery Convention, Monograph 26, Symposium on Quality Issues and HACCP. Stockholm, 1997, pp 95–107.
89. S Hoffer, C Jimenez-Misas, A Lundin. Luminescence 14:255–261, 1999.
90. MJ Harber, W Asscheer. J Antimicrob Chemother 3:35–41, 1997.
91. H Hanberger, E Svensson, M Nilsson, LE Nilsson, EG Horsten, R Maller. J Antimi-crob Chemother 31:245–260, 1993.

92. I Codita, C Popescu Roum. Arch Microbiol Immunol 55:323–331, 1996.
93. V Loimaranta, J Tenovuo, L Koivisto, M Karp. Antimicrob Agents Ch 42:1906–1910, 1998.
94. S Trajkovska, S Dzhekova-Stojkova, S Kostovska. Anal Chim Acta 290:246–248, 1994.
95. LE Nilsson, Ö Molin, S Ånséhn. J Biolum Chemilum 3:101–104, 1989.
96. A Decool, V Goury, A Tibi, S Gibaud, F Vincent, JC Darbord. Anal Chim Acta 255:423–425, 1991.
97. S Girotti, E Ferri, ML Cascione, A Orlandini, L Farina, S Nucci, F Di Graci, R Budini. Anal Biochem 192:350–357, 1991.
98. HN Nguyen, BU Sevin, HE Averette, J Perras, D Donato, M Penalver. Gynecol Oncol 45:185–191, 1992.
99. BU Sevin, HE Averette, J Perras, D Donato, M Penalver. Cancer Suppl 71:1613–1620, 1993.
100. A Steren, BU Sevin, J Perras, R Angioli, HN Nguyen, L Guerra, O koechli, HE Averette. Gynecol Oncol 48:252–258, 1993.
101. RD Petty, RA Sutherland, EM Hunter, IA Cree. J Biolum Chemilum 10:29–34, 1995.
102. PE Andreotti, TT Thonthwaite, IS Morse In: Bioluminescence Chemiluminescence: Current Status. New York: Wiley, 1991, pp 417–420.
103. M Dellian, S Walenta, F Gamarra, GEH Kuhnle, W Mueller-Klieser, AE Goetz. Br J Cancer 68:26–31, 1993.
104. A Guadano, E de la Pena, A Gonzalez-Coloma, JF Alvarez. Mutagenesis 14:411–415, 1999.
105. A Koop, PH Cobbold. J Biochem 295:165–170, 1993.
106. S Forman, J Kas, F Fini, M Steinberg, T Ruml. J Biochem Mol Toxic 13:11–15, 1999.
107. V Schultz, I Sussman, K Bokvist, K Tornheim. Anal Biochem 215:302–304, 1993.
108. CK Browers, AP Allshire, PH Cobbold. J Mol Cell Cardiol 24:213–218, 1992.
109. PG Arthur, PW Hochachka. Anal Biochem 227:281–284, 1995.
110. BJ Tuovila, FC Dobbs, PA LaRock, BZ Siegel. Appl Environ Microbiol 53:2749–2753, 1987.
111. V Prodi, S Girotti, S Agostini, E Ferri. Quad Med Lav Med Riabil 1:203–204, 1995.
112. IW Stewart, G Leaver, SJ Futter. J Aerosol Sci 28:511–523, 1997.
113. C Sorlini, E Cesarotti. Ann Microbiol 41:267, 1991.
114. M Arretxe, JM Heap, N Christofi. Inc Environ Toxicol Water Qual 12:23–29, 1997.
115. SJ Saul, E Zomer, D Puopolo, SE Charm. J Food Protect 59:306–311, 1996.
116. T Reemtsma, A Putschew, M Jekel. Waste Manag 19:181–188, 1999.
117. WJ Simpson, JRM Hammond. J Biolum Chemilum 6:97–106, 1991.
118. M Velazquez, JM Feirtag. J Food Protect 60:799–803, 1997.
119. T Kamidate, T Kinkou, H Watanabe. Anal Biochem 244:62–66, 1997.
120. A Lundin, A Rickardsson, A Thore. Anal Biochem 75:611–620, 1976.
121. A Lundin. In: A Szalay, LJ Kricka, PE Stanley eds. Bioluminescence and Chemiluminescence: Status Report. Chichester: Wiley, 1993, pp 291–295.
122. N Kajiyama, T Masuda, H Tatsumi, E Nakano. Biochim Biophys Acta 1120:228–232, 1992.

123. IA Lundovskik, EI Dementieva, NN Ugarova. Biochemistry 63:691–696, 1998.
124. T Higashi, A Isomoto, I Tyuma, E Kakishita, M Uomoto, K Nagai. Thromb Haemost 53:65–69, 1985.
125. K Miayi, T Miyagi, N Ashida, Y Narizuka, K Taniguchi, H Tatsumi, K Ikonaoka, S Ida, T Oura. J Endocrinol 46:761–766, 1998.
126. T Mosees, R Geiger, W Miska. J Biolum Chemilum 4:213–218, 1995.
127. W Miska, R Geiger. Clin Chem Biochem 25:23–30, 1987.
128. R Hauber, R Geiger. J Biolum Chemilum 4:367–372, 1989.
129. A-S Rhedin, U Tidefelt, K Jönsson, A Lundin, C Paul. Leukemia Res 17:271–276, 1993.
130. A Lundin. In: PE Stanley, R Smither, WJ Simpson. A Practical Guide to Industrial Uses of ATP-Luminescence in Rapid Microbiology. Lingfield: Cara Technology, 1997, pp 55–62.
131. WD Griffits, GAL DeCosemo. J Aerosol Sci 8:1425–1458, 1994.
132. S Girotti, F Zanetti, E Ferri, S Stampi, F Fini, S Righetti, A Roda, R Budini. Ann Chim 88:291–298, 1998.
133. A Moller, K Gustafsson, JK Jansson. FEMS Microbiol Ecol 15:193–206, 1994.
134. SR Ford, MS Hall, FR Leach. J Biolum Chemilum 7:185–193, 1992.
135. WM Barnes. Proc Natl Acad Sci USA 87:9183–9187, 1990.
136. C Aflafo. Biochemistry 29:4758–4766, 1990.
137. E Kobatake, T Niimi, T Haruyama, Y Ikariyama, M Aizawa. Biosens Bioelectron 10:601–605, 1995.
138. Y Ikariyama, S Nishiguchi, T Koyama, E Kobatake, M Aizawa, M Tsuda, T Nakazawa. Anal Chem 69:2600–2605, 1997.
139. S Subramani, M DeLuca. In: J Setlow, A Hollaendr, eds. Genetic Engineering-Principles and Methods, Vol 10. New York: Plenum Press, 1985, pp 75–89.
140. TJ Sweeney, V Mailander, AA Tucker, AB Olomu, W Zhang, Y Cao, RS Negrin, CH Contag. Proc Natl Acad Sci USA 96:12044–12049, 1999.
141. J Yang, DB Thomason. BioTechniques 15:848–850, 1993.
142. S Lee, M Suzuki, E Tamiya, I Karube. Anal Chim Acta 244:201–206, 1991.
143. M Virta, J Lampinen, M Karp. Anal Chem 67:667–669, 1995.
144. A Cebolla, ME Vazquez, AJ Palomares. Appl Environ Microbiol 61:660–668, 1995.
145. M Virta, S Lineri, P Kankaanpaa, M Karp, K Peltonen, J Nuutila, Esa-Matti Lilius. Appl Environ Microbiol 64:515–519, 1998.
146. JH Axelrod, A Honigman. AIDS Res Hum Retroviruses 20:759–767, 1999.
147. PV Dunlap, U Mueller, TA Lisa, KS Lundberg. J Gen Microbiol 138:115–123, 1992.
148. MA DeLuca, WD McElroy. Methods Enzymol 57:125–234, 1978.
149. A Roda. S Girotti, S Ghini, G Carrea. J Biolum Chemilum 4:423–435, 1989.
150. MH Vatn, B Thusus, A Norheim. J Cancer 3:366–369, 1990.
151. JC Makemson. J Bacteriol 172:4725–4727, 1990.
152. K Tanaka, E Ishikawa. Anal Lett 23:241–253, 1990.
153. JC Makemson, JW Hastings. J Biolum Chemilum 6:131–136, 1991.
154. DA Lane, D Nadeau. J Biochem Bioph Meth 17:107–118, 1988.
155. P Cristol, S Alameddine, B Terouanne, JC Nicolas, F Arnal, B Hedon, A Crastes de Paulet. Ann Biol Clin 47:167–180, 1989.

156. PJ Holt, NC Bruce, CR Lowe. Anal Chem 68:1877–1882, 1996.
157. FJ Romero, W Mueller-Klieser. J Biolum Chemilum 13:263–266, 1998.
158. LC Welch, A Bryant, T Kealey. Anal Biochem 176:228–233, 1989.
159. OW Paschen, G Mies, O Kloiber, KA Hossmann. J Cerebr Blood F Met 5:465–468, 1985.
160. PG Arthur, JC Kent, PE Hartmann. Anal Biochem 176:449–456, 1989.
161. PG Arthur, PW Hochachka. Anal Biochem 227:281–284, 1995.
162. NL Jocoy, DJ Butcher. Spectrosc Lett 31:1589–1597, 1998.
163. A Frantz, S Salla, C Redbrake. Graef Arch Clin Exp Ophthalmol 236:61–64, 1998.
164. MJ Thompson, LM Kennaugh, TM Casey, PG Arthur. Anal Biochem 269:168–173, 1999.
165. NS Kudryasheva, IY Kudinova, EN Esimbekova, VA Kratasyuk, DI Stom. Chemosphere 38:751–758, 1999.
166. VA Kratasyuk, VA Vetrova, NS Kudryasheva. Luminescence 14:193–195, 1999.
167. GSAB Stewart. Letters Appl Microbiol 10:1–8, 1990.
168. KK Kwan, BJ Dutka, SS Rao, D Liu. Environ Pollut 65:323–332, 1990.
169. AA Bulich, KK Tung, G Scheibner. J Biolum Chemilum 5:71–77, 1990.
170. LI Sweet, DE Travers, PG Mieier. Environ Toxicol Chem 16:2187–2189, 1997.
171. S Ulitzur, A Matin, C Fraley, E Meighen. Curr Microbiol 35:336–342, 1997.
172. TSC Sun, HM Stahr. J AOAC Int 76:893–898, 1993.
173. BT Johnson, ER Long. Environ Toxicol Chem 17:1099–1106, 1998.
174. H Heida, R van der Oost. Water Sci Technol 34:109–116, 1996.
175. A Fernandez, C Tejedor, F Cabrera, A Chordi. Water Res 29:1281–1286, 1995.
176. L Guzzella, S Galassi. Sci Total Environ (Suppl):1217–1226, 1993.
177. AS Jarvis, ME Honeycutt, VA McFarland, AA Bulich, HC Bounds. Ecotoxicol Environ Safety 33:193–200, 1996.
178. H Brouwer, T Murphy, L McArdle. Environ Toxicol Chem 9:1353–1358, 1990.
179. E Argese, C Bettiol, AV Ghirardini, M Fasolo, G Giurin, PF Ghetti. Environ Toxicol Chem 17:1005–1012, 1998.
180. A Carlson-Ekvall, GM Morrison. Environ Toxicol 14:17–22, 1995.
181. C Blaise, R Forghani, R Legault, J Guzzo, MS Dubow. Biotechniques 16:932–937, 1994.
182. MC Newman, JT McCloskey. Environ Toxicol Chem 15:275–281, 1996.
183. KW Thomulka, JH Lange. Ecotox Environ Safe 32:201–204, 1995.
184. MW Jacobs, JJ Delfino, G Bitton. Environ Toxicol Chem 11:1137–1143, 1992.
185. MG Shettlemore, KJ Bundy. J Biomed Mater Res 45:395–403, 1999.
186. K Rathinam, PV Mohanan. J Biomat Appl 13:166–171, 1998.
187. KW Thomulka, JH Lange. Fresenius Envir Bull 4:731–736, 1995.
188. DL Campbell, LA Lawton, KA Beattie, GA Codd. Environ Toxicol Water Qual 9:71–77, 1994.
189. JM Ribo. Environ Toxicol Water Qual 12:283–294, 1997.
190. JC Makemson, NR Fulayfil, LV Ert. J Biolum Chemilum 13:147–156, 1998.
191. EK Rodicheva, IN Trubachev, SE Medvedeva, OI Egorova, L Yu Shitova. J Biolum Chemilum 6:293–299, 1993.
192. S Lee, K Sode, K Nakanishi, JL Marty, E Tamiya, I Karube. Biosens Bioelectron 7:273–277, 1992.

193. N Kudryasheva, V Kratasyuk, E Esimbekova, E Vetrova, E Nemtseva, I Kudinova. Field Anal Chem Technol 2:277–280, 1998.
194. KB Konstantinov, P Dhurjati, T Van Dyk, W Majarian, R LaRossa. Biotechnol Bioeng 42:1190–1198, 1993.
195. S Ulitzur. J Biolum Chemilum 13:365–369, 1998.
196. WC Fuqua, SC Winans, EP Greenberg. J Bacteriol 176:269–275, 1994.
197. NBA Illarionov, VM Blinov, AP Donchenko, MV Protopopova, VA Karginov, NP Mertvetsov, JI Gitelson. Gene 86:89–91, 1990.
198. S Ulitzur, J Kuhn. J Biolum Chemilum 4:317–325, 1989.
199. S Duncan, LA Glover, K Killham, JI Prosser. Appl Environ Microbiol 60:1308–1316, 1994.
200. D Fearnside, I Caffoor. Environ Toxicol Water Qual 13:347–357, 1998.
201. AG Prest, MK Winson, JRM Hammond, GSAB Stewart. Lett Appl Microbiol 24:355–360, 1997.
202. L Geiselhart, M Osgood, DS Holmes. Ann NY Acad Sci 53–60, 1991.
203. GM Sinclair, GI Paton, AA Meharg, K Killham. FEMS Microbiol Lett 174:273–278, 1999.
204. G Palmer, R McFadzean, K Killham, A Sinclair, GI Paton. Chemosphere 36:2683–2697, 1998.
205. DJ Silcock, RN Waterhouse, LA Glover, JI Prosser, K Killham. Appl Environ Microbiol 58:2444–2448, 1992.
206. M Virta, M Karp, P Vuorinen. Antimicrob Agents Ch 38:2775–2779, 1994.
207. A Guzzo, MS DuBow. FEMS Microbiol Rev 14:369–374, 1994.
208. R Rychert, M Mortimer. Environ Toxicol Water Qual 6:415–421, 1991.
209. SP McGrath, B Knight, K Killham, S Preston, GI Paton. Environ Toxicol Chem 18:659–663, 1999.
210. B Applegate, C Kelly, L Lackey, J McPherson, S Kehrmeyer, F-M Menn, P Bienkowski, G Sayler. J Ind Microbiol Biotechnol 18:4–9, 1997.
211. EM Boyd, AA Meharg, J Wright, K Killham. Environ Toxicol Chem 17:2134–2140, 1998.
212. KE Gustavson, A Svenson, JM Harkin. Environ Toxicol Chem 17:1917–1921, 1998.
213. S Sousa, C Duffy, H Weitz, LA Glover, E Bär, R Henkler, K Killham. Environ Toxicol Chem 17:1039–1045, 1998.
214. AC Layton, B Gregory, TW Schultz, GS Sayler. Ecotox Environ Safe 43:222–228, 1999.
215. S Ramanathan, WP Shi, BP Rosen, S Dautnert. Anal Chem 69:3380–3384, 1997.
216. J Cai, MS DuBow. Biodegradation 8:105–111, 1997.
217. LM Hudson, J Chen, AR Hill, MW Griffiths. J Food Protect 60:891–897, 1997.
218. D Vanderlelie, L Regniers, B Borremans, A Provoost, L Verschave. Mutat Res 389:279–290, 1997.
219. P Rettberg, C Baumstark-Khan, K Bandel, LR Ptitsyn, G Horneck. Anal Chim Acta 387:289–296, 1999.
220. W Beyer, R Bohm. Microbiol Res 151:407–419, 1996.
221. DJ Min, JD Andrade, RJ Stewart. Anal Biochem 270:133–139, 1999.
222. PR Coulet, LJ Blum. In: DL Wise, LB Wingard, eds. Biosensors with Fiber Optics. Clifton, NJ:Humana Press, 1991, pp 293–324.

223. AR Ribeiro, RM Santos, LM Rosario, MH Gil. J Biolum Chemilum 13:371–378, 1998.
224. LJ Blum, SM Gautier. In: LJ Blum, PR Coulet eds. Biosensor Principles and Applications. New York: Marcel Dekker, 1991, pp 213–247.
225. CU Matrubutham, GS Sayler. In: A Mulchandani, KR Rogers eds. Methods in Biotechnology, Vol 6. Totowa, NJ: Humana Press, 1998, pp 249–255.
226. LJ Blum, SM Gautier, PR Coulet. J Biotechnol 31:357–368, 1993.
227. SM Gautier, PE Michel, LJ Blum. Anal Lett 27:2055–2069, 1994.
228. S Girotti, E Ferri, S Ghini, R Budini, G Carrea, R Bovara, S Piazzi, R Merighi, A Roda. Talanta 40:425–430, 1993.
229. K Kurkijarvi, T Vierijoki, T Korpela. Ann NY Acad Sci 585:394–403, 1990.
230. BA Barshop, DT Adamson, DC Vellom, F Rosen, BL Epstain, JE Seegmiller. Anal Biochem 197:266–272, 1991.
231. S Girotti, E Ferri, S Ghini, P Rauch, G Carrea, R Bovara, A Roda, MA Giosuè, G Gangemi. Analyst 118:849–853, 1993.
232. SD Clerc, RA Jewsbury, MG Mortimer, J Zeng. Anal Chim Acta 339:225–229, 1997.
233. S Girotti, M Muratori, F Fini, EN Ferri, G Carrea, M Koran, P Rauch. Eur Food Res Technol 210:90–94, 2000.
234. PE Michel, SM Gautier, LJ Blum. Enzyme Microb Technol 21:108–116, 1997.
235. PE Michel, SM Gautier, LJ Blum. Anal Chim Acta 360:89–99, 1998.
236. A Roda, S Girotti, B Grigolo, S Ghini, G Carrea, R Bovara, I Zini, R Grimaldi. Biosens Bioelectron 6:21–29, 1991.
237. A Roda, M Pazzagli, LJ Kricka, PE Stanley. Bioluminescence and Chemiluminescence: Perspectives for the 21st Century. New York: Wiley, 1999, pp 357–483.
238. KV Wood, YA Lam, HH Seliger, WD McElroy. Science 244:700–702, 1989.
239. KV Wood, MG Gruber. Biosens Bioelectron 11:207–214, 1996.
240. J Knight. In: A Roda, M Pazzagli, LJ Kricka, PE Stanley, eds. Bioluminescence and Chemiluminescence: Perspectives for the 21st Century. New York: J Wiley, 1999, pp 251–254.
241. M Mitani, S Sakaki, Y Koinuma, Y Toya, M Kosugi. Anal Sci 11:1013–1015, 1995.
242. MP Skatchov, D Sperling, U Hink, E Anggard, T Munzel. Biochem Biophys Res Commun 248:382–386, 1998.
243. JD Andrade, M Lisonbee, D Min. In: AK Campbell, LJ Kricka, PE Stanley eds. Bioluminescence and Chemiluminescence: Fundamentals and Applied Aspects. New York: Wiley, 1994, pp 371–374.
244. JM Kendall, MN Badminton. Tibtech 16:216–224, 1998.
245. DJ Blyth, SJ Poynter, DA Russell. Analyst 121:1975–1978, 1996.
246. O Shimomura, S Inouye. Protein Expres Purif 16:91–95, 1999.
247. DS Goldfarb. Cell 70:185–188, 1992.
248. S Lizano, S Ramanathan, A Feltus, A Witkowski, S Daunert. Methods Enzymol 279:296–303, 1997.
249. A Witkowski, S Ramanathan, S Daunert. Anal Chem 66:1837–1840, 1994.
250. B Galvan, TK Christopoulos. Anal Chem 68:3545–3550, 1996.
251. CL Crofcheck, AL Grosvenor, KW Anderson, JK Lumpp, DL Scott, S Daunert. Anal Chem 69:4768–4772, 1997.

252. DS Sgoutas, TE Tuten, AA Verras, A Love EG Barton. Clin Chem 41:1637–1643, 1995.
253. JK Actor, FS Nolte, DF Smith. In: A Roda, M Pazzagli, LJ Kricka, PE Stanley, eds. Bioluminescence and Chemiluminescence: Perspectives for the 21st Century. New York: Wiley, 1999, pp 83–86.
254. JK Actor, T Kuffner, CS Dezzutti, RL Hunter, JM McNicholl. J Immunol Methods 211:65–77, 1998.
255. CW Cody, DC Prasher, WM Westler, FG Prendergast, WW Ward. Biochemistry 32:1212–1218, 1993.
256. DC Prasher, VK Eckenrode, WW Ward, FG Prendergast, MJ Cormier. Gene 11: 229–233, 1992.
257. M Ikawa, K Kominami, Y Yoshimura, K Tanaka, Y Nishimune, M Okabe. Dev Growth Diff 37:455–459, 1995.
258. J Haselhoff, B Amoa. Trends Genet 11:328–329, 1995.
259. M Chalfie, Y Tu, G Euskirchen, WW Ward, DC Prasher. Science 263:802–805, 1994.
260. S Eriksson, E Raivio, JP Kukkonen, K Eriksson, K Lindquist. J Virol Meth 59: 127–133, 1996.
261. Z Boldogkoi, F Erdelyi, A Sik, TF Freund, I Fodor. Luminescence 14:69–74, 1999.
262. R Tombolini, A Unge, ME Davey, FJ Debruijn, JK Jansson. FEMS Microbiol Ecol 22:17–28, 1997.
263. AA Arrage, DC White. In: JW Hastings, LJ Kricka, PE Stanley, eds. Bioluminescence and Chemiluminescence: Molecular Reporting with Photons. New York: Wiley, 1997, pp 383–386.
264. BA Illarionov, VS Bondar, VA Illarionova, ES Vysotski. Gene 153:273–274, 1995.
265. LA Frank, ES Vysotski. In: JW Hastings, LJ Kricka, PE Stanley, eds. Bioluminescence and Chemiluminescence: Molecular Reporting with Photons. New York: Wiley, 1997, pp 439–442.
266. LA Frank, SA Efimenko, AI Petunin, ES Vysotski. In: A Roda, M Pazzagli, LJ Kricka, PE Stanley, eds. Bioluminescence and Chemiluminescence: Perspectives for the 21st Century. New York: Wiley, 1999, pp 111–114.
267. WW Lorenz, RO McCann, M Longiaru, MJ Cormier. Proc Natl Acad Sci USA 88:4438–4442, 1991.
268. Y Wang, G Wang, DJ O'Kane, AA Szalay. In: JW Hastings, LJ Kricka, PE Stanley, eds. Bioluminescence and Chemiluminescence: Molecular Reporting with Photons. New York: Wiley, 1997, pp 419–422.
269. G Grentzmann, JA Ingram, PJ Kelly, RF Gesteland, JF Atkins. RNA 4:479–486, 1998.
270. J Stables, S Scott, S Brown, C Roelant, D Burns, MG Lee, S Rees. J Recept Signal Transduct Res 19:395–410, 1999.
271. I Minko, SP Holloway, S Nikaido, M Carter, OW Odom, CH Johnson, DL Herrin. Mol Gen Genet 262:421–425, 1999.

11

The Role of Organized Media in Chemiluminescence Reactions

José Juan Santana Rodríguez
University of Las Palmas de G.C., Las Palmas de G.C., Spain

1. INTRODUCTION

Organized media have been extensively applied in various analytical methodologies to enhance their sensitivity and selectivity [1–6]. The success of such applications is due to the fact that organized systems can be employed to change the solubility and microenvironment of analytes and reagents and to control the reactivity, equilibrium, and pathway of chemical or photochemical processes among other effects [1, 2, 7]. These properties of organized media can also be utilized to facilitate analytical chemiluminescence (CL) measurements [8–13]. Advantages cited include elimination of solubility problems [8, 9], improved sensitivity [9–13], increased selectivity [9, 13] better precision [13], and a less strict pH requirement for observation of efficient chemiluminescence [9, 13].

In the following sections the most important features of the organized media that are most frequently used in chemiluminescent reactions (micellar media and cyclodextrins) will be summarized as well as their influence on various chemiluminescent systems, including their corresponding applications in chemical analysis.

2. ORGANIZED MEDIA

2.1 Micellar Media

2.1.1 Definitions and Characteristics

Surfactants (a contraction of the term *surface-active agent*) are substances one of whose properties is that of being adsorbed in surfaces or interfaces of the system and altering the free energy of these surfaces (or interfaces). Here, the term *interface* refers to the union between two inmiscible phases, while the term *surface* refers to an interface in which one of the phases is a gas, generally air.

Free surface energy is the minimum work required to create an interface. The surface tension of a liquid is the free surface energy per unit of area in the bond between the liquid and the air. Thus, the minimum work required to create an interface is:

$$W_{min} = \gamma \Delta A$$

where γ is the surface tension and ΔA the increase in area between phases.

The surfactant usually acts by reducing the free surface energy, though on occasion it will increase it. The molecules present on the surface have greater energy than those found in the interior of a medium, as the latter interact more strongly in such a way that their energy is lowered. Therefore, a certain amount of work will be required to carry a molecule from the interior to the surface.

Surfactants have a molecular structure characteristic, called amphipathic, consisting of a group that has little affinity for the solvent, called *hydrophobic*, when the solvent is water, and a group with a high affinity for the solvent, called *hydrophilic*, in an aqueous solution. When the surfactant is dissolved in water, the presence of the hydrophobic group in the interior of the water causes a distortion, thereby bringing about an increase in the free energy of the system. This means that the work required to carry a surfactant molecule to the surface is less than the work required to transfer a water molecule. Therefore, the surfactant will concentrate on the surface. This brings about a decrease in the work required to increase the surface unit area (surface tension). On the other hand, the presence of the hydrophilic group impedes the surfactant from being completely expelled from the solvent, which would require the dehydration of this group. The amphipathic structure of the surfactant causes not only concentration of surfactant on the surface and reduction of the surface tension of the water, but also orientation of the molecule with its hydrophilic group in the aqueous phase and its hydrophobic group toward the exterior [14].

The hydrophobic group is usually a hydrocarbon chain and is generally called the *tail*, while the hydrophilic group, called the *head*, is an ionic group and very polar. Depending on the nature of the hydrophilic group, the surfactants are classified as:

1. *Anionic*: the *head* group is negatively charged such as, for example, alkylcarboxilic or sulfonic acids salts.
2. *Cationic*: the *head* group is positively charged such as, for example, quaternary ammonium salts.
3. *Nonionic*: the surfactant has no groups with charge such as, for example, monoglycerides of long-chain fatty acids or polyoxyethylenated alkylphenols.
4. *Zwitterionic*: in the surfactant there are positively and negatively charged groups such as, for example, long-chain amino acids.

Table 1 shows examples of these four types of surfactants and their corresponding structures.

In diluted solutions (generally at concentrations less than 10^{-4} M), the surfactants are generally found to be monomers, though there can also be dimers, trimers, etc. If the concentration of surfactant in solution increases, it can reach a point where a process of aggregation occurs and many of the physicochemical properties undergo change. The colloidal aggregate is given the name *micelle* and its shape varies depending on the surfactant and the medium. The concentration at which these changes occur is known as the *critical micelle concentration* (cmc).

It was mentioned above that when substances that contain hydrophobic groups are dissolved in water the free energy of the system increases. To make this free energy minimal, the substance positions itself with its hydrophobic part toward the exterior of the solvent. However, there is another way to minimize this energy, which consists of the aggregation of these surface-active molecules (surfactants) in clusters (micelles), with their hydrophobic groups facing toward the interior of the cluster and the hydrophilic groups toward the solvent. Thus, the formation of micelles is an alternative mechanism to adsorption in the interfaces to avoid contact between the hydrophobic groups and the water, and so decrease the free energy of the system [14].

In spite of the fact that keeping the hydrophobic groups from being in contact with the water can produce a decrease in the free energy of the system,

Table 1 Name, Structure, and Abbreviation of Some Surfactants

Name	Molecular structure	Abbreviation
Anionic		
Sodium dodecylsulfate (sodium laurylsulfate)	$CH_3(CH_2)_{11}$-O-$SO_3 - Na^+$	SDS
Sodium hexadecylsulfate (sodium cetylsulfate)	$CH_3(CH_2)_{15}$-O$SO_3 - Na^+$	SHS
Sodium dodecylsulfonate	$CH_3(CH_2)_{11}$-$SO_3 - Na^+$	SDDS
Sodium dodecylbenzene-sulfonate	$CH_3(CH_2)_{11}$-C_6H_4-$SO_3 - Na^+$	SDBS
Cationic		
Hexadecyltrimethylammonium bromide (cetyltrimethyl-ammonium bromide)	$CH_3(CH_2)_{15}N^+(CH_3)_3.Br^-$	HTAB
Hexadecyltrimethylammonium chloride (cetyltrimethyl-ammonium chloride)	$CH_3(CH_2)_{15}N^+(CH_3)_3Cl^-$	HTAC
Dodecyltrimethylammonium bromide (lauryltrimethyl-ammonium bromide)	$CH_3(CH_2)_{11}N^+(CH_3)_3.Br^-$	DTAB
Hexadecylpyridinium chloride (cetylpyridinium chloride)	$CH_3(CH_2)_{15}$-$C_6H_4N^+Cl^-$	HPC
Hexadecylpyridinium bromide (cetylpyridinium bromide)	$CH_3(CH_2)_{15}$-$C_6H_4N^+Br^-$	HPB
Dioctadecyldimethylammonium chloride		DODAC

$CH_3(CH_2)_{17}$ —— $\overset{CH_3}{\underset{CH_3}{N^+}}$ —— Cl^-

$CH_3(CH_2)_{17}$

Nonionic

Polyoxyethylene(8)dodecanol	$CH_3(CH_2)_{11}$-O-$(CH_2$-$CH_2O)_8H$	$C_{12}E_8$
Polyoxyethylene(23)dodecanol	$CH_3(CH_2)_{11}$-O-$(CH_2$-$CH_2O)_{23}H$	Brij 35
Polyoxyethylene(9.5)-t-octyl-phenol	$(CH_3)_3C$-CH_2-$C(CH_3)_2$-C_6H_4-$(OCH_2CH_2)_{9.5}H$	Triton X-100
Polyoxyethylene(7.5)nonyl-phenyl ether	$CH_3(CH_2)_8$-C_6H_4-$(OCH_2CH_2)_{7.5}H$	PONPE 7.5
Octylglucoside	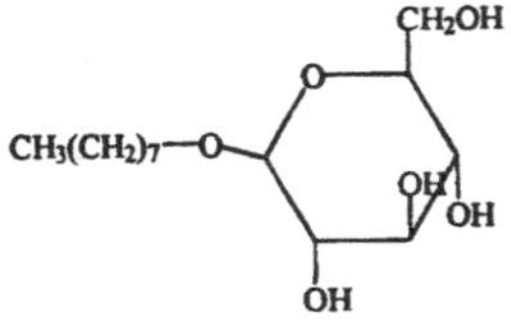	OG

Zwitterionic

Decyldimethylammonium acetate (*n*-decylbetaine)	$CH_3(CH_2)_9(CH_3)_2N^+CH_2COO^-$	DeDAA
N-Dodecyl-*N*,*N*-dimethyl-ammonium-3-propane-1-sulfonate (*n*-dodecylsultaine)	$CH_3(CH_2)_{11}(CH_3)_2N^+(CH_2)_3SO_3^-$	SB-12
N-Hexadecyl-*N*,*N*-dimethyl-ammonium-3-propane-1-sulfonate (*n*-hexadecylsultaine)	$CH_3(CH_2)_{15}(CH_3)_2N^+(CH_2)_3SO_3^-$	SB-16

the surfactant molecule can lose its freedom on finding itself in the micelle. More-over, an electrostatic repulsion can occur in the ionic surfactants between mole-cules of surfactants in micelles and surfactants that are free to have the same charge. These forces increase the free energy of the system and resist the forma-tion of micelles. As a consequence, if in a certain case micelles are formed, the concentration of surfactant of which they are produced depends on the equilib-rium between the factors that favor the formation of micelles and those that resist it.

The structure of the micelles remains an object of study and controversy. A series of models has been proposed that attempt to explain the experimental evidence. In this study the chronological order of publication of the four most important models has been considered.

The classical model, as shown in Figure 1, assumes that the micelle adopts a spherical structure [2, 15–17]. In aqueous solution the hydrocarbon chains or the hydrophobic part of the surfactants from the *core* of the micelle, while the ionic or polar groups face toward the exterior of the same, and together with a certain amount of counterions form what is known as the *Stern* layer. The remain-der of the counterions, which are more or less associated with the micelle, make up the *Gouy-Chapman* layer. For the nonionic polyoxyethylene surfactants the structure is essentially the same except that the external region does not contain counterions but rather rings of hydrated polyoxyethylene chains. A micelle of

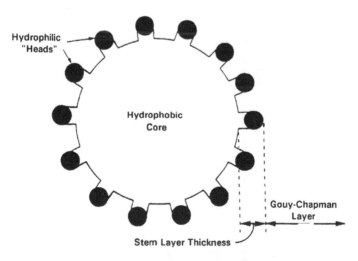

Figure 1   Typical cross-sectional schematic representing the classical view of an aque-ous micelle. Counterions are not shown. (From Ref. 2 with permission.)

this type has a radius that is approximately equal to the length of the extended hydrocarbon chain and usually fluctuates between 3 and 6 nm [3].

This model has been used for years for its simplicity, but it presents several contradictions and for this reason other researchers have proposed different alternatives. Thus, it has been noted that there is contact between the water and the tails of the surfactants that make up the micelle. This can occur in three different ways:

If there were penetration of the water in the core of the micelle.
If, even if there were no penetration of the water in the interior of the core of the micelle, a part of the hydrocarbon chains were exposed to the micelle-water interface.
If a combination of the above two situations were to occur.

The model proposed by Menger et al. (Fig. 2) shows two extreme conformations, one in which the hydrocarbon chains are fully extended and another in which they are folded [18, 19]. The surface of *Menger's micelle* is less defined than in the classical model and the surfactants that form the micelle are randomly orientated. The water can penetrate and enter in contact with the hydrophobic part of the surfactants. This model, apart from being more acceptable from an esteric point of view, gives a better explanation than the classical model of a series of experimental results such as viscosity, polarity, or kinetics.

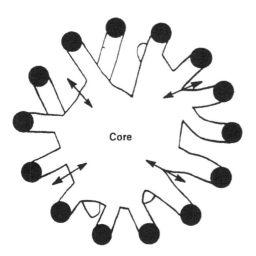

Figure 2 A possible representation of the Menger micelle in which the hydrocarbon "tails" are extended. Counterions are not shown. (From Ref. 2 with permission.)

The other two, more recent, micellar models consist of a core that has barely any contact with the water (as in the classical model), and part of the hydrocarbon chains of the surfactants, as well as the heads of these, are exposed toward the surface, with which they are in contact with the water (as in Menger's model).

Fromherz's model considers a spherical micelle where the surfactants are arranged in parallel forming a packaging without tensions and without contact with the water, in which the heads of the surfactants are as separated as possible [20]. The surfactant chains, in the region of the heads, are bent to lower electrostatic repulsion as much as possible. Figure 3 shows a cross-section of this model of micelle.

In the model of micelle proposed by Dill et al. [21], the hydrocarbon chains of the surfactants are more randomly distributed, bearing in mind statistical considerations (Fig. 4). A considerable number of hydrocarbon chains are exposed to the water at the surface.

The models of Fromherz and Dill allow the solubilization of hydrophobic solutes near the surface of the micelle and explain how these solutes (in addition to part of the hydrocarbon chains of the surfactants) can be in contact with the water when they are associated with the micelle. However, the debate concerning the structure of the clusters is not yet finished and research on the subject continues.

The micelles are spherical, but when the concentration of surfactant increases, the shape of the ionic micelles changes following the spherical sequence: cylindrical-hexagonal-laminar [22]. In the case of nonionic micelles the shape

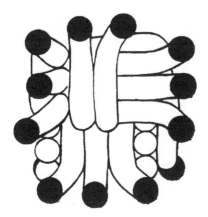

Figure 3   A possible cross-sectional view of the Fromherz micelle. This model is constructed using cylindrical sticks to represent the surfactant molecules. Counterions are not shown. (From Ref. 2 with permission.)

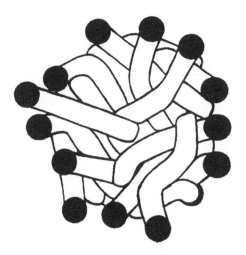

Figure 4 A cross-sectional schematic of the Dill model for the micelle. Counterions are not shown. (From Ref. 2 with permission.)

changes directly from spherical to laminar as the concentration of surfactant increases.

It was mentioned previously that the narrow range of concentrations in which sudden changes are produced in the physicochemical properties in solutions of surfactants is known as critical micelle concentration. To determine the value of this parameter the change in one of these properties can be used; so normally electrical conductivity, surface tension, or refraction index can be measured. Numerous cmc values have been published, most of them for surfactants that contain hydrocarbon chains of between 10 and 16 carbon atoms [1, 3, 7]. The value of the cmc depends on several factors such as the length of the surfactant chain, the presence of electrolytes, temperature, and pressure [7, 14]. Some of these values of cmc are shown in Table 2.

It is important to point out that, in general, the micelles composed of nonionic surfactants usually have a lower cmc and higher aggregation numbers than the analogous ionic micelles. This is partly due to the absence of electrostatic repulsion between the heads of the nonionic surfactants. However, in the ionic micelles these repulsion tend to limit the aggregation number and the cmc.

Most of the applications of the micelles in aqueous medium are based on the association or solubilization of solutes. The interactions between both can be electrostatic, hydrophobic, or, more normally, a combination of both effects [23, 24]. It was thought initially that hydrophobic solutes dissolve in the core of the micelle in the same way in which they would do so in an organic solvent, but

Table 2 Critical Micelle Concentration of Some Surfactants (aqueous solutions at 25°C)

Surfactant	cmc (M)
Anionic	
Sodium decylsulfate	3.3×10^{-2}
Sodium dodecylsulfate	8.1×10^{-3}
Sodium hexadecylsulfate	5.2×10^{-4}
Sodium dodecylsulfonate	9.8×10^{-3}
Cationic	
Dodecyltrimethylammonium bromide	1.5×10^{-2}
Hexadecyltrimethylammonium bromide	9.0×10^{-4}
Hexadecyltrimethylammonium chloride	1.3×10^{-3}
Hexadecylpiridinium chloride	9.0×10^{-4}
Nonionic	
Polyoxyethylene(8)dodecanol	7.0×10^{-5}
Polyoxyethylene(9.5)-t-octylphenol	2.3×10^{-4}
Polyoxyethylene(23)dodecanol	1.0×10^{-4}
Octylglucoside	2.5×10^{-2}
Zwitterionic	
n-Dodecylsultaine	1.2×10^{-3}
n-Hexadecylsultaine	1.0×10^{-4}
n-Decylbetaine	$(1.0–2.1) \times 10^{-2}$
n-Dodecylbetaine	1.5×10^{-3}

Source: From Ref. 7 with permission.

this is probably a simplification. In accordance with the structure of the micelle of Menger, Fromherz, or Dill, the interaction between many hydrophobic solutes and the micelle can be somewhat similar to a surface adsorption phenomenon where hydrophobic and electrostatic reactions are important. This would explain not only the apparent presence of solutes that are relatively nonpolar near the surface of the micelle [23], but also the fact that some polar solutes have a higher solubility in micellar solutions than in water or organic solvents. There are solutes, such as aliphatic hydrocarbons, that are dissolved in the core of the micelle and other solutes, such as dodecanol, that have a part of their structure integrated in the core of the micelle. It is therefore believed that the micelle has at least two interaction sites: a hydrophobic one in the core and another more polar one near the surface of the micelle [15, 25].

Thus far reference has only been made to the so-called *normal micelles* that are formed in polar solvents such as water. However, when the surfactants are dissolved in organic nonpolar solvents, the hydrophilic groups are found in the interior of the aggregate, while the hydrophobic chains extend toward the

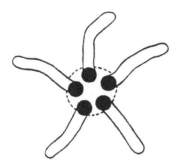

Figure 5 Model of a reversed micelle in a nonpolar organic solvent from which all possible water and impurities have been removed. (From Ref. 2 with permission.)

nonpolar phase (Fig. 5) [26]. The accepted term for this type of aggregate, the structure and properties of which are generally different from those of normal micelles, is *reversed micelle*. The interior core of the reversed micelle, i.e., the micellar interface and the inner aqueous phase, provides a unique and versatile reaction field. Depending upon its water content (which also dictates the size of the aggregate), the microscopic polarity, the local concentration (proximity), and the mobility of substrates (microviscosity) can vary markedly and, therefore, the chemical reactions can be controlled as required [26–28].

Finally, and apart from the importance of micelles in the solubilization of chemical species, mention should also be made of their intervention in the displacement of equilibria and in the modification of kinetics of reactions, as well as in the alteration of physicochemical parameters of certain ions and molecules that affect electrochemical measurements, processes of visible-ultraviolet radiation, fluorescence and phosphorescence emission, flame emission, and plasma spectroscopy, or in processes of extraction, thin-layer chromatography, or high-performance liquid chromatography [2–4, 29–33].

2.2 Cyclodextrins

2.2.1 General Characteristics

Cyclodextrins are cyclic oligosaccharides formed by the connection of individual glucopyranose units through α-1,4-glycosidic oxygen bridges. These compounds form inclusion complexes with appropriate molecules, producing changes in the photophysical properties of these molecules.

The three most used cyclodextrins are known as α, β, and γ, and these contain six, seven, and eight units of glucopyranose, respectively. The structure of the cyclodextrins is in the shape of a truncated cone, the cavity of which has an average diameter of 0.57, 0.78, and 0.95 nm for α, β, and γ cyclodextrins, respectively [34]. The solute molecules that have the correct dimensions can interact with the cyclodextrin cavity, which is nonpolar, and form inclusion complexes.

The cyclodextrins are stable bodies in aqueous solution, unlike the micelles, which are transitory and are in a state of dynamic equilibrium with the monomer surfactants. However, in many aspects the inclusion of analytes in the cyclodextrin cavity is reminiscent of the solubilization of hydrophobic molecules in micelles in aqueous solution.

The reduced polarity and the protection supplied by the cyclodextrin cavity have a strong and often favorable influence on the properties of the included solute. Cyclodextrins have been used extensively in luminescence techniques due to the fact that many analytes, on forming complexes with them, undergo an enhancement of their luminescent efficiency compared to that seen in an aqueous medium [35–37]. The factors responsible for this increase are the protection of the analyte molecule in the complex with the cyclodextrin from quenching of the water molecules or of other molecules present in solution and, moreover, the viscosity increase in the cyclodextrin cavity with a corresponding reduction in quenching caused by oxygen. As a consequence, the cyclodextrin cavity can protect the excited state of the analyte from nonradiative processes and from quenching that normally occur in aqueous solution.

3. INFLUENCE OF THE PRESENCE OF ORGANIZED MEDIA ON CHEMILUMINESCENT REACTIONS: ANALYTICAL APPLICATIONS

3.1 Micellar Media

3.1.1 Normal Micelles

In recent years different types of surfactants have been used, in concentrations above their cmc, forming normal micelles, to improve different aspects of chemiluminescent reactions. Though the choice of the best surfactant depends on the characteristics of the chemiluminescent reaction, the surfactants most used have been the cationic (fundamentally quaternary ammonium salts) and to a lesser degree the anionic, the nonionic, and the zwiterionic compounds.

3.1.1.1   Cationic Surfactants

Alkyltrimethylammonium surfactants, such as hexadecyltrimethylammonium chloride (HTAC) [6, 9, 38–41], hexadecyltrimethylammonium hydroxide (HTAH) [38, 42], hexadecyltrimethylammonium bromide (HTAB) [43–46], tetradecyltrimethylammonium bromide (TTAB) [47], pentadecyltriethylammonium bromide (PTAB) [48], stearyltrimethylammonium chloride (STAC) [46], cetylethyldimethylammonium bromide (CEDAB) [46], didodecyldimethylammonium bromide (DDDAB) [46], and dioctadecyldimethylammonium bromide (DODAB) [41], have been widely used in various chemiluminescent reactions.

HTAC and HTAH have been used as surfactants in the chemiluminescence reaction of lucigenin (10,10'-dimethyl-9,9'-biacridinium dinitrate) with biological reductants (such as fructose, glucose, ascorbic and uric acid) or hydrogen peroxide [38].

Lucigenin produces a weak light in an alkaline solution, and the intensity of its light is substantially increased by the addition of hydrogen peroxide [49]. As to the reaction mechanism, it has been postulated [50, 51] that lucigenin is oxidized to form a peroxide, which is then decomposed to yield an excited state of N-methylacridine, as shown in Figure 6.

When biological reductants (sugars) are present, the rate-limiting step in the CL reaction between these agents and lucigenin apparently involves a base-catalyzed ketoenol tautomerization in which the sugars (in keto form) are converted to 1,2-enediols (enol form of parent sugar) as shown in Figure 7 for glucose and fructose [52–56]. The 1,2-enediol anion tautomers, which are stronger reducing agents than the parent sugars, subsequently react with lucigenin in a series of rapid steps that lead to CL emission [54].

The results indicate that the enhancement produced by HTAH is greater than that produced by HTAC in CL intensity in the presence of biological reductants. This is due to enhanced micellar catalysis of the rate-limiting step of the lucigenin-reductant CL reaction in HTAH micelles compared to that possible in HTAC. Figure 8 shows the influence of HTAH concentration on CL intensity of lucigenin-glucose and lucigenin-fructose systems. The presence of HTAH as a micellar medium improves the sensitivity and precision for lucigenin CL assay of biological reductants. For example, the detection limits for fructose were 2.3, 7.0, and 9.8 mg/L for the process with HTAH, HTAC, and an aqueous solution,

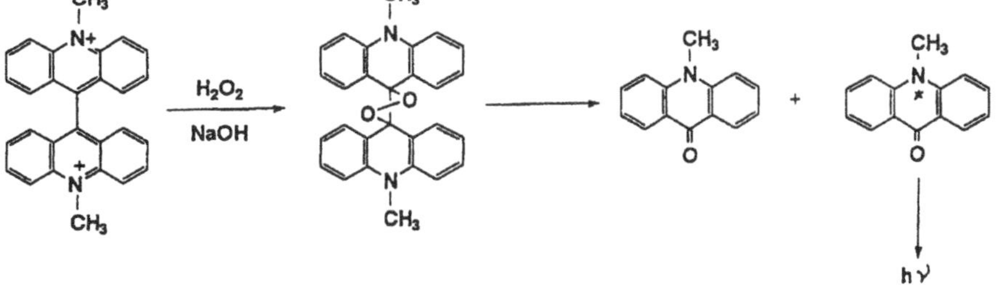

Figure 6   Chemiluminiscent reaction of lucigenin. (From Ref. 49 with permission.)

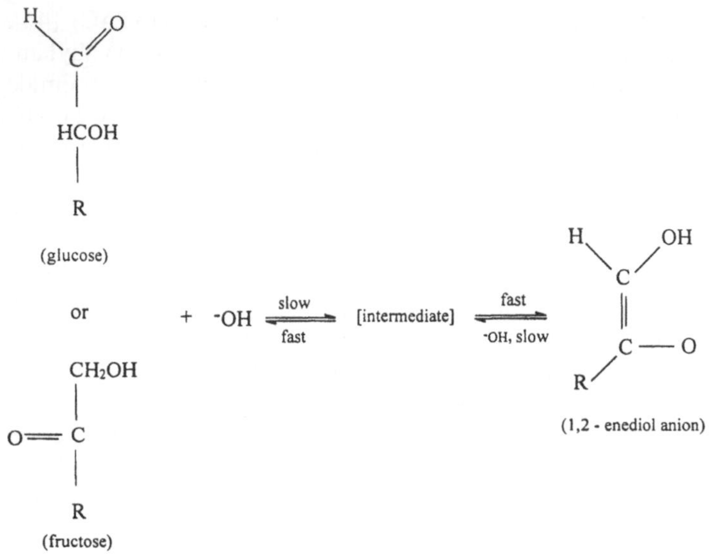

$$R = -CH(OH)CH(OH)CH_2OH$$

Figure 7   Ketoenol tautomerization for glucose and fructose.

respectively. With respect to the lucigenin-hydrogen peroxide CL system, both micellar media produced similar effects in the observed CL.

Since most CL assays require basic conditions, the use of a HTAH micellar medium compared to HTAC offers the advantage of providing the micelle-forming surfactant (HTA$^+$ required for the enhancement of CL intensity) and the hydroxide ion (required for efficient CL).

HTAC cationic micelles also markedly enhance the CL intensity of fluorescein (FL) in the oxidation of hydrogen peroxide catalyzed by horseradish peroxidase (HRP) [39]. However, no CL enhancement was observed when anionic micelles of sodium dodecyl sulphate (SDS) or nonionic micelles of polyoxyethylene (23) dodecanol (Brij-35) were used (Fig. 9). CL enhancement is attributed to the electrostatic interaction of the anionic fluorescein with the HTAC micelles. The local concentration of fluorescein on the surface of the micelle increases the efficiency of the energy transferred from the singlet oxygen (which is produced in the peroxidation catalyzed by the HRP) to fluorescein. This chemiluminescent enhancement was applied to the determination of traces of hydrogen peroxide. The detection limit was three times smaller than that obtained in aqueous solution.

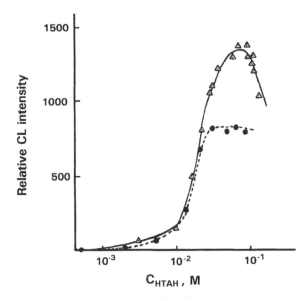

Figure 8 Dependence of the relative CL intensity upon the final molar concentration of HTAH observed from the reaction of 3.3×10^{-4}M lucigenin with (Δ, solid line) 200 mg/L glucose and ($\bullet$, dashed line) 48 mg/L fructose. (From Ref. 38 with permission.)

Riehl et al. also characterized the CL system lucigenin-hydrogen peroxide-N-methylacridone in the presence of different cationic surfactants such as HTAC, 3-(N-dodecyl-N,N-dimethylammonio) propane-1-sulfonate, and DODAB [41]. Enhancement factors (ratio between CL intensity in the presence of organized medium and CL intensity in the absence of organized medium) of CL intensity were found of 3.4, 2.5, and 1.6, respectively. The alterations in CL intensity are explained in terms of the effect of the different surfactants on the rate of the reaction and on excitation efficiency.

The CL enhancement of the lucigenin reaction with catecholamines in the presence of HTAH micelles was used for determination of dopamine, norepinephrine, and epinephrine [42]. However, the presence of an anionic surfactant, SDS, inhibits the CL of the system. The aforementioned CL enhancement in the presence of HTAH can be explained in the following way: the deprotonated forms of the catecholamines are expected to be the principal species present in aqueous alkaline solution due to the dissociation of the catechol hydroxyl groups, and to react with lucigenin to produce CL. The anionic form of the catecholamines and the hydroxide ion interact electrostatically with and bond to the cationic micelle, to which the lucigenin also bonds. Therefore, the effective concentration of the

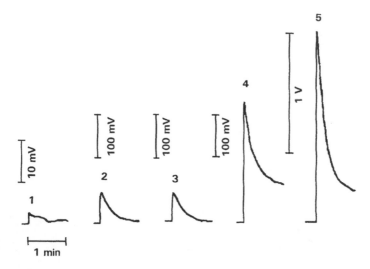

Figure 9 CL response curves from the oxidation of H_2O_2 with sodium hypochlorite in the presence of fluorescein and surfactants. (1) Aqueous system without fluorescein; (2) aqueous system; (3) $C_{SDS} = 2.0 \times 10^{-2}$ M; (4) $C_{Brij-35} = 4.2 \times 10^{-3}$ M; (5) $C_{HTAC} = 3.0 \times 10^{-3}$ M; $C_{H2O2} = 1.5 \times 10^{-4}$ M; $C_{NaOCl} = 2.0 \times 10^{-3}$ M; $C_{fluorescein} = 2.7 \times 10^{-4}$ M; $C_{NaOH} = 0.05$ M. (From Ref. 39 with permission.)

catecholamines, hydroxide ion, and lucigenin is higher than the stoichiometric concentration in water.

HTAB has been used, on the one hand, to increase the CL intensity of the reaction of 2,6,7-trihydroxy-9-(4'chlorophenyl)-3-fluorene with hydrogen peroxide in alkaline solution, in the presence of traces of Co(II) as a catalyst [43]. As a consequence, a CL method has been established for determination of ultratraces of Co(II). On the other hand, HTAB micelles sensitize the CL oxidation of pyrogallol with N-bromosuccimide in an alkaline medium [44], while anionic and nonionic surfactants inhibit the CL intensity of this reaction (Table 3). This sensitized process allows the determination of pyrogallol by flow injection in an interval of 5×10^{-7}–3×10^{-5} M.

The CL reaction of luminol (5-amino-2,3-dihydro-1,4-phthalazinedione) (1) is one of the more commonly used nonenzymatic CL reactions and has been extensively studied [49, 57–59]. It is well known that the luminol CL reaction is catalyzed by many kinds of substances, e.g., ozone, halogen-Fe complex, hemin, hemoglobin, persulfate, and oxidized transition metals. The most acceptable scheme is shown in Figure 10. Luminol forms a six-membered ring of peroxide (3) from a diazaquinone intermediate (2) and then, by the decomposition of 3, N_2 gas and the S_1-excited state of the phthalate dianion are produced, yielding

Table 3 Effect of Surfactant Concentration on CL Intensity with 1.0×10^{-5} M Pyrogallol

	Relative CL intensity		
Concentration (M)	HTAB	SDS	Triton X-100
0.0	30	30	30
5.0×10^{-5}	44	26	24
1.0×10^{-4}	55	20	18
5.0×10^{-4}	83	13	11
1.0×10^{-3}	120	9	8
5.0×10^{-3}	100	6	4

Source: From Ref. 44 with permission.

emission of light. It is certain that the dianion of 2-aminophtalic acid is an emitter. White et al. [58] confirmed this fact by the following experimental results: (a) luminol yielded 2-aminophtalic acid in 90% yield by the reaction with NaOH and O_2 in dimethylsulfoxide; (b) the fluorescence spectrum of aminophthalic dianion agreed well with the resultant CL spectrum; (c) in the reaction system using $^{18}O_2$, the consumed oxygen was all found in the carbonyl group of the dianion and 1 mol N_2 gas was yielded simultaneously.

This CL reaction is most efficient at pH values in the range 10.5–11. This pH limitation is the main handicap of the system when coupled to enzymatic reactions at neutral pH. Abdel-Latif and Guilbault [45] have used HTAB to increase the CL intensity and to save the pH incompatibility of the luminescent

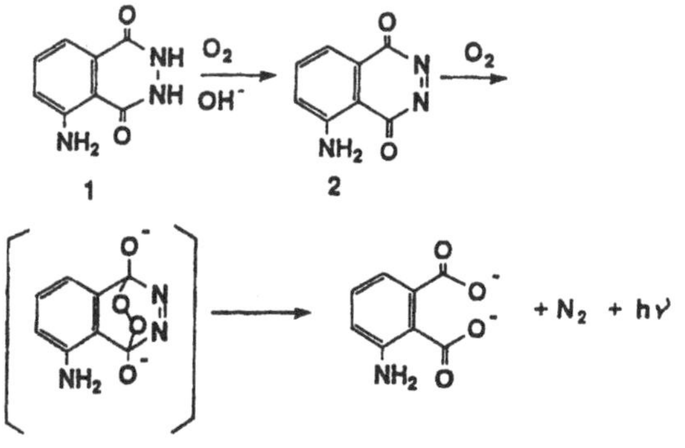

Figure 10 Chemiluminescent reaction of luminol. (From Ref. 49 with permission.)

reaction of luminol when it binds to the enzyme reaction glucose/glucose oxidase at neutral pH. The results show the reliability of the simultaneous and efficient operation of both reactions with pH 7.5–8.5, allowing the determination of glucose in an interval of 3×10^{-7}–3×10^{-4} M with a relative standard deviation of 3.8%.

Other cationic surfactants such as TTAB, DTAB, DODAB, STAC, CEDAB, and DDDAB have been used in CL reactions with less frequency. Thus, tetradecyltrimethylammonium bromide [TTAB] has been used to increase the sensitivity of the method to determine Fe(II) and total Fe based on the catalytic action of Fe(II) in the oxidation of luminol with hydrogen peroxide in an alkaline medium [47]. While other surfactants such as HTAB, hexadecylpiridinium bromide (HPB), Brij-35, and SDS do not enhance the CL intensity, TTAB shows a maximum enhancement at a concentration of 2.7×10^{-2} M (Fig. 11). At the same time it was found that the catalytic effect of Fe(II) is extremely efficient in the presence of citric acid. With regard to the mechanism of the reaction, it is thought that Fe(II) forms an anionic complex with citric acid, being later concentrated on the surface of the TTAB cationic micelle. The complex reacts with the hydrogen peroxide to form hydroxy radical or superoxide ion on the

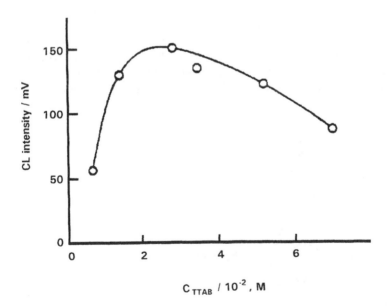

Figure 11 Effect of TTAB concentration on the CL intensity in the presence of citric acid. $C_{Fe(II)} = 1.0 \times 10^{-7}$ M; $C_{citric\ acid} = 1.0 \times 10^{-2}$ M; $C_{luminol} = 1.0 \times 10^{-5}$ M; $C_{H2O2} = 1.0 \times 10^{-2}$ M; pH = 9.3 (borate buffer). (From Ref. 47 with permission.)

surface of the micelle. In this way the speed of the CL reaction increases, with the CL intensity being enhanced.

Sukhan has used PTAB cationic micelles to enhance the CL reaction of 4-diethylaminophthalohydrazide with oxygen and Co(II) in the presence of fluorescein as sensitizer [48]. This enhancement is mainly due to electron-excited energy transfer from the donor (4-diethylaminophthalohydrazide) to the acceptor (fluorescein). The addition of fluorescein combined with the presence of PTAB reduces the detection limit of Co(II) by a factor of 6. The method was successfully applied in the determination of Co in tap water samples.

Finally, Yamada and Suzuki made a comparative study of the use of DDAB, HTAB, STAC, and CEDAB to improve the sensitivity and selectivity of the determination of ultratraces of Cu(II) by means of the CL reaction of 1,10-phenanthroline with hydrogen peroxide and sodium hydroxide, used as detection in a flow injection system [46]. Of the four cited surfactants it was found that CEDAB causes the greatest enhancement of the chemiluminescent signal (Fig. 12) (an enhancement factor of 140 with respect to the absence of surfactant).

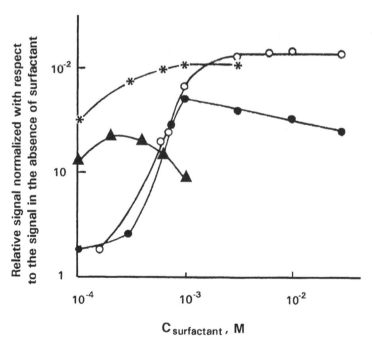

Figure 12 CL enhancement versus surfactant concentration. (▲) DDAB; (●) HTAB, (*)STAC; (○) CEDAB. $C_{Cu(II)} = 2.0 \times 10^{-8}$ M; $C_{H2O2} = 5\%$; $C_{1,10\text{-phenanthroline}} = 6.0 \times 10^{-5}$ M; $C_{NaOH} = 0.1$ M. (From Ref. 46 with permission.)

The CL enhancement may be interpreted as follows: 1,10-phenantroline is repelled from the cationic micellar interface owing to its charge; it therefore seems that 1,10-phenantroline exists as a nonionic species in the alkaline solution and migrates to the micellar surfaces, i.e., to the Stern layer, which is significantly less polar than water. On the other hand, superoxide radical anion $^{\cdot}O_2^{-}$, the oxidant in the present system, which is produced from the hydrogen peroxide decomposition catalyzed by the copper–1,10-phenantroline complex in the alkaline solution, migrates easily on the positively charged micellar surface. Consequently, in the hydrophobic environment of the Stern layer the nonionic species may react more effectively with $^{\cdot}O_2^{-}$ to form a dioxetane intermediate that decomposes via an exoenergic route producing an emitter. The CL enhancement is therefore attributed to higher excitation efficiency of the dioxetane decomposition in the less polar environment of the Stern layer. In the presence of CEDAB a detection limit of 0.3 pg is obtained, which is much lower than the detection limit obtained by other CL methods and is comparable to that obtained by methods that employ atomic spectrometry. At the same time, it was found that the established method for determination of Cu(II) using CEDAB is highly selective.

3.1.1.2   Anionic Surfactants

From an analytical point of view, the use of anionic surfactants as enhancers of CL reactions is most limited. One of the most recent examples is the use of sodium dodecylbenzene sulfonate (SDBS) as a CL enhancer of the system $Ru(bpy)_3^{2+} - SO_3^{2-} - KBrO_3$ (bpy = 2,2'-bipyridyl) [60]. The authors of this work propose the following mechanism for the chemiluminescent system:

$$SO_3^{2-} + BrO_3^{-} \rightarrow {}^{\cdot}SO_3^{-}$$
$$Ru(bpy)_3^{2+} + BrO_3^{-} \rightarrow Ru(bpy)_3^{3+}$$
$$Ru(bpy)_3^{3+} + {}^{\cdot}SO_3^{-} \rightarrow Ru(bpy)_3^{2+*} + {}^{\cdot}SO_5^{-}$$
$${}^{\cdot}SO_3^{-} + O_2 \rightarrow {}^{\cdot}SO_5^{-}$$
$${}^{\cdot}SO_5^{-} + SO_3^{2-} \rightarrow SO_5^{2-} + {}^{\cdot}SO_3^{-}$$
$$SO_5^{2-} + SO_3^{2-} \rightarrow 2SO_4^{2-}$$
$$Ru(bpy)_3^{2+*} \rightarrow Ru(bpy)_3^{2+} + h\nu$$
$${}^{\cdot}SO_3^{-} \rightarrow {}^{3}SO_2^{*}$$
$${}^{3}SO_2^{*} \rightarrow SO_2 + h\nu,$$

where $Ru(bpy)_3^{2+*}$ and ${}^{3}SO_2^{*}$ are excited state and triplet state of both species, respectively. When this system was studied in the presence of different anionic, cationic, and nonionic surfactants, it was found that SDBS produces the highest enhancement (Table 4). This is due to the fact that $Ru(bpy)_3^{2+}$ exists as a cationic complex in aqueous solution, where it is surrounded by the anionic surfactant SDBS, which prevents extinction of CL by the water and increases the excited-

Table 4 Effect of Different Surfactants on the Chemiluminiscent System $Ru(bpy)_3^{2+}$ -SO_3^{2-} -$KBrO_3$

Surfactant	Intensity/mV	Surfactant	Intensity/mV
Water	3.8	Triton X-100	4.3
SDBS	126	TPB	7.7
Tween-20	7.9	HPB	5.6
Tween-40	7.7	HTAB	3.5
Tween-80	4.2		

TPB, tetradecylpyridine bromide.
Source: From Ref. 60 with permission.

state lifetime of the complex $Ru(bpy)_3^{2+}$. This chemiluminescent system in the presence of SDBS has been successfully applied to determination of sulfite in sugar.

However, surfactants have been used not only to enhance the signal of CL systems, but also to avoid problems of solubility in these systems. Thus, Klopf and Nieman have used SDS, at a submicellar concentration, to solubilize the product (N-methylacridone) of the CL reaction of lucigenin, due to its insolubility in water [9]. In this way the appearance is avoided of solid deposits in the observation cell and in other components of the flow system.

3.1.1.3 Nonionic Surfactants

One of the nonionic surfactants most used as an enhancer of chemiluminescent reactions is Brij-35. This surfactant increases the reaction of lucigenin with catecholamines by a factor of 2.6 compared with the CL intensity in an aqueous medium [42]. This enhancement can be explained in the following way: it is known that oxygen from the polyoxyethylene chains in Brij-35 can react with sodium ion to form an oxonium ion, by which means the polyoxyethylene chains act as an oxonium cation. In this way the increase in CL intensity due to Brij-35 can be attributed to the same effect described for the micelles of a cationic surfactant.

Dan et al. found that Brij-35 also produces an enhancement of the CL intensity of the reaction of *bis*[N-[2-(N'-methyl-2'-pyridiniumyl)ethyl]-N-[(trifluoromethyl)sulfonyl]]oxamide with hydrogen peroxide in the presence of some fluorophors [40]. To be precise, the CL intensity increased by a factor of 130 for 8-aniline-1-naphthalenesulfonic acid (ANS) and 5.6 for rhodamine B (RH B), compared to the CL intensity in the absence of surfactant. This leads to an increase of 2–3 orders of magnitude in the linear dynamic ranges and a more precise determination of these analytes. However, improvement of the detection limits is limited owing to the increase in background noise when using a micellar medium.

When the influence was studied of different surfactants on the CL intensity of the reaction of lucigenin with isoprenaline, it was found that while cationic surfactants such as HTAH and HTAB and anionic surfactants such as SDS decrease the CL signal, the presence of Brij-35 increases the signal by a factor of 2.1 compared to that obtained in an aqueous medium [61]. As a result, a quite sensitive analytical method has been established for determination of isoprenaline, using Brij-35 as a CL enhancer. Application of the method has been satisfactorily verified with the determination of isoprenaline in pharmaceutical preparations.

3.1.1.4 Zwitterionic Surfactants

One of the few zwitterionic surfactant used in CL reactions is N-dodecyl-N,N-dimethyl-ammonium-3-propane-1-sulfonate (SB-12). Particularly, SB-12 has been assayed in the study of the CL reaction of lucigenin with various biological reductants [10]. The results show that SB-12 enhances CL intensity of the lucigenin-glucose and lucigenin-fructose systems by factors of 3.0 and 1.5, respectively, compared to the intensity obtained in aqueous medium. In these conditions detection limits were found for-both analytes of 0.7×10^{-4} and 2.5×10^{-5} M, respectively.

3.1.2 Reversed Micelles

As mentioned earlier, reversed micelles have different properties from normal micelles. These properties have the potential to favorably affect the sensitivity and other analytical aspects of CL reactions. Thus, reversed micelles have been used to prolong the duration of the observed CL of various oxalate ester (or acid)–hydrogen peroxide–sensitizer reaction systems for application as chemical light sources [62].

The CL system luminol–hydrogen peroxide was characterized by Hoshino and Hinze in HTAC reversed micelles, formed in a 6:5 (v/v) chloroform-cyclohexane mixture [63]. The results indicate that such a CL system can be used from an analytical point of view in a pH interval of 7.8–9.0 without the need to add a catalyst or a co-oxidant. In these conditions an analytical method was established for determination of hydrogen peroxide that, apart from supplying much milder conditions compared to the usual situation in an aqueous medium, is also acceptably precise and reproducible.

Studies have also been made on the coupling of the CL luminol–hydrogen peroxide in HTAB reversed micellar medium detection system with enzyme reactions [64]. The use of HTAB reversed micellar medium permits the simultaneous performance of both reactions, enzyme and CL detection, at a mild pH in the

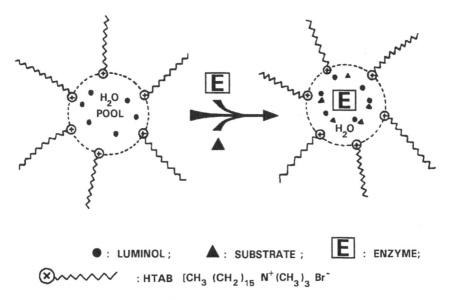

● : LUMINOL ;   ▲ : SUBSTRATE ; $\boxed{E}$: ENZYME;

$\otimes\!\sim\!\sim\!\sim$: HTAB $[CH_3 (CH_2)_{15} N^+(CH_3)_3 \; Br^-$

Figure 13 Schematic representation of the coupled substrate-enzymatic-luminol CL reaction system in a reversed micellar medium. (From Ref. 64 with permission.)

absence of a co-oxidant or catalyst. Figure 13 shows a scheme of the coupled substrate-enzymatic-luminol CL detection reaction system in a reversed micellar medium.

Based on these results, a simple and unique determination of 14 L-amino acids and glucose as substrates was developed. Thus, the calibration graph for a representative amino acid, L-phenylalanine was linear in the concentration range 1.0×10^{-6}–2×10^{-8} M with a relative standard deviation of 5.78% and a correlation coefficient of 0.9974. The detection limit obtained was 1.05×10^{-8} M. In the case of glucose the calibration graph was linear in the concentration range 2.7×10^{-6}–2.7×10^{-8} M with a relative standard deviation of 4.27% and a correlation coefficient of 0.9980. The detection limit was 2.7×10^{-8} M. The method was successfully applied to the determination of glucose in human blood serum.

3.2 Cyclodextrins

Grayeski and Woolf have studied the influence of the presence of cyclodextrins in chemiluminescent systems [65, 66]. They have found that aqueous solutions of α-, β-, and γ-cyclodextrins enhance the CL intensity of the reaction of lucigenin with hydrogen peroxide [65]. The greatest enhancement was observed for a solution 1×10^{-2} M of β-cyclodextrin, presumably due to restrictions of size in accommodating the corresponding species. Different experiments show that the CL enhancement can be attributed to an increase in the excitation efficiency and in the rate of the reaction through the inclusion of a reaction intermediate in the cyclodextrin cavity.

The same authors studied the CL of 4,4'-[oxalylbis(trifluoromethylsulfonyl)imino]*bis*[4-methylmorphilinium trifluoromethane sulfonate] (METQ) with hydrogen peroxide and a fluorophor in the presence of α, β, γ, and heptakis 2,6-di-*O*-methyl β-cyclodextrin [66]. The fluorophors studied were rhodamine B (RH B), 8-aniline-1-naphthalene sulfonic acid (ANS), potassium 2-*p*-toluidinylnaphthalene-6-sulfonate (TNS), and fluorescein. It was found that TNS, ANS, and fluorescein show CL intensity enhancement in all cyclodextrins, while the CL of rhodamine B is enhanced in α- and γ-cyclodextrin and reduced in β-cyclodextrin medium. The enhancement factors were found in the range of 1.4 for rhodamine B in α-cyclodextrin and 300 for TNS in heptakis 2,6-di-*O*-methyl β-cyclodextrin. The authors conclude that this enhancement could be attributed to increases in reaction rate, excitation efficiency, and fluorescence efficiency of the emitting species. Inclusion of a reaction intermediate and fluorophore in the cyclodextrin cavity is proposed as one possible mechanism for the observed enhancement.

4. CONCLUSIONS

In accordance with the above, it is clear that the organized media may play an important role in the development of CL reactions. This role may be shown in the improvement of the sensitivity, precision, and selectivity of many CL reactions, due principally to the change of the microenvironment of the CL system. Organized media can alter the microviscosity, local pH, polarity, reaction pathway or rate, etc. This situation allows application of these organized media to determination of organic and inorganic analytes in different kinds of matrices using CL reactions. A summary of these applications is shown in Table 5.

Moreover, the organized systems can be used to smooth the required experimental conditions of CL reactions, above all pH conditions. In this way, it is possible to carry out CL reactions more easily, as well as to develop simultaneously these reactions with other kinds of reactions, such as enzymatic ones, that are of great biochemical interest.

Finally, it is possible to use these media to solubilize some products of CL reactions, which due to their low solubility in aqueous medium can produce interferences in the development of CL reaction as well as in the detection of CL signal. Obviously, this leads to an improvement in the precision of the CL reaction.

Table 5 Analytical Applications of CL Systems in the Presence of Organized Media[a]

Analyte	CL system	Organized medium	Analytical parameters	Applications	Ref.
Ascorbic acid	Lucigenin–ascorbic acid Alkaline medium	HTAC	LDR[b]: 0.5×10^{-5}–0.2×10^{-3} mol.L^{-1} Slope: 1.447; CC[c]: 0.9996 LOD[d]: 1.0×10^{-6} mol.L^{-1}	—	10
Ascorbic acid	Lucigenin–ascorbic acid Alkaline medium	SDS	LDR: 8.5×10^{-5}–2.0×10^{-3} mol.L^{-1} Slope: 0.951; CC: 0.9961 LOD: 5.0×10^{-5} mol.L^{-1}	—	10
Glucose	Lucigenin-glucose Alkaline medium	HTAC	LDR: 1.3×10^{-4}–4.5×10^{-3} mol.L^{-1} Slope: 0.664; CC: 0.9933 LOD: 0.8×10^{-4} mol.L^{-1}	—	10
Glucose	Lucigenin-glucose Alkaline medium	SDS	LDR: 1.3×10^{-4}–4.5×10^{-3} mol.L^{-1} Slope: 0.788; CC: 0.9944 LOD: 0.9×10^{-4} mol.L^{-1}	—	10
Glucose	Lucigenin-glucose Alkaline medium	SB-12	LDR: 0.6×10^{-4}–2.6×10^{-3} mol.L^{-1} Slope: 0.783; CC: 0.9959 LOD: 0.7×10^{-4} mol.L^{-1}	—	10
Uric acid	Lucigenin–uric acid Alkaline medium	HTAC	LDR: 0.6×10^{-5}–5.0×10^{-4} mol.L^{-1} Slope: 0.639; CC: 0.9958 LOD: 2.9×10^{-6} mol.L^{-1}	—	10

Table 5 Continued

Analyte	CL system	Organized medium	Analytical parameters	Applications	Ref.
Glucuronic acid	Lucigenin–glucuronic acid Alkaline medium	SDS	LDR: 3.5×10^{-4}–1.0×10^{-2} mol.L^{-1}; Slope: 0.887; CC: 0.9981; LOD: 8.4×10^{-5} mol.L^{-1}	—	10
Fructose	Lucigenin-fructose Alkaline medium	HTAC	LDR: 1.2×10^{-4}–1.3×10^{-3} mol.L^{-1}; Slope: 1.100; CC: 0.9961; LOD: 4.0×10^{-5} mol.L^{-1}	—	10
Fructose	Lucigenin-fructose Alkaline medium	SB-12	LDR: 1.2×10^{-4}–1.3×10^{-3} mol.L^{-1}; Slope: 0.934; CC: 0.9955; LOD: 2.5×10^{-5} mol.L^{-1}	—	10
Fructose	Lucigenin-fructose Alkaline medium	SDS	LDR: 0.6×10^{-4}–1.0×10^{-3} mol.L^{-1}; Slope: 1.052; CC: 0.9998; LOD: 2.0×10^{-5} mol.L^{-1}	—	10
Hydrogen peroxide	Lucigenin–hydrogen peroxide Alkaline medium	SB-12	LDR: 1.0×10^{-6}–1.0×10^{-3} mol.L^{-1}; Slope: 0.9998, Intercep: 6.18; CC: 0.9998; LOD: 7.5×10^{-7} mol.L^{-1}	—	13
Hydrogen peroxide	Lucigenin–hydrogen peroxide Alkaline medium	HTAC	LDR: 3.0×10^{-6}–2.0×10^{-3} mol.L^{-1}; Slope: 0.98; Intercep: 5.66; CC: 0.9999; LOD: 9.5×10^{-7} mol.L^{-1}	—	13

Analyte	CL system	Medium	Analytical parameters		Ref.
Hydrogen peroxide	Lucigenin–hydrogen peroxide Alkaline medium	Brij-35	LDR: 2.0×10^{-6}–2.0×10^{-3} mol.L^{-1}; Slope: 0.99; Intercep: 5.86 CC: 0.9999 LOD: 7.4×10^{-7} mol.L^{-1}	—	13
Hydrogen peroxide	Lucigenin–hydrogen peroxide Alkaline medium	β-CD	LDR: 2.0×10^{-6}–6.0×10^{-4} mol.L^{-1}; Slope: 0.99; Intercep: 6.62 CC: 0.9989 LOD: 8.0×10^{-6} mol.L^{-1}	—	13
Lucigenin	Lucigenin–hydrogen peroxide Alkaline medium	HTAC	LDR: 5.0×10^{-7}–5.0×10^{-3} mol.L^{-1}; Slope: 0.71; Intercep: 5.14 CC: 0.9998	—	13
Lucigenin	Lucigenin–hydrogen peroxide Alkaline medium	Brij-35	LDR: 5.0×10^{-7}–5.0×10^{-4} mol.L^{-1}; Slope: 0.92; Intercep: 6.58 CC: 0.9984 LOD: 2.9×10^{-7} mol.L^{-1}	—	13
Lucigenin	Lucigenin–hydrogen peroxide Alkaline medium	β-CD	LDR: 5.0×10^{-7}–3.0×10^{-4} mol.L^{-1}; Slope: 0.82; Intercep: 6.33 CC: 0.9984 LOD: 1.2×10^{-7} mol.L^{-1}	—	13
Fructose	Lucigenin–fructose	HTAH	LDR: 6.5–84 mg.L^{-1}; Slope: 12.57; Intercep: 361.5 CC: 0.9997; RSD^e: 1.6% LOD: 2.3 mg.L^{-1}	—	37
Hydrogen peroxide	Fluorescein-H_2O_2 pH: 7.0	HTAC	LDR: 5.0×10^{-9}–1.0×10^{-6} mol.L^{-1}; Slope: 0.63	—	38

Table 5 Continued

Analyte	CL system	Organized medium	Analytical parameters	Applications	Ref.
Dansyl chloride	METQ[f]· H_2O_2–dansyl chloride; pH: 7.0	Brij-35	LDR: 0.6×10^{-6}–3.0×10^{-6} mol.L⁻¹ Slope: 0.15 RSD[g]: 3.43–15.2% LOD[h] 4.0×10^{-3} mol.L⁻¹	—	40
Dansyl chloride	METQ- H_2O_2–dansyl chloride; pH: 7.0	HTAC	LDR: 3.0×10^{-6}–10.0×10^{-6} mol.L⁻¹ Slope: 2.31 RSD[g]: 2.24–5.05% LOD[h]: 2.0×10^{-3} mol.L⁻¹	—	40
Rhodamine B (RH B)	METQ- H_2O_2–RH B pH: 7.0	Brij-35	LDR: 5.5×10^{-8}–1.1×10^{-4} mol.L⁻¹ Slope: 4.11 RSD: 6.4–11.5% LOD[h]: 2.0×10^{-3} mol.L⁻¹	—	40
Rhodamine B (RH B)	METQ-H_2O_2–RH B pH 7.0	HTAC	LDR:5.5×10^{-8}–5.5×10^{-5} mol.L⁻¹ Slope: 1.13 RSD[g]: 4.75–14.5% LOD[h]: 4.0×10^{-3} mol.L⁻¹	—	40
8-anilino-1-naphthalenesulfonic acid (ANS)	METQ-H_2O_2-ANS pH: 7.0	Brij-35	LDR:5.5×10^{-8}–5.5×10^{-5} mol.L⁻¹ Slope: 9.94; RSD[h]:0.46–6.10% LOD[h]: 3.0×10^{-3} mol.L⁻¹	—	40

Analyte	System	Surfactant	Analytical data		Ref.
8-anilino-1-naphthalenesulfonic acid (ANS)	METQ-H_2O_2-ANS pH: 7.0	HTAC	LDR: $5.5 \times 10^{-8} - 5.5 \times 10^{-5}$ mol.L^{-1}; Slope: 3.74; RSD^g: 1.74–2.26%; LOD^h: 2.0×10^{-3} mol.L^{-1}	—	40
Albumin–dansyl chloride	METQ-H_2O_2-albumin–dansyl chloride pH: 7.0	Brij-35	LDR:0.7×10^{-7}–8.7×10^{-7} mol.L^{-1}; Slope: 3.80; RSD^g: 5.86–6.84%; LOD^h: 6.0×10^{-3} mol.L^{-1}	—	40
Albumin–dansyl chloride	METQ-H_2O_2-albumin–dansyl chloride pH: 7.0	HTAC	LDR:0.7×10^{-7}–8.7×10^{-7} mol.L^{-1}; Slope: 14.2; RSD^g: 1.87–2.05%; LOD^h: 2.0×10^{-3} mol.L^{-1}	—	40
Avidin–RH B	METQ-H_2O_2-avidin-RHB; pH: 7.0	Brij-35	LDR:3.6×10^{-7}–3.6×10^{-5} mol.L^{-1}; Slope: 0.62; RSD^g: 0.20–17.4%; LOD^h: 4.0×10^{-3} mol.L^{-1}	—	40
Avidin–RH B	METQ-H_2O_2-avidin-RHB; pH: 7.0	HTAC	LDR:3.6×10^{-7}–3.6×10^{-5} mol.L^{-1}; Slope: 0.36; RSD^g: 3.30–5.93%; LOD^h: 2.0×10^{-3} mol.L^{-1}	—	40
H_2O_2-ANS	METQ-H_2O_2-ANS; pH: 7.0	Brij-35	LDR: 1.1×10^{-3}–5.5×10^{-2} mol.L^{-1}; Slope: 4.42; RSD^g: 2.54–2.75%; LOD^h: 3.0×10^{-3} mol.L^{-1}	—	40

Table 5 Continued

Analyte	CL system	Organized medium	Analytical parameters	Applications	Ref.
H_2O_2-ANS	METQ-H_2O_2-ANS; pH: 7.0	HTAC	LDR: 5.5×10^{-5}–5.5×10^{-2} mol.L^{-1} Slope: 1.80; RSD^g: 1.32–6.46% LOD^h: 4.0×10^{-3} mol.L^{-1}	—	40
H_2O_2-RH B	METQ-H_2O_2-RH B; pH: 7.0	Brij-35	LDR: 5.5×10^{-5}–5.5×10^{-2} mol.L^{-1} Slope: 2.27 RSD^g: 2.35–4.06% LOD^h: 2.0×10^{-3} mol.L^{-1}	—	40
H_2O_2-RH B	METQ-H_2O_2-RH B; pH: 7.0	HTAC	LDR: 5.5×10^{-5}–5.5×10^{-2} mol.L^{-1} Slope: 5.56 RSD^g: 3.42–5.22% LOD^h: 2.0×10^{-3} mol.L^{-1}	—	42
Dopamine	Lucigenin-dopamine	Brij-35	LDR: 5.0×10^{-7}–1.0×10^{-3} mol.L^{-1} Slope: 0.88 RSD: 2.5% LOD: 5.0×10^{-7} mol.L^{-1}	—	42
Dopamine	Lucigenin-dopamine	HTAH	LDR: 3.0×10^{-7}–1.0×10^{-3} mol.L^{-1} Slope: 0.60 RSD: 2.5% LOD: 3.0×10^{-7} mol.L^{-1}	—	42

Analyte	System	Surfactant	Analytical figures	Sample	Ref.
Norepinephrine	Lucigenin-norepinephrine	Brij-35	LDR: 5.0×10^{-7}–1.0×10^{-3} mol.L^{-1} Slope: 0.84; RSD: 2.5% LOD: 5.0×10^{-7} mol.L^{-1}	—	42
Norepinephrine	Lucigenin-norepinephrine	HTAH	LDR: 5.0×10^{-7}–1.0×10^{-3} mol.L^{-1} Slope: 0.68 RSD: 2.5% LOD: 5.0×10^{-7} mol.L^{-1}	—	42
Epinephrine	Lucigenin-epinephrine	Brij-35	LDR: 5.0×10^{-8}–1.0×10^{-4} mol.L^{-1} Slope: 1.40 RSD: 2.5% LOD: 5.0×10^{-8} mol.L^{-1}	—	42
Epinephrine	Lucigenin-epinephrine	HTAH	LDR: 5.0×10^{-8}–1.0×10^{-4} mol.L^{-1} Slope: 1.40 RSD: 2.5% LOD: 5.0×10^{-8} mol.L^{-1}	—	42
Co(II)	2,6,7-trihydroxy-9-(4-chlorophenyl)-3-fluorone–H_2O_2-Co(II) Alkaline medium	HTAB	LDR: 0.5–6.0 ng.mL^{-1} Slope: 13.05 RSD: 2.1% LOD: 0.07 ng.mL^{-1}	Water (river, lake)	43
Pyrogallol	N-bromosuccini-mide(NBS)-pyrogallol Akaline medium and hydroxylammonium chloride	HTAB	LDR: 5.0×10^{-7}–3.0×10^{-5} mol.L^{-1} RSD[f]: 2.1–1.6% LOD: 2.0×10^{-7} mol.L^{-1}	—	44
Glucose	Luminol-glucose–glucose oxidase pH: 7.25	HTAB	LDR: 3.0×10^{-7}–3.0×10^{-4} mol.L^{-1} RSD: 3.8% LOD: 1.0×10^{-7} mol.L^{-1}	Blood serum	45

Table 5 Continued

Analyte	CL system	Organized medium	Analytical parameters	Applications	Ref.
Hydrogen peroxide	Luminol-H_2O_2-peroxidase pH: 7.25	HTAB	LDR: 2.4×10^{-8}–1.2×10^{-4} mol.L^{-1} RSD: 2.6%	—	45
Cu(II)	1,10-phenantroline-H_2O_2-Cu(II) Alkaline medium	CEDAB	LOD: 0.3 pg	—	46
Fe(II)	Luminol-H_2O_2-Fe(II) pH: 9.3 and citric acid	TTAB	LDR: 5.0×10^{-9}–1.0×10^{-6} mol.L^{-1} RSD: 3.0% LOD: 2.0×10^{-9} mol.L^{-1}	Natural waters	47
Fe(II) + Fe(III)	Luminol- H_2O_2-Fe(II) pH: 9.3 and citric acid	TTAB	LDR: 5.0×10^{-9}–1.0×10^{-6} mol.L^{-1} RSD: 1.5%; LOD: 1.0×10^{-9} mol.L^{-1}	Human hair River water	47
Co(II)	4-diethylaminophthalohydrazide-O_2-Co(II); pH: 11.6 and fluorescein	PTAB	LDR: 5.9×10^{-3}–5.9 ng.mL^{-1} LOD: 1.0×10^{-3} ng.mL^{-1}	Tap water	48
Sulfite	Ru(bpy)$_3^{2+}$-SO_3^{2-}-$KBrO_3$ Acid medium	SDBS	LDR: 2.5×10^{-8}–9.5×10^{-5} mol.L^{-1} Slope: 3.1×10^7 RSD: 4.6% LOD: 3.8×10^{-9} mol.L^{-1}	Sugar	60

Isoprenaline	Lucigenin-isoprenaline Alkaline medium	Brij-35	LDR: 5.0×10^{-8}–1.0×10^{-5} mol.L^{-1} Slope: 0.94 RSD: 0.97% LOD: 5.0×10^{-8} mol.L^{-1}	Pharmaceutical preparations	61
Hydrogen peroxide	Luminol-H$_2$O$_2$-3-aminophthalate pH: 7.8–9.1	HTAC (reversed micelles)	LDR: 6.4×10^{-9}–6.4×10^{-7} mol.L^{-1} RSD[g]: 0.9–12.3% LOD[h]: 4.4×10^{-6} mol.L^{-1}	—	63
Glucose	Luminol-H$_2$O$_2$-glucose–glucose oxidase pH: 8.5	HTAB (reversed micelles)	LDR: 2.7×10^{-8}–2.7×10^{-6} mol.L^{-1} Slope: 1.58 RSD: 4.27% LOD: 2.7×10^{-8} mol.L^{-1}	Human blood serum Soda	64
L-Phenylalanine	Luminol-H$_2$O$_2$-L-phenylalanine–L-aminoacid oxidase pH: 8.5	HTAB (reversed micelles)	LDR: 2.0×10^{-8}–1.0×10^{-6} mol.L^{-1} Slope:1.41 RSD: 5.78% LOD: 1.05×10^{-8} mol.L^{-1}	—	64

[a] Data obtained from respective references with corresponding permission.
[b] Linear dynamic range.
[c] Correlation coefficient.
[d] Limit of detection.
[e] Relative standard deviation.
[f] *Bis*-[*N*-[2-(*N'*-methyl-2'-piridinumy)ethyl]-*N*-[(trifluoromethyl)sulfonyl]].
[g] RSD for upper concentration of linear dynamic range—RSD for low concentration of linear dynamic range.
[h] Limit of detection given in molarity of original analyte sample.

REFERENCES

1. WL Hinze. In: KL Mittal, ed. Solution Chemistry of Surfactants. New York: Plenum, 1979, Vol 1, pp 79–127.
2. DW Armstrong. Sep Purif Methods 14:213–304, 1985.
3. LJ Cline Love, JG Harbarta, JG Dorsey. Anal Chem 56:1132A–1148A, 1984.
4. E Pelizzetti, E Pramauro. Anal Chim Acta 169:1–29, 1985.
5. WL Hinze, HN Singh, Y Baba, NG Harvey. Trends Anal Chem 3:193–199, 1984.
6. ME Díaz García, A Sanz-Medel. Talanta 33:255–275, 1988.
7. E Pramauro, E Pelizzetti. In: SG Weber, ed. Surfactants in Analytical Chemistry. Applications of Organized Amphiphilic Media. Amsterdam: Wilson & Wilson's, 1996, pp 93–498.
8. WL Hinze, N Srinivasan, TK Smith, S Igarashi, H Hoshino. Advances in Multidimensional Luminescence. Tokyo: JAI Press, 1991, Vol 1, pp 149–206.
9. LL Klopf, TA Nieman. Anal Chem 56:1539–1542, 1984.
10. WL Hinze, TE Riehl, HN Singh, Y Baba. Anal Chem 56:2180–2191, 1984.
11. M Kato, M Yamada, S Susuki. Anal Chem 56:2529–2534, 1984.
12. M Yamada, S Susuki. Anal Lett 17:251–263, 1984.
13. CL Malehorn, TE Riehl, WL Hinze. Analyst 111:941–947, 1986.
14. MJ Rosen. Surfactants and Interfacial Phenomena, 2nd ed. New York: Wiley, 1989, pp 1–6, 108–136.
15. P Mukerjee, JR Cardinal. J Phys Chem 82:1620–1627, 1978.
16. C Tanford. J Phys Chem 78:2469–2479, 1974.
17. F Israelchivili, V Luzatti. J Phys Chem 71:3320–3330, 1967.
18. FM Menger, BJ Boyer. J Am Chem Soc 102:5936–5938, 1980.
19. FM Menger, JM Bonicamp. J Am Chem Soc 103:2140–2141, 1981.
20. P Fromherz. Chem Phys Lett 77(3):460–465, 1981.
21. KA Dill, DE Koppel, RS Cantor, JD Dill, D Bendedouch, SH Chen. Nature 309: 42–45, 1984.
22. D Bendedouch, SH Chen, WC Koeler. J Phys Chem 87:2621–2628, 1983.
23. JH Fendler, LK Patterson. J Phys Chem 74:4608–4609, 1970.
24. DW Armstrong, GY Stine. J Am Chem Soc 105:2962–2964, 1983.
25. GL Mcintire, DM Chiappardi, RL Casselberry, HN Blount. J Phys Chem 86:2632–2640, 1982.
26. JH Fendler. Acc Chem Res 9:153–161, 1976.
27. PL Luisi, BE Straub, eds. Reverse Micelles. New York: Plenum, 1984.
28. PL Luisi. Angew Chem, Int Ed Engl 24:439–528, 1985.
29. ME Díaz García, A Sanz Medel. Talanta 33:255–264, 1986.
30. WL Hinze. Ann Chim 77:167–207, 1987.
31. R von Wandruszka. Crit Rev Anal Chem 23(3):187–215, 1992.
32. G Ramis Ramos, MC García Alvarez-Coque, A Berthod, JD Winefordner. Ana Chim Acta 208:1–19, 1988.
33. A Sanz-Medel, MM Fernández, M De La Guardia, JL Carrión. Anal Chem 58: 2161–2166, 1986.
34. RP Frankewich, KN Thimmaiah, WL Hinze, Anal Chem 63:2924–2933, 1991.

35. RA Femia, S Scypinski, LJ Cline Love. Environ Sci Technol 19:155–159, 1985.
36. Y Kusumoto. Chem Phys Lett 136(6):535–538, 1987.
37. L Szente, J Szejtli. Analyst 123:735–741, 1998.
38. A Ingvarsson, CL Flurer, TE Riehl, KN Thimmaiah, JM Williams, WL Hinze. Anal Chem 60:2047–2055, 1988.
39. T Segawa, H Ishikawa, T Kamidate, H Watanabe. Anal Sci 10:589–593, 1994.
40. N Dan, ML Lau, ML Grayeski. Anal Chem 63:1766–1771, 1991.
41. TE Riehl, CL Malehorn, WL Hinze. Analyst 111:931–939, 1986.
42. T Kamidate, K Yoshida, T Kaneyasu, T Segawa, H Watanabe. Anal Sci 6:645–649, 1990.
43. Z Xie, F Zhang, Y Pan. Analyst 123:273–275, 1998.
44. A Safavi, MR Baezzat. Anal Chim Acta 368:113–116, 1998.
45. MS Abdel-Latif, GG Guilbault. Anal Chim Acta 221:11–17, 1989.
46. M Yamada, S Suzuki. Anal Lett 17(A4):251–263, 1984.
47. K Saitoh, T Hasebe, N Teshima, M Kurihara, T Kawashima. Anal Chim Acta 376: 247–254, 1998.
48. VV Sukhan. J Anal Chem USSR 46(12):1696–1700, 1991.
49. K Nakashima, K Imai. In: SG Schulman ed. Molecular Luminescence Spectroscopy. Methods and Applications: Part 3. New York: Wiley, 1993, pp 1–23.
50. KW Lee, LA Singer, KD Legg. J Org Chem 41:2685–2688, 1976.
51. R Maskiewicz, D Sogah, T Bruice. J Am Chem Soc 101:5347–5354, 1979.
52. T Riley, FA Long. J Am Chem Soc 84:522–526, 1962.
53. N Nath, MP Singh. J Phys Chem 69:2038–2043, 1965.
54. RL Veazey, H Nekimken, TA Nieman. Talanta 31:603–606, 1984.
55. YZ Lai. Carbohyd Res 28:154–157, 1973.
56. MP Singh, AK Singh, V Tripathi. J Phys Chem 82:1222–1225, 1978.
57. EH White, DF Roswell, OC Zafiriou. J Org Chem 34:2462–2468, 1969.
58. EH White, OC Zafiriou, HH Kagi, JHM Hill. J Am Chem Soc 86:940–941, 1964.
59. MM Rauhut, AM Semsel, BG Roberts. J Org Chem 31:2431–2436, 1966.
60. F Wu, Z He, H Meng, Y Zeng. Analyst 123:2109–2112, 1998.
61. AA Alwarthan, HA Al-Lohedan, ZA Issa. Anal Lett 29(9):1589–1602, 1996.
62. ML Cohen, FJ Arthan, SS Tsang. Eur Pat Appl EP 96, 749, 28 Dec 1983, 21 pp. Chem Abst 1984, 100, 18297o.
63. H Hoshino, WL Hinze. Anal Chem 59:496–504, 1987.
64. S Igarashi, WL Hinze. Anal Chim Acta 225:147–157, 1989.
65. ML Grayeski, EJ Woolf. J Lumin 33:115–121, 1985.
66. EJ Woolf, ML Grayeski. J Lumin 39:19–27, 1987.

12

Chemiluminescence in Flow Injection Analysis

Antony C. Calokerinos and Leonidas P. Palilis
University of Athens, Athens, Greece

1. INTRODUCTION

Chemiluminescence (CL) is usually generated by fast reactions and hence the phenomenon can be followed only when the chemical reaction is initiated in front of the light detector. A plethora of analytical research on CL was carried out with modified fluorimeters [1] that allowed mixing of reagents and injection of the analyte through a syringe maintaining the lighttightness of the unit. The air-segmented continuous-flow system is not appropriate for CL measurements since the time between mixing of the reagents and passing the final solution through the debubbler before measurement is sometimes longer than the time required for maximum CL intensity. Nevertheless, a plethora of analytes have been determined by nonsegmented continuous-flow manifolds [2] and some of them are summarized in Table 1.

The great boost of analytical CL appeared soon after the discovery of flow injection analysis (FIA) by Ruzicka and Hansen [3]. The speed with which the solutions of reagents can be supplied to the detector proved to be the best for CL reactions. Various mixing coils were investigated and this was the beginning of an avalanche of research on CL [4].

In this chapter the basic principles and various designs of flow injection chemiluminometers will be presented.

2. BASIC PRINCIPLES OF FLOW INJECTION ANALYSIS

FIA [5–7] is a version of continuous-flow analysis based on a nonsegmented flowing stream into which highly reproducible volumes of sample are injected, carried through the manifold, and subjected to one or more chemical or biochemical reactions and/or separation processes. Finally, as the stream transports the final solution, it passes through a flow cell where a detector is used to monitor a property of the solution that is related to the concentration of the analyte as a

Table 1 Selected Applications of Continuous Flow Manifolds with CL Detection

Analyte	Comments	L.o.D.	Ref.
Acetaminophen	Ce(IV) CL, drugs	0.07 µg/mL	18
Amiloride, streptomycin	N-bromosuccinimide CL, drugs	0.16, 1.61 µg/mL	19
Ammonium	N-bromosuccinimide-dichlorofluorescein (enhancer) CL, fertilizers	0.032 µg/mL	20
Dihydralazine, rifampicin, rifamycin SV	N-bromosuccinimide CL, drugs	1.23, 0.0017, 0.0005 µg/mL	21
Carbon dioxide	Luminol-Co(II) phthalocyanine CL, air, human breath	1.5 ppm	22
Corticosteroids	Enhancement of Ce(IV)-SO$_3^{2-}$ CL	0.02–0.3 µg/mL	23
Formaldehyde	Gallic acid-H$_2$O$_2$-OH$^-$ CL, air samples	10 ppb	24
Furancarboxylic acid	H$_2$O$_2$-OH$^-$ CL, serum	1 µM	25
Isoniazid	N-bromosuccinimide CL, drugs	0.024 µg/mL	26
Pyrogallol	H$_2$O$_2$-formaldehyde enhancer	6 nM	27
Sulfite, sulfur dioxide	Ce(IV)-3-(cyclohexylamino)propanesulfonic acid (sensitizer), air	0.02 µg/mL	2
Tertiary amines	ClO$^-$-OH$^-$ CL, fluorescein sensitizer, fish samples	2.7–7.7 nmol	28
Tetracyclines	N-bromosuccinimide CL, drugs, honey samples	0.005–0.4 µg/mL	29
Tetracyclines	Lucigenin or [Fe(CN)$_6$]$^{3-}$ CL	0.04–0.80 µg/mL	30
Thiamine	[Fe(CN)$_6$]$^{3-}$ CL, drugs	9 µM	31
Quinine, quinidine	Enhancement of Ce(IV)-SO$_3^{2-}$ CL, drugs	0.64, 1.6 µg/mL	32
Urushiol	Uranine-OH$^-$ CL, chinese urushi	10 µM	33

function of time. A typical recording has the shape of a peak (Fig. 1). The height
of the peak (H) is related to the concentration of the analyte. The area (*A*) under
this peak or the width (*W*ᵢ) of the peak at a fixed height can also used for correlation with the concentration of the analyte. The time interval between sample
injection and recording of the maximum value of the peak is called residence
time (*T*). During this time interval, all necessary physical and chemical processes
occur to generate the analytical signal. Wash time (*t*w) is defined as the time
interval between the maximum value of the peak height until the removal of the
sample zone from the detection area. The dispersion or dilution of the analyte or
its reaction product can be controlled through the geometrical and hydrodynamic
features of the flow system. Neither physical equilibrium (homogeneous flow)
nor chemical equilibrium (completion of reaction) is reached by the time of signal
detection. In a well-defined flow injection system the time required for completion of a single measurement ranges between 20 and 60 s.

Analytical measurement by CL is very sensitive to a variety of experimental
factors and even slight variations of these factors affect extensively the emitted
radiation. Thus, analytical CL measurements require highly reproducible mixing
of the reagents necessary for the chemiluminescent reaction and in such a way
to ensure repetition of the whole procedure at different times. A suitable observation cell (flow cell) that allows detection of the CL emission at an appropriate
time after mixing of analyte with reagents and initiation of the reaction is also
needed. Therefore, combining FIA with CL is a powerful tool to exploit the
advantages of both these techniques.

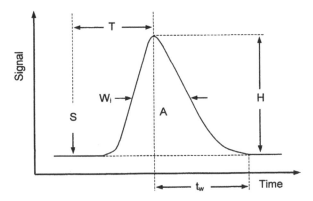

Figure 1   Schematic diagram of a typical FIA peak. *S*, time of sample introduction; *T*,
residence time; *H*, peak height; *A*, area under peak; *W*ᵢ, peak width at fixed height; *t*w,
wash time.

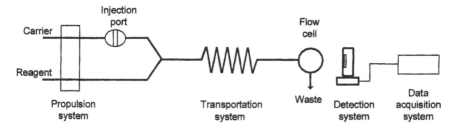

Figure 2 Schematic diagram of flow injection chemiluminometer.

3. A BASIC FLOW INJECTION SYSTEM FOR CHEMILUMINESCENCE MEASUREMENTS

A simple flow injection manifold for CL measurements is depicted in Figure 2. The main components of this manifold are the following:

The propulsion system, which establishes and controls accurately the flow of one or more solutions containing the reagents needed for initiating the CL reaction or, in some cases, merely acting as carrier for the sample that will be introduced later in this stream.

The injection system, which introduces the sample into the flowing stream.

The transportation system, into which dispersion of the sample into the carrier occurs and sometimes other analytical procedures may also happen, such as extraction, osmosis, ion exchange and non-CL reactions, before the final CL reaction is initiated.

The flow cell, into which the CL reaction takes place (throughout or, at least, the major percentage of it) and radiation is emitted.

The detection system, which detects radiation and transduces it into electrical signal.

The data acquisition system, which handles the signal received by the CL detector.

4. PRINCIPLES OF FLOW INJECTION ANALYSIS WITH CHEMILUMINESCENCE DETECTION

To understand how FIA functions with respect to CL detection, it is necessary to examine first, what happens when a sample is introduced into a flowing stream and second, how the rate of the CL reaction affects the emitted radiation.

4.1 Sample Dispersion

A very important concept in understanding the theoretical background of FIA is the dispersion of the sample. Dispersion is expressed as the dispersion coefficient, D, which is defined as follows:

$$D = \frac{C_0}{C_{max}} = \frac{H_0 K_1}{H K_2} = \frac{A_0 K_3}{A K_4}$$

where:

> C_0 is the initial concentration of the sample (before the effect of dispersion takes place) while C_{max} is the maximum concentration of the sample during the observation time (when the dispersion has taken place),
> H_0 or A_0 is the height or area of the peak that would be measured by the detector if no dispersion had taken place,
> H or A is the height or area of the peak during the observation time,
> K_1, K_2, K_3, and K_4 are ratio factors. If the concentration of the sample is within the linear range of the detector, then the ratio factor K_1 is equal to K_2 and K_3 is equal to K_4.

When a certain volume of a sample is instantaneously introduced into the stream of the carrier solution, the sample is carried away by the carrier not as a compact plug but gradually mixes with it. Mixing of the sample with the flowing solution is achieved by the action of two different mechanisms of dispersion: axial dispersion (parallel to the direction of the flow) due to the continuous flow of the stream, and radial dispersion (in vertical direction to the flow) due to diffusion.

Measurements by FIA occur under conditions where laminar flow predominates over turbulent flow (Fig. 3, a and b) and hence a parabolic profile of the concentration of analyte solution inside the carrier stream is developed. The layers of the analyte that are adjacent to the inner surface of the transportation tube flow slowly owing to the friction forces developed between these two different

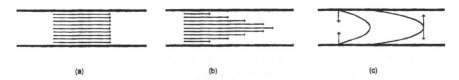

(a)   (b)   (c)

Figure 3 Schematic diagram of transportation of sample via the stream of carrier by (a) turbulent flow, (b) laminar flow, and (c) radial diffusion of the sample in response to radial concentration gradient, within the flowing stream.

surfaces. As the distance of the layer from the inner surface of the tube increases, the flow rate of the layer increases due to reduction of the friction forces. Hence, a concentration gradient is developed, which builds up a radial diffusion to equilibrate the different concentrations (Fig. 3c). Radial mass transfer ensures that sample and carrier are mixed and also reduces spreading out of the sample into the stream, maintaining the dilution of the sample in low level and allowing for relative short time intervals between successive sample injections.

A flow injection system can operate under low (D = 1–3), intermediate (D = 3–10), or high (D > 10) dispersion. For example, if D = 2, then the sample has been subjected to 1 : 1 dilution with the carrier before analytical measurement. Generally, if D = n, then the sample is subjected to 1: (n − 1) dilution by the carrier prior to measurement. The distribution of the sample in the flowing stream at various times after injection into the stream of the carrier is depicted in Figure 4. D increases with the distance covered by the sample zone through the transportation tube and the recorded peak becomes flat, even though the enclosed peak area usually remains constant. Thus, if peak height is used as the parameter for analytical measurement, then the sensitivity is lowered as D increases. Furthermore, sample throughput decreases with increasing D and hence flat peaks with

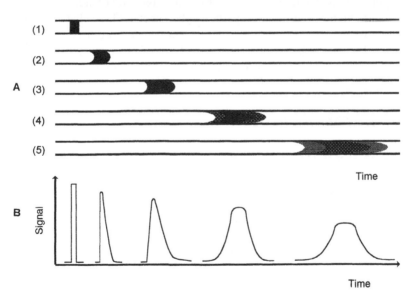

Figure 4 Sample dispersion (A) and signal profiles (B) at different distances within the tube (2–5) after injection of the sample into the stream (1), assuming that the detector is able to follow the sample zone as it flows.

high risk of overlapping are recorded. Under typical operating conditions, the zone of the sample is not completely symmetrical and peaks might be tailed. Flow injection chemiluminometers usually operate under low or medium dispersion.

Dispersion depends on several experimental factors, such as volume of sample injected, length, inner diameter, and geometrical configuration to the space of the transportation tube [and the reactor(s), if there is any], and flow rate of the carrier. If all other variables are constant, then dispersion depends on the following parameters [8]:

Volume of sample: If the volume of the sample increases, then dispersion decreases but, at the same time, the width of the peak increases and hence the measurement rate is reduced.

Geometrical characteristics of flow components: If the length or the internal diameter of the transportation tube increases, then dispersion increases. Dispersion also increases if a reactor is placed between the injection point and the flow cell. However, dispersion greatly depends on the geometry of the tube and/or the reactor used. Lower dispersion is achieved by using single bed string reactors such as the one shown in Figure 5a. With this reactor, broadening of the peaks is reduced to a minimum and this feature permits the introduction of several samples with very low time interval between injections, without any risk of sample overlapping. A shortcoming of this reactor is the high ratio of area to volume, which promotes high pressures and superficial phenomena. These characteristics are desirable when immobilized enzymes or solid reagents are used online. A knitted tube (Figure 5b) has similar attributes to the former reactor, although dispersion is generally higher. A mediocre dispersion is achieved when a coiled

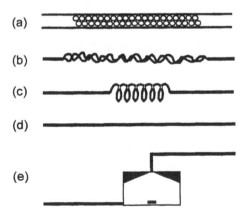

Figure 5   Schematic diagrams of reactors used in FIA in order of increasing dispersion: (a) single bed string reactor, (b) knitted tube, (c) coiled tube, (d) straight tube, and (e) external mixing chamber with stirring.

or a straight open tube is used (Fig. 5, c and d). Because this level of dispersion is usually acceptable in CL measurements and provided that these tubes are widespread, they are very widely used. Maximum dispersion can be achieved if an external mixing chamber (Fig. 5e) is used. The peaks recorded when this reactor is used are unsymmetrical and they decline very slowly to the baseline. This characteristic reduces greatly the number of measurements per unit of time. As a conclusion, it should be pointed out that when designing a reactor to be incorporated into the flow manifold, the main requirements are maximum mixing of the sample with the reagents while the width of the peaks is maintained as small as possible.

Flow rate: If the flow rate decreases, then dispersion increases and hence the residence time of the sample within the manifold increases.

4.2 Kinetics of Chemiluminescence Reaction

The chemiluminogenic reaction is initiated as soon as the analyte and the reagent(s) mix within the flow manifold. Therefore, the emission intensity is time dependent as the reaction proceeds and the excited intermediate or product is formed. If the analyte and reagent(s) were mixed in a chamber and the emitted radiation was monitored as a function of time, then a profile similar to that in Figure 6a would be recorded. The emission intensity increases, attains a maximum value, and then declines back to the baseline (for more details, see Chapter 8). This emission profile is the result of the change in the reaction rate versus time as the analyte and the reagents are consumed. Although all CL reactions generally follow the same pattern as for the shape of this emission profile, the time required by the reaction to generate maximum intensity, which then declines to the baseline, varies extensively from less than a few seconds to several hours. Except for the nature of the analyte and CL reagents, this time interval also depends on some factors that affect the chemical environment such as temperature, solvent, ionic strength, pH, and the presence of other species. Since these factors affect greatly the rate of the CL reaction and thus the emission intensity, they should be under strict control to achieve reproducible results.

4.3 Chemiluminescence and Flow System

When maximum peak height is recorded, neither the CL reaction is complete nor the flow system has reached a physical equilibrium (particularly as for the mixing of the streams). Thus a flow system with CL detection should incorporate several features, discussed below:

Number of flow streams: The number of reagent and/or carrier streams is a very important aspect when designing a flow manifold. In contrast to other detection techniques in FIA, with CL detection there is a very important limitation

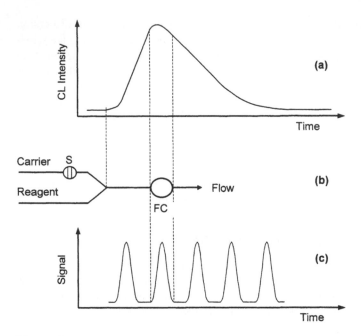

Figure 6 CL emission profile with the time (a) after mixing of reagents, initiation of reaction, entrance and exit from the flow cell (FC) of the chemiluminescent solution (b) and typical peaks recorded after successive injections of the same analyte into the manifold (c).

that should be encountered. This is the need for total exclusion of ambient light from the region of the detector; otherwise a high and uncontrollable noise signal will develop. Although it may seem easy to exclude ambient light from the CL detector, in many cases the transportation tubes used might act as optical fibers and lead light directly to the detector. Thus, for practical reasons and whenever it is possible, the sample is usually injected into a stream of carrier, which then merges into the stream of the CL reagent to initiate the reaction. This configuration also depends on the rate of the CL reaction. If the reaction is relatively slow, the sample can be injected directly into the stream of CL reagent, since the rate of the reaction ensures that maximum intensity will not be emitted before the analytical stream enters into the flow cell. If the reaction rate is high, the sample should be introduced into a separate stream and mix with the CL reagent near the inlet of the flow cell.

Volume of flow cell: Ideally, the solution of reactants should enter into the flow cell as soon as the radiation has such a value to generate maximum intensity while residing within the cell and start declining as soon as the solution flows out of the cell. On the other hand, the photosensitive area of the light detector is limited, e.g., about 2–3 cm² for side-window PMTs. Furthermore, the construction of the flow cell also depends on glass-blowing limitations. Hence, the volume of the flow cell is usually controlled by practical limitations and cannot be easily changed according to the requirements of a single CL reaction.

Flow rate: The limitations associated with the volume of flow cell can be overcome by accurately controlling the flow rate of each stream entering into the manifold. This experimental parameter controls the residence time of the chemiluminescent solution within the cell and can be easily optimized by the operator. Flow rates are directly proportional to the rate of the CL reaction. As the rate of the reaction increases, the flow rate should be increased but, at the same time, consumption of reagents increases. The flow rate also affects the shape and the height of the peak as well as the measurement rate (number of sample or standard solutions injected per hour).

Typical flow rates in FIA vary between 0.5 and 5.0 ml/min per channel, although higher values have also been used. Most of the published work on CL with FIA is based on equal flow rates for every stream entering the manifold. Nevertheless, if different rates must be used, deterioration of repeatability and reproducibility might appear due to incomplete mixing and anomalous hydrodynamic characteristics. These problems can be avoided if the general rule that the ratio of the fastest to the slowest flow rate should not exceed the value of 3 or 4 is followed.

Volume of analyte injected: As the volume of sample and standard solution injected increases, the peak height and duration increase. Nevertheless, if the volume increases too much, dispersion of the carrier into the analyte is limited and the concentration gradient within the flow is distorted resulting in very high dispersion values at the edges and very low values in the center. As a result, double peaks are recorded that are not appropriate for analytical measurement.

4.4 Optimization of Flow Injection Chemiluminescence Measurements

The target for optimization in FIA with CL detection is to adjust all experimental factors in such a way so that the detector views as much radiation as possible while the chemiluminescent solution flows through the cell. Hence the kinetics of the flow and detector system should be monitored to match the kinetics of the reaction and generate maximum intensity inside the cell. The effect of experimental variables on the CL signal cannot be exactly predicted in advance and there is not enough theoretical background to support any suggestion.

Some empirical observations can reduce the number of experimental factors that must be investigated. CL reactions require low or medium dispersion to achieve high detection sensitivity. Hence since the predominant feature is the time interval between mixing the sample with the reagent(s) and production of maximum emission intensity, optimization should be focused on this parameter. This time interval depends on the flow rate of each stream of the manifold and the length and diameter of the transportation tube between the mixing point and the flow cell. Subsequently, the detector sensitivity can be maximized by controlling these parameters so that the observation time window is near the peak of the CL emission intensity (Fig. 6a). If the volume of the flow cell and the geometrical characteristics of the tube from mixing to the cell are constant, then a change in the flow rate of solutions will alter the position and the width of the observation window. In addition, an extensive study of the effect of concentration of each reagent used will probably help to increase the sensitivity even more.

In some cases, the use of the stopped-flow technique might improve sensitivity. BL reactions or CL reactions with very low rate might benefit from this technique. By stopping the pump when the sample zone has reached the flow cell, it is possible to extend the time allowed for the analyte to react with the immobilized enzyme or with the reagents within the cell. In this way, sample dispersion is kept within low levels and sensitivity remains high. Implementation of this technique presupposes that the propulsion system has the ability to stop instantaneously and reproducibly to entrap always the same portion of sample zone within the flow cell. For very slow CL reactions, stopping of the flow can be done before the sample zone reaches the flow cell and for a time interval enough to generate maximum radiation and then the flow is reestablished for analytical measurement. Flow injection systems that operate under computer control of both the injection system and the propulsion system can be used for this purpose.

5. INSTRUMENTATION

The instrumentation used for FIA with CL detection is usually simple and is composed of the components depicted in Figure 2. These components are readily assembled to form the analytical manifold, although there are also commercially available flow injection systems with CL detection. Spectrophotometric or fluorimetric flow injection systems can often be used for CL measurements after some modifications.

The components required for a dedicated flow injection chemiluminometer are briefly discussed.

5.1 Propulsion Units

Ideally, the flow rate should be constant and perfectly reproducible to achieve reproducible residence time and maintain constant dispersion throughout the sys-

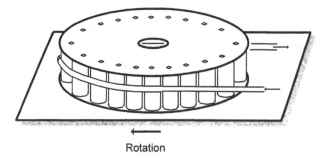

Rotation

Figure 7 Schematic diagram of the drum of a peristaltic pump showing the rollers, which squeeze the tubes. The arrows show the direction of the flow.

tem. For this purpose, peristaltic pumps are commonly used. These types of pumps are subjected to pulsation, which depends on the diameter of the roller used to squeeze the tubes (Fig. 7) and the distance between them. Pumps that operate with high-frequency, low-amplitude pulses and/or with pulse dumber are satisfactory. The number of channel (streams) used varies from one to 16, but usually is four.

The flow rate can be altered either by changing the diameter of tubes or by changing the rotation speed of the roller, provided that all other parameters are kept constant. Modern pumps can be controlled by microcomputer and different flow rates can be established during one analytical measurement. The pump is usually located before the injection port but sometimes it can be placed after the detection cell to reduce the effect of pulsation.

Alternatively, but not commonly, propulsion of the solutions can be achieved by gravity-based units, which rely on the difference in height between the solution(s) reservoir(s) and the flow lines of the manifold. Also, gas-pressure units, which rely on the action of pressure by an inert gas inside the vessels, which contain the solutions, can be used. Both these uncommon propulsion units yield pulse-free flow but require periodic refilling of solution reservoirs and adjustment of the desired flow rate is very difficult.

5.2 Flow Lines, Connectors, and Intermediate Reaction Systems

Tubes made by polyvinylchloride (PVC), Teflon, silicone, or any other similar material (with internal diameter between 0.1 and 2.5 mm) are usually used to transfer analytical streams along the manifold. A shortcoming of most tubes is that they are not suitable to transfer all kinds of organic solvents. They also require periodic replacement depending on the time and conditions of operation. Long-time operation alters the initial length and diameter of the tube and hence the flow rate of the tube changes. This problem is more intense when the propulsion system operates at high rotation speed. A gradual change in the flow rate would reduce sensitivity, which is not easily noticeable and may lead to erroneous results.

In some applications, additional components acting as reactors for specific chemical pretreatment are incorporated within the flow manifold. Typical examples are ion-exchange microcolumns for preconcentration of the analyte or removal of interferences and redox reactors, which are used either to convert the analyte into a more suitable oxidation state or to produce online an unstable reagent. Typical examples of online pretreatment are given in Table 2. Apart from these sophisticated reactors, a simple and frequently used reactor is a delay coil (see also Fig. 4), which may be formed by knitting a segment of the transfer line. This coil allows slow CL reactions to proceed extensively and enter into the flow cell at the time required for maximum radiation. The position of the reactors within the manifold is either before or after the injection port depending on the application.

The connectors used in FIA to join the tubes when the analytical system uses more than one stream usually have the shape of those in Figure 8. These connectors provide adequate mixing of the several analytical streams, while dispersion remains in relatively low levels, to achieve high measured values. The connector in Figure 8d is normally used for very fast CL reactions and allows mixing of the reagents with the analyte in front of the light detector. The choice of the connector depends mainly on the CL reaction rate (Table 3).

5.3 Sample Introduction Unit

The purpose of the injection unit is to insert accurately and reproducibly a very small volume of analyte into the flowing stream. This volume is usually within the range of 25–250 μL and very rarely less or more. Injection of the solution into the flowing stream of carrier or reagent should not cause any disturbance to flow of the solution and should be done fast enough to allow a high sampling rate. These requirements are fulfilled by volume-based injection units, such as the typical six-port rotary valve shown in Figure 9, which are based on the entrapment of a certain volume of the sample inside a loop. This is the most commonly used insertion unit and can be controlled either manually or by microcomputer via a pneumatic actuator for instance. By changing the loop with a lager or a smaller one, the volume of the solution injected can be altered. When operating the sample valve, it is essential to avoid the presence of air bubbles as they can modify the flow pattern and affect the mixing process with a subsequent change of the analytical signal.

Table 2 Typical Examples of Online Pretreatment in Flow Injection CL

Analyte	Comments	L.o.D.	Ref.
Anions	Column packed with anionic luminol derivative bound to a strong anion-exchange resin, displaced derivative–H_2O_2-$[Fe(CN)_6]^{3-}$ CL	$1.0\ \mu M\ SO_4^{2-}$–$5.0\ \mu M$ Br^-	34
Catecholamines	Online electrogeneration of ClO-, inhibition of luminol CL	0.6–1 nM	35
Citrate	Photochemical reaction, luminol CL, Pharmaceuticals, soft drinks	0.2 μM	36
Cobalt	8-Quinolinol column, gallic acid–H_2O_2 CL, seawater	10 ng/L	37
Chromium	Anion exchange column, luminol-H_2O_2 CL, speciation	0.05 μg/L Cr(III) 0.1 μg/L Cr(VI)	38
Cyanide	Resin with immobilized luminol and Cu^{2+}, Tap and waste water	2 ng/mL	39
Dopa	Online electrogeneration of ClO-, inhibition of luminol CL	0.8 ng/mL	40
Etamsylate	Online electrogeneration of ClO-, inhibition of luminol CL	0.6 ng/mL	41
Glucose	Immobilization of glucose oxidase, peroxyoxalate-H_2O_2 CL	0.05 μM	42
Glucose, cholesterol	Immobilized glucose oxidase or cholesterol oxidase to H_2O_2, luminol CL	Not reported	43
Glutamate, glutamine	Coimmobilized glutamate oxidase + glutaminase, Luminol-H_2O_2 CL	0.1 μM glutamate, 1 μM glutamine	44
Hydrazine	Ion-exchange columns to remove interfering metal ions, hydrazine-NaClO CL	0.5 nM	45
Hydrogen peroxide	Removal of interferences by anion and cation resin, luminol CL, rainwater	35 nM	46
Inorganic anions	Anion-exchange column, KOH-HNO_3 CL, water	70–100 ng/mL	47
Iron	8-Hydroxyquinoline column, luminol-H_2O_2 CL, seawater, on board analysis	0.021 nM	48
Iron	8-Hydroxyquinoline column, luminol CL, seawater	40 pM	49
Iron	Preconcentration on 8-hydroxyquinoline-5-sulfonic acid column, luminol-H_2O_2 CL	2 pg/mL	50
Isoniazid	Online electrogeneration of BrO^-, sensitization of luminol CL	7 ng/mL	51, 52
Nitrate	Photochemical activation, luminol CL, natural waters	70 nM	53
Persatin	Online electrogeneration of BrO^-, inhibition of luminol CL	4 μg/L	54
Pesticides	Immobilized choline oxidase/peroxidase, luminol-H_2O_2 CL	4 μg/L aldicarb 0.75 μg/L paraoxon	55
Vanadium	Potassium dichromate-KI reactor, luminol CL	0.7 nM	56
Vanadium	Jones reductor, lucigenin CL, tapwater, river water, industrial wastewater	0.3 ng/mL	57
Xanthine hypoxanthine	Immobilization of peroxidase or xanthine oxidase. luminol-H_2O_2 CL	2 and 5 μM	58

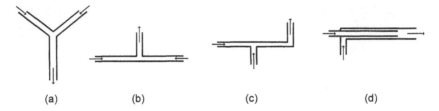

Figure 8 Schematic diagram of connectors used to join different streams in FIA: (a) Y-shaped, (b, c) T-shaped, and (d) concentric tubes. The arrows show the direction of the flow.

A disadvantage of injection via rotary valves is that while the volume of the sample that is injected is very small, the procedure for filling the loop of the valve requires a high volume of sample to flow through. This may reduce application of this technique in cases where there is only a limited quantity of sample available.

5.4 Flow Cell

The flow cell is the most important component of a flow injection manifold for CL measurements since maximum radiation should be generated while the solution is flowing in front of the detector. Other attributes of the flow cell are the small "dead volume" of the cell to allow fast and effective washing between injections

Table 3 Typical Examples of Connectors Used in Flow Injection with CL Detection

Design	Analyte	Comments	L.o.D.	Ref.
Concentric	Urea	NaBrO CL, urine and natural water	90 nM	59
T piece	Cinchona alkaloids	Ce(IV)-S^{2-} CL	0.01–0.6 µg/mL	60
T piece	Serotonin	KMnO$_4$ CL	0.5 µM	61
T piece	Choline	Choline oxidase column, luminol-H$_2$O$_2$ CL	0.1 µM	62
T piece	ATP	Molybdovanadophosphoric acid-luminol CL	10 nM	63
T piece	Isoprenaline	Lucigenin CL	0.05 µM	64
T piece	Nitroprusside	Luminol-H$_2$O$_2$ CL	0.05 µg/mL	65
Y piece	Iron	Luminol CL, seawater	40 pM	49

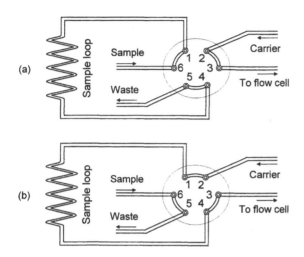

Figure 9 Schematic diagram of typical six-port rotary injection valve at (a) filling and (b) emptying position. (1) Sample-loop inlet, (2) carrier inlet, (3) outlet to the flow cell, (4) sample-loop outlet, (5) outlet to waste, and (6) sample inlet.

as well as accurate and reproducible control of dispersion and/or mixing of the reagents within the construction. A variety of flow cells used in FIA with CL detection are shown in Figure 10.

The first design of flow cell was based on mixing of reagents by blowing air into the solution but the presence of bubbles disturbed severely the CL radiation [9] (Fig. 10a). A modification of this design allows mixing by introducing the reagents from opposite sides of the cell [4] (Fig. 10b) and it was the first used in FIA with CL detection. Another design of flow cell incorporates a movable back plate to adjust the volume of the cell [10] (Fig. 10c). Nevertheless, the most widely used flow cell is the flat spiral of glass tube (Fig. 10d), introduced by Burguera et al. in 1980 [11]. The design is very simple and can be constructed very easily. It can be placed very close to the light-sensitive area of the detector and, therefore, maximize light intensity. A shortcoming of the cylindrical shape of this flow cell is that only a very thin layer of solution emits light opposite to the detector while light is also emitted to other directions, which are not viewed by the detector.

A recent development of the "technology" of the flow cells is the fountain cell [12] (Fig. 10e). An advantage of this cell for CL is that it has a relatively large surface, where the CL reaction takes place, and more light is emitted, but it must be placed horizontally; otherwise the gravity will influence mixing and dispersion of the flowing solution.

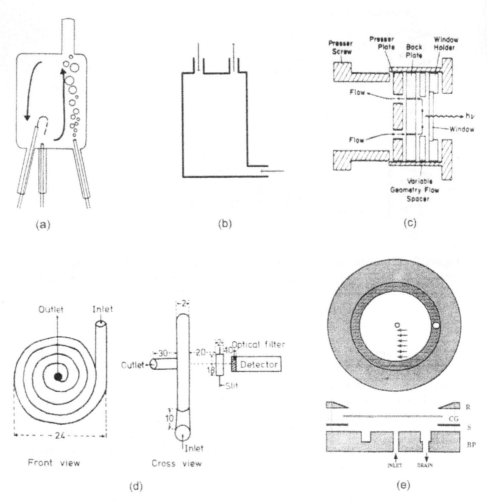

Figure 10 Schematic diagrams of flow cells: (a) flow cell using air bubbles for better stirring (reprinted with permission from Ref. 9), (b) flow cell using counterbalancing inlet flows, (c) flow cell with changeable volume (reprinted with permission from Ref. 10), (d) coiled flow cell (reprinted with permission from Ref. 11), and (e) fountain cell (reprinted with permission from Ref. 12). The arrows show the direction of the flow.

Flow cells may also act as reactors. In BL, enzymes may be immobilized inside the cell either by chemical bonding on the inner surface or by entrapping the enzyme as a heterogeneous system by mechanical ways. This approach has the advantage of low consumption of expensive reagents and enhancement of their stability, which is usually low. Many bioluminescent reactions have utilized the benefit of this process. The flow cell is also used as a reactor in the case of electrogenerated chemiluminescence (ECL) when used with FI manifolds. Some of these applications are included in Table 4.

5.5 Detector

Any device that can monitor radiation can be used as detector. In most cases, end-on and side-on photomultiplier tubes (PMT) are used. The end-on PMT has the benefit of circular photocathode, which views more effectively the radiation emitted from the circular flow cells. Nonetheless, the side-on PMTs are more commonly used due to their lower cost. The signal from the PMT is current and, hence, in most applications a current-to-voltage converter is required to convert the current to voltage, which is then monitored.

Light detection can also be achieved by semiconductor photodiodes or by photodiode array detectors. Their sensitivity, so far, is lower than that of PMTs but they possess the great advantages of much smaller dimensions and lower demand on power supply. These features make them attractive, especially for the construction of portable chemiluminometers. The sensitivity of these detectors

Table 4 Applications of Electrogenerated Chemiluminescence FIA

Analyte	Comments	L.o.D.	Ref.
Amino acids	$[Ru(bpy)_3]^{2+}$ ECL	0.1 pM–0.4 nM	66
Aminobutyl-*N*-ethylisoluminol	Luminol ECL, immunoassays	6 fmol	67
Chlorine species	Rhodamine 6G, electrochemically modified CL	90 μg/L	68
Glucose	H_2O_2 sensor, luminol ECL	60 pmol	69
Hydrogen peroxide	Luminol ECL	30 pmol	70, 71
Luminol	Luminol ECL	10 nM	72
Methyl linoleate hydroperoxide	Luminol ECL	0.3 nmol	73
Primary amines	Derivatization with divinyl sulfone, $[Ru(bpy)_3]^{2+}$ ECL	1–30 pmol	74

is expected to improve considerably within the next few years and probably in the near future chemiluminometers will depend solely on them.

Apart from the obvious dependence of the output of a given detector to the concentration of the chemiluminescent species, several other factors also affect the output [13]:

The area of the photosensitive surface of the detector, A.
The distance of the flow cell from the photosensitive area of the detector, d.
The three-dimensional shape of the cell.
The use of mirrors.
The use of lenses.

The first two factors are very important and it has been found that the detected radiation, L, is related to the intensity of emission, I, by the expression

$$L = I A/d^2$$

Therefore, the sensitivity of the measurement is greatly improved by decreasing the distance of the flow cell from the detector. Practically, the flow cell should be positioned as close as possible to the photosensitive area of the detector.

The effect of the other factors is less significant but sensitivity can still be improved if, for example, mirrors are used to focus the radiation on the detector.

5.6   Data Acquisition System

In the early years of FIA, the signal was followed by recorders and the height of the recorded peak was used as the analytical parameter. Although today recorders are still in use, the progress in computer technology has led to their wide use in handling the analytical signal produced by the detector. They can also be used to control the whole system of the flow injection manifold, including control of the injection and propulsion system. Thus, apart from the height, the area of the recorded signal is now used in the interpretation of the received data, which in some cases extends the linear range of the measurement. Integrators may also be used to record the analytical signal.

6.   NEW CHEMILUMINESCENCE REACTIONS AND FLOW INJECTION ANALYSIS

When a reaction is under investigation to establish possible chemiluminogenic properties, a batch chemiluminometer is preferable to be used (Fig. 11). This system can reveal the emission profile of the reaction and provide useful information about the kinetics of the reaction. It is suitable for reactions of all rates even

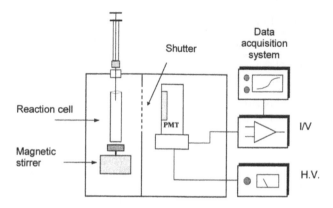

Figure 11   Schematic diagram of a batch chemiluminometer. H.V., high-voltage power supply; I/V, current-to-voltage converter.

though in cases of extremely fast reactions the rising part of the profile would not be representative of the reaction rate. By using the optimization data from batch systems, it is possible to predict with good accuracy, many of the experimental parameters for establishing the same reaction within a flow system, saving valuable time and reagents [14]. In general, if both systems are available, the batch system should be used for understanding the reaction and the flow system for applying the reaction.

7. FLOW INJECTION ANALYSIS VERSUS SEGMENTED FLOW ANALYSIS AND SEPARATION TECHNIQUES

Two independent groups of scientists developed FIA in the middle of seventies: Ruzicka and Hansen in Denmark [3] and Stewart et al. in the United States [15] and since then it has been developed rapidly. The first group developed the method using primarily instrumentation normally associated with segmented flow analyzers (SFA). In contrast, the second group based their initial work on HPLC components. These two different origins of FIA are responsible for several features of the technique related to SFA and /or HPLC that may characterize FIA as a hybrid of SFA and HPLC.

The main differences between SFA and FIA lie in the absence of air bubbles, which are used in SFA to prevent carryover. This shortcoming is not associated with FIA because the hydrodynamic conditions and the geometry of the system provide sufficient mixing of the solutions without the risk of sample over-

Table 5 Selected Applications of FIA with CL Detection

Analyte	Comments	L.o.D.	Ref.
Amidopyrine	$KMnO_4$ CL	30 µg/L	75
Ascorbic acid	$KMnO_4$ CL	0.5 µM	76
Beta-lactam antibiotics	Luminol CL	1–100 ng/5 µL	77
Bilirubin	ClO^- or $NBS-OH^-$ CL	0.05 or 0.075 µg/mL	14
Captopril	$Ce(IV)-H^-$ CL, sensitizer rhodamine 6G	10 pmol	78
Cefadroxil	$KMnO_4$ CL, sensitizer quinine	0.05 µg/mL	79
Cinchona alkaloids	$Ce(IV)-S^{2-}$ CL	0.01–0.6 µg/mL	60
Codeine	$KMnO_4$ CL	0.2 µM	80
Ciprofloxacin	$Ce(IV)$ CL	0.27 mg/L	81
Epinephrine, norepinephrine, dopamine, L-dopa	$KMnO_4$ CL	0.03–0.05 µg/L	82
Ergonovine maleate	$[Fe(CN)_6]^{3-}$ CL	0.07 ppb	83
Folic acid	$Ce(IV)-SO_3^{2-}$ CL, sensitizer rhodamine B	10 nM	84
Hydrochlorothiazide	$Ce(IV)$ CL, sensitizer rhodamine 6G	0.15 µM	85
Imipramine, chlorpromazine	$KMnO_4$ CL	50, 20 µM	86
Isoprenaline	$Lucigenin-OH^-$-surfactant CL	0.05 µM	64
Medazepam	$KMnO_4$ CL	18.5 µM	87
Methotrexate	$Formaldehyde-KMnO_4$ CL	3.4 nM	88
Morphine	$KMnO_4$ CL	0.05 µM	89
Penicillamine	$Ce(IV)-H^-$ CL, sensitizer quinine	15 pmol	90
Phenothiazines	$Ce(IV)-H^+$ CL, sensitizer rhodamine B	0.01–0.1 µg/mL	91
Promethazine	Addition of glyoxal, $KMnO_4$ CL	35 nM	92
Pyridoxine	$Luminol-H_2O_2$ CL	10 µg/mL	93
Ranitidine	$Tris$(2,2′-bipyridyl)ruthenium(III)-H_2SO_4 CL	0.6 µM	94
Salbutamol	$KMnO_4$ CL	25 nM	94
Steroids	$Ce(IV)-SO_3^{2-}$ CL, sensitization	0.013–4 µg/mL	95
Steroids	$BrO_3^- -SO_3^{2-}$ CL, sensitization	0.1–0.4 µg/mL	96
Tetracyclines	$KMnO_4$ CL	0.4–0.6 µg/mL	97
Tetracyclines	Treatment with H_2O_2 + persulfate, catalyst Cu(II)	0.1–0.01 nmol	98
Tiopronin	$Ce(IV)$ CL, sensitizer rhodamine 6G and quinine	0.036 µM	99, 100
Vitamin B_{12}	Acidification to release Co(II), $luminol-H_2O_2$ CL	0.35 µg/L	101

lapping. The absence of air bubbles makes the measurement procedure less sophisticated in the instrumentation used and allows for better sensitivity, reproducibility, and precision. In addition, it reduces significantly the interval between mixing of the solutions and detection of the signal, which is very important for the usually fast-rated CL reactions. Thus, an extremely wide variety of organic analytes have been measured by CL utilizing flow injection manifolds (Table 5).

In relation to separation techniques (mainly HPLC), FIA has the disadvantage that it is not able to separate the sample to its constituents and if the sample contains more than one analyte, then unless the kinetics of the reactions are totally different, it is not possible to measure them. Nonetheless, by placing a separator (e.g., an ion-exchange column) before the flow cell, it might be possible to eliminate these problems. Nevertheless, the scope of each technique is different. Separation techniques are aimed at separating and measuring the constituents of a mixture, while FIA is oriented toward the rapid determination of one or two species in a large number of samples. Also, the FIA instrumentation is much simpler than the corresponding for a liquid chromatographic technique.

8. RECENT FLOW INJECTION ANALYSIS VERSIONS

Progress in developing new flow injection techniques has led to sequential injection analysis (SIA). This technique is relatively new [16] and utilizes simpler instrumentation without reducing the precision comparing to classical flow systems. SIA, contrary to FIA, does not push the analytical stream continuously in the same direction, but the direction and the rate of the flow are altered according to the progress of the analytical procedure. SIA utilizes only one flowing stream, which can move in both directions. Initially the propulsion system pumps a washing solution, followed by the sample and finally the solution of the reagent. All these solutions are introduced via a special injection valve into the same tube, which serves as a medium for the reaction, the incubation, and finally for the detection of the analytical signal. This valve ensures that this successive introduction of different solutions inside the tube is synchronized with the rate of the pump. After the reaction process has been completed, the pump alters the direction of the flow and thus the solutions are flown out of the tube and then the system is ready for the next measurement sequence. Mixing of the reagent(s) with the sample is done by diffusion of one into the other. The magnitude and the time of mixing depend on the residence time of solutions at different sections of the tube, as well as by the direction of the flow and the flow rate. The aim is to obtain the maximum signal possible when the solutions exit the tube and pass through the detector, in the final stage of the analytical procedure. SIA has already been used with CL detection [17] (Table 6).

Table 6 Examples of Sequential Injection Analysis with CL Detection

Analyte	Comments	L.o.D.	Ref.
Digoxin	Immunoreactor, immobilized antibodies, acridinium ester–H_2O_2 CL	0.2 ng/mL	102
Glucose	Immobilized glucose oxidase, luminol-H_2O_2 CL	15 μM	103
Lactic acid	Immobilized lactate oxidase, luminol-H_2O_2 CL	5 mg/L	104
Morphine	$KMnO_4$ CL	0.00003%	105
		25 nM	106

9. CONCLUSIONS

Combination of FIA with CL detection offers several advantages for CL measurements. Some of them are, briefly:

Small analysis time: The typical time needed for a complete sequence of a measurement in FIA is usually less than 60 s. This feature makes FIA very attractive when a large number of samples have to be measured.

Automation: CL measurements can be performed very rapidly and with minimal human participation if the flow system is carefully designed. Several reagents can be added to the sample if multiple flow lines are used and at predetermined times to fulfill the best condition for maximum CL sensitivity. Introduction of special devices into the flow line, which allow procedures otherwise time-consuming such as solvent extraction or ion exchange, improve substantially the sensitivity and selectivity of the technique.

Precision: FIA measurements typically show low relative standard deviations (RSD) on replicate measurements, mainly due to the definite and reproducible way of sample introduction. This is a very important feature especially for CL, which is very sensitive to several environmental factors and sensitivity relies greatly on the rate of the reaction.

Versatility: Another benefit that derives from the fact that CL reagents are continuously mixed in front of the detector, regardless of the presence of the analyte, is implementation of the analytical procedure even in cases of reaction where the reagents produce a low background emission. This happens because this background emission can be regarded as the baseline since it is constant with the time and, hence, it will not interfere with the analytical signal produced by the analyte.

On the other hand, there are a few shortcomings in the use of FIA with CL detection:

Kinetics of CL reaction: CL reactions with complex kinetics are not always easily handled. The correlation of peak height with concentration of analyte is not always linear since the characteristics and limitations in the design of the flow cell allow only a fraction of the emitted radiation to be viewed by the detector.

Physical properties of solutions: Hydrodynamic problems and disturbance of the flow profile might appear when solutions with different properties, such as density or viscosity, are introduced into the manifold.

Fundamental research: Flow systems are not suitable for fundamental CL research, as they do not allow observation of the whole emission profile of a reaction. In addition, in cases of very fast or very slow CL reactions a flow system might lead to wrong conclusions. Therefore, a fast or a slow chemiluminogenic reaction might be missed if the flow system is not well established.

REFERENCES

1. JL Burguera, A Townshend. Talanta 26:795–798, 1979.
2. II Koukli, EG Sarantonis, AC Calokerinos. Analyst 113:603–608, 1988.
3. J Ruzicka, EH Hansen. Anal Chim Acta 78:145–157, 1975.
4. JL Burguera, A Townshend. Proc. Anal Div Chem Soc 16:262–264, 1979.
5. A Townshend. In Encyclopedia of Analytical Science. 1st ed. London: Academic Press Limited, 1995, pp. 1299–1318.
6. M Valcárcel, MD Luque de Castro. Flow-Injection Analysis, Principles and Applications. 1st ed. Chichester: Ellis Horwood, 1987, pp. 40–98.
7. J Ruzicka, EH Hansen. Anal Chim Acta 114:19–44, 1980.
8. JT Vanderslice, KK Stewart, AG Rosenfeld, DJ Higgs. Talanta:28 11–18, 1981.
9. WR Seitz, DM Hercules. Anal Chem 44:2143–2149, 1972.
10. S Stieg, TA Nieman. Anal Chem 50:401–404, 1978.
11. JL Burguera, A Townshend, S Greenfield. Anal Chim Acta 114:209–214, 1980.
12. KM Scudder, CH Pollema, J Ruzicka. Anal Chem 64:2657–2660, 1992.
13. DG Bullock, RA Bunce, TJN Carter. Anal Lett 12:841–854, 1979.
14. LP Palilis, AC Calokerinos, N Grekas. Anal Chim Acta 333:267–275, 1996.
15. KK Stewart, GR Beecher, PE Hare. Anal Biochem 70:167, 1976.
16. J Ruzicka, GD Marshall. Anal Chim Acta 237:329–343, 1990.
17. DJ Tucker, B Toivola, CH Pollema, R Ruzicka, GD Christian. Analyst 119:975–979, 1994.
18. II Koukli, AC Calokerinos, TP Hadjiioannou, Analyst 114:711–714, 1989.
19. SA Halvatzis, AM Mihalatos, LP Palilis, AC Calokerinos. Anal Chim Acta, 290:172–178, 1994.
20. SA Halvatzis, MM Timotheou-Potamia. Talanta 40:1245–1254, 1993.
21. SA Halvatzis, MM Timotheou-Potamia, TP Hadjiioannou. Anal Chim Acta 272:251–263, 1993.
22. ZH Lan, A Mottola. Anal Chim Acta 329:305–310, 1996.
23. II Koukli, AC Calokerinos. Analyst 115:1553–1557, 1990.
24. Y Maeda, XC Hu, S Itou, M Kitano, N Takenaka, H Bandow, M Munemori. Analyst 119:2237–2240, 1994.

25. M Ishii, K Itoh, Y Yoshihiro, T Nakamura. Anal Chim Acta 299:269–275, 1994.
26. SA Halvatzis, MM Timotheou-Potamia, AC Calokerinos. Analyst 115:1229–1234, 1990.
27. NJ Kearney, ES Bridgeland, RA Jewsbury, ND Martin, SJ Kelly, SR Korn. Anal Commun 33:241–243, 1996.
28. M Cobo, M Silva, D Perez-Bendito. Anal Chim Acta 356:51–59, 1997.
29. SA Halvatzis, MM Timotheou-Potamia, AC Calokerinos. Analyst 118:633–637, 1993.
30. AB Syropoulos, AC Calokerinos. Anal Chim Acta 255:403–411, 1991.
31. N Grekas, AC Calokerinos. Talanta 37:1043–1048, 1990.
32. II Koukli, AC Calokerinos. Anal Chim Acta 236:463–468, 1990.
33. M Ishii, M Takimoto, T Miyakoshi, T Nakamura. Anal Sci 11:79–83, 1995.
34. MM Cooper, SR Spurlin. Anal Lett 19:2221–2230, 1986.
35. CX Zhang, JH Huang, ZJ Zhang, M Aizawa. Anal Chim Acta 374:105–110, 1998.
36. T Perez Ruiz, C Martinez Lozano, V Tomas, OVal. Analyst 120:471–475, 1995.
37. S Hirata, Y Hashimoto, M Aihara, G Vitharana-Mallika. Fresenius J Anal Chem 355:676–679, 1996.
38. HG Beere, P Jones. Anal Chim Acta 293:237–243, 1994.
39. JH Lu, W Qin, ZJ Zhang, ML Feng, YJ Wang. Anal Chim Acta 304:369–373, 1995.
40. JC Huang, CX Zhang, ZJ Zhang. Fenxi. Huaxue 27:41–43, 1999.
41. CX Zhang, JC Huang, ML Feng, ZJ Zhang. Anal Lett 31:1917–1928, 1998.
42. P Van Zoonen, I De Herder, C Gooijer, NH Velthorst, RW Frei, E Kuentzberg, G Guebitz. Anal Lett 19:1949–1961, 1986.
43. G Xie, Y Huang, G Huang. Fenxi Ceshi Tongbao 9:30–34, 1990.
44. G Blankenstein, F Preuschoff, U Spohn, KH Mohr, MR Kula. Anal Chim Acta 271:231–237, 1993.
45. AT Faizullah, A Townshend. Anal Proc 22:15–16, 1985.
46. W Qin, ZJ Zhang, HH Chen. Int J Environ Anal Chem 66:191–200, 1997.
47. H Sakai, T Fujiwara, T Kumamaru. Anal Chim Acta 331:239–244, 1996.
48. JTM De Jong, J den Das, U Bathmann, MHC Stoll, G Kattner, RF Nolting, HJW de Baar. Anal Chim Acta 377:113–124, 1998.
49. AR Bowie, EP Achterberg, RFC Mantoura, PJ Worsfold. Anal Chim Acta 361:189–200, 1998.
50. AA Alwarthan, KAJ Habib, A Townshend. Fresenius J Anal Chem 337:848–851, 1990.
51. X Zheng, Z Zhang. Analyst 124:763–766, 1999.
52. JC Huang, CX Zhang, ZJ Zhang. Fresenius J Anal Chem 363:126–128, 1999.
53. L Renmin, L Daojie, S Ailing, L Guihua. Talanta 42:437–440, 1995.
54. XW Zheng, ZJ Zhang. Fenxi Huaxue 27:145–148, 1999.
55. A Roda, P Rauch, E Ferri, S Girotti, S Ghini, G Carrea, R Bovara. Anal Chim Acta 294:35–42, 1994.
56. WH Li, ZL Wang, JZ Li, ZJ Zhang. Fenxi Shiyanshi 17:41–44, 1998.
57. ZJ Zhu, JR Lu. Fenxi Shiyanshi 13:65–67, 1994.
58. U Spohn, F Preuschoff, G Blankenstein, D Janasek, MR Kula, A Hacker. Anal Chim Acta 303:109–120, 1995.

59. X Hu, N Takenaka, M Kitano, H Bandow, Y Maeda, M Hattori. Analyst 119:1829–1833, 1994.
60. M Feng, G Zhang, Z Zhang. Anal Lett 31:2537–2548, 1998.
61. NW Barnett, BJ Hindson, SW Lewis. Anal Chim Acta 362:131–139, 1998.
62. M Yaqoob, A Nabi, M Masoom Yasinzai. J Biolumin Chemilumin 12:135–140, 1997.
63. T Fujiwara, K Kurahashi, T Kumamaru. Anal Chim Acta 349:159–164, 1997.
64. AA Alwarthan, HA Al Lohedan, ZA Issa. Anal Lett 29:1589–1602, 1996.
65. AA Alwarthan. Talanta 41:1683–1688, 1994.
66. X Chen, M Sato. Fenxi Kexue Xuebao 14:278–282, 1998.
67. K Arai, K Takahashi, F Kusu. Anal Chem 71:2237–2240, 1999.
68. GP Irons, GM Greenway. Analyst 120:477–483, 1995.
69. CA Marquette, LJ Blum. Anal Chim Acta 381:1–10, 1999.
70. S Sakura, H Imai. Anal Sci 4:9–12, 1988.
71. D Janasek, U Spohn, D Beckmann. Sens Actuators B51:107–113, 1998.
72. X Chen, M Sato. J Flow Injection Anal 12:54–65, 1995.
73. S Sakura, J Terao. Anal Chim Acta 262:59–65, 1992.
74. K Uchikura, M Kirisawa, A Sugii. Anal Sci 9:121–123, 1993.
75. YH He, JX Du, ML Feng, JR Lu. Fenxi Shiyanshi 18:60–62, 1999.
76. IB Agater, RA Jewsbury. Anal Chim Acta 356:289–294, 1997.
77. H Kubo, M Saitoh, S Murase, T Inomato, Y Yoshimura, H Nakazawa. Anal Chim Acta 389:89–94, 1999.
78. ZD Zhang, WRG Baeyens, XR Zhang, G Van der Weken. J Pharm Biomed Anal 14:939–945, 1996.
79. FA Aly, NA Alarfaffj, AA Alwarthan. Talanta 47:471–478, 1998.
80. TJ Christie, RH Hanway, DA Paulls, A Townshend. Anal Proc 32:91–93, 1995.
81. YD Liang, JZ Li, ZJ Zhang. Fenxi Huaxue 25:1307–1310, 1997.
82. NT Deftereos, AC Calokerinos, CE Efstathiou. Analyst 118:627–632, 1993.
83. Y Fuster Mestre, B Fernandez Band, L Lahuerta Zamora, J Martinez Calatayud. Analyst 124:413–416, 1999.
84. AA Alwarthan. Anal Sci 10:919–922, 1994.
85. J Ouyang, WRG Baeyens, J Delanghe, G Van der Weken, AC Calokerinos. Talanta 46:961–968, 1998.
86. JL Lopez Paz, A Townshend. Anal Commun 33:31–33, 1996.
87. SM Sultan, AM Almuaibed, A Townshend. Fresenius J Anal Chem 362:167–169, 1998.
88. YH He, YY Xue, ML Feng, JR Lu. Fenxi Huaxue 26:1136–1138, 1998.
89. NW Barnett, DG Rolfe, TA Bowser, TW Paton. Anal Chim Acta 282:551–557, 1993.
90. ZD Zhang, WRG Baeyens, XR Zhang, G Van der Weken. Analyst 121:1569–1572, 1996.
91. FA Aly, NA Alarfaj, AA Alwarthan. Anal Chim Acta 358:255–262, 1998.
92. YY Xue, YH He, ML Feng, JR Lu. Fenxi Huaxue 27:427–429, 1999.
93. AA Alwarthan, FA Aly. Talanta 45:1131–1138, 1998.
94. NW Barnett, BJ Hindson, SW Lewis. Anal Chim Acta 384:151–158, 1999.
95. NT Deftereos, AC Calokerinos. Anal Chim Acta 290:190–200, 1994.

96. AB Syropoulos, EG Sarantonis, AC Calorkerinos. Anal Chim Acta 239:195–202, 1990.
97. Z Li, M Feng, J Lu, Z Gong, H Jiang. Anal Lett 30:797–807, 1997.
98. ZR Zhang, WRG Baeyens, A Van den Borre, G Van der Weken, AC Calokerinos, SG Schulman. Analyst 120:463–466, 1995.
99. T Perez Ruiz, C Martinez Lozano, WRG Baeyens, A Sanz, MT San Miguel. J Pharm Biomed Anal 17:823–828, 1998.
100. YN Zhao, WRG Baeyens, XR Zhang, AC Calokerinos, K Nakashima, G Van Der Weken. Analyst 122:103–106, 1997.
101. W Qin, ZJ Zhang, HJ Liu. Anal Chim Acta 357:127–132, 1997.
102. D Dreveny, J Michalowski, R Seidl, G Guebitz. Analyst 123:2271–2276, 1998.
103. X Liu, EH Hansen. Anal Chim Acta 326:1–12, 1996.
104. RW Min, J Nielsen, J Villadsen. Anal Chim Acta 312:149–156, 1995.
105. NW Barnett, CE Lenchan, SW Lewis, DJ Tucker, KM Essery. Analyst 123:601–605, 1998.
106. NW Barnett, SW Lewis, DJ Tucker. Fresenius J Anal Chem 355:591–595, 1996.

13

Gas-Phase Chemiluminescence Detection

James E. Boulter and John W. Birks
University of Colorado, Boulder, Colorado

1. INTRODUCTION

The use of fire, being the first chemical reaction controlled by humans, is an important marker for the beginning of civilization. The thermal and visible radiation of fire was used to great benefit for warmth, cooking, nighttime visibility, and warding off predators. For millennia, our ancestors must have marveled at the luminescence and other unique properties of fire; indeed the Greeks considered fire one of four basic elements comprising all matter. Chemiluminescent emissions from flames served as one of the first analytical tools used for qualitative identification of specific elements, and flame emission spectrometry is still one of the most sensitive and selective means for quantification of alkali metals and alkaline earths. The discovery of the ability to produce gas-phase chemiluminescence without the assistance of the thermal energy provided by flames led to many new analytical applications. This chapter focuses primarily on the analytical utility of these so-called "cold" chemiluminescent reactions, where reactants are formed and molecular excitation is achieved without the assistance of the thermal energy associated with a flame.

Chemiluminescence techniques have several inherent advantages, including high sensitivity, high selectivity, and simplicity of instrumentation, making them the methods of choice for select applications. The high quantum efficiency of photon detection, in combination with a dark background in the absence of any analyte, can result in very low limits of detection. Additionally, because very few chemical reactions produce intense chemiluminescence in the UV/Vis region, the selectivity of chemiluminescence is often very high. For example, nitric oxide is measured in the atmosphere at low parts per trillion (pptv) levels in the presence of hundreds of other compounds at higher concentrations without any separation step, based on its chemiluminescent reaction with ozone [1–4]. Finally, the instrumentation for chemiluminescence is simple, robust, and inexpensive, making it well suited for both field and laboratory applications.

2. CHARACTERISTICS OF GAS-PHASE CHEMILUMINESCENCE

As in the condensed phase, gas-phase chemiluminescence consists of a chemical reaction forming an excited-state product that then undergoes one or more relaxation processes to attain its ground state. In theory, any reaction sufficiently exo-ergic to generate a significant fraction of products in an excited state has the potential to be chemiluminescent. In such a reaction, the excess energy is parti-tioned into a combination of translational, rotational, vibrational, and electronic states of the products. Emission of a visible or ultraviolet photon, which can be detected with a high quantum efficiency, generally requires that the reaction form a product in an excited electronic state. Vibrational overtone emission such as that of HF produced in the reaction of fluorine with certain sulfur compounds is a rare exception to this rule [5–7]. Once a product molecule is formed in an excited electronic state, the mechanism by which it relaxes to its ground state also contributes to the chemiluminescent character of the reaction. Excess energy within a molecule may be redistributed by vibrational relaxation, internal conver-sion, intersystem crossing, fluorescence, or phosphorescence [8]. In gas-phase chemiluminescence detection, radiative emission is usually competitive with non-radiative processes with the result that both the quantum yield of the reaction and the emission spectrum vary with physical conditions such as bath gas compo-sition, temperature, and pressure.

3. REACTOR DESIGN AND OPTIMIZATION

As shown schematically in Figure 1, a gas-phase chemiluminescence detector consists of a reaction chamber, inlets for the analyte and reagent gas streams, a vacuum pump to lower the pressure in the reaction chamber (typically to a few torr), and a transducer such as a photomultiplier tube (PMT) to monitor the light produced in the reaction. The reagent gas, usually present in large excess, reacts with a trace concentration of analyte to produce an excited product that subse-

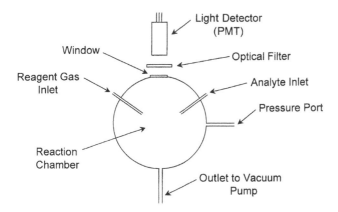

Figure 1 Schematic diagram of a generalized gas-phase chemiluminescence detector.

quently emits light. Of course, the actual design and optimization of a chemiluminescence instrument is somewhat more complicated [9]. For example, any stray
light entering the reaction chamber provides a radiation background, so the reaction chamber must be well sealed from ambient light and from any light produced
in the reagent gas source, such as a microwave discharge, if atomic or radical
species serve as the reagent gas. The pressure within the reaction chamber depends on the flow rates of the analyte and reagent gas streams and on the speed
of the vacuum pump. Finally, the reaction chamber volume affects both the signal
intensity and response time of the detector.

The interdependence of the flow rates, pumping speed, total pressure, reaction chamber volume, residence time, reaction rate, quantum yield, and chemiluminescence signal is given by simple equations if one assumes a large excess of
reagent gas and plug-flow conditions, characteristic of most gas-phase chemiluminescence detectors in the pressure range 1–10 torr. As an example, we will
consider the measurement of the analyte NO using the well-known chemiluminescent reaction between nitric oxide and ozone, the mechanism of which includes the following steps [10]:

$$NO + O_3 \rightarrow NO_2^* + O_2 \tag{1a}$$

$$NO + O_3 \rightarrow NO_2 + O_2 \tag{1b}$$

$$NO_2^* + M \rightarrow NO_2 + M \tag{2}$$

$$NO_2^* \rightarrow NO_2 + h\nu \tag{3}$$

where M is the bath gas, and the rate constant for Reaction (2) depends on its
concentration and identity. The rate-limiting step for this mechanism is the reac-
tion of NO with O_3, the total rate constant ($k_1 = k_{1a} + k_{1b}$) of which is 1.8×10^{-14} cm^3 molec^{-1} s^{-1} at 298 K [11]. The chemiluminescence efficiency, defined
as the fraction of reactions of NO with O_3 that result in the production of a
photon, is determined by two factors, the branching ratio for the formation of
the electronically excited product NO_2^*, and the quantum efficiency for radiative
deactivation of NO_2^*. The branching ratio (fraction of reactions that produce
NO_2^*) is given by

$$\phi_{b.r.} = \frac{k_{1a}}{k_{1a} + k_{1b}} \tag{E1}$$

and the chemiluminescence quantum efficiency (fraction of NO_2^* molecules that
emit a photon) is given by

$$\phi_{q.e.} = \frac{k_3}{k_3 + k_2[M]} \tag{E2}$$

The fraction of NO molecules that react during the residence time, t_{res}, within the reactor is given by,

$$\frac{[NO]_0 - [NO]}{[NO]_0} = 1 - \exp(-k_1[O_3]t_{res}) \tag{E3}$$

where $[NO]_0$ is the concentration of NO in the absence of O_3, and $[NO]$ is the concentration of NO in the presence of O_3. The light intensity, I, expressed in photons s^{-1}, is given by the flow rate of analyte molecules into the reaction chamber multiplied by the fraction of molecules that react during the residence time within the reactor, the branching ratio to form excited-state products, and the quantum efficiency for emission of light:

$$I = X_{NO}F_{NO}\phi_{b.r.}\phi_{q.e.}\{1 - \exp(-k_1[O_3]t_{res})\} \tag{E4}$$

where F_{NO} is the flow rate of the gas stream containing NO in units of molec s^{-1}, and X_{NO} is the mixing ratio (mole fraction) of NO in the gas stream. The residence time is given by

$$t_{res} = \frac{V}{S} \tag{E5}$$

where V is the volume of the reaction chamber in units of L and S is the pumping speed in units of L s^{-1} after considering the conductance of the connecting tubing and any intervening valves. Of course, the measured signal in photon counts s^{-1} will be reduced from Eq. (E4) by the collection efficiency (fraction of emitted photons falling on the detector) and quantum efficiency of the PMT or other detector averaged over the emission spectrum.

Note that increasing the flow rate of the analyte NO results in an increased signal according to Eq. (E4), but a complication arises in that the pressure (and therefore $[M]$) also increases. For a fixed pumping speed, the residence time within the reaction chamber is constant, and the pressure in the reaction cell will increase linearly with increasing flow rate of gas into the cell:

$$P = \frac{FRT}{N_A S} = \frac{FRT t_{res}}{N_A V} \tag{E6}$$

where F is the total flow rate (sum of analyte and reagent streams) into the cell in molec s^{-1}, R is the molar gas constant, T is the cell temperature, and N_A is Avogadro's number. Thus, as the throughput of analyte increases, the pressure also increases. The corresponding increase in molecular concentration ($[M] = N_A P/RT$) results in quenching of NO_2^* and a decrease in the chemiluminescence quantum efficiency, according to Eq. (E2). These competing effects result in the existence of an optimal flow rate for the gas stream containing the analyte, above which no improvement in signal is obtained. Ultimately, the sensitivity is limited by reactor considerations such as the size of the vacuum pump and the flow rates of the analyte and reagent gas streams.

Optimization strategies and a number of generalized limitations to the design of gas-phase chemiluminescence detectors have been described based on exact solutions of the governing equations for both exponential dilution and plug-flow models of the reaction chamber by Mehrabzadeh et al. [12, 13]. However, application of this approach requires a knowledge of the reaction mechanism and rate coefficients for the rate-determining steps of the chemiluminescent reaction considered.

4. GAS-PHASE CHEMILUMINESCENT REACTIONS

Typically, intense chemiluminescence in the UV/Vis spectral region requires highly exothermic reactions such as atomic or radical recombinations (e.g., S + S + M $\rightarrow$ S$_2$* + M) or reactions of reduced species such as hydrogen atoms, olefins, and certain sulfur and phosphorus compounds with strong oxidants such as ozone, fluorine, and chlorine dioxide. Here we review the chemistry and applications of some of the most intense chemiluminescent reactions having either demonstrated or anticipated analytical utility.

4.1 Chemiluminescent Reactions of O$_3$

The most widely used gas-phase chemiluminescence reagent is ozone. Analytically useful chemiluminescence signals are obtained in the reactions of ozone with NO, SO, and olefins such as ethylene and isoprene, but many other compounds also chemiluminesce with ozone. Ozone is conveniently generated online at mixing ratios of $\approx$1–5% by electrical discharge of air or O$_2$ at atmospheric pressure [14].

4.1.1   O$_3$ + NO Chemiluminescence

The chemiluminescence associated with the reaction of NO with O$_3$ is perhaps the best known and most analytically useful gas-phase chemiluminescent reaction [15]. The mechanism of the reaction is summarized by Reactions (1–3), given above. The exothermicity of the reaction, 200 kJ mol^{-1}, corresponds to the short-wavelength cutoff of the reaction [16]. Emission is observed in the range $\approx$600–3000 nm, as shown in Figure 2. Although the spectroscopy of NO$_2$ is still not well understood, the emission appears to originate in a combination of the ^{2}B$_1$ and ^{2}B$_2$ states [17]. As seen in Figure 2, only a small fraction of emission occurs below 800 nm where photons can be detected with high quantum efficiency. Furthermore, only a small fraction of the reaction of NO with O$_3$ results in ex-

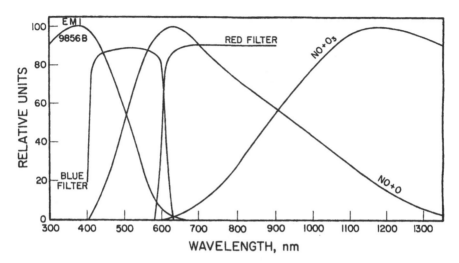

Figure 2 Comparison of NO + O_3 and NO + O chemiluminescence spectra with blue and red optical filter transmissions and the response of a blue-sensitive photomultiplier tube. Note the blue shift of the NO reaction with O compared to that with O_3 and that the addition of a blue filter effectively removes emission from NO + O_3, while the red filter effectively removes emission from NO + O.

cited-state NO_2 [10]. Assuming that the emitting species was $^2B_2$, Clough and Thrush estimated the rate constant for formation of electronically excited-state molecules to be given by $k_{1a} = 1.26 \times 10^{-12} \exp(-2100/T)$ cm³ molec⁻¹ s⁻¹ [10]. Comparing this with the now well-established rate constant for the overall reaction of $k_1 = 2.0 \times 10^{-12} \exp(-1400/T)$ cm³ molec⁻¹ s⁻¹ [11], the fraction of reactions that produce excited states is only $\approx 6.0\%$ at room temperature.

The radiative lifetime of NO_2^* is calculated from its integrated absorption spectrum to be 0.3 μs [18], but the actual measured lifetime is at least two orders of magnitude longer. This anomaly, which was investigated by numerous research groups [17], has been attributed to coupling of an excited $^2B_2$ electronic state to high vibrational levels of the $^2A_1$ ground state [19–22]. Those states with a larger fraction of $^2B_2$ character show shorter lifetimes.

4.1.2 O_3 + SO Chemiluminescence

The chemiluminescent reaction of SO with ozone is the basis of the sulfur chemiluminescence detector (SCD) [23] discussed later in this chapter,

$$SO + O_3 \rightarrow SO_2^* + O_2 \tag{4}$$

Emission is thought to occur from at least four excited states of SO_2, $^3B_1$, $^1A_2$, $^1B_1$, and $^1B_2$, which have radiative lifetimes of 8 ms, 30 μs, 600 μs, and 30 ns, respectively [17]. The emission extends from just beyond 260 nm, closely corresponding to the exothermicity of the reaction (445 kJ/mol), and extends to ≈480 nm [24–26], as in Figure 3, which shows the emission spectra of SO^*_2 formed as an intermediate in the reaction of CH_3SH with O_3. (In this figure, a glass window present in the chemiluminescence apparatus imposes the apparent lower wavelength limit.) Under most conditions, the emission spectrum consists of a pseudocontinuum with very little band structure. In the presence of helium or a nonquenching gas such as SO_2, a simple band structure occurs between 380 and

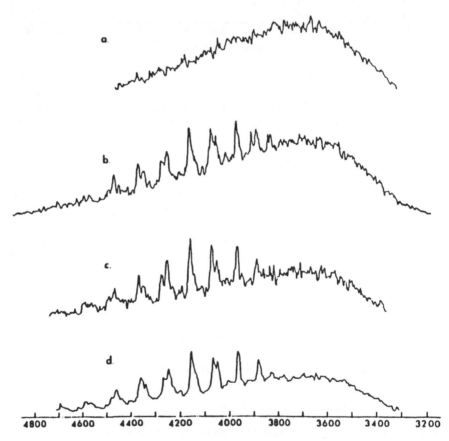

Figure 3 $SO_2(^3B_1)$ emission spectra obtained in the reaction of methane thiol with O_3 at 620 mtorr in He; (a) 9 mtorr O_3, 3 mtorr CH_3SH; (b) 15 mtorr O_3, 7 mtorr CH_3SH; (c) 20 mtorr O_3, 7 mtorr CH_3SH; (d) 30 mtorr O_3, 15 mtorr CH_3SH. (Reprinted with permission from Ref. 27. Copyright 1981 American Chemical Society.)

470 nm, similar to that seen in Figure 3. These bands originate in the $^3B_1$ state, probably as a result of intersystem crossing from the $^1A_1$ state [27]. Additionally, as a result of strong vibronic coupling between the $^1A_2$ and $^1B_1$ states, the radiative properties of SO_2^* are quite complex.

The rate constant for formation of excited-state SO_2 was measured by Halstead and Thrush to be $1.7 \times 10^{-12} \exp(-2100/T)$ cm^3 molec^{-1} s^{-1} [28–31], while the overall rate constant for the loss of SO is $3.6 \times 10^{-12} \exp(-1100/T)$ cm^3 molec^{-1} s^{-1} [11]. Thus, at room temperature, SO_2^* constitutes $\approx 1.6\%$ of the product channels. This figure can only be considered approximate, however, since mixing of states within the singlet manifold affects the interpretation of the rate constant for formation of the excited state [32].

Electronically excited SO_2 has been found to be the principal emitter in the multistep reactions of a number of reduced sulfur compounds such as H_2S, CH_3SH, $(CH_3)_2S$, CS_2, and thiophene with ozone [25, 27, 33, 34]. Unfortunately, the mechanisms of these complex reactions are not understood.

4.1.3 Chemiluminescent Reactions of O_3 with Hydrocarbons

At room temperature ozone produces chemiluminescence in its reactions with unsaturated hydrocarbons such as alkenes, alkynes, and aromatics, and at elevated temperatures even alkanes chemiluminescence with O_3 [35]. These reactions are highly complex, involving many reaction steps and numerous emitters. The simplest of these reactions is with ethylene, where emission is observed from $CH_2O(^1A_2)$ in the wavelength region 350–520 nm, vibrationally excited $OH(X^2\Pi$, $v \leq 9)$ at 700–1100 nm, and the (0,0) band of electronically excited $OH(A^2\Sigma^+)$ at 306 nm [14, 36]. Additional phosphorescence from the triplet states of glyoxal, $(CHO)_2(^3A_u \rightarrow {}^1A_g)$, and methyl glyoxal, $(CH_3CO)_2(^3A_u \rightarrow {}^1A_g)$, are observed for substituted olefins [36, 37]. These reactions begin with addition of ozone across the double bond to form an ozonide,

$$\tag{5}$$

Decomposition of this intermediate, the "Criegee split," forms an aldehyde, which is electronically excited formaldehyde in the case of a terminal alkene, and a diradical,

$$\tag{6}$$

It has been shown that abstraction of an α or β hydrogen from the ozonide also can occur. In fact, α-hydrogen abstraction can occur up to 2.5 times faster than Criegee fragmentation [36]. As an example, the proposed mechanism for the reaction of *cis*-2-butene with ozone is shown in Figure 4.

Vibrational overtone bands within the ground electronic state of OH, known as Meinel bands, result from the reaction of H atoms with ozone,

$$H + O_3 \rightarrow OH(X^2\Pi, v \leq 9) + O_2 \tag{7}$$

Channels involving Criegee splitting and α-hydrogen abstraction can produce H atoms, as seen in Figure 4. Herron and Huie reported that $\approx 9\%$ of the total reaction produces H atoms by the pathways shown [38]. The $H + O_3$ reaction, for which $\Delta H°_{298} = -322$ kJ mol^{-1}, is not sufficiently exothermic to produce electronically excited $OH(A^2\Sigma^+)$, which requires an excitation energy of 390 kJ mol^{-1}. The excitation energy could be provided from energy pooling of vibrationally excited OH or possibly through the reaction [39]

$$CH + O_2 \rightarrow CO + OH(A^2\Sigma^+) \tag{8}$$

which can provide up to 665 kJ mol^{-1}. This proposed reaction involves an unfavorable four-center intermediate, and the source of CH is unknown. However,

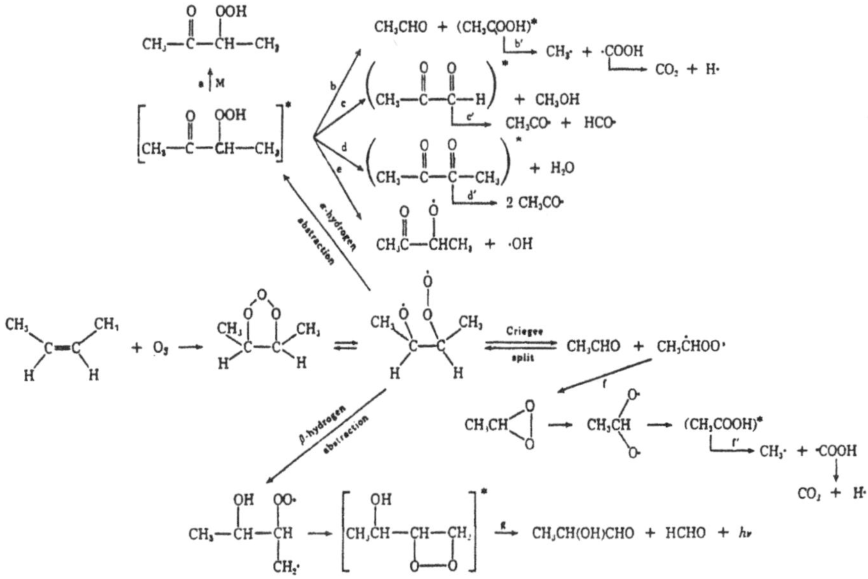

Figure 4 Proposed reaction mechanism for the reaction of *cis*-2-butene with O_3. (Reprinted with permission from Ref. 36. Copyright 1974 American Chemical Society.)

emission from CH radicals has been observed in reactions of ozone with aromatics, allene, and acetylene, as reviewed by Toby [14].

As the reaction temperature is increased, chemiluminescence is observed in the reactions of ozone with aromatic hydrocarbons and even alkanes. Variation of temperature has been used to control the selectivity in a gas chromatography (GC) detector [35]. At room temperature, only olefins are detected; at a temperature of 150°C, aromatic compounds begin to exhibit a chemiluminescent response; and at 250°C alkanes respond, giving the detector a nearly universal response similar to a flame ionization detector (FID). The mechanisms of these reactions are complex and unknown. However, it seems likely that oxygen atoms produced in the thermal decomposition of ozone may play a significant role, as may surface reactions with O_3 and O atoms.

4.1.4 Other Chemiluminescent Reactions of O_3

Chemiluminescence associated with the reactions of ozone with phosphine (PH_3), arsine (AsH_3), and stibine (SbH_3) have been investigated by Fraser et al. [40, 41]. The reaction mechanisms are very complex. For example, in the $AsH_3/O_3/O_2$ system, a 24-step mechanism was required for the computer simulation [41]. The emissions consist of a continuum in the wavelength range 360–700 nm, tentatively assigned to emission from the $^2B_1$ state of AsO_2, analogous to NO_2, and a strongly banded ultraviolet emission corresponding to the transition $AsO(A^2\Sigma^+ \rightarrow X^2\Pi)$. The relative contribution of these two emissions depends on the presence or absence of O_2 and of externally added oxygen atoms. The $PH_3/O_3/O_2$ system produced only analogous PO_2^* emissions, while the $SbH_3/O_3/O_2$ reaction resulted in emissions of SbO from the $B^2\Sigma^+$ and $A^2\Pi_{1/2,3/2}$ states to the ground $X^2\Pi_{1/2,3/2}$ state in the wavelength regions 340–450 nm and 450–680 nm, respectively [41].

Electronically excited metal oxides have been produced in the reactions of ozone with molecular beams of metal atoms, A,

$$A + O_3 \rightarrow AO^* + O_2 \tag{9}$$

These studies have allowed the spectroscopic identification of a number of electronically excited states of the metal oxides, but there appear to have been no analytical applications of the reactions to date. The emitting states, as summarized by Toby [14], are $CaO(A^1\Pi)$, $SrO(A^1\Pi)$, $PbO(a^3\Sigma^+, b^3\Sigma^+)$, $ScO(C^2\Pi)$, $YO(C^2\Pi)$, $FeO(C^1)$, $AlO(A^2\Pi_i, B^2\Sigma^+)$, and $BaO(A^1\Sigma^+, D^1\Sigma^+)$. Nickel carbonyl reacts with ozone to produce chemiluminescence from an excited electronic state of NiO, which is probably produced in the Ni + O_3 reaction [42, 43].

Tin, antimony, and selenium hydrides, produced by the borohydride reduction technique, were found to chemiluminesce with ozone in an analytical detection scheme. Limits of detection were 35, 10, and 110 ng of Sn, Sb, and Se, respectively. Bismuth and mercury hydrides gave a nonlinear chemiluminescent response, while lead, germanium, and tellurium did not chemiluminesce with ozone under the conditions used [44].

4.2 Chemiluminescent Reactions of O Atoms

Although not frequently used in analytical applications, oxygen atoms are even stronger oxidants than ozone and produce chemiluminescence when reacted with a wide range of analytes. Chemiluminescent reactions of oxygen atoms were first studied in the "air afterglow" in the early 1900s by Lord Rayleigh, who passed air through a low-pressure electrical discharge and observed a faint green-yellow glow just downstream [45].

A background emission underlies all chemiluminescence spectra obtained in reactions of O atoms. This emission results from the recombination of O atoms,

$$O + O + M \rightarrow O_2{}^* + M \tag{10}$$

Emissions from five different excited electronic states of O_2 throughout the visible spectral region of 400–800 nm have been identified [17], placing a practical limit on the concentration of O atoms that can be used as a chemiluminescence reagent since the background emission increases quadratically in O atom concentration.

A common way to generate oxygen atoms is to flow either pure oxygen or oxygen dilute in a noble gas such as He or Ar through a microwave discharge [46, 47]; this produces mostly ground-state $O(^3P)$ atoms but also some $O(^1D)$. The species $O(^1D)$ is 189.5 kJ/mol more energetic than $O(^3P)$ and, as a result, is more reactive and can populate higher excited states in its chemiluminescent reactions than does $O(^3P)$. However, producing a constant concentration of $O(^1D)$ in a flow stream is extremely difficult due to wall deactivation and its rapid reactions with water vapor to produce OH radicals. Therefore, in a detector making use of oxygen atoms, it is generally best to remove $O(^1D)$ by providing adequate residence time between the discharge source and the reaction chamber. Significant quantities of $O_2(^1\Delta_g)$, the lowest bound electronic state of oxygen, also are produced in microwave discharges of O_2. At high concentrations, $O_2(^1\Delta_g)$ can provide a background emission at 634 and 703 nm via the "dimol" emission process in which two singlet oxygen molecules cooperate to emit a single photon [48].

4.2.1 O + NO Chemiluminescence

The associative reaction of oxygen atoms with nitric oxide produces the yellow-green chemiluminescence in the "air afterglow," easily seen by the naked eye. The reaction has long been used to measure the concentrations of O atoms in kinetics experiments [49–51] and is so bright that it has been used to visualize and photograph flow properties such as the mixing of gas streams in flow reactors. Kinetics studies have shown that the first step of the reaction mechanism is truly second order and reversible. A weakly bound electronically excited molecule, (NO_2^*), can undergo radiative emission to the ground state or be stabilized by collision with a third body M to form a more strongly bound excited-state molecule, NO_2^* [52]. The latter species may emit light or be collisionally quenched:

$$NO + O \leftrightarrow (NO_2^*) \tag{11}$$

$$(NO_2^*) + M \rightarrow NO_2^* + M \quad \text{(high-pressure regime)} \tag{12a}$$

$$(NO_2^*) \rightarrow NO_2 + h\nu \quad \text{(low-pressure regime)} \tag{12b}$$

$$NO_2^* + M \rightarrow NO_2 + M \tag{13}$$

$$NO_2^* \rightarrow NO_2 + h\nu \tag{14}$$

Chemiluminescence is believed to arise from the $^2B_1$ and the $^2B_2$ electronic states, as discussed above for the reaction of NO with ozone [17]. The primary emission is in a continuum in the range $\approx 400\text{--}1400$ nm, with a maximum at ≈ 615 nm at 1 torr. This emission is significantly blue-shifted with respect to chemiluminescence in the NO + O_3 reaction ($\lambda_{max} \approx 1200$ nm), as shown in Figure 2, owing to the greater exothermicity available to excite the NO_2 product [52]. At pressures above approximately 1 torr of O_2, the chemiluminescence reaction becomes independent of pressure with a second-order rate coefficient of 6.4×10^{-17} cm^3 molec^{-1} s^{-1}. At lower pressures, however, this rate constant decreases and then levels off at a minimum of 4.2×10^{-18} cm^3 molec^{-1} s^{-1} near 1 mtorr, and the emission maximum blue shifts to ≈ 560 nm [52]. These results are consistent with the above mechanism in which the fractional contribution of (NO_2^*) to the emission spectrum increases as the pressure is decreased, therefore decreasing the rate at which (NO_2^*) is deactivated to form NO_2^*. Additionally, the radiative lifetime and emission spectrum of excited-state NO_2 vary with pressure, as discussed above for the NO + O_3 reaction [19–22].

4.2.2 O + SO Chemiluminescence

As with SO + O_3, discussed above, it is believed that sulfur monoxide reacts with ground-state O atoms to form SO_2 in either the $^3B_1$ or $^1B_2$ electronic states, although the $^1A_2$ and $^1B_1$ may also contribute [24, 28, 29]. Standard third-order chemiluminescence reaction mechanisms and kinetics are thought to apply:

$$SO + O + M \rightarrow SO_2^* + M \tag{15}$$

$$SO_2^* + M \rightarrow SO_2 + M \tag{16}$$

$$SO_2^* \rightarrow SO_2 + h\nu \tag{17}$$

The emission spectrum consists of a series of weak bands starting at about 220 nm and then growing into a continuum from about 240 to 400 nm, with a maximum at approximately 270 nm as shown in Figure 5. Halstead and Thrush estimated that $\approx$65% of the emission occurs from the $^1B_2$ state, $\approx$15% from the $^3B_1$, and $\approx$20% from a combination of the $^1A_2$ and $^1B_1$ states [24, 28, 29] with a rate constant of 2×10^{-31} cm^6 molec^{-2} s^{-1} using argon as the bath gas at 300 K [53]. As with the reaction of SO + O$_3$ discussed above, collisional coupling results in a radiative lifetime that is pressure dependent.

It is important to note that the SO + O chemiluminescence reaction can be surface mediated in addition to occurring in the gas phase. Here, the oxygen atoms are weakly bound to the surface and react with SO diffusing to the surface:

$$\text{surface}{-}\text{O} + \text{SO} \rightarrow \text{SO}_2{}^* + \text{surface} \tag{18}$$

The emission maximum is red-shifted from about 340 nm to 380 nm and shows a more banded structure than in the homogeneous reaction [17]. This emission spectrum more closely approximates the spectrum of SO + O$_3$ than that of the homogeneous SO + O reaction, indicative of energy loss to the surface [31].

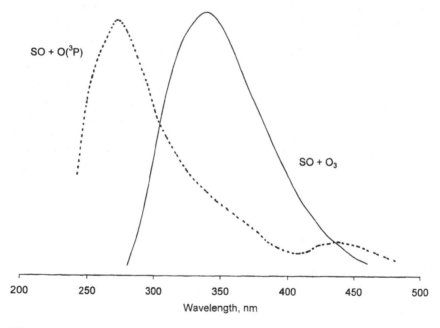

Figure 5 Comparison of SO + O and SO + O$_3$ chemiluminescence spectra. Note the blue shift of the SO + O spectrum as compared to that of SO + O$_3$. (Adapted from Refs. 25, 152).

4.3 Chemiluminescent Reactions of Active Nitrogen

The use of active nitrogen as a chemiluminescence reagent results in an almost universal analysis technique for hydrocarbons; however, it also is possible to selectively detect organometallics, halogens, and compounds containing oxygen, phosphorus, or sulfur. Active nitrogen is generated in a similar manner to that for oxygen atoms and was first reported in the "nitrogen afterglow" by E. P. Lewis in 1904 [54]. Passing a stream of pure, ground-state $N_2(X^1\Sigma_g^+)$ through a low-pressure microwave or electrical discharge generates a population of ground-state N (^{4}S) atoms and N_2 molecules excited to the $A^3\Sigma_u^+$ and $B^3\Pi_g$ states. It is the ability to readily react with or to transfer energy to a wide variety of molecules that gave this reaction mixture the name "active nitrogen." Nitrogen atoms readily recombine to form nitrogen in either the $A^3\Sigma_u^+$ state or higher $B^3\Pi_g$ state, and molecules in the $B^3\Pi_g$ state can decay radiatively to the lower $A^3\Sigma_u^+$ state. Because the radiative transition of this triplet state to the singlet ground state is spin forbidden, the $A^3\Sigma_u^+$ state is metastable with a lifetime on the order of seconds. The $B^3\Pi_g \rightarrow A^3\Sigma_u^+$ transition produces light in the yellow portion of the visible spectrum as a background to whatever chemiluminescent signal is sought above it [55]. The relative contributions of N atoms and N_2* to chemiluminescent reactions depend on the source and pressure. At reduced pressure, high-energy discharges such as microwave discharges produce higher concentrations of N atoms than electrical discharges, while active nitrogen produced at atmospheric pressure consists almost entirely of $N_2(A^3\Sigma_u^+)$ [39].

4.3.1 Chemiluminescent Reactions of Active Nitrogen with Hydrocarbons

The chemiluminescent products of the reactions of active nitrogen with organics produce cyanogen radicals (CN) in an excited $B^2\Sigma^+$ state, which radiatively decay to the $X^2\Sigma^+$ ground state. The chemiluminescence spectrum consists of a number of banded features between 350 and 420 nm, corresponding to the $\Delta v = -1, 0$, and $+1$ band progressions, as seen in Figure 6. The most intense feature is the $\Delta v = 0$ sequence in the narrow wavelength range of 383–388 nm. The source of CN* responsible for these emissions is not known. The reaction of CH radicals with N atoms has adequate exothermicity to produce CN*:

$$CH + N \rightarrow CN + H \qquad \Delta H^0_{298} = -413 \text{ kJ mol}^{-1} \qquad (19)$$

but the first step in the production of CH by successive hydrogen abstractions by N atoms is highly endothermic (e.g., $\Delta H^0_{298} = +87$ kJ mol^{-1} for N + C_2H_6).

Low-pressure discharges tend to produce much more intense chemiluminescence from alkenes than from alkanes. This has been attributed to the addition of N to the double bond, as illustrated for ethylene,

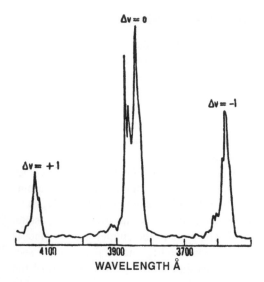

Figure 6 $CN(B^2\Sigma^+ \rightarrow X^2\Sigma^+)$ emission in the reaction of active nitrogen with hydrocarbons. (Reprinted with permission from Ref. 58. Copyright 1979 American Chemical Society.)

$$\begin{array}{c} \overset{H}{\underset{H}{\diagdown}}C=C\overset{H}{\underset{H}{\diagup}} \quad + \quad N \quad \longrightarrow \quad H-\overset{\overset{\displaystyle N}{\diagup\diagdown}}{C}-\overset{}{\underset{H}{\overset{|}{C}}}-H \end{array} \qquad (20)$$

It has been proposed that this reaction intermediate could decompose to produce HCN and CH_3 [55]. Chemiluminescence from alkanes can be greatly enhanced by addition of HCl. The proposed explanation is that energy transfer from active nitrogen dissociates HCl to produce chlorine atoms, which have rapid hydrogen-atom abstraction reactions with alkanes,

$$N + N \text{ (or } N_2^*) + HCl \rightarrow N_2 + H + Cl \qquad (21)$$

$$Cl + RH \rightarrow HCl + R \qquad (22)$$

This effect of adding HCl to the reaction cell has been used to advantage in distinguishing alkanes from alkenes and oxygenated hydrocarbons in a GC detector utilizing active nitrogen [56].

In atmospheric pressure discharges, where concentrations of N atoms are greatly reduced and possibly absent, the discrimination between alkanes and alkenes is not observed and an alternative means of initiating the reaction is required. In the absence of significant concentrations of N, the mechanism may consist of collisional energy transfer from $N_2(A^3\Sigma_u^+)$, which lies 593 kJ mol^{-1}

above the ground state, to the organic molecule, followed by fragmentation of the carbon skeleton [57]. Again, the complex reaction mechanism leading to CN and/or CN* is not known, but CN* could be formed via collisional energy transfer from N_2^* to ground-state CN.

4.3.2 Other Chemiluminescent Reactions of Active Nitrogen

In addition to the ability to react nonspecifically with hydrocarbons, active nitrogen can readily participate in energy transfer reactions with volatile organometallic compounds, leading to atomic emission from the metal atom. By use of appropriate optical filters, selective detection of elements such as aluminum, lead, tin, and mercury has been achieved in the presence of large excesses of organics [58].

Molecules containing oxygen, sulfur, and phosphorus react with active nitrogen to produce excited states of NO, SN, and PN radicals, respectively [59]. Electronically excited OH also is observed in reactions of oxygenated compounds [56]. A disadvantage of selective detection of oxygen compounds using active nitrogen is that oxygen impurities also react with active nitrogen to form NO*. Finally, the reactions of active nitrogen with inorganic compounds containing boron, chlorine, bromine, and iodine have been found to chemiluminesce in characteristic spectral regions [57].

4.4 Chemiluminescent Reactions of F and F_2

Being the most electronegative element and one of the strongest oxidants known, rich chemiluminescence is associated with reactions of F atoms, while F_2 is surprisingly selective in its gas-phase reactions. Fluorine atoms can be generated by flowing a stream of F_2 dilute in a noble gas through a microwave electrical discharge or, for those less courageous, SF_6 or CF_4 may be used as the source gas [60]. Because fluorine atoms react with silica at the high temperatures of the discharge, it is necessary to use a ceramic (alumina) discharge tube. The plasma produces a stream containing a mixture of F and F_2; by providing adequate residence time in the connecting tubing, F atoms can recombine to form F_2 for greater selectivity. Fluorine atoms may also be produced by multiphoton dissociation of SF_6 using a highly focused infrared laser [61]. This approach has the advantage of producing the F atoms at the point where they react, rather than upstream, so that fewer atoms are lost to recombination on the vessel walls. Fluorine can be removed after the chemiluminescence cell and prior to entering the vacuum pump using a charcoal trap to produce nontoxic CF_4, and the reaction product HF may be removed with an ascarite trap. Thus, it is possible to produce F/F_2 online from innocuous starting materials and then remove it without any significant health risk.

4.4.1   Reactions of F Atoms with Hydrocarbons

The reactions of fluorine atoms with hydrocarbons are similar to those of active nitrogen in that they provide an essentially universal response. Fluorine atoms abstract H atoms from hydrocarbons at near-collisional reaction rates. Reactions with fluorine are highly exothermic, forming strong H—F (≈ 570 kJ mol^{-1}) and C—F (≈ 485 kJ mol^{-1}) bonds while breaking much weaker C—H (≈ 414 kJ mol^{-1}) and C—C (≈ 368 kJ mol^{-1}) bonds. The hydrogen abstraction reaction

$$F + RH \rightarrow HF^\dagger + R \tag{23}$$

is exothermic by about 156 kJ mol^{-1}, sufficient to produce vibrationally excited HF† in levels up to v = 4. The (3,0) vibrational overtone band may be detected at 880 nm for most species containing a C—H bond and many other compounds containing hydrogen [62]. This is not a favorable wavelength for sensitive detection using photomultiplier tubes, and in order to be selective for hydrocarbons, emissions from electronically excited C$_2$(d$^3\Pi_g$) at 470 nm and CH(A$^2\Delta$) at 431 nm may be monitored. The detailed mechanisms that produce these species are unknown but must involve many reaction steps. The same emissions are observed in fluorine/hydrocarbon flames [63]. The excited state of diatomic carbon, C$_2$(d$^3\Pi$), has been observed in many systems, ranging from solar spectra to organic flames. It may be formed in the disproportionation reaction of CH radicals [64]:

$$CH + CH \rightarrow C_2^* + H_2 \tag{24}$$

The chemiluminescent response of hydrocarbons reacting with an excess of fluorine atoms produced by multiphoton dissociation of SF$_6$ is linear below 50 parts per million (ppmv) [61].

4.4.2   Reactions of F$_2$ with Sulfur Compounds

Certain reduced sulfur compounds such as thiols, sulfides, disulfides, and trisulfides, specifically those having the functionality

$$
\begin{array}{c}
R_3 \\
| \\
R_2-C-S-R_1 \\
| \\
H
\end{array}
\tag{25}
$$

exhibit intense chemiluminescence with molecular fluorine, providing the basis of the fluorine-induced chemiluminescence detector (FCLD) [65]. Analogous selenium and tellurium compounds also respond with high sensitivity [66, 67]. Interestingly, the gases H$_2$S, CS$_2$, OCS, and SO$_2$ do not provide significant responses in the FCLD, and the detector exhibits selectivities of $>10^7$ against alkanes. Compounds with weak C—H bonds, such as alkenes (allylic hydrogens), toluene, xylenes (benzylic hydrogens), aldehydes, and halocarbons, provide relatively weak responses [65]. Molecular fluorine also exhibits intense chemiluminescence with phosphines, alkyl phosphines, and monophosphinate esters [68].

Chemiluminescence is observed from several different emitting species, depending on the analyte and reaction conditions. Vibrational overtone bands of HF in the wavelength region of ≈500–900 nm are observed under nearly all conditions and are often the dominant spectral feature, the (3,0), (4,0) (5,1), and (6,2) bands being the most intense, while for some reaction conditions emissions from levels up to $v = 8$ are observed [63]. It is likely that hydrogen atoms are produced in the reaction and form vibrationally excited HF in the reaction reported by Mann et al. [62]:

$$H + F_2 \rightarrow HF^\dagger(v \leq 9) + F \tag{26}$$

While investigating the potential for an instrument to measure atmospheric dimethyl sulfide (DMS) [69], discussed below, Hills et al. investigated the possibility of adding H_2 to the reaction cell to provide chemical amplification of the chemiluminescence signal via the catalytic chain reaction:

$$F + H_2 \rightarrow HF + H \tag{27}$$

$$H + F_2 \rightarrow HF^\dagger(v \leq 9) + F \tag{28}$$

$$\underline{HF^\dagger \rightarrow HF + h\nu} \tag{29}$$

$$\text{Net: } H_2 + F_2 \rightarrow 2\,HF + h\nu \tag{30}$$

In this case, chemiluminescence was monitored using a red-sensitive PMT to detect emissions from $HF^\dagger$. A factor-of-six enhancement in sensitivity to 1.1 parts per billion (ppbv) DMS was obtained. This is consistent with the fact that, based on the rate constant for the $H + F_2$ rate determining step [Reaction (28)], the reaction can cycle approximately 7 times during the cell residence time and confirms the observation by Turnipseed and Birks [7] that F atoms are produced in the $F_2 + DMS$ reaction.

In the reactions of F_2 with excess DMS and methane thiol, strong phosphorescence from triplet-excited thioformaldehyde, $CH_2S(^3A_2)$, dominates the chemiluminescence, while for conditions of excess F_2, emission is predominantly that of $HCF(A^1A'' \rightarrow X^1A')$ [5, 6, 70] as shown in Figure 7. The kinetics and mechanism of the reaction of $(CH_3)_2S$ with F_2 were investigated by Turnipseed and Birks using the flow tube technique with simultaneous detection by mass spectrometry and observation of the emission spectrum using a diode array spectrometer [7]. As DMS was titrated by addition of F_2, a product having a m/e ratio of 80, corresponding to the mass of $H_2C=SFCH_3$, increased in proportion to the loss of DMS at m/e = 62. As shown in Figure 8, once the endpoint of the titration

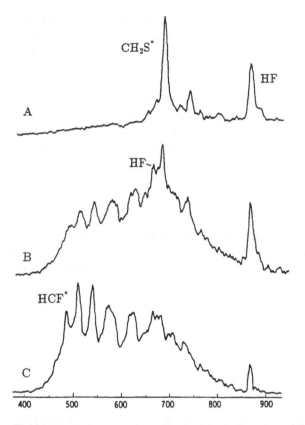

Figure 7 Emission spectra obtained in the reaction of DMS with F_2. (A) 155 mtorr CH_3SCH_3, 30 mtorr F_2, 570 mtorr He; (B) 30 mtorr CH_3SCH_3, 20 mtorr F_2, 350 mtorr He; (C) 30 mtorr CH_3SCH_3, 50 mtorr F_2, 900 mtorr He. (Reprinted from RJ Glinski, EA Mishalanie, JW Birks, "Molecular emission spectra in the visible and near IR produced in the chemiluminescent reactions of molecular fluorine with organosulfur compounds," *Journal of Photochemistry* 37:223, 1987, with permission from Elsevier Science.)

was reached, the ion current at m/e = 80 began to decline with increasing F_2 concentration. In agreement with the observations of Glinski et al. [6], the emission spectrum was found to change from CH_2S^* to HCF* and possibly some $HF^\dagger$ near the titration endpoint. The rate constant for the reaction of F_2 with DMS was found to be $(1.6 \pm 0.5) \times 10^{-11}$ cm^3 molec^{-1} s^{-1}, which is extremely fast for the reaction of two closed-shell molecules [7]. By contrast, the rate constant for the F_2 reaction with H_2S, which does not respond in the FCLD, was found to have an upper limit of $\leq 6.4 \times 10^{-16}$ cm^3 molec^{-1} s^{-1}. It was argued that the

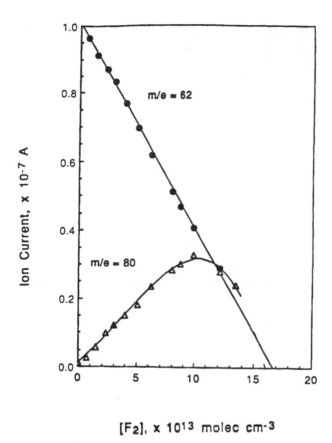

Figure 8 Changes in the DMS ($m/e = 62$) reactant and CH_3SFCH_2 ($m/e = 80$) product ion currents as a function of reaction time using the flow tube technique to study the reaction of DMS with F_2. (Reprinted with permission from Ref. 7. Copyright 1991 American Chemical Society.)

fast rate of the F_2 + DMS reaction, occurring about once in every six collisions, may be due to the formation of a charge-transfer complex,

$$CH_3SCH_3 \ + \ F_2 \ \longrightarrow \ \underset{H_3C\diagup\overset{|}{\underset{}{S}}\diagdown CH_3}{\overset{F\text{---}F^-}{\vdots}} \tag{31}$$

which is calculated, using the electron affinity for F_2 and the ionization potential of DMS, to have a radius of 2.63 Å. By comparison, the critical radius for forming a charge transfer complex with H_2S is 1.94 Å. Turnipseed and Birks showed that the critical radius for forming a charge-transfer complex is highly correlated with the relative sensitivity of the FCLD to various compound classes [7].

Once the DMS-F_2 complex is formed, atomic fluorine may be eliminated, analogous to the proposed mechanism for the reaction of F_2 with olefins [71]:

$$
\begin{array}{c}
F-\!\!-\!\!-F^{\,-} \\
\vdots \\
S^+ \\
H_3C^{\diagup \quad \diagdown} CH_3
\end{array}
\quad\longrightarrow\quad
H_3C-\underset{F}{\overset{F}{\underset{|}{S}}}-CH_3 \;\; + \;\; F
\qquad (32)
$$

This radical could then eliminate an H atom to form a C=S bond, also analogous to olefinic systems [71, 72], to form the species observed at m/e = 80,

$$
H_3C-\underset{\cdot}{\overset{F}{\underset{|}{S}}}-CH_3 \quad\longrightarrow\quad H_3C-\overset{F}{\overset{|}{S}}\!=\!CH_2 \;\; + \;\; H
\qquad (33)
$$

Subsequent reactions of the H and F atoms produced in these reactions may be responsible for HF^+ formation via Reactions (27–28) above and for CH_2S^* by means of reactions such as

$$H + CH_3SCH_3 \rightarrow CH_3 + CH_3SH \qquad (34)$$

$$CH_3SH + F \rightarrow CH_3S + HF \qquad (35)$$

$$CH_3S + H \rightarrow H_2 + CH_2S^* \qquad (36)$$

$$CH_3S + F \rightarrow HF + CH_2S^* \qquad (37)$$

The reactions of H and F with CH_3S are exothermic by 259 and 393 kJ mol^{-1}, respectively, and either could be responsible for formation of $CH_2S(A^3\Sigma)$ [7].

Methylated sulfur compounds produce intense emission from HCF*, while the corresponding ethyl-substituted compounds produce only trace or no HCF* [6]. Emission from HF^+ was much stronger for the ethyl-substituted than for the methyl-substituted sulfides [6]. This suggests that methyl radicals, formed perhaps in Reaction (34), lead to the formation of HCF*. Emission of HCF* has been identified in F_2/CH_4 flames [70] where it is attributed to the association reaction

$$CH + F + M \rightarrow HCF^* + M \qquad (38)$$

Under excess F_2 conditions, the reaction with DMS and other organosulfur compounds may produce CH as well, and at the low pressures (≈ 1 torr) of the FCLD, a more favorable reaction may be,

$$CH + F_2 \rightarrow HCF^* + F \qquad (39)$$

which has a more than adequate exothermicity of 326 kJ mol^{-1} for ground-state products.

Carbon disulfide produces weak chemiluminescence relative to organo-sulfur compounds bearing hydrogen atoms. The absence of H atoms in CS_2 precludes emissions from $HF^\dagger$, CH_2S^*, or HCF^*. A banded emission in the wavelength region $\approx 600-900$ nm was tentatively assigned to FCS [6] but later identified as arising from an excited electronic state of SF_2 [73].

4.4.3 Reactions of F_2 with Selenium and Tellurium Compounds

Intense chemiluminescence also accompanies the reaction of F_2 with organo-selenium and -tellurium compounds [67]. HCF and selenoformaldehyde emissions are observed in the reaction of F_2 with dimethyl diselenide, $(CH_3Se)_2$ [66]. Analogous to the reaction of DMS, the emission spectrum changes from that of CH_2Se^* phosphorescence and some $HF^\dagger$ under conditions of excess $(CH_3Se)_2$, to HCF^* under conditions of excess F_2. Selenium and tellurium compounds have even lower ionization potentials than the corresponding sulfur compounds, so they also are expected to form charge transfer complexes with F_2 [7] in an manner analogous to Reaction (31).

4.4.4 Reactions of F_2 with Phosphorus Compounds

Phosphine, alkyl phosphines, and monophosphinate esters also produce intense chemiluminescence with F_2 [68]. Again, these compounds have low ionization potentials and are expected to form charge transfer complexes with F_2 [7]. The emission spectra obtained in the reaction of trimethyl phosphine (TMP) with excess F_2 have contributions from HCF^*, $HF^\dagger$, and an unidentified broad-band emission as shown in Figure 9. The relative contributions of HCF^* and the un-

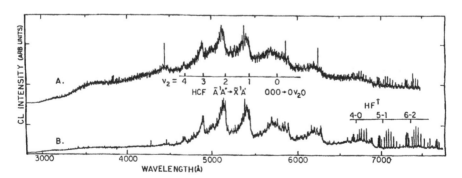

Figure 9 Emission spectra obtained in the reaction of TMP with F_2. (A) 8 mtorr TMP, 7 mtorr F_2, 59 mtorr He; (B) 20 mtorr TMP, 47 mtorr F_2, 423 torr He. (Reprinted from Ref. 68, with permission from JW Birks.)

known emitter change with the F_2/TMP ratio, with the relative amount of HCF* increasing as this ratio increases. The HCF*/HF† ratio remains constant within a factor of two as the F_2/TMP ratio is varied from ≈ 1 to ≈ 5. Interestingly, the emission spectrum for triethyl phosphine (TEP) contains HF† and the broad-band emitter, but no HCF*. This is the same trend observed for the sulfide reaction system discussed above and provides further evidence of the participation of methyl radicals in the formation of HCF*.

4.4.5   Reactions of F and F_2 with Iodine Compounds

Birks et al. reported chemiluminescence from the $A^3\Pi_0{}^+$ and $B^3\Pi_1$ states of IF in the reaction of F_2 with I_2 and suggested that the reaction kinetics were consistent with a four-center reaction forming the products IF* + IF [74]. In a series of molecular beam studies, it was shown that the reaction actually forms a collision complex that decomposes to form two sets of products, IF* + IF and I_2F + F [75–77]:

$$I_2 + F_2 \rightarrow [I_2F \cdot F] \rightarrow IF^* + IF \text{ (minor channel)} \tag{40}$$

$$\rightarrow I_2F + F \text{ (major channel)} \tag{41}$$

Those experiments suggest that chemiluminescence in the earlier work by Birks et al. was due principally to the secondary reaction

$$F + I_2F \rightarrow IF^* + IF \tag{42}$$

The FCLD, described above for S, Se, Te, and P compounds, actually was first demonstrated for alkyl iodides [60]. Fluorine atoms (and almost certainly F_2 through atom recombination) were produced by flowing a 1:1 mixture of SF_6 and He through a microwave discharge. Limits of detection were of the order of 1 μg. Ironically, the detector was later found to be about five orders of magnitude more sensitive to organosulfur compounds in a study of possible interferences [65]. The formation of IF* is thought to occur through an expanded octet intermediate RIF, analogous to the $I_2 + F_2$ reaction, which has been positively identified in the reaction of methyl iodide with F atoms [78]:

$$F + RI + M \rightarrow RIF + M \tag{43}$$

$$F + RIF \rightarrow RF + IF^* \tag{44}$$

4.5   Other Chemiluminescent Reactions

Other reactants that have been used to generate chemiluminescent reactions useful for chemical analysis include atomic sodium to detect halocarbons and chlorine dioxide to detect H_2S and CH_3SH.

4.5.1 Reactions of Atomic Sodium with Halocarbons

One fairly obscure chemiluminescent technique uses atomic sodium vapor to detect organic compounds containing more than one halogen atom. A piece of sodium metal is heated to approximately 400°C, enough to produce a significant vapor pressure of Na atoms in the reaction cell [79]. The mechanism, as it is hypothesized, produces sodium halides and a double bond in the carbon skeleton where the halogens were located, when structurally possible. Formation of the C=C double bond provides sufficient excess energy to populate high vibrational levels of the sodium halide:

$$R—CXY + Na \rightarrow R—CX + NaY \tag{45}$$

$$R—CX + Na \rightarrow R{=}C + NaX^\dagger \tag{46}$$

Here, X and Y are halogens and R represents the remainder of the carbon skeleton. The reaction proceeds particularly well for the halogens chlorine and bromine, but much less so for fluorine, as the C—F bond is much stronger than either C—Cl or C—Br. However, the excited sodium halide, $NaX^\dagger$, is not the emitting species. Another collision between this molecule and another vapor-phase sodium atom is needed:

$$NaX^\dagger + Na(3^2S) \rightarrow NaX + Na^*(3^2P) \tag{47}$$

The resulting doublet sodium atom can then emit light at the sodium D-line at 589 nm. With so many steps requiring atomic sodium, its concentration, which increases with temperature, significantly affects the chemiluminescent yield. Therefore, the temperature of the solid sodium must be tightly controlled. The excited sodium halide is readily quenched by ambient air molecules, so the system works best at reduced pressure.

   Limits of detection as low as 9.1×10^{-14} g for $C_2H_4Cl_2$ have been measured with a relative standard deviation of $\sim$2% and a linear dynamic range of 4–5 orders of magnitude [79]. Weak interferences were found for oxygen- and nitrogen-containing organics. This technique also produces a chemiluminescent response for compounds with a single carbon and/or halogen by an unknown mechanism, and has been applied to the detection of atmospheric N_2O at sub-ppm levels through a mechanism similar to Reactions (45–47) [80].

4.5.2 Reactions of OClO with Sulfur Compounds

The chemiluminescent reaction with chlorine dioxide provides a highly sensitive and highly selective method for only two sulfur compounds, hydrogen sulfide and methane thiol [81]. As in the flame photometric detector (FPD), discussed below, atomic sulfur emission, $S_2(B^3\Sigma_u^- \rightarrow X^3\Sigma_g^-)$ is monitored in the wavelength range 250–400 nm, as shown in Figure 10. Although population of S_2^* is thought to proceed through the recombination of sulfur atoms, the reaction mechanism is not understood; however, the overall stoichiometry of the reaction is

$$2 H_2S + OClO \rightarrow S_2^* + \frac{1}{2} Cl_2 + 2 H_2O \tag{48}$$

Surprisingly, despite requiring two analyte molecules to produce one S_2 molecule, the kinetics of the chemiluminescent reaction are first order with respect to the sulfur compound. This can be explained if every H_2S or CH_3SH molecule is consumed in the reaction and every S atom recombines to form S_2, through the use of an excess of OClO to maintain pseudo-first-order reaction conditions [81]. The limit of detection for this analysis was found to be ≈ 3 ppbv for H_2S.

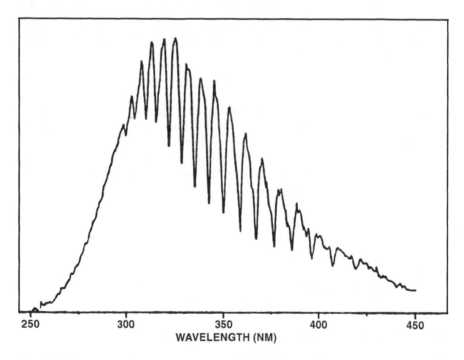

Figure 10 Chemiluminescence spectrum of S_2 observed in the selective reaction of OClO with H_2S. Note that this is the same excited-state species observed by the FPD. (Reprinted with permission from Ref. 81. Copyright 1982 American Chemical Society.)

5. APPLICATIONS TO GAS CHROMATOGRAPHY

Gas chromatography is one of the most powerful analytical techniques available for chemical analysis. Commercially available chemiluminescence detectors for GC include the FPD, the SCD, the thermal energy analysis (TEA) detector, and nitrogen-selective detectors. Highly sensitive detectors based on chemiluminescent reactions with F_2 and active nitrogen also have been developed.

5.1 Requirements for GC Detection

A number of demands are placed on detectors applied to gas chromatography owing to the specific nature of this separation technique. The time resolution must be greater than approximately 10 Hz in order not to degrade the resolution of the separation. Ideally, the sensitivity should be sufficient to detect quantities as little as picograms of an analyte as it elutes from the column but should have a linear response up to nearly microgram quantities. The detector must be able to accommodate flow rates of typically 1 $cm^3 min^{-1}$ of gas at atmospheric pressure or slightly higher and at temperatures ranging up to 400°C. It must be stable, robust, and simple enough for routine use. Often, the design must be adapted to meet these requirements. For example, it may be necessary to reduce the volume of the detection cell at the expense of sensitivity to shorten the response time. A vacuum pump may be used to optimize the sensitivity and/or reduce the residence time in the detector. Either the GC effluent can be thermally equilibrated before reaching the detector or the detector temperature held constant at some temperature above the oven temperature. The mobile phase can have additional constituents in addition to helium if necessary for detection. Computers can easily correct for some of the complexities inherent to some detectors, such as a nonlinear response.

5.2 Flame Photometric Detector

The most commonly used and widely marketed GC detector based on chemiluminescence is the FPD [82]. This detector differs from other gas-phase chemiluminescence techniques described below in that it detects chemiluminescence occurring in a flame, rather than "cold" chemiluminescence. The high temperatures of the flame promote chemical reactions that form key reaction intermediates and may provide additional thermal excitation of the emitting species. Flame emissions may be used to selectively detect compounds containing sulfur, nitrogen, phosphorus, boron, antimony, and arsenic, and even halogens under special reaction conditions [83, 84], but commercial detectors normally are configured only for sulfur and phosphorus detection [85–87]. In the FPD, the GC column extends to the sample inlet where it is mixed with oxygen (or air) and with hydrogen fuel prior to the burner head, and before entering the chemiluminescence cell. Fuel-rich, hydrogen/oxygen flames are generally preferred, but the combustion mixture must be optimized for specific analytes. As is common in chemiluminescence analysis, the greatest limitation to sensitivity of the FPD is noise associated with the background signal, which arises primarily from other flame emissions. To optimize the limit of detection, the PMT is positioned, and lenses are used to view a region of the flame where the signal-to-background ratio is greatest, while filters are employed to reduce background contributions of flame emissions.

5.2.1   Sulfur Detection by FPD

The most common application of the FPD to routine GC is detection of organo-sulfur compounds in complex matrices. Examples include quantification of sulfur compounds in petrochemical feedstocks [88, 89], measurements of sulfur-containing pesticides such as malathion and parathion in environmental samples [90, 91], and detection of sulfur compounds in foods and beverages, especially beer, where they are strong contributors to flavor and fragrance [92]. The FPD also detects inorganic sulfur compounds such as sulfates, sulfites, thiosulfates, and thiocyanates [87, 93]. The trace gases SO_2, H_2S, OCS, CS_2, CH_3SH, and $(CH_3)_2S$ may be detected in the atmosphere, but require a preconcentration technique [94, 95].

Despite occurring in a very complex flame environment, the mechanism for sulfur detection in the FPD is largely understood. Sulfur compounds are fully combusted in the flame, resulting in the formation of sulfur atoms, among other sulfur species. These atoms recombine to form S_2 in its $B^3\Sigma_u^-$ electronically excited state, which relaxes to the $X^3\Sigma_g^-$ ground state by emission of light in a series of bands with maximum intensities at 284 and 294 nm [96, 97], as shown in Figure 10 for the chemiluminescent reaction between OClO and H_2S.

$$S(^3P) + S(^3P) + M \rightarrow S_2(B^3\Sigma_u^-) + M \tag{49}$$

$$S_2(B^3\Sigma_u^-) \rightarrow S_2(X^3\Sigma_g^-) + h\nu \tag{50}$$

Because the reaction forming S_2^* is second order in sulfur atoms, the signal is approximately quadratic in concentration of the sulfur analyte. Although the detector may be linearized electronically by taking the square root of the signal, the power dependence on sulfur analyte concentration has been observed to vary from 1.5 to 2.0, depending on flame conditions [85, 87]. Because of this, calibration curves must still be constructed over the entire range for which compounds are to be quantified. The detector can also be linearized chemically at the expense of its dynamic range by adding a large excess of a sulfur compounds such as SF_6 to the burner gas [98–100], bringing the kinetics into a pseudo-first-order regime in sulfur. This chemical linearization has the additional benefit of improving the limit of detection by one to two orders of magnitude [87]. The FPD suffers from quenching by high concentrations of coeluting hydrocarbons. It has been suggested that hydrocarbons may enhance the formation of CS at the expense of S atoms, thereby reducing the sensitivity [101]. Reported detection limits for a chemically linearized FPD have been determined to be 2–10 pg S s^{-1} [87].

5.2.2 Detection of Other Elements by FPD

In addition to sulfur compounds, the FPD has been used for the measurement of organic compounds containing atoms such as phosphorus, nitrogen, boron, arsenic, antimony, and even chlorine [93]. Organophosphorus compounds are of particular interest owing to their common use as pesticides and chemical warfare agents. A modification to the FPD design, discussed below, known as pulsed flame photometric detection (PFPD) has additionally detected hydrocarbons and organic compounds containing tin, germanium, selenium, silicon, iron, and manganese [102]. The mechanism of detection for organophosphorus compounds is through the formation of PO, which subsequently reacts with H atoms in the fuel-rich flame to produce HPO*:

$$H + PO + M \rightarrow HPO* + M \tag{51}$$

emitting light at $\approx$526 nm [86]. Nitrogen-containing organics are detected in an analogous manner, generating HNO*, which emits light at $\approx$690 nm [93]. Unlike the nonlinear response to sulfur, HPO and HNO give a response proportional to analyte concentration. However, the response is not directly proportional to the number of P or N atoms, as structural and compositional differences between compounds can alter both the chemiluminescence efficiency and the emission spectra of the products [86].

Emissions from BO$_2$* [103], AsO*, and SbO* [103, 104] in an FPD flame have been used to detect organics or highly reduced species containing B, As, and Sb, respectively. These metal atoms react to form the same excited-state metal oxides discussed in their reactions with ozone above. These analytes have limits of detection measured to be approximately 50 ppbv, 10 ppbv, and 20 ppbv, respectively [93].

5.2.3 Improvements to FPD Design

Two improvements to the FPD design have been made in recent years. First, the addition of an oxygen-rich burner upstream of the FPD oxidizes hydrocarbons to CO and CO$_2$ and thereby eliminates the hydrocarbon interference within the analytical flame [105, 106]. This increases the sensitivity of the detector to approximately 20 pg S s^{-1} for a nonlinearized FPD [102]. In a second improvement, the PFPD was developed to significantly reduce the background and increase the sensitivity to all detectable atoms [107]. This detector is now commercially available from Varian Instruments (San Fernando, CA). In this design, the burner is constructed to generate a noncontinuous flame, which is reignited at a frequency of about 1–10 Hz. This periodic interruption allows for the acquisition of the time-dependent emissions from the various excited-state species present in the detector. As each of these has differing fluorescence lifetimes on the order of milliseconds, they can be differentiated in the time domain between flame pulses. Partially because the flame emission background is absent from the signal and other coeluting interferences may be eliminated, this technique has lowered the detection limits for sulfur, phosphorus, and nitrogen to 0.2, 0.01, and 2 pg s^{-1}, respectively. Chemically linearizing the sulfur detection resulted in a detection limit of ≈ 30 fg S s^{-1} [108]. Additionally, carbon can be detected at levels near 60 pg s^{-1} with this technique based on the emissions of both C_2^* and CH^* species [102].

5.3 Fluorine-Induced Chemiluminescence Detection of S, Se, Te, and P Compounds

In the FCLD [65], F_2 is reacted with the effluent of a GC in an evacuated cell at a pressure of ≈ 1 torr. Chemiluminescence is monitored using a red-sensitive PMT in conjunction with a band-pass filter that isolates wavelengths in the range 660–740 nm, as in the generic chemiluminescence detector shown in Figure 1. A commercial version of this detector was manufactured as the Model 300 SCD by Sievers Research (now Ionics-Sievers, Boulder, CO). It was later replaced by the Model 350 SCD, which is based on SO + O_3 chemiluminescence, described below. In the Model 300 instrument, fluorine was generated online by means of a high-frequency electrical discharge of SF_6. The instrument was found to be linear over at least three orders of magnitude and displayed a limit of detection of ≈ 1 pg S s^{-1} for a wide range of mercaptans, sulfides, disulfides, and trisulfides. Cyclic sulfur compounds such as thiophene were detected with about one order of magnitude less sensitivity. The detector has been successfully interfaced to high-performance liquid chromatography (HPLC) [109] and supercritical fluid chromatography (SFC) [110], in addition to GC.

The gases H_2S, CS_2, OCS, and SO_2 do not provide significant responses in the FCLD [65], apparently because their ionization potentials are too large to form long-lived charge-transfer complexes, and because they do not contain β hydrogens that can be eliminated to form the C=S bond, as in Reaction (33). The inability to detect these compounds is a disadvantage for some analytical applications (but can be an advantage for the highly selective detection of atmospheric DMS, discussed below). The detector also responds analogously with high sensitivity to organo-selenium and -tellurium [67] compounds and to phosphines, alkyl phosphines, and phosphinate esters [68]. As previously mentioned, the FCLD exhibits selectivities of $>10^7$ against alkanes, but compounds with weak C—H bonds also provide weak responses. This lack of adequate selectivity was the ''Achilles heel'' for this detector from a commercial standpoint, since one of the largest markets for sulfur-selective detectors is for the analysis of petrochemical feedstocks for sulfur compounds that can foul catalysts. These feedstocks often contain relatively high concentrations of olefins.

The FCLD has been particularly useful as a highly sensitive means of measuring volatile sulfur, selenium, and tellurium compounds in studies of bacterial methylation [67, 111–114]. Because of its high sensitivity to phosphinate esters [68], the FCLD could potentially serve as a monitor for nerve gases.

5.4 Sulfur Chemiluminescence Detector

The sulfur chemiluminescence detector, invented by Benner and Stedman [115, 116], combusts the GC effluent in a fuel-rich H_2/O_2 flame. The combustion products are reacted with ozone at low pressure ($\approx 1-10$ torr), and the chemiluminescence is detected by a PMT in combination with an optical filter to select for SO_2* [Reaction (4)] chemiluminescence, as in the generic chemiluminescence detector shown in Figure 1. Commercial versions of this detector, manufactured by Ionics-Sievers and Antek Instruments (Houston, TX), combust the GC effluent with a H_2/O_2 mixture in a ceramic or quartz furnace resistively heated to 800–1000°C, although the original SCD manufactured by Sievers Instruments used a ceramic capillary to sample the exhaust gas from the flame of an FID [117]. The SCD is linear over four or more orders of magnitude and exhibits a nearly equimolar response to all sulfur compounds [115, 116]. Because potential interferents such as olefins are combusted to form products (CO_2 and H_2O) that do not react with ozone, the detector is highly selective. An example of the selective detection of trace sulfur compounds in a hydrocarbon fuel is given in Figure 11. The limit of detection of the SCD has been reported to be as low as 25 fg S s^{-1} [118]. The SCD has advantages over the FCLD for sulfur-compound detection in that it responds to all sulfur compounds and exhibits much higher selectivity. In comparison to the FPD, the SCD is more sensitive and exhibits a linear response.

Although there has been some dispute among the manufacturers about the mechanism of detection, it is clear from the scientific evidence that the signal is derived from the reaction of SO with O_3 [Reaction (4) discussed above]. Martin and Glinski [119] have identified SO_2* as the emitting species. That the SO_2* is derived from the reaction of the SO intermediate with O_3 has been confirmed by flow-tube kinetics studies in which the flame or furnace effluents were reacted with O_3 and NO_2 [23, 120]. The rate constants for reaction of the chemiluminescent intermediate with O_3 and with NO_2 agree with the literature values for those rate constants and with the rate constants measured in the same apparatus using SO generated in a microwave discharge of SO_2/He. One proviso, however, found

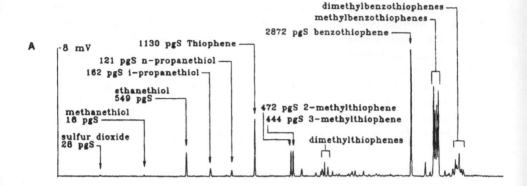

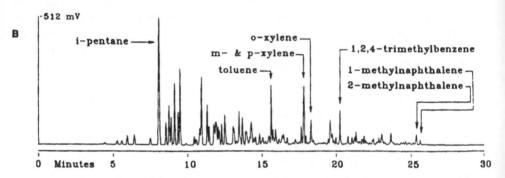

Figure 11 SCD chromatogram of catalytic cracked (FCC) gasoline, demonstrating the selectivity for sulfur compounds. (A) SCD response; (B) FID response. (Reprinted from *American Laboratory* 23(3):117, 1991. Copyright 1991 by International Scientific Communications, Inc.)

by Burrow and Birks [120], is that with the higher sulfur concentrations used in the kinetics experiments, the SO was derived from some other chemical species, X, which itself reacts rapidly with O_3 to form SO:

$$X + O_3 \rightarrow SO + \text{products} \qquad (52)$$

Based on kinetics considerations, it was possible to rule out a large number of $H_xS_yO_z$ species, including S atoms, which are known to react rapidly with ozone. The most likely candidate for X is S_3 formed in the association of S atoms with S_2, a reaction that could occur within the transfer line linking the furnace with the chemiluminescence chamber [120]. Watson and Birks (unpublished results,

1998) demonstrated that sulfur vapor, which contains S_3, does chemiluminesce with ozone.

5.5 Thermal Energy Analysis and Nitrogen-Selective Detectors

The NO + O_3 chemiluminescent reaction [Reactions (1–3)] is utilized in two commercially available GC detectors, the TEA detector, manufactured by Thermal Electric Corporation (Saddle Brook, NJ), and two nitrogen-selective detectors, manufactured by Thermal Electric Corporation and Antek Instruments, respectively. The TEA detector provides a highly sensitive and selective means of analyzing samples for N-nitrosamines, many of which are known carcinogens. These compounds can be found in such diverse matrices as foods, cosmetics, tobacco products, and environmental samples of soil and water. The TEA detector can also be used to quantify nitroaromatics. This class of compounds includes many explosives and various reactive intermediates used in the chemical industry [121]. Several nitroaromatics are known carcinogens, and are found as environmental contaminants. They have been repeatedly identified in organic aerosol particles, formed from the reaction of polycyclic aromatic hydrocarbons with atmospheric nitric acid at the particle surface [122–124]. The TEA detector is extremely selective, which aids analyses in complex matrices, but also severely limits the number of potential applications for the detector [125–127].

Both N-nitroso and nitro groups are thermally labile and can be pyrolyzed to liberate nitric oxide. Because of the low thermal stability and low volatility of the compounds usually detected with this technique, GC cannot always be used for their separations. An analyte simply may not survive the temperature gradients required to elute it [121]. To circumvent this problem, the TEA detector has also been interfaced to HPLC with some success [128, 129]. The HPLC column effluent is discharged through a restricting orifice and nebulized into the heated detection cell, where the mobile phase is evaporated and the analysis proceeds as with GC. The GC/TEA design possesses the advantage of simplicity, consisting only of a heated tube or catalyst bed interfaced to an evacuated cell supplied with a flow of ozone and viewed through an optical filter by a PMT as in Figure 1.

The name thermal energy analysis refers to the amount of heat energy required to break a bond in an analyte to produce the detected NO. The detector can be made more or less selective, depending on the pyrolysis conditions used. Although the simplest design uses only a heated tube as the pyrolysis chamber, these detectors have more often made use of catalysts to detect a wider variety of compounds at lower temperatures [121]. While many catalysts have been investigated, WO_3 and $W_{20}O_{58}$ have been shown to be especially effective [125]. Other catalysts investigated included platinum and nickel alloy oxides to achieve similar results [130, 131]. For a simple pyrolysis tube made of quartz or ceramic, the N—NO bond in N-nitroso compounds cleaves at temperatures in the range 200–300°C with few interferences from other nitrogen-containing compounds (with the exception of some organic nitrites). At much higher pyrolytic zone temperatures, up to approximately 600°C, the signal for nitrosamines begins to decrease and those for nitroaromatics, nitroalkanes, and nitroamines become significant. For each compound class, detection limits on the order of picograms of analyte have been determined [121].

Analogously, nitrogen-selective detectors have been developed to detect all nitrogen-containing organics [132]. This is accomplished by adding oxygen to the column effluent upstream of the pyrolysis tube. At catalyst temperatures near 800–1000°C, any nitrogen present in an analyte is converted to NO, while all carbon and hydrogen is oxidized to CO_2 and water, respectively. The NO is then detected in the same manner as the TEA detector described above. Limits of detection have been demonstrated to be in the picogram range for organic analytes containing at least one nitrogen [121]. The response to nitrogen in compounds containing differing numbers of nitrogen atoms is not always equimolar. For example, the ratio of responses to N-nitrosodimethylamine and pyridine is 2:1, not 3:1 as expected from their nitrogen molar ratios [121].

5.6 Redox Chemiluminescence Detector

The redox chemiluminescence detector (RCD), developed by Nyarady et al. [133], and marketed for a short time by Sievers Research, selectively detects compounds capable of reducing NO_2 to NO at a heated gold surface. The NO is detected by chemiluminescence with ozone [Reactions (1–3)], as in the TEA detector and the nitrogen-selective detectors described above. The RCD detects most compounds containing oxygen, nitrogen, sulfur, or other reactive functional groups, and the selectivity can be tuned by varying the NO_2 concentration or the temperature of the gold catalyst bed. Atmospheric gases that respond include CO, H_2O_2, SO_2, H_2S, OCS, CS_2, and H_2. Figure 12 compares FID and RCD chromatograms of jet fuel containing 10 ppmv of the antioxidant BHT (2, 6-di-$tert$-butyl-4-methylphenol), illustrating the high degree of selectivity that can be achieved with this detector. The RCD also has been adapted for detection in SFC and HPLC [134, 135]. The principal difficulty encountered with the RCD is the tendency for the gold catalyst to become poisoned; it is often necessary to recondition the catalyst by flowing oxygen through the bed at high temperature. Perhaps the main reason that the RCD did not become widely accepted is that it responds to a wide range of compounds rather than having an element-specific or functional-group-specific response. As a result, little chemical information is provided for unknown chromatographic peaks. The detector could be useful for

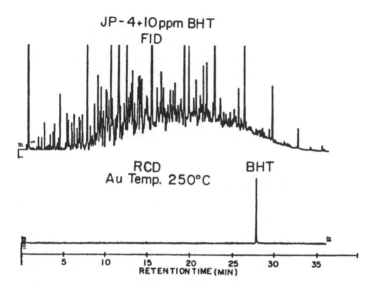

Figure 12 FID and RCD chromatograms of JP-4 jet fuel containing 10 ppm of BHT, demonstrating the high degree of selectivity against hydrocarbons. (Reprinted with permission from Ref. 153. Copyright 1985 American Chemical Society.)

some applications, however, provided the irreproducibility associated with the catalyst is solved.

5.7   Active Nitrogen Detectors

There has been considerable interest in the application of active nitrogen as a universal GC detector [56–58, 136]. Melzer and Sutton et al. named this detection technique metastable transfer emission spectroscopy (MTES) after the metastable N_2^* component of active nitrogen. An advantage of the use of active nitrogen, as discussed above, is the ability to distinguish between paraffins and olefins by selectively detecting olefins with the atomic component of active nitrogen [Reaction (20)], present in low-pressure discharges, and then doping the column effluent with HCl [Reactions (21,22)] to detect both simultaneously. The system can also be applied to the detection of organometallics in GC separations with high specificity [58]. More recently, Rice et al. applied an active nitrogen detection scheme at ambient pressure, the atmospheric pressure active nitrogen (APAN) detector, for nonselective detection of both hydrocarbons and organometallics [57]. Although clearly useful, detection by active nitrogen does not appear to have sufficient advantages to supplant the FID as a universal detector, and GC is unlikely to compete with atomic absorption or atomic emission as a practical method for metal analysis. The main advantage of active nitrogen detection is flexibility; this one detector, when properly configured, can perform a variety of analyses that would otherwise require a number of individual detectors.

6. APPLICATIONS TO ATMOSPHERIC RESEARCH

Chemiluminescence is the method of choice for measurements of a few select atmospheric species, especially the oxides of nitrogen. Where chemiluminescence is sufficiently selective that it can be used without prior separation, the high inherent sensitivity of the technique can be exploited. Although chemiluminescence requires no light source, the necessity of a vacuum pump to create low pressures and gas cylinders to provide reagent gases often makes the instruments large and heavy. The NO_x and NO_y instruments described below, for example, weigh up to several hundred pounds. An additional difficulty associated with the application of gas-phase chemiluminescence to atmospheric measurements is that the detector response changes with reactor pressure, while the atmospheric pressure changes with altitude. Nevertheless, there are several atmospheric measurements that are best made by chemiluminescence.

6.1 "NO_x Box" Detector for NO, NO_2, and NO_y

The most widely used chemiluminescent reaction in atmospheric field studies is the reaction of NO with O_3 [Reactions (1–3)]. The so-called "NO_x box" measures both NO and the sum of NO and NO_2, which is defined as NO_x [1, 2, 4]. The concentration of NO_2 is obtained by difference. First, NO is measured by mixing air with ozone and measuring the chemiluminescence signal with a red-sensitive PMT and optical filter to discriminate against interferences from alkenes and reduced sulfur compounds. The air is then diverted through a photochemical reactor prior to entering the reaction chamber where a fraction, on the order of 50%, of the NO_2 is photolyzed to form NO. The sensitivity to NO and efficiency of photolysis of NO_2 are periodically measured by means of standards. Limits of detection for integration times of the order of 1 s are in the range 1–10 pptv [4, 137, 138].

The measurement of the total reactive oxides of nitrogen, NO_y (= NO + NO_2 + NO_3 + $2 N_2O_5$ + HNO_2 + HNO_3 + HNO_4 + $ClONO_2$ + organic nitrogen compounds + particulate nitrate), has been of great interest in the atmosphere for several years. It has been found that all these compounds may be quantitatively converted to NO by either passing air through a heated molybdenum tube (which has a catalytically active oxide surface) or mixing CO or H_2 with the air stream in a heated gold tube in a manner similar to that of the RCD described above [137–139]. The NO produced may then be quantified using a NO_x box.

New laboratory data on conversion efficiencies for a wide range of gas-phase odd-nitrogen species and a review of the status of NO_y measurements were published recently [140]. Also, measurements of NO_y have been subject to recent scrutiny [140, 141]. During the field experiment known as PEM-West (fall of 1991), for example, two NO_y instruments operating aboard a DC-8 aircraft showed significant disagreement. Also, the measured NO_y was greater than the sum of its separately measured components, often by a factor of two or more. This discrepancy, which has been observed several times before, has come to be known as the "missing NO_y" [141] and is of great interest to atmospheric chemists. It is especially important to be able to account for all of the reactive nitrogen species in the atmosphere. Although the chemiluminescence measurements of NO and NO_2 are now well established, measurements of NO_y are still problematic. Especially poorly established is the efficiency of conversion of particulate nitrate to NO in the catalytic converters.

6.2 Fast Ozone Detector

Ozone is usually measured in the atmosphere by UV absorbance at 254 nm with a response time of the order of several seconds. Faster measurements are desired for aircraft measurements owing to the high speed of the aircraft, and measurements of ozone flux using the eddy covariance technique require a method having a response time of up to ≈ 10 Hz [142]. This can be accomplished by using the chemiluminescent reaction of O_3 with a large excess of either NO, as in the detection of NO by O_3 [Reactions (1–3)], or ethylene [Reactions (5–8)] [36, 143–146]. A fast-response detector based on the heterogeneous, gas-solid reaction of ozone with a fluorescent dye such as rhodamine B may also be used and has the advantage of being extremely lightweight [147–149]. The latter method requires frequent calibration, however, which can be accomplished with the slower UV absorbance instrument. Miniaturized UV ozone sondes weighing less than 1 kg have been developed recently and may be coupled with fast-response ozone detectors [150].

6.3 Isoprene Detector

Isoprene, the most abundant hydrocarbon emitted to the atmosphere by plants, can also be measured using ozone chemiluminescence. As discussed above, alkenes react with ozone to produce formaldehyde in its $^1A_2$ electronic state, in addition to several other chemiluminescent products. In a fast isoprene detector manufactured by Hills Scientific (Boulder, CO), the chemiluminescence is detected using a blue-sensitive PMT to maximize the sensitivity for isoprene detection while minimizing any interference from the NO + O_3 reaction, which occurs in the red and near-IR spectral regions [151]. The fully optimized reaction cell is evacuated to approximately 300 torr with an ozone partial pressure of approximately 22 torr. Under these conditions, and with a 5.4 s integration time, the detection limit was found to be 400 pptv, but 5 ppbv could be measured as fast as 1 Hz. Although it might be expected that other common biogenic compounds such as pinene, limonene, and various other monoterpenes would chemiluminesce under the same conditions, investigations indicated that the light produced from their reactions with ozone was virtually nondetectable. It is known that these compounds readily react with ozone in the atmosphere, but the exact nature of their failure to chemiluminesce in the isoprene detector is not understood. Significant interferences include propene, and to a lesser extent, ethene, dimethyl sulfide, 3-butene-2-one, and 2-methylpropenal. The instrument is able to measure isoprene emissions from a single leaf and to measure eddy correlation fluxes at frequencies up to 1 Hz.

6.4   Dimethyl Sulfide Detector

The chemiluminescent reaction with F_2 was demonstrated to be sufficiently sensitive and selective to measure dimethyl sulfide, emitted by oceanic phytoplankton, in the marine boundary layer without a prior chromatographic separation [69] and with a detector response fast enough to measure DMS fluxes using the eddy covariance technique [142]. Limits of detection (S/N = 1) of 39, 12, and 4 pptv DMS were demonstrated for 0.1-, 1-, and 10-s integration times, respectively. Whereas all organosulfur compounds produce vibrational overtone emission from HF^+, only methylated sulfur compounds produce HCF* [6]. Emission of HCF* occurs in the wavelength range 500–700 nm, which is blue-shifted from HF^+ emission at 670–900 nm (Fig. 7). Therefore, selectivity for HCF* is achieved by using a blue-sensitive PMT.

7.   CONCLUDING REMARKS AND FUTURE TRENDS

Gas-phase chemiluminescence has been demonstrated to provide highly sensitive detection for a wide range of compound classes with selectivities ranging from very specific, as in the detection of only H_2S and CH_3SH with OClO, to nearly universal, as in the reactions of active nitrogen at atmospheric pressure. The use of a combustion chamber to achieve element selectivity by converting all compounds containing that element to a single surrogate analyte (e.g., NO for nitrogen compounds and SO for sulfur compounds) has proven especially beneficial in gas chromatography. There are a number of highly efficient chemiluminescence reactions that have yet to be fully exploited in analytical chemistry. Chemiluminescence of NO and SO with O atoms, for example, holds the potential of greater sensitivity than the corresponding reactions with ozone, provided contributions from background emissions can be minimized.

The chemical characterization of aerosol particles currently is of great interest in the field of atmospheric chemistry. A major goal is the development of a method for continuous elemental analysis of aerosols, especially for the elements C, N, and S. Chemiluminescence reactions described in this chapter have adequate sensitivity and selectivity for such analyses. In fact, considering that a 1-μm-diameter particle has a mass of $\approx 0.5-1.0$ pg, online analysis of single aerosol particles should be achievable, especially for larger particles.

During the past decade, single fluorescent molecules have been detected by repeatedly exciting them to fluorescence within a high-intensity laser beam, and the method has been applied to problems ranging from studies of the dynamics of single molecules to rapid sequencing of DNA. Although the concept of single-molecule detection is now well established, it has not yet been applied to atmospheric chemistry. Few, if any, atmospheric molecules have large enough extinction coefficients and fluorescence quantum yields to be detected as single molecules by laser-induced fluorescence. Although highly speculative at this point, it may be possible to detect single molecules in the gas phase based on initiation of chain reactions, such as the $F_2 + H_2$ reaction, that produce chemiluminescence. If successful, this would, of course, provide the ultimate sensitivity for gas-phase detection.

REFERENCES

1. A Fontijn, AJ Sabadell, RJ Ronco. Anal Chem 42:575–579, 1970.
2. DW Joseph, CW Spicer. Anal Chem 50:1400–1403, 1978.
3. DW Fahey, G Hubler, DD Parrish, EJ Williams, RB Norton, BA Ridley, HB Singh, SC Liu, FC Fehsenfeld. J Geophys Res 91:9781–9793, 1986.
4. BA Ridley, FE Grahek. J Atmos Ocean Technol 7:307–311, 1990.
5. RJ Glinski, JN Getty, JW Birks. Chem Phys Lett 117:359–364, 1985.
6. RJ Glinski, EA Mishalanie, JW Birks. J Photochem 37:217–231, 1987.
7. AA Turnipseed, JW Birks. J Phys Chem 95:6569–6574, 1991.
8. JW Birks. In: JW Birks, ed. Chemiluminescence and Photochemical Reaction Detection in Chromatography. New York: VCH Publishers, 1989, pp 1–37.
9. DM Steffenson, DH Stedman. Anal Chem 46:1704–1709, 1974.
10. PN Clough, BA Thrush. Chem Commun 783–784, 1966.
11. WB DeMore, SP Sander, DM Golden, RF Hampson, MJ Kurylo, CJ Howard, AR Ravishankara, CE Kolb, MJ Molina. Chemical Kinetics and Photochemical Data for Use in Stratospheric Modeling, JPL Publication 97-4. Pasadena: NASA JPL, 1997.
12. AA Mehrabzadeh, RJ O'Brien, TM Hard. Anal Chem 55:1660–1665, 1983.

13. AA Mehrabzadeh, RJ O'Brien, TM Hard. Rev Sci Instrum 54:1712–1718, 1983.
14. S Toby. Chem Rev 84:277–285, 1984.
15. JC Greaves, DJ Garvin. J Chem Phys 30:348–349, 1959.
16. H Harrison, DM Scattergood. Upper Limits for Chemiluminescence from Single Collisions of O_3 with NO, CO, H_2S, and CS_2, Boeing Scientific Research Laboratories Document D1-82-0480. Seattle: Boeing Scientific Research Laboratories, 1966.
17. RD Kenner, EA Ogryzlo. In: JG Burr, ed. Chemi- and Bioluminescence. New York: Marcel Dekker, 1985, pp 45–185.
18. AE Douglas. J Chem Phys 45:1007–1015, 1966.
19. VM Donelly, F Kaufman. J Chem Phys 66:4100–4110, 1977.
20. VM Donelly, F Kaufman. J Chem Phys 68:5671, 1978.
21. VM Donelly, DG Keil, F Kaufman. J Chem Phys 71:659–673, 1979.
22. DG Keil, VM Donelly, F Kaufman. J Chem Phys 73:1514–1520, 1980.
23. RL Benner, DH Stedman. Appl Spectrosc 48:848–851, 1994.
24. CJ Halstead, BA Thrush. Photochem Photobiol 4:1007–1013, 1965.
25. H Akimoto, JJ Finlayson, JN Pitts. Chem Phys Lett 12:199–202, 1971.
26. A Kaldor, W Braun, MJ Kurylo. J Chem Phys 61:2496–2499, 1974.
27. RJ Glinski, JA Sedarski, DA Dixon. J Phys Chem 85:2440–2443, 1981.
28. MAA Clyne, CJ Halstead, BA Thrush. Proc Roy Soc Lond A 295:355–362, 1966.
29. CJ Halstead, BA Thrush. Proc Roy Soc Lond A 295:363–379, 1966.
30. CJ Halstead, BA Thrush. Proc Roy Soc Lond A 295:380–398, 1966.
31. A McKenzie, BA Thrush. Proc Roy Soc Lond A 308:133–140, 1968.
32. EKC Lee, GL Lopin. In: SH Lin, Ed. Radiationless Transitions. New York: Academic Press, 1980, pp 1–425.
33. S Glavas, S Toby. J Phys Chem 79:779–782, 1975.
34. JS Gaffney, DJ Spandau, TJ Kelly, RL Tanner. J Chromatogr 347:121–127, 1985.
35. W Bruening, FJM Concha. J Chromatogr 112:253–265, 1975.
36. BJ Finlayson, JN Pitts, R Atkinson. J Am Chem Soc 96:5356–5367, 1974.
37. U Schurath, H Gusten, R-D Penzhorn. J Photochem 5:33–40, 1976.
38. JT Herron, RE Huie. J Am Chem Soc 99:5430–5435, 1977.
39. AA Turnipseed. In: JW Birks, Ed. Chemiluminescence and Photochemical Reaction Detection in Chromatography. New York: VCH Publishers, 1989, pp 39–69.
40. ME Fraser, DH Stedman, MJ Henderson. Anal Chem 54:1200–1201, 1982.
41. ME Fraser, DH Stedman. J Chem Soc Faraday Trans I 79:527–542, 1983.
42. DH Stedman, DA Tammaro. Anal Lett 9:81–89, 1976.
43. DH Stedman, DA Tammaro, DK Branch, R Pearson. Anal Chem 51:2340–2342, 1979.
44. K Fujiwara, Y Watanabe, K Fuwa, JD Winefordner. Anal Chem 54:125–128, 1982.
45. RJ Strutt. Proc Phys Soc 23:66, 1910.
46. FC Fehsenfeld, KM Evenson, HP Broida. Rev Sci Instrum 36:294–298, 1965.
47. JW Carnahan. Am Lab 31–36, 1983.
48. A Khan, M Kasha. J Am Chem Soc 92:3293–3300, 1970.
49. JW Birks, B Shoemaker, TJ Leck, DM Hinton. J Chem Phys 65:5181–5185, 1976.
50. AP Ongstad, JW Birks. J Chem Phys 81:3922–3930, 1984.
51. AP Ongstad, JW Birks. J Chem Phys 85:3359–3368, 1986.

52. KH Becker, W Groth, D Thran. Chem Phys Lett 6:583–586, 1970.
53. A Sharma, JP Padur, P Warneck. J Phys Chem 71:1602, 1967.
54. EP Lewis. Phys Rev 18:125–128, 1904.
55. AN Wright, CA Winkler. Active Nitrogen. New York: Academic Press, 1968, pp 1–602.
56. JE Melzer, DG Sutton. Appl Spectrosc 34:434–437, 1980.
57. GW Rice, JJ Richard, AP D'Silva, VA Fassel. Anal Chem 53:1519–1522, 1981.
58. DG Sutton, KR Westburg, JE Melzer. Anal Chem 51:1399–1401, 1979.
59. JT Clay, TM Niemczyk. Spectroscopy 2:36, 1987.
60. RH Getty, JW Birks. Anal Lett 12:469–476, 1979.
61. SR Spurlin, ES Yeung. Anal Chem 57:1223–1227, 1985.
62. DE Mann, BA Thrush, DR Lide, JJ Ball, N Acquista. J Chem Phys 34:420–431, 1961.
63. WH Duewer, DW Setser. J Chem Phys 58:2310–2320, 1973.
64. M VanPee, KD Cashin, RJ Mainiero. Combust Flame 31:187–196, 1978.
65. JK Nelson, RH Getty, JW Birks. Anal Chem 55:1767–1770, 1983.
66. RJ Glinski, EA Mishalanie, JW Birks. J Am Chem Soc 108:531–532, 1986.
67. TG Chasteen, GM Silver, JW Birks, R Fall. Chromatographia 30:181–185, 1990.
68. TG Chasteen, R Fall, JW Birks, HR Martin, RJ Glinski. Chromatographia 31:342–346, 1991.
69. AJ Hills, DH Lenschow, JW Birks. Anal Chem 70:1735–1742, 1998.
70. RI Patel, GW Stewart, K Castleton, JL Gole, JR Lombardi. Chem Phys 52:461–468, 1980.
71. K Shobatake, JM Parson, YT Yee, SA Rice. J Chem Phys 59:1416–1426, 1973.
72. JM Parson, YT Lee. J Chem Phys 56:4658–4666, 1972.
73. RJ Glinski, CD Taylor. Chem Phys Lett 155:511–512, 1989.
74. JW Birks, SD Gabelnick, HS Johnston. J Mol Spectrosc 57:23–46, 1975.
75. JJ Valentini, MJ Coggiola, YT Lee. J Am Chem Soc 98:853–854, 1976.
76. JJ Valentini, MS Coggiola, YT Lee. Disc Faraday Soc 62:232–245, 1977.
77. CC Kahler, YT Lee. J Chem Phys 73:5122–5130, 1980.
78. JM Farrar, YT Lee. J Am Chem Soc 96:7570–7572, 1974.
79. M Yamada, A Ishiwada, T Hobo, S Suzuki, S Araki. J Chromatogr 238:347–356, 1982.
80. S Araki, S Suzuki, M Yamada, H Suzuki, T Hobo. J Chromatogr Sci 16:249–253, 1978.
81. SR Spurlin, ES Yeung. Anal Chem 54:318–320, 1982.
82. SS Brody, JE Chaney. J Gas Chromatogr 4:42–46, 1966.
83. PT Gilbert. Anal Chem 38:1920–1922, 1966.
84. WL Crider. Anal Chem 41:534–537, 1969.
85. T Sugiyama, Y Suzuki, T Takeuchi. J Chromatogr 77:309–316, 1973.
86. S Sass, GA Parker. J Chromatogr 189:331–349, 1980.
87. SO Farwell, CJ Barinaga. J Chromatogr Sci 24:483–494, 1988.
88. DA Ferguson, LA Luck. Chromatographia 12:197–203, 1979.
89. AR Baig, CJ Cowper, PA Gibbons. Chromatographia 16:297–300, 1982.
90. WA Aue. J Chromatogr Sci 13:329–333, 1975.
91. R Greenhaigh, MA Wilson. J Chromatogr 128:157–162, 1976.

92. M Dressler. Selective Gas Chromatographic Detectors. Amsterdam: Elsevier, 1986, pp 1–319.
93. DH Stedman, ME Fraser. In: JG Burr, ed. Chemi- and Bioluminescence. New York: Marcel Dekker, 1985, pp 439–468.
94. RS Braman, JM Ammons, JL Bricker. Anal Chem 50:992–994, 1978.
95. J Godin, J Cluet, C Boudene. Anal Chem 51:2100–2102, 1979.
96. RW Fair, BA Thrush. Trans Faraday Soc 65:1208, 1969.
97. WC Swope, YP Lee, HF Schaefer. J Chem Phys 70:947–953, 1979.
98. WL Crider, RW Slater. Anal Chem 41:531–533, 1969.
99. JM Zehner, RA Simonaitis. J Chromatogr Sci 14:348–350, 1976.
100. G Marcalin. J Chromatogr Sci 15:560–562, 1977.
101. S Fredriksson, A Cedergren. Anal Chim Acta 100:429–438, 1978.
102. S Cheskis, E Atar, A Amirav. Anal Chem 65:539–555, 1993.
103. EJ Sowinski, IH Suffet. J Chromatogr Sci 9:632–634, 1971.
104. R Belcher, SL Bogdanski, SA Ghonaim, A Townshend. Anal Chim Acta 109:229–239, 1979.
105. PL Patterson. Anal Chem 50:345–348, 1978.
106. PL Patterson, RL Howe, A Abu-Shumays. Anal Chem 50:339–344, 1978.
107. E Atar, S Cheskis, A Amirav. Anal Chem 63:2061–2064, 1991.
108. A Amirav, H Jing. Anal Chem 67:3305–3318, 1995.
109. EA Mishalanie, JW Birks. Anal Chem 54:918–923, 1986.
110. WT Foreman, CL Shellum, JW Birks, RE Sievers. J Chromatogr 465:23–33, 1989.
111. U Brunner, TG Chasteen, P Ferloni, R Bachofen. Chromatographia 40:399–403, 1995.
112. V Stalder, N Bernard, KW Hanselmann, R Bachofen, TG Chasteen. Chromatographia 303: 91–97, 1995.
113. H Gürleyük, VV Fleet-Stalder, TG Chasteen. Appl Organomet Chem 11:471–483, 1997.
114. V VanFleet-Stalder. J Photochem Photobiol 43:193–203, 1998.
115. RL Benner, DH Stedman. Anal Chem 61:1268–1271, 1989.
116. RL Benner, DH Stedman. Environ Sci Technol 24:1592–1596, 1990.
117. NG Johansen, JW Birks. Am Lab 23:112–119, 1991.
118. RL Shearer. Anal Chem 64:2192–2196, 1992.
119. HR Martin, RJ Glinski. Appl Spectrosc 46:948–952, 1992.
120. PL Burrow, JW Birks. Anal Chem 69:1299–1306, 1997.
121. RL Shearer, RE Sievers. In: JW Birks, ed. Chemiluminescence and Photochemical Reaction Detection in Chromatography. New York: VCH Publishers, 1989, pp 71–97.
122. J Jager. J Chromatogr 152:575–578, 1978.
123. JN Pitts, KAV Cauwenberghe, D Grosjean, JP Schmid, DR Fitz, WL Belser, GB Knudson, PM Hynds. Science 202:515–519, 1978.
124. CY Wang, M-S Lee, CM King, PO Warner. Chemosphere 9:83–87, 1980.
125. DH Fine, D Lieb, F Rufeh. J Chromatogr 107:351–357, 1975.
126. DH Fine, DP Roundbehler. J Chromatogr 109:271–279, 1975.
127. DH Fine, F Rufeh, D Lieb, DP Roundbehler. Anal Chem 47:1188–1191, 1975.
128. PE Oettinger, F Huffman, DH Fine, D Lieb. Anal Lett 8:411–414, 1975.

129. C Ruhl, J Reusch. J Chromatogr 328:362–366, 1985.
130. DH Fine, F Rufeh, B Gunther. Anal Lett 6:731–733, 1973.
131. DH Fine, F Rufeh, D Lieb. Nature (Lond) 247:309–310, 1974.
132. HV Drushel. Anal Chem 49:932–939, 1977.
133. SA Nyarady, RM Barkley, RE Sievers. Anal Chem 57:2074–2079, 1985.
134. JJ DeAngelis, RM Barkley, RE Sievers. J Chromatogr 441:125–134, 1988.
135. WT Foreman, RE Sievers, B Wenclawiak. Fres J Anal Chem 330:231–234, 1988.
136. WB Dodge, RO Allen. Anal Chem 53:1279–1286, 1981.
137. MJ Bollinger, RE Sievers, DW Fahey, FC Fehsenfeld. Anal Chem 55:1980–1986, 1983.
138. DW Fahey, CS Eubank, G Hubler, FC Fehsenfeld. J Atmos Chem 3:435–468, 1985.
139. LC Anderson, DW Fahey. J Phys Chem 94:644–652, 1990.
140. DAV Kliner, DC Daube, JD Burley, SC Wofsy. J Geophys Res 102:10,759–10,776, 1997.
141. DR Crosley. J Geophys Res 101:2049–2052, 1996.
142. DH Lenschow. In: PA Matson RC Harriss, Eds. Biogenic Trace Gases: Measuring Emissions from Soil and Water. American Chemical Society, Washington, D.C., 1995, pp 126–163.
143. GW Nederbragt, AVD Horst, JV Duijn. Nature (Lond) 206:87, 1965.
144. F Chisaka, S Yanagihara. Anal Chem 54:1015–1017, 1982.
145. RJ O'Brien, TM Hard, AA Mehrabzadeh. Environ Sci Technol 17:560–562, 1983.
146. DH Stedman, DL McElwee, GJ Wendel. In: A Fontijn, ed. Gas-Phase Chemiluminescence and Chemi-ionization. New York: Elsevier Science Publishing, 1985, pp 345–355.
147. E Hilsenrath, L Seiden, P Goodman. J Geophys Res 74:6873–6880, 1969.
148. JA Hodgeson, KJ Krost, AE O'Keeffe, RK Stevens. Anal Chem 42:1795–1797, 1970.
149. E Hilsenrath, PT Kirschner. Rev Sci Instrum 51:1381–1384, 1980.
150. JA Bognar, JW Birks. Anal Chem 68:3059–3062, 1996.
151. AJ Hills, PR Zimmerman. Anal Chem 62:1055–1060, 1990.
152. L Herman, J Akriche, H Grenat. J Quant Spectrosc Radiat Transfer 2:215–224, 1962.
153. W Worthy. Chem Eng News 63:42–44, 1985.

14
Chemiluminescence Detection in Liquid Chromatography

Naotaka Kuroda, Masaaki Kai, and Kenichiro Nakashima
Nagasaki University, Nagasaki, Japan

1. INTRODUCTION

For the analysis of biological and environmental samples, separation analyses have an important role in the determination of analytes in complex matrices. High-performance liquid chromatography (HPLC) has currently become dominant as a potential tool for the separation of a wide range of analytes in the diverse fields of analysis. As detection systems for HPLC, the ultraviolet-visible light absorption and, to a lesser extent, refractive index (RI) detectors have been employed most commonly. Although these detection techniques are very universal and conveniently applicable, in general they lack sensitivity. In the case of trace analyses, fluorescence (FL) and electrochemical (EC) detections are utilized owing to their relatively high sensitivity and selectivity. When ultratrace quantities of an analyte in a complex sample matrix (e.g., biologically active compounds such as hormones in body fluids) are to be determined, highly sensitive chemiluminescence (CL) and laser-induced fluorescence (LIF) detections are often very powerful. CL-based techniques as a means of detection for HPLC have been developed since the 1980s. Although it is not as universal as FL detection, CL detection is rapidly growing owing to the very low detection limits, wide linear working ranges, and relatively simple instrumentations.

In this chapter, developments and applications of the CL detection methods in HPLC are reviewed.

2. FEATURES OF CL DETECTION IN HPLC

In the CL detection method, the excitation of a molecule is achieved via a chemical reaction that is generally an oxidation process. That is, an exciting light source is not required; thus, the CL is not accompanied by any scattering light and source instability. This permits a large signal-to-noise ratio (S/N), which finally provides an increase in sensitivity.

Most of the light emission in CL methods are in the visible region (the limiting factor for the occurrence of CL is that the energy required for luminescence in the visible region lies between 44 and 77 kcal/mol) [1]. To apply this CL emission to a detection system in HPLC, generation of CL by a postcolumn reaction of analytes in an eluent with reagent(s) is required. Therefore, excellent efficiency of the CL reaction is desired for HPLC. Although CL techniques permit sensitive detection of analytes owing to the reason described above, undesirable emission of CL derived from impurities in solvents and reagents sometimes interferes with the highly sensitive detection. When compared to FL detection, CL detection is generally more sensitive but requires an additional pump(s) to deliver postcolumn CL reagent(s) resulting in an increase of the running cost.

3. CL REACTIONS USED FOR HPLC

In spite of the various CL reactions reported so far, only a few are used as practical tools for CL detection in HPLC [2, 3].

Representative CL compounds are shown in Figure 1. CL reactions such as the ones using luminol (5-amino-2,3-dihydro-1,4-phthalazinedione), lucigenin (*N, N'*-dimethyl-9,9'-bisacridinium dinitrate), lophine (2,4,5-triphenylimidazole), and aryloxalates are well known. Luminol, lucigenin, lophine, and their derivatives directly produce emission of light in the process accompanying their decomposition, and the CL intensities are increased employing various catalysts. On the other hand, another chemiluminescent reaction using aryloxalates, which involves an energy transfer reaction, is called peroxyoxalate chemiluminescence (PO-CL). PO-CL is the emission of light produced by a chemical reaction of aryloxalate, hydrogen peroxide, and a fluorophore.

In these CL reactions, as the total emission of light is proportional to the concentration of the various substrates associated with the CL reaction, the substrates can be determined sensitively. Understanding of the principle of each CL

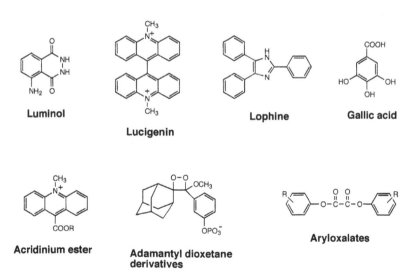

Figure 1   Representative CL compounds.

reaction is therefore very important for the determination of an analyte participating in a given CL reaction.

3.1   CL Reaction of Luminol

Luminol derivatives produce emission of light by oxidation with oxygen and hydrogen peroxide under alkaline conditions. By utilizing this reaction, peroxides such as hydrogen peroxide and lipid hydroperoxides can be determined after HPLC separation. Metal ions [e.g., iron(II), cobalt(II), etc.] catalyzing the luminol CL reaction can also be determined.

Some luminol derivatives have been developed as CL labeling reagents. Analytes prelabeled with luminol derivatives are separated by HPLC, mixed with postcolumn reagents such as hydrogen peroxide and an alkaline solution of potassium hexacyanoferrate (III), and then detected by a CL detector. Highly sensitive determination is possible by optimizing the conditions to increase the CL reaction efficiency for each analyte.

3.2   CL Reaction of Lucigenin

Although acridinium derivatives including lucigenin are well-known CL compounds, few methods have been described on the use of these compounds for detection in HPLC. Differing from other CL reagents, lucigenin produces CL by the reaction with organic reductants as well as with hydrogen peroxide. Therefore, the HPLC determination of reductants i.e., ascorbic acid, glucose, etc., can be performed by using lucigenin as a postcolumn CL reagent.

3.3   CL Reaction of Aryloxalate

A CL reaction of aryloxalate is the so-called peroxyoxalate chemiluminescence (PO-CL) reaction. PO-CL is one of the most efficient and versatile CL systems for HPLC. In this system, aryloxalates or oxamides react with hydrogen peroxide to produce a high-energy intermediate 1,2-dioxetanedione. This intermediate transfers its energy to a coexisting fluorophore, and then emission of light is observed from the excited fluorophore (Fig. 2). The total emission of light is proportional to the concentration of each substrate (i.e., oxalates, hydrogen peroxide, fluorophores, or basic catalysts). Therefore, these compounds can be determined by a suitable CL detection system. By using the catalytic effects of bases in this CL reaction, various amines can also be determined by HPLC. Many compounds can be used in the PO-CL reaction system, most of which can be detected sensitively. This is one of the major reasons why the PO-CL reaction has been exclusively used for postcolumn CL detection in HPLC.

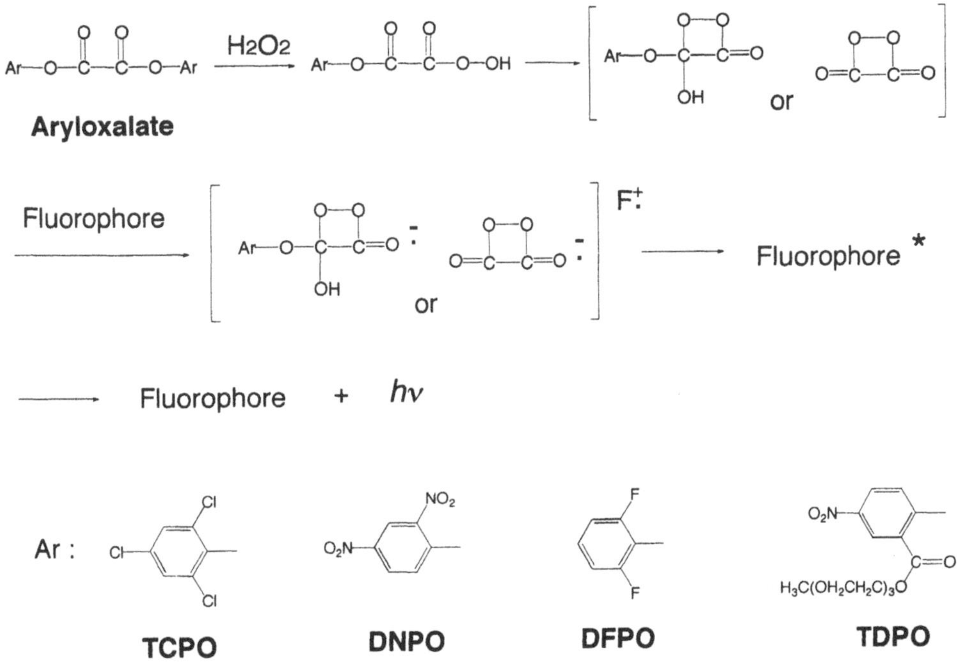

Figure 2 Aryloxalates and the PO-CL reaction. TCPO, *bis*(2,4,6-trichlorophenyl)oxalate; DNPO, *bis*(2,4-dinitrophenyl)oxalate; DFPO, *bis*(2,6-difluorophenyl)oxalate; TDPO, *bis*[2-(3,6,9-trioxadecyloxycarbonyl)-4-phenyl]oxalate.

4. CONSTRUCTION OF HPLC-CL DETECTION SYSTEMS

An HPLC-CL detection system is constructed by considering the conditions for HPLC separation, efficiency of CL reaction, and stabilities of reagents. The three systems shown in Figure 3 are most widely employed.

    System A is very simple and used for delivering a single solution of postcolumn reagent(s). This system is used in a case; after separation, hydrogen peroxide is determined by using a mixture of an aryloxalate and a fluorophore for postcolumn CL reaction.

    System B is the most widely used for CL detection after an HPLC separation. In this system, two pumps are required for delivering the reagent solutions in the following cases: (1) the solutions for CL reaction are first combined and then mixed with an eluent; (2) CL reaction conditions (e.g., pH, water and organic solvent contents, and salt concentration) need to be optimized before mixing with the CL reagent.

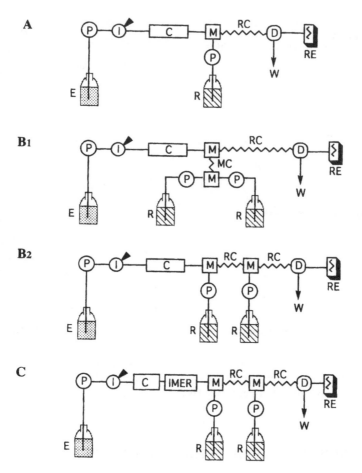

Figure 3 CL detection systems in combination with HPLC. P, pump: I, injector; C, column; M, mixing tee; D, detector; RC, reaction coil; MC, mixing coil; RE, recorder; E, eluent; R, reagent; W, waste; IMER, immobilized enzyme reactor.

System C is used when an immobilized enzyme reactor (IMER) is introduced into system B. The analyte(s) separated by HPLC is converted to a suitable species for CL detection with an IMER, and then mixed with the CL reagent. In this system, a buffer solution as a mobile phase and an ion-exchange-type column are preferable for an enzyme reaction.

More complicated systems will be required when a gradient elution is incorporated for the separation of analyte(s). However, more simple systems are preferable in view of operations, maintenance, and running costs of analysis.

5. DESIGN TO INCREASE THE EFFICIENCY OF CL REACTIONS

To achieve a sensitive and selective determination of analyte(s), resolution of HPLC separation and sensitivity of CL detection are the major critical factors. Many attempts to increase the sensitivity, namely to improve the efficiency of the CL reaction, have been made.

5.1 Devices

5.1.1 Mixing Device

In the postcolumn CL reaction, the thorough and rapid mixing of the column eluate with reagent solution(s) is an important factor to obtain a stable baseline and reproducible peak heights. In general, a T-shaped joint or other mixing devices for gradient elution are employed as mixing devices for the postcolumn CL reaction. However, several mixing devices to improve the CL reaction efficiency have also been proposed [4–7].

As shown in Figure 4, a mixing device with a small vessel, in which two flowing solutions are caused to rotate (A), was developed [6]. Another rotating flow mixing device (B) having three directional inlets, in which three solutions are mixed and flowed out through an outlet at the center of the top of the vessel, was also reported [7]. By using these devices, the CL reaction conditions were investigated from the standpoint of the improvement on the signal-to-noise (S/N) ratio and peak broadening.

5.1.2 Reaction Coil

In general, the CL reaction for HPLC requires rapid reaction accompanied by intense CL. Taking into account the sensitivity, it is preferable to measure CL at the time that the maximum CL intensity is observed. However, it should be considered that a large background signal may be observed simultaneously. Therefore, the time for measuring CL should be set to give the maximum S/N ratio. The adjustment of length and diameter of the reaction coil that is introduced between a mixing device and a detector is effective to obtain an optimized CL reaction time.

5.1.3 Electronic Device to Improve S/N Ratio

Electronic devices that improve the S/N ratio by filtering electronic signals from the CL detector are commercially available. Devices using both the analog and digital technologies are currently obtainable. These devices are inserted between a detector and a recorder, and treat electronic signals from a detector to give a

(A)   **(B)**

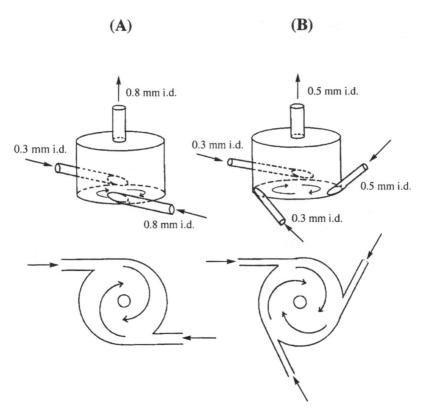

Figure 4 Rotating flow mixing devices having two (A) and three (B) directional inlets.

chromatogram with an improved S/N ratio. Typical results are shown in Figure 5 [8]. The S/N ratio increased 5–10 times by using this device for the detection of hydrogen peroxide in PO-CL.

5.1.4 Flow Cell and Flow Rate

A spiral coil of Teflon, or a spiral groove on a stainless steel surface covered with quartz, is used as a flow cell in the CL detector. These cells are placed in front of a photomultiplier tube (PMT), which detects photons emitted by the CL reaction. The cell volumes are generally in the range of 60–120 μL.

A joining part of the spiral coil or the quartz part that is in contact with stainless steel in the flow cell might not withstand high levels of pressure (generally smaller than 10 kg/cm^2); hence careful operation to prevent excess flow is

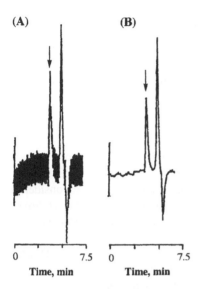

Figure 5 Chromatograms of hydrogen peroxide obtained without (A) and with (B) a signal cleaner. HPLC conditions: peak (2.5 pmol of hydrogen peroxide on column); eluent, 50 mM imidazole-HNO$_3$ (pH 7.0)/CH$_3$CN (=87:13, v/v); postcolumn CL reagent, 0.6 mM TDPO in CH$_3$CN/ 0.5 μM in CH$_3$CN (=50:50, v/v); column, TSK-gel ODS-80TM (250 × 4.6 mm id). (From Ref. 8.)

required. It should be noted that precipitation of salts in a mobile phase or postcolumn CL reagents might increase the pressure to cause breakage of the flow-cell. A suitable flow rate of a mobile phase and postcolumn CL reagent(s) must be determined under a user-specified pressure to prevent damage of the flow cell by considering a mixing efficiency and a CL reaction yield.

5.1.5 Pump

The use of the pumps capable of delivering pulseless flows is desired for the mobile phase and CL reagent(s) to obtain a stable and reproducible chromatogram. For this purpose, a dual-piston pump with minor pump pulsation or a syringe-type pump without pulsation are preferable. Pumps for an HPLC separation with high-performance ability are often used to deliver postcolumn CL reagent(s). However, these pumps are rather expensive and plural pumps will be required when delivering many reagent solutions. To overcome this, attempts to effectively delivering two reagent solutions by using an inexpensive dual-head reciprocating piston pump with a short stroke (4.9 μl per stroke) were made [9].

5.2   Optimization of Reaction Conditions

To achieve highly sensitive detection, optimization of various factors affecting the CL reaction is required. Reaction temperature, pH, solvent, nature of CL compounds, and coexisting compounds such as a catalyst and an enhancer affect the CL reaction yield.

5.2.1   Choice of the CL reagent

The choice of a CL reagent has a serious influence on the detectability of analytes. Therefore, applicability of a CL reagent to a given analyte and method, stability, and ease of preparation as well as CL reaction efficiency should be considered when selecting a reagent. Although novel CL reagents such as adamantyl dioxetane derivatives have been developed, only a few of them can be combined with HPLC.

Some derivatives of luminol and acridine are currently used for the HPLC-CL detection. They are prepared by changing a substituent or introducing a new functional group into the luminol or acridine skeleton to give better CL efficiency or reactivity with analyte(s).

In the PO-CL system, efforts to increase the solubility of aryloxalate in the mobile phases employed for HPLC and to improve the fluorescence quantum yield of the fluorophore have been made. Many fluorophores have been developed for the PO-CL system and applied to various analytes [10].

5.2.2   Temperature

The fluctuation of the temperature in the mixing device, reaction coil, and flow cell affects the CL reaction velocity and emission duration thus affecting the sensitivity and reproducibility. Therefore, it is preferable to keep a constant temperature [11].

5.2.3   Solvent

The nature of the employed organic solvents affects the efficiency of the PO-CL reaction. Organic solvents such as acetonitrile and methanol utilized frequently for reversed-phase HPLC are well suited for a PO-CL reaction. Generally, a higher content of these solvents in the mobile phase gives more intense CL. Effects of organic solvents' nature on the stability of postcolumn reagents were investigated and it was found that a mixture of hydrogen peroxide and *bis*(2,4,6-trichlorophenyl)oxalate (TCPO) was most stable in acetonitrile. When *bis*[4-nitro-2-(3,6,9-trioxadecylcarbonyl)phenyl] oxalate (TDPO) and hydrogen peroxide dissolved in a mixture of ethyl acetate and acetonitrile were used, the addition of phthalic acid esters to this solution decreased the noise level of CL [7].

5.2.4   Catalyst

Many compounds are known to catalyze CL reactions. As these catalysts act at trace amounts, the CL reaction can be applied to determination of these substances. Since most of the CL reactions concern oxidation reactions, compounds catalyzing the oxidation process have been well investigated.

For example, peroxidase catalyzes the reaction of luminol derivatives with hydrogen peroxide and results in an increase of the CL reaction velocity and CL intensity. Therefore, intense CL can be obtained from the analyte labeled with luminol derivatives after HPLC separation, followed by reaction with peroxidase.

On the other hand, several oxidases are known to generate hydrogen peroxide, acting as an oxidant in the CL system, from corresponding substrates. IMERs in which the oxidases are immobilized on adequate supporting materials such as glass beads have been developed. IMERs are often used for flow injection with CL detection of uric acid and glucose, and are also applicable to the CL determination of acetylcholine, choline, polyamines, enzyme substrates, etc., after online HPLC separation.

5.2.5   Enhancer

As compounds exhibiting enhancing effects on CL reactions, a variety of phenols, e.g., firefly luciferin and 6-hydroxybenzothiazole derivatives [12, 13], 4-iodophenol [14], 4-(4-hydroxyphenyl)thiazole [15], 2-(4′-hydroxy-3′-methoxy-benzylidene)-4-cyclopentene-1,3-dione (KIH-201) [16], and 2-(4-hydroxyphenyl)-4,5-diphenylimidazole (HDI) and 2-(4-hydroxyphenyl)-4,5-di(2-pyridyl)imidazole (HPI)[17] (Fig. 6A), and phenylboronic acid derivatives, e.g., 4-phenylylboronic acid [18], 4-iodophenylboronic acid [19], and 4-[4,5-di(2-pyridyl)-1 H-imidazol-2-yl]phenylboronic acid (DPPA) [20] (Fig. 6B), in the luminol/hydrogen peroxide/peroxidase system are well known. Rhodamine B and quinine are used as sensitizers in the CL-emitting reaction between cerium (IV) and thiol compounds. This CL reaction was successfully applied to the sensitive determination of various thiol drugs [21–32].

Fluorophores having lower oxidation potentials also enhance the PO-CL reaction. The addition of these compounds to the postcolumn reagent is very effective so as to increase the sensitivity. The attempt to enhance CL intensity using micelles of surfactants has also been reported [33], but it has yet not been applied to HPLC.

5.2.6   Buffer and pH

Ion species present in the CL reaction mixture sometimes affect the CL intensity. In the PO-CL system, phosphate buffer has a tendency to decrease the CL intensity, but borate and imidazole buffers provide more intense CL.

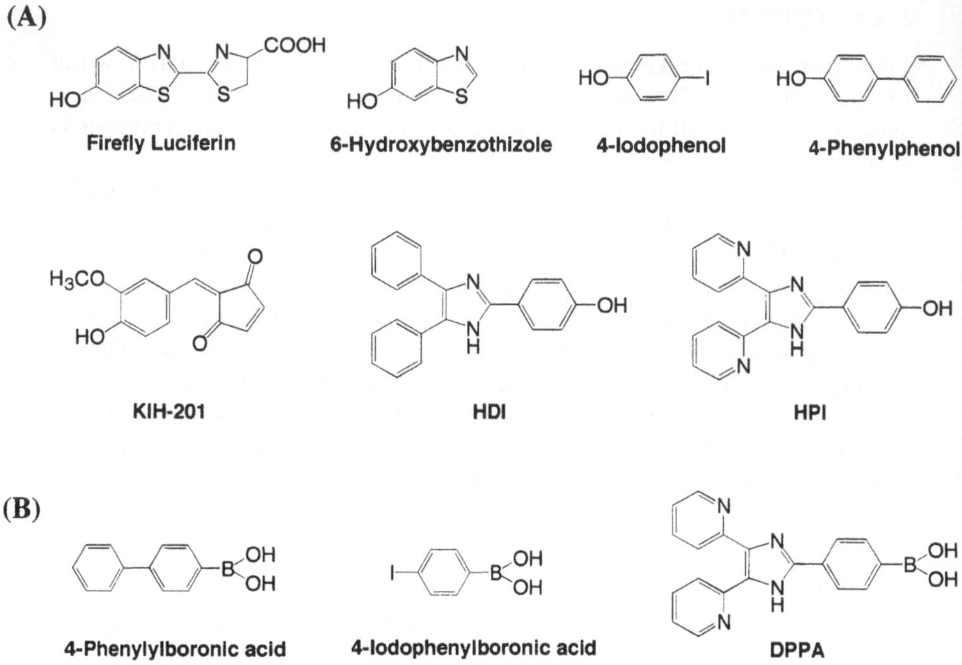

(A)

Firefly Luciferin   6-Hydroxybenzothizole   4-Iodophenol   4-Phenylphenol

KIH-201   HDI   HPI

(B)

4-Phenylylboronic acid   4-Iodophenylboronic acid   DPPA

Figure 6   Representative (A) phenol-type and (B) phenylboronic acid-type enhancers for luminol/hydrogen peroxide/peroxidase system. KIH-201, 2-(4′-hydroxy-3′-methoxy-benzylidene)-4-cyclopentene-1,3-dione; HDI, 2-(4-hydroxyphenyl)-4,5-diphenylimidazole; HPI, 2-(4-hydroxyphenyl)-4,5-di(2-pyridyl)imidazole; DPPA, 4-[4,5-di(2-pyridyl)-1*H*-imidazol-2-yl]phenylboronic acid).

In general, halogen ions have a decreasing effect upon the CL intensity. Salt concentration and buffer pH are also known to affect the CL reaction. The choice of buffer and its pH suited to each CL reaction system are hence very important to obtain intense CL.

6.   APPLICATIONS

6.1   HPLC-CL Detection Using Luminol Derivatives

The structures of luminol derivatives used for HPLC-CL detection are shown in Figure 7A. Analytes labeled with luminol derivatives can be detected using hydrogen peroxide and potassium hexacyanoferrate(III) under alkaline conditions after HPLC separation (Table 1). For example, ibuprofen in saliva [34], saturated

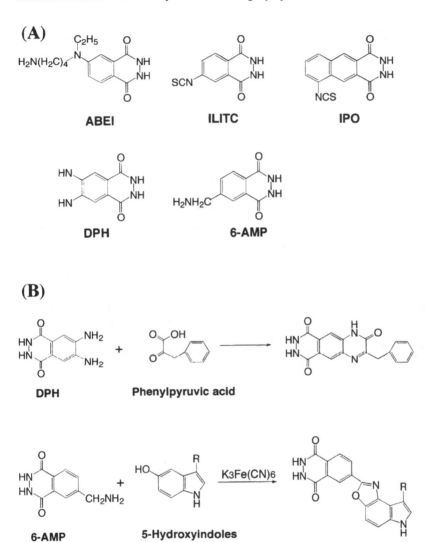

Figure 7 (A) Luminol-type CL reagents and (B) derivatization reactions for DPH and 6-AMP.

Table 1 Application of Luminol Derivatives as CL Labeling Reagents to HPLC

Analyte	Labeling reagent	Separation column	Detection limit (on column)	Ref.
Ibuprofen	ABEI	PLRP-S (150 × 4.6 mm id)	85 fmol	34
Cholic acid	ABEI	Spherical C_{18} (150 × 4.6 mm id)	20 fmol	35
Eicosapentaenoic acid	ABEI	Nova Pak C_{18} (150 × 3.9 mm id)	200 fmol	36
Methamphetamine	ABEI	Shimpack CLC-C_{18} (150 × 6.0 mm id)	20 fmol	37, 38
Amino acids	ILITC	Bio-Sil ODS-5s (150 mm)	10 fmol (average)	39
Primary and secondary amines	IPO	TSK gel ODS-120T (150 × 4.6 mm id)	30–120 fmol (primary amines) 0.8–3 fmol (secondary amines)	40
Maprotiline	IPO	TSK gel ODS-80 (150 × 4.6 mm id)	1.5 fmol	41
α-Keto acids and α-dicarbonyl compounds	DPH	TSK gel ODS-120T (250 × 4.6 mm id)	4–50 fmol (α-keto acids) 1.1–300 fmol (α-dicarbonyl compounds)	45, 46
3α, 5β-Tetrahydroaldosterone	DPH	TSK gel ODS-120T (250 × 4.6 mm id)	1.5 fmol	48
Dexamethasone	DPH	TSK gel ODS-120T (250 × 4.6 mm id)	8 fmol	49
5-Hydroxyindoles	6-AMP	Capcell Pak C_{18} (250 × 4.6 mm id)	0.7–4 fmol	50

and unsaturated fatty acids [35, 36], and central nervous stimulants [37, 38] could be detected at fmol levels using *N*-(4-aminobutyl)-*N*-ethylisoluminol (ABEI) as a labeling reagent. For labeling of primary and secondary amines, ABEI is used with *N, N'*-disuccinimidyl carbodiimide (DSC) as a condensing reagent. 2-Chloro-1-methylpyridinium iodide (CMPI) and 3,4-dihydro-2*H*-pyrido[1,2-α]pyrimidin-2-one are employed as condensing reagents to label carboxylic acids with ABEI (Fig. 8). 4-Isothiocyanatophthalhydrazide (ILITC) [39] and 6-isothiocyanatobenzo[g]phthalazine-1,4 (2*H*,3*H*)-dione (IPO) have been developed for the labeling of amines [40]. Femtomol levels of amino acids and amines could be detected by use of these reagents. IPO has also been applied to analysis of maprotiline, a widely used antidepressant, in human plasma [41].

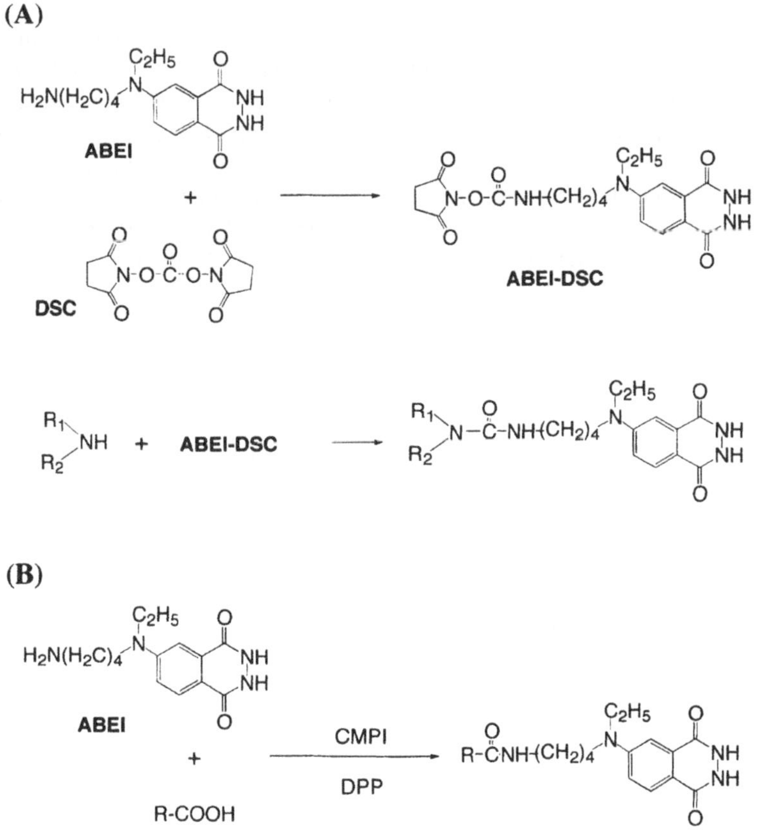

(A)

(B)

Figure 8 Labeling reaction of ABEI with (A) primary and secondary amines, and (B) carboxylic acids.

Hydrogen peroxide and peroxides can also be detected utilizing the CL from luminol derivatives. Determination of peroxides is very important to obtain information on human diseases such as a atherosclerosis and aging, and also on food spoilage. Lipid hydroperoxides such as phosphatidylcholine hydroperoxide in rat tissues separated by HPLC were successfully determined by using the CL generated by the reaction of luminol with cytochrome c or microperoxidase [42]. An HPLC determination of hydrogen peroxide with a cation-exchange gel column was examined and a detection limit of 4 pmol was obtained. This method was successfully applied to determination of hydrogen peroxide in coffee drinks [43].

An IMER immobilizing 3 α-hydroxysteroid dehydrogenase was employed for HPLC determination of cholic acid and a detection limit of 2 pmol was achieved [44]. This approach has the advantage of permitting a repeatable use of the enzyme.

Methods measuring CL via the luminol-type reaction products from analytes have been proposed. Derivatization reactions of analytes with luminol-type reagents are shown in Figure 7B. α-Keto acids including phenylpyruvic acid and α-dicarbonyl compounds were reacted with 4,5-diaminophthalhydrazide (DPH) under different conditions to give corresponding chemiluminescent luminol-type compounds. The detection limits of eight biologically important α-keto acids and five α-dicarbonyl compounds (phenylglyoxal, diacetyl, 2,3-pentanedione, 2,3-hexanedione, and 3,4-hexanedione) were in the range of 4–50 fmol [45] and 1.1–300 fmol per injection [46], respectively. N-Acetylneuraminic acid (NANA) derivatized with DPH was determined with a detection limit of 9 fmol [47]. DPH has also been extended to the sensitive determination of 3α, 5β-tetrahydroaldosterone [48] and dexamethasone in human plasma [49]. A new luminol-type reagent, 6-aminomethylphthalhydrazide (6-AMP), was synthesized as a CL derivatization reagent for 5-hydroxyindoles (Fig. 7B) [50]. The detection limits for 5-hydroxyindoles were in the range of 0.7–4 fmol (S/N = 3) (Fig. 9). The advantages of these methods using DPH and 6-AMP are high sensitivity and selectivity due to the unique derivatization reaction.

6.2 HPLC-CL Detection Using Lucigenin Derivatives

Reducing sugars can be determined using lucigenin; oxidized products of reducing sugars with sodium periodate react with lucigenin to generate an intense CL. This phenomenon is based on the reaction of lucigenin with the α-hydroxycarbonyl group. Therefore, compounds such as glyceraldehyde, cortisol, phenacylalcohols, and phenacylesters, which contain an α-hydroxycarbonyl group in their structure, are determined sensitively by this system. Detection limits of corticosteroids and p-nitrophenacyl esters were reported to be ca. 0.5 pmol per injection [51].

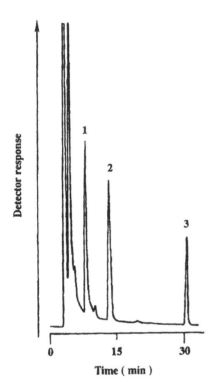

Figure 9 Chromatogram of 5-hydroxyindoles derivatized with 6-AMP. Peaks (2.5 pmol each on column): 1 = 5-hydroxytryptophan; 2 = serotonin; 3 = 5-hydroxyindole-3-acetic acid. (From Ref. 50.)

An HPLC-CL determination of environmentally important chlorophenols was reported by using 10-methyl-9-acridinium carboxylate as a CL label. A two-step derivatization was used to produce the CL derivatives (Fig. 10). Following the separation under reversed-phase conditions, the CL reaction was performed by the base-catalyzed postcolumn oxidation. The quantum efficiency was dependent on the species of analytes. The detection limit of chlorophenols (S/N = 3) ranged from 300 amol to 1.25 fmol per injection (Fig. 11) [52].

6.3 HPLC-PO-CL Detection

The nature of the aryloxalate used in a PO-CL system is an important factor in view of the produced CL intensity. Several aryloxalates, e.g., *bis*(2,4,6-trichlorophenyl)oxalate (TCPO), *bis*(2,4-dinitrophenyl)oxalate (DNPO), *bis*(2,6-difluorophenyl)oxalate (DFPO), *bis*(pentafluorophenyl)oxalate (PFPO), and *bis*[2-

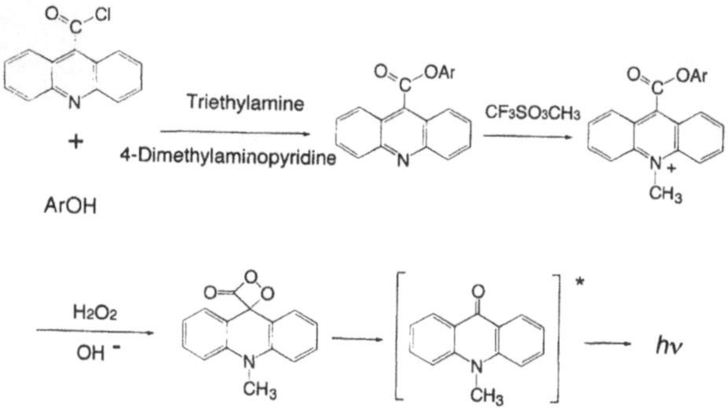

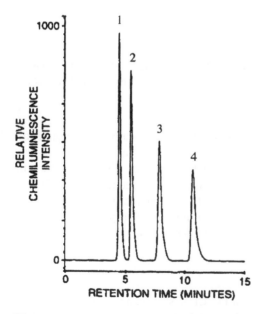

Figure 10 CL derivative of chlorophenol and its CL reaction.

Figure 11 Chromatogram of chlorophenols. Peaks (40 fmol each on column): 1 = chlo-rophenol; 2 = 4-chlorophenol; 3 = 2,4-dichlorophenol; 4 = 2,4,6-trichlorophenol. (From Ref. 52.)

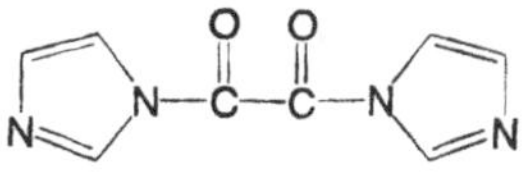

ODI

Figure 12   Structure of ODI.

(3,6,9-trioxadecyloxycarbonyl)-4-phenyl]oxalate (TDPO), were evaluated on their properties [53–55]. Among them, TCPO, TDPO, and DNPO are the most frequently used. Recently, 1,1'-oxalyldiimidazole (ODI) was applied to determine hydrogen peroxide, and the results showed that ODI was about 10 times more sensitive than TCPO (Fig. 12) [56].

In the PO-CL system, the compounds showing native fluorescence or that fluoresce after chemical derivatization can be detected. As examples of the PO-CL detection of native fluorescence compounds, dipyridamole and benzydamine in rat plasma [57] and fluphenazine [58] have been reported; in the former method, the detection limits of dipyridamole and benzydamine were 345 pM and 147 nM in plasma, respectively. Diamino- and aminopyrenes were sensitively determined using TCPO and their detection limits were in the sub-fmol range [59]. Carcinogenic compounds such as 1- nitropyrene and its metabolites, can also be determined by the HPLC-PO-CL system. Nonfluorescent nitropyrenes were converted into the corresponding fluorescent aminopyrenes by online reduction on a Zn column followed by detection; 2–50-fmol detection limits were achieved in the determination of ethanol extracts from airborne particulates (Fig. 13) [60].

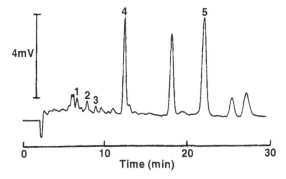

Figure 13   Chromatogram of an airborne particulate sample. Peaks: 1 = 1,6-dinitropyrene; 2 = 1,8-dinitropyrene; 3 = 1,3-dinitropyrene; 4 = 2-fluoro-7-nitrofluorene; 5 = 1-nitropyrene. (From Ref. 60.)

Table 2 Application of Fluorescence Derivatization Reagents to HPLC with PO-CL Detection

Analyte	Fluorophore[a]	Aryloxalate[b]	Detection limit (on column)	Ref.
Catecholamines	Fluorescamine	TCPO	25 fmol	61
Histamine	Fluorescamine	TCPO	9.0 pmol	62
Amino acids	DNS-Cl	TCPO	10 fmol	63
	DNS-Cl	TCPO	2–5 fmol	64
	DNS-Cl	TCPO	0.2 fmol	67
Methamphetamine	DNS-Cl	TCPO	4 fmol	68
Mexiletine	DNS-Cl	TDPO	1.0 fmol	69
Amphetamines	DNS-Cl	TCPO	3–4 fmol	70
	NDA	TCPO	0.2 fmol	70
Methamphetamines	DNS-Cl	TCPO	10–30 fmol	71
	NDA	TCPO	0.3–1.5 fmol	71
Catecholamines	NDA	DNPO	1 fmol	72
Methamphetamines	DBD-F	TDPO	25–133 fmol	73
Amino acids	DBD-F	TDPO	37 fmol (Ile), 73 fmol (Ala)	74
Epinephrine	DBD-F	TDPO	80 fmol	74
Metoprolol	DBD-F	TDPO	5.9 fmol	75
Primary and secondary amines	Luminarin 1	TCPO	15–100 fmol	76

Prostaglandin E$_2$	Luminarin 4	TCPO	32 fmol	77
Arachidonic acid metabolite	DMQPH	TCPO	500 amol	78
Saturated fatty acid	HCPI	TCPO	12–18 fmol	79
Biological thiols	DBPM	TCPO	7–113 fmol	80
Oxosteroids	DNS-H	TDPO	2–4 fmol	81
Hyaluronic acid	DNS-H	TDPO	100 fmol	82
Glycosaminoglycans	DNS-H	TDPO	100 fmol	83
Medroxyprogesterone	DBD-H	TDPO	8.7 fmol	84
Propentofylline	DBD-H	TDPO	30 fmol	85
Malondialdehyde	DETBA	TDPO	20 fmol	86
Fluoropyrimidines	DCIA	TCPO	20–40 fmol	87
Catecholamines	Ethylenediamine	TDPO	1 fmol	88
Catecholamines	m-CED	TDPO	40–120 amol	90

[a] DNS-Cl = dansyl chloride, NDA = naphthalene-2,3-dicarboxaldehyde, DBD-F = 4-(N,N-dimethylaminosulphonyl)-7-fluoro-2,1,3-benzoxadiazole, DMQPH = 6,7-dimethoxy-1-methyl-2(1H)-quinoxalinone-3-proprionylcarboxylic acid hydrazide, HCPI = 2-(4-hydrazinocarbonylphenyl)-4,5-diphenyli-midazole, DBPM = N-[4-(6-dimethylamino-2-benzofuranyl)phenyl]maleimide, DNS-H = dansyl hydrazine, DBD-H = 4-(N,N-dimethylaminosulphonyl)-7-hydrazino-2,1,3-benzoxadiazole, DETBA = 1,3-diethyl-2-thiobarbituric acid, DCIA = 7-dimethylamino-3-{4-[(iodoacetyl)amino]phenyl}-4-methylcou-marin, m-CED = 1, 2-bis(3-chlorophenyl)ethylenediamine.

[b] TCPO = bis(2,4,6-trichlorophenyl)oxalate, TDPO = bis[2-(3,6,9-trioxadecyloxycarbonyl)-4-phenyl]oxalate, DNPO = bis(2,4-dinitrophenyl)oxalate.

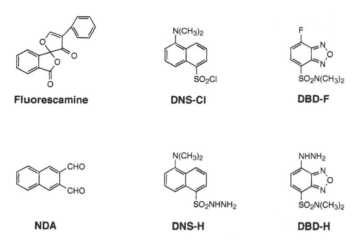

Figure 14 Representative fluorescence derivatization reagents used for PO-CL detection. DNS-Cl, dansyl chloride; DBD-F, 4-(*N,N*-dimethylaminosulphonyl-7-fluoro-2,1, 3-benzoxadiazole; NDA, naphthalene-2,3-dicarboxaldehyde; DNS-H, dansyl hydrazine; DBD-H, 4-(*N,N*-dimethylaminosulphonyl-7-hydrazino-2,1,3-benzoxadiazole).

For analysis of nonfluorescent compounds, many fluorescence derivatization reagents can be used for HPLC with PO-CL detection (Table 2). Representative fluorescence derivatization reagents are shown in Figure 14. Dansyl derivatives such as dansyl chloride (DNS-Cl) and dansyl hydrazine (DNS-H) are often utilized as fluorescence labeling reagents. DNS-Cl has been utilized for PO-CL detection of amino acids [7, 63–66]. A highly sensitive microbore HPLC-PO-CL detection technique for DNS amino acids was also investigated and the detection limits obtained were 0.8–1.7 fmol [91]. Determination of methamphetamine and its metabolites was investigated and determination limits as low as 10–30 fmol were obtained [71]. For labeling of the carbonyl group, DNS-H is widely used. Oxosteroids and oxobile acid ethyl esters [81], hyaluronic acid (HA) in blood plasma [82], HA, chondroitin sulfate, and dermatan sulfate [83] were successfully determined using DNS-H.

Femtomol levels of detection limits were also achieved in the determination of stimulant amines with the benzofurazan derivative 4-(*N,N*-dimethylaminosulphonyl)-7-fluoro-2,1,3-benzoxadiazole (DBD-F) [73]. DBD-F was successfully applied to the PO-CL detection of amino acids and epinephrine [74] and a β-blocker, metoprolol [75]. 4-(*N,N*-Dimethylaminosulphonyl)-7-hydrazino-2,1,3-benzoxadizole (DBD-H) has also been used for PO-CL determination of a neuronal cell protective compound, propentofylline. The method was applied for the first time to determine propentofylline concentration in the dialysate obtained from the rat hippocampus [85].

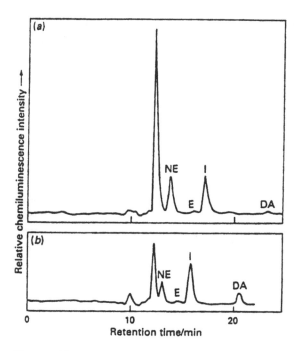

Figure 15 Chromatograms of catecholamines obtained from (a) human and (b) Sprague-Dawley rat plasma. Peaks: NE = norepinephrine; E = epinephrine; I = 3,4-dihydroxybenzylamine; DA = dopamine. (From Ref. 88.)

Naphthalene-2,3-dicarboxaldehyde (NDA), which reacts with primary amines to give highly fluorescent cyanobenz[f]isoindole (CBI) derivatives, has also been used for determination of amphetamine-related compounds [70], dopamine, and norepinephrine [72]; sub-fmol to fmol amounts of these were detected by HPLC with the PO-CL detection.

A highly sensitive method was developed for determination of plasma catecholamines. These analytes were derivatized online with ethylenediamine to give the corresponding intensely fluorescent compounds, and then detected by the PO-CL reaction [88]; the detection limit for all the catecholamines obtained was 1 fmol (S/N = 2) (Fig. 15). Detection limits of amol levels (S/N = 3) for catecholamines were obtained by the precolumn derivatization method using 1,2-diarylethylenediamine derivatives as fluorescence derivatization reagents [89, 90]. In these methods, TDPO permitting a highly sensitive determination of catecholamines was used as the aryloxalate.

Hydrogen peroxide was determined with fmol detection limits by using rhodamine 6G and pyrimidopyrimidine derivatives as fluorescent enhancers. The method employing the latter reagent was applied to cola drinks [92]. Sensitive determination of hydrogen peroxide was also reported using ODI in a PO-CL system. In this system, solid-phase detection reactor immobilizing lophine derivatives were introduced and a 10-fmol detection limit for hydrogen peroxide was obtained [93]. Phospholipids separated by preparative HPLC were converted to give hydrogen peroxide by FIA with an IMER, in which phospholipase D and choline oxidase were immobilized, and then determined [94]. Certain amines are known to promote the PO-CL reaction, which was utilized for determination of polyamines in tomatoes [95].

Efforts to find and develop new fluorophores being efficiently chemically excited in the PO-CL reaction were carried out. Pyrimido[5,4-*d*]pyrimidines together with several fluorescent compounds were evaluated [96]; 2,6-*bis*[di-(2-hydroxyethyl)amino]-4,8-dipiperidinopyrimido[5,4-*d*]pyrimidine (Dipyridamole) and 2,4,6,8-tetrathiomorpholinopyrimido[5,4-*d*]pyrimidine (TMP) gave intense CL, the signals being larger than with any other commercially available fluorescent compound tested (Fig. 16). These pyrimido[5,4-*d*]pyrimidines were also applied to a PO-CL photographic assay of hydrogen peroxide and glucose by

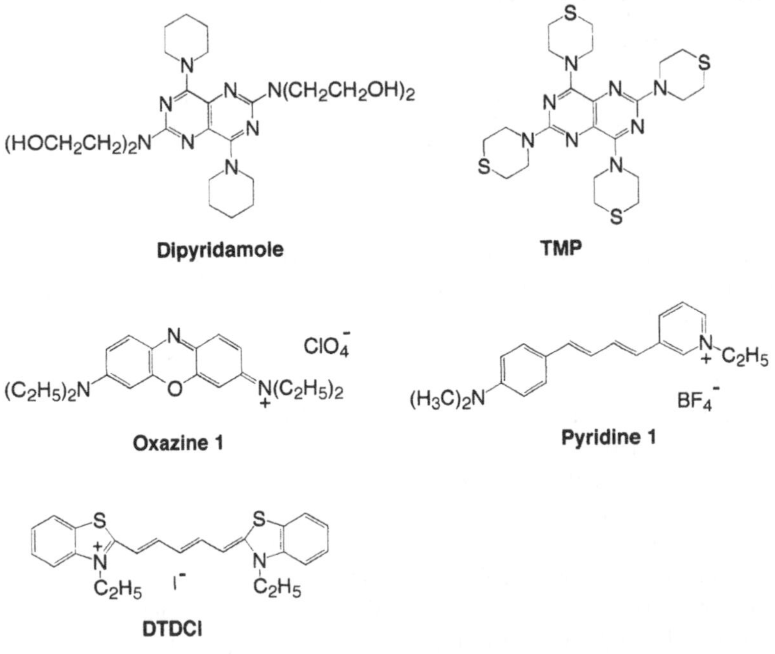

Figure 16 Fluorophores with efficient chemical excitation in the PO-CL reaction. TMP, 2,4,6,8-tetrathiomorpholinopyrimido[5,4-*d*]pyrimidine; DTDCI, 3,3'-diethylthiadicarbocyanine iodide.

using a water-soluble oxamide, 4,4′-oxalyl-*bis*[(trifluoromethylsulfonyl)imino]-trimethylene-*bis*(4-methylmorpholinium)trifluoromethanesulfonate (MPTQ) [97]. The highly sensitive detection of near-infrared (near-IR) fluorescent dyes using HPLC with PO-CL detection was examined [98]. These dyes are assumed to be suitable for PO-CL detection owing to their low singlet excitation energy. The detection limits for methylene blue, pyridine 1, oxazine 1 and 3,3′-diethylthiadi-carbocyanine iodide (DTDCI) were 120, 27, 31, and 0.19 fmol on column, respectively. DTDCI was found to be the preferred structure for PO-CL detection and its sensitivity was 250 times that obtained by HPLC with conventional fluorescent detection (Fig. 16).

6.4 HPLC-CL Detection Using Ruthenium Complex

A unique CL reagent, *tris*(2,2′-bipyridyl)ruthenium(II) [Ru(bpy)3^{2+}] for the post-column CL reaction, was applied to HPLC detection. The oxidative-reduction reaction scheme of CL from Ru(bpy)$_3^{2+}$ is shown in Figure 17. When the production of light following an oxidation of Ru(bpy)$_3^{2+}$ to Ru(bpy)$_3^{3+}$ at an electrode surface is measured, this CL reaction is termed electrogenerated chemiluminescence (ECL). The CL intensity is directly proportional to the amount of the reductant, that is, the analyte.

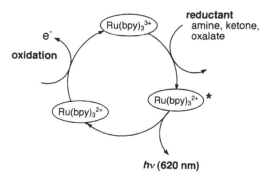

Figure 17 Ru(bpy)$_3^{2+}$ and its CL reaction.

Table 3 Drug Analysis by HPLC with CL Detection Using Ruthenium Complex

Drug	Separation column	Detection	Detection limit (on column)	Ref.
Antihistamines	Asahipak ODP-50 C$_{18}$ (150 × 4.6 mm id)	ECL	0.09–0.21 µg/mL (8–16 pmol)	99
Anticholinergic drugs	PRP-1 (150 × 2 mm id) or Deltabond octyl silica (150 × 2 mm id)	ECL	0.1–1 µg/mL (0.4–3 pmol)	100
Erythromycin	Bioanalytical Unijet C$_{18}$ (150 × 1 mm id)	ECL	7.4 ng/mL (50 fmol)	101
Erythromycin derivative (EM523)	Inertsil ODS-3 (150 × 4.6 mm id)	CL (oxidizing reagent and light irradiation)	1 ng/mL plasma 10 ng/mL urine	102

Since the order of increasing CL intensity for alkyl amines reacted with $Ru(bpy)_3^{2+}$ is tertiary amines > secondary amines > primary amines, pharmaceutical compounds bearing a tertiary amine function (e.g., antihistamine drugs [99], anticholinergic drugs [100], erythromycin [101], and its derivatives [102]) have been sensitively determined after HPLC separation (Table 3). The method was applied to the detection of D- and L-tryptophan (Trp) after separation by a ligand-exchange HPLC [103]. The detection limits for D- and L-Trp were both 0.2 pmol per injection. Oxalate in urine and blood plasma samples has also been determined by a reversed-phase ion-pair HPLC (Fig. 18) [104]. Direct addition of

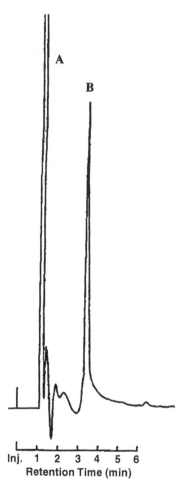

Figure 18 Chromatogram of oxalate in blood plasma. Peaks: A = amino acids; B = oxalate. (From Ref. 104.)

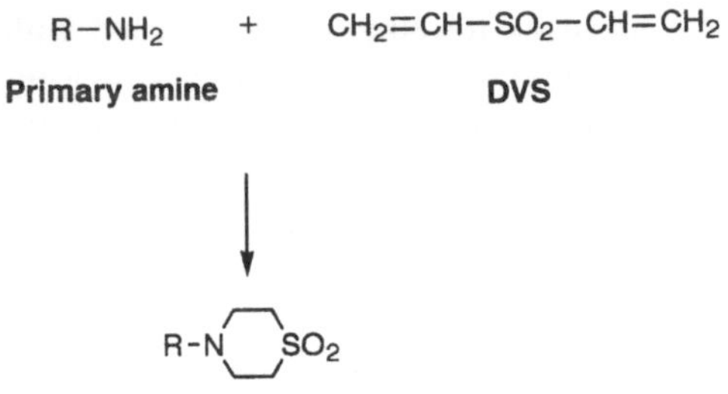

R—NH$_2$ +   CH$_2$=CH—SO$_2$—CH=CH$_2$

Primary amine   **DVS**

R-N⟨ ⟩SO$_2$

Figure 19 Derivatization of primary amines with DVS.

Ru(bpy)$_3^{2+}$ to the mobile phase was investigated and compared with conventional postcolumn Ru(bpy)$_3^{2+}$ addition. The detection limit using oxalate standards with Ru(bpy)$_3^{2+}$ in the mobile phase was below 0.1 μM, which was significantly superior to the postcolumn technique. The mobile-phase addition method allowed the instrumentation to be simplified and reduced band broadening caused by postcolumn mixing.

Amino acids labeled with DNS-Cl were determined using the Ru(bpy)$_3^{2+}$ CL reaction after HPLC separation with a reversed-phase column [104, 105]. DNS derivatives are expected to produce intense CL owing to their secondary and tertiary amino groups. The detection limit for DNS-Glu was 0.1 μM (2 pmol/injection). Although underivatized amino acids could be detected by Ru(bpy)$_3^{2+}$ CL, the DNS derivatives showed improved detection limits by three orders of magnitude [105]. An approach to convert primary amines to tertiary amines was also reported [106]. In this method, divinyl sulfone (DVS) was used for a cycloaddition reaction of primary amines (Fig. 19). The DVS derivatives after HPLC separation were sensitively detected (e.g., detection limits for propylamine and 3-aminopentane were 30 and 1 pmol, respectively).

6.5 Other CL Detection Methods for HPLC

Adenine, guanine, and their nucleos(t)ides are known to react with glyoxal derivatives to give chemiluminescent compounds [107–110]. Structures of the chemiluminescent species of both nucleic acid bases are still unknown, but the possible pathway of the derivatization reaction with phenylglyoxal to produce CL is shown in Figure 19. The derivatization products exhibit intense CL in an alkaline medium in the presence of the aprotic polar solvent N,N'-dimethylformamide (DMF). Guanine-containing compounds separated by reversed-phase chromatography were detected with CL and their detection limits range from 4 to 53 pmol (Fig. 20) [110].

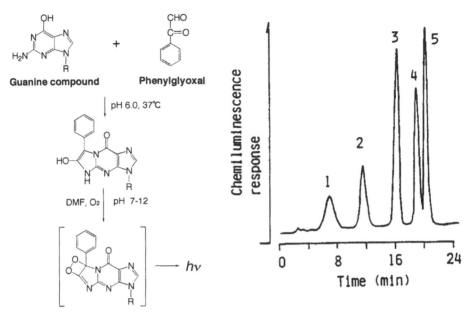

Figure 20 Possible pathway of the CL reaction between phenylglyoxal and guanine compound, and chromatogram of guanine compounds. Peaks: 1 = GTP; 2 = GMP; 3 = cGMP; 4 = guanosine; 5 = deoxyguanosine. (From Ref. 109.)

Methods for determination of thiol drugs (i.e., captopril [21–25], penicillamine [26–28], hydrochlorothiazide [24, 25, 29, 30], and tiopronin [31, 32]) have been developed. These methods are based on CL from a cerium (IV) oxidation system sensitized by adequate fluorophores such as quinine and rhodamine B. By using HPLC-coupled CL-flow-injection analysis method, tiopronin and its metabolite 2-mercaptopropionic acid in human urine were sensitively determined with the detection limits of 0.8 and 1 μM, respectively [32].

7. CONCLUSIONS

Research on the application of CL detection in HPLC is progressing rapidly owing to its great sensitivity and the simplicity of instrumentation. These advantages are also well suited for miniaturized separation techniques such as capillary liquid chromatography, capillary electrophoresis, and capillary electrochromatography, which will be further extended. Most of the CL detection systems for HPLC so far reported utilize long-standing CL compounds and reactions, and a few methods have been developed based on new principles. We expect that the efforts will be focused not only on applications but also on the development of new CL reactions and reagents. It is expected that in the near future, CL detection techniques will provide a wider range of applications in the fields of life sciences, environmental science, and other areas.

REFERENCES

1. LJ Kricka, GHG Thorpe. Analyst 108:1274–1296, 1983.
2. PJM Kwakman, UATh Brinkman. Anal Chim Acta 266:175–192, 1992.
3. K Nakashima, K Imai. Molecular Luminescence Spectroscopy, Part 3. New York: Wiley, 1993, pp 1–23.
4. RW Frei, L Michel, W Santi. J Chromatogr 126:665–677, 1976.
5. S Katz, WW Pitt Jr, G Jones Jr. Clin Chem 19:817–820, 1973.
6. S Kobayashi, K Imai. Anal Chem 52:1548–1549, 1980.
7. M Sugiura, S Kanda, K Imai. Biomed Chromatogr 7:149–154, 1993.
8. K Nakasima. Bunseki 7:518–524, 1996.
9. K Hayakawa, N Imaizumi, M Miyazaki. Biomed Chromatogr 5:148–152, 1991.
10. K Imai. Chromatogr Sci 48:359–379, 1990.
11. N Hanaoka. J Chromatogr 503:155–165, 1990.
12. TP Whitehead, GHG Thorpe, TJN Carter, C Croucutt, LJ Kricka. Nature (Lond) 305:158–159, 1983.
13. GHG Thorpe, LJ Kricka, E Gillespie, S Moseley, R Amess, N Baggett, TP Whitehead. Anal Biochem 145:96–100, 1985.
14. GHG Thorpe, LJ Kricka. Methods Enzymol 133:331–353, 1986.
15. M Ii, H Yoshida, Y Aramaki, H Masuya, T Hada, M Terada, M Hatanaka, Y Ichimori. Biochem Biophys Res Commun 193:540–545, 1993.
16. H Hori, T Fujii, A Kubo, N Pan, S Sako, C Tada, T Matsubara. Anal Lett 27: 1109–1122, 1994.
17. N Kuroda, R Shimoda, M Wada, K Nakashima. Anal Chim Acta 403:131–136, 2000.
18. LJ Kricka, X Ji. J Biolumin Chemilumin 10:49–54, 1995.
19. LJ Kricka, M Cooper, X Ji. Anal Biochem 240:11–125, 1996.
20. N Kuroda, K Kawazoe, H Nakano, M Wada, K Nakashima. Luminescence 14: 361–364, 1999.
21. XR Zhang, WRG Baeyens, G Van der Weken, AC Calokerinos, K Nakashima. Anal Chim Acta 303:121–125, 1995.
22. Z Xinrong, WRG Baeyens, G Van der Weken, AC Calokerinos, K Nakashima. J Pharm Biomed Anal 13:425–429, 1995.
23. ZD Zhang, WRG Baeyens, XR Zhang, G Van der Weken. J Pharm Biomed Anal 14:939–945, 1996.
24. J Ouyang, WRG Baeyens, J Delanghe, G Van der Weken, W Vandaele, D De Keukeleire, AM Garcia-Campaña. Anal Chim Acta 386:257–264, 1999.

25. J Ouyang, WRG Baeyens, J Delanghe, G Van der Weken, D De Keukeleire, W Vandaele, AM Garcia-Campaña, AC Calokerinos. Biomed Chromatogr 12:160–161, 1998.
26. ZD Zhang, WRG Baeyens, XR Zhang, G Van der Weken. Analyst 121:1569–1572, 1996.
27. ZD Zhang, WRG Baeyens, XR Zhang, G Van der Weken. Biomed Chromatogr 9:287–288, 1995.
28. ZD Zhang, WRG Baeyens, XR Zhang, Y Zhao, G Van der Weken. Anal Chim Acta 347:325–332, 1997.
29. J Ouyang, WRG Baeyens, J Delanghe, G Van der Weken, AC Calokerinos. Talanta 46:961–968, 1998.
30. J Ouyang, WRG Baeyens, J Delanghe, G Van der Weken, D De Keukeleire, AC Calokerinos. Biomed Chromatogr 12:162–163, 1998.
31. Y Zhao, WRG Baeyens, XR Zhang, AC Calokerinos, K Nakashima, G Van der Weken. Analyst 122:103–106, 1997.
32. Y Zhao, WRG Baeyens, XR Zhang, AC Calokerinos, K Nakashima, G Van der Weken, A Van Overbeke. Chromatographia 44:31–36, 1997.
33. K Ishii, M Yamada. Bunnseki:452–459, 1994.
34. OM Steijger, H Lingeman, UAT Brinkman, JJM Holthuis, AK Smilde, DA Doornbos. J Chromatogr 615:97–110, 1993.
35. T Kawasaki, M Maeda, A Tsuji. J Chromatogr 328:121–126, 1985.
36. H Yuki, Y Azuma, N Maeda, H Kawasaki. Chem Pharm Bull 36:1905–1908, 1988.
37. K Nakashima, K Suetsugu, S Akiyama, K Yoshida. J Chromatogr 530:154–159, 1990.
38. K Nakashima, K Suetsugu, K Yoshida, K Imai, S Akiyama. Anal Sci 7:815–816, 1991.
39. SR Spurlin, MM Cooper. Anal Lett 19:2277–2283, 1986.
40. J Ishida, N Horike, M Yamaguchi. Anal Chim Acta 302:61–676, 1995.
41. J Ishida, N Horike, M Yamaguchi. J Chromatogr B 669:390–396, 1995.
42. JR Zhang, AR Cazers, BS Lutzke, ED Hall. Free Radical Bio Med 18:1–10, 1995.
43. T Miyazawa, S Lertsiri, K Fujimoto, M Oka. J Chromatogr A 667:99–104, 1994.
44. M Maeda, S Shimada, A Tsuji. J Chromatogr 515:329–335, 1990.
45. J Ishida, M Yamaguchi, T Nakahara, M Nakamura. Anal Chim Acta 231:1–6, 1990.
46. J Ishida, S Sonezaki, M Yamaguchi. J Chromatogr 598:203–208, 1992.
47. J Ishida, T Nakahara, M Yamaguchi. Biomed Chromatogr 6:135–140, 1992.
48. J Ishida, S Sonezaki, M Yamaguchi, T Yoshitake. Analyst 117:1719–1724, 1992.
49. J Ishida, S Sonezaki, M Yamaguchi, T Yoshitake. Anal Sci 9:319–322, 1993.
50. J Ishida, T Yakabe, H Nohta, M Yamaguchi. Anal Chim Acta 346:175–181, 1997.
51. M Maeda, A Tsuji. J Chromatogr 352:213–220, 1986.
52. TJ Novak, ML Grayeski. Microchem J 50:151–160, 1994.
53. K Honda, K Miyaguchi, K Imai. Anal Chim Acta 177:103–110, 1985.
54. K Imai, H Nawa, M Tanaka, H Ogata. Analyst 111:209–211, 1986.
55. K Nakashima, K Maki, S Akiyama, WH Wang, Y Tsukamoto, K Imai. Analyst 114:1413–1416, 1989.
56. M Stigbrand, E Ponten, K Irgum. Anal Chem 66:1766–1770, 1994.
57. A Nishitani, Y Tsukamoto, S Kanda, K Imai. Anal Chim Acta 251:247–253, 1991.

58. B Mann, ML Grayeski. Biomed Chromatogr 5:47–52, 1991.
59. K Hayakawa, R Kitamura, M Butoh, N Imaizumi, M Miyazaki. Anal Sci 7:573–577, 1991.
60. K Hayakawa, N Terai, PG Dinning, K Akutsu, Y Iwamoto, R Etoh, T Murahashi. Biomed Chromatogr 10:364–350, 1996.
61. S Kobayashi, J Sekino, K Honda, K Imai. Anal Biochem 112:99–104, 1981.
62. DL Walters, JE James, FB Vest, HT Karnes. Biomed Chromatogr 8:207–211, 1994.
63. S Kobayashi, K Imai. Anal Chem 52:424–427, 1980.
64. K Miyaguchi, K Honda, K Imai. J Chromatogr 303:173–176, 1984.
65. WRG Baeyens, J Bruggeman, B Lin. Chromatographia 27:191–193, 1989.
66. WRG Baeyens, J Bruggeman, C Dewaele, B Lin, K Imai. J Biolumin Chemilumin 5:13–23, 1990.
67. K Miyaguchi, K Honda, K Imai. J Chromatogr 316:501–505, 1984.
68. K Hayakawa, N Imaizumi, H Ishikura, E Minogawa, N Takayama, H Kobayashi, M Miyazaki. J Chromatogr 515:459–466, 1990.
69. A Nishitani, S Kanda, K Imai. Biomed Chromatogr 6:124–127, 1992.
70. K Hayakawa, K Hasegawa, N Imaizumi, OS Wong, M Miyazaki. J Chromatogr 464:343–352, 1989.
71. K Hayakawa, Y Miyoshi, H Kurimoto, Y Matsushima, N Takayama, S Tanaka, M Miyazaki. Biol Pharm Bull 16:817–821, 1993.
72. T Kawasaki, K Imai, T Higuchi, OS Wong. Biomed Chromatogr 4:113–117, 1990.
73. K Nakashima, K Suestsugu, K Yoshida, S Akiyama, S Uzu, K Imai. Biomed Chromatogr 6:149–154, 1992.
74. S Uzu, K Imai, K Nakashima, S Akiyama. Biomed Chromatogr 5:184–185, 1991.
75. S Uzu, K Imai, K Nakashima, S Akiyama. Analyst 116:1353–1357, 1991.
76. H Kouwatli, J Chalom, M Tod, R Farinotti, G Mahuzier. Anal Chim Acta 266:243–249, 1992.
77. M Tod, M Prevot, J Charon, R Farinotti, G Mahuzier. J Chromatogr 542:295–306, 1991.
78. BW Sandmann, ML Grayeski. J Chromatogr B 653:123–130, 1994.
79. G-L Duan, K Nakashima, N Kuroda, S Akiyama. J Clin Pharm Sci 4:22–28, 1995.
80. K Nakashima, C Umekawa, S Nakatsuji, S Akiyama, RS Givens. Biomed Chromatogr 3:39–42, 1989.
81. K Imai, S Higashidate, A Nishitani, Y Tsukamoto, M Ishibashi, J Shoda, T Osuga. Anal Chim Acta 227:21–27, 1989.
82. H Akiyama, T Toida, T Imanari. Anal Sci 7:807–809, 1991.
83. H Akiyama, S Shidawara, A Mada, H Toyoda, T Toida, T Imanari. J Chromatogr 579:203–207, 1992.
84. S Uzu, K Imai, K Nakashima, S Akiyama. J Pharm Biomed Anal 10:979–984, 1992.
85. Y Hamachi, MN Nakashima, K Nakashima. J Chromatogr B 724:189–194, 1999.
86. K Nakashima, M Nagata, M Takahashi, S Akiyama. Biomed Chromatogr 6:55–58, 1992.
87. S Yoshida, K Urakami, M Kito, S Takeshima, S Hirose. Anal Chim Acta 239:181–187, 1990.
88. S Higashidate, K Imai. Analyst 117:1863–1868, 1992.

89. GH Ragab, H Nohta, M Kai, Y Ohkura. Anal Chim Acta 298:431–438, 1994.
90. GH Ragab, H Nohta, M Kai, Y Ohkura, K Zaitsu. J Pharm Biomed Anal 13:645–650, 1995.
91. K Miyaguchi, K Honda, T Toyo'oka, K Imai. J Chromatogr 352:255–260, 1986.
92. K Nakashima, M Wada, N Kuroda, S Akiyama, K Imai. J Liq Chromatogr 17:2111–2126, 1994.
93. E Ponten, P Appelblad, M Stigbrand, K Irgum, K Nakashima. Fresenius J Anal Chem 356:84–89, 1996.
94. M Wada, K Nakashima, N Kuroda, S Akiyama, K Imai. J Chromatogr B 678:129–136, 1996.
95. M Katayama. Meiji Yakka Daigaku Kenkyuu Kiyo 24:43–48, 1994.
96. K Nakashima, K Maki, S Akiyama, K Imai. Biomed Chromatogr 4:105–107, 1990.
97. K Nakashima, S Kawaguchi, RS Givens, S Akiyama. Anal Sci 6:833–836, 1990.
98. K Kimoto, R Gohda, K Murayama, T Santa, T Fukushima, H Homma, K Imai. Biomed Chromatogr 10:189–190, 1996.
99. JA Holeman, ND Danielson. J Chromatogr A 679:277–284, 1994.
100. JA Holeman, ND Danielson. J Chromatogr Sci 33:297–302, 1995.
101. JS Ridlen, DR Skotty, PT Kissinger, TA Nieman. J Chromatogr B 694:393–400, 1997.
102. H Monji, M Yamaguchi, I Aoki, H Ueno. J Chromatogr B 690:305–313, 1997.
103. K Uchikura, M Kirisawa. Anal Sci 7:971–973, 1991.
104. DR Skotty, W-Y Lee, TA Nieman. Anal Chem 68:1530–1535, 1996.
105. W-Y Lee, TA Nieman. J Chromatogr A 659:111–118, 1994.
106. K Uchikura, K Kirisawa, A Sugii. Anal Sci 9:121–123, 1993.
107. N Kuroda, K Nakashima, S Akiyama. Anal Chim Acta 278:275–278, 1993.
108. N Sato, K Shirakawa, K Sugihara, T Kanamori. Anal Sci 13:59–65, 1997.
109. N Kuroda, K Nakashima, S Akiyama, N Sato, N Imi, K Shirakawa, A Uemura. J Biolumin Chemilumin 13:25–19, 1998.
110. M Kai, Y Ohkura, S Yonekura, M Iwasaki. Anal Chim Acta 287:75–81, 1994.

15

Chemiluminescence Detection in Capillary Electrophoresis

Ana M. García-Campaña
University of Granada, Granada, Spain

Willy R. G. Baeyens
Ghent University, Ghent, Belgium

Norberto A. Guzman
The R. W. Johnson Pharmaceutical Research Institute, Raritan, New Jersey

1. INTRODUCTION

In the last decade, capillary electrophoresis (CE) has become one of the most powerful and conceptually simple separation techniques for the analysis of complex mixtures. The main reasons are its high resolution, relatively short analysis times, and low operational cost when compared to high-performance liquid chromatography (HPLC). The ability to analyze ultrasmall volume samples in the picoliter-to-nanoliter ranges makes it an ideal analytical method for extremely volume-limited biological microenvironments.

Thanks to the efforts of a continuously increasing number of research groups, CE has by now been accepted as a highly efficient separation technique for qualitative purposes, with about 1500 CE-related documents appearing annually in analytical journals. However, CE has not yet been fully established as a quantification method, mainly due to the predominance of HPLC techniques applied in standard, validated analytical protocols.

Analytical techniques that involve the measurement of chemiluminescence (CL), though less commonly encountered in the literature, can be applied for quite sensitive measurements but often suffer from a lack of selectivity. Although the first observation of CL was made by Radziszewski in 1877—lophine (2,4,5-triphenylimidazol) emitting green light when reacting with oxygen in alkaline medium—investigations on the analytical use of CL have mainly been performed since the 1970s for gas-phase and from the 1980s onward for liquid-phase reactions. In addition, the rapid development of immobilization techniques has considerably enhanced the applications of chemiluminometry especially in flow injection analysis (FIA) and in liquid chromatographic systems [1].

The number of reactions producing CL cited in the literature is increasing each year, being analytically applied in chemical, biomedical, food, environmental, and toxicological disciplines [2–4]. In combination with HPLC separations several CL reactions have been used, among others, peroxyoxalate, firefly luciferase, lucigenin, and luminol. The most commonly used CL system for postcolumn detection in conventional and in microcolumn LC is the peroxyoxalate reaction [5]. The main inconvenience is related to the use of organic solvents that may cause precipitation problems with reversed-phase eluents due to the low water solubility of the average oxalate esters. In comparison with other CL reactions and in the presence of a suitable fluorophore, highest quantum efficiencies may be reached with the latter system and a more wide range of fluorophores can be analyzed.

Applying different CL systems, continuous-flow CL-based detection of several analytes has been widely applied for determination of several biological compounds and drugs. This technique has already become a highly sensitive method of detection in FIA, in liquid and gas chromatography, and in immunoassays [6–12].

With the purpose of increasing the specificity of analysis employing CL-based detection devices, several methods have been applied, including separation techniques. In this way, the combination of CL as a detection method with CE as prior separation methodology has provided a powerful analytical tool in recent years, offering excellent analytical sensitivity and selectivity and allowing the resolution and quantification of various analytes in a complex mixture. As a matter of fact, until the 1990s, chemiluminometric detection was not applied after capillary electrophoretic separation, but fast developments from some important research groups were recently noticed. Due to the advantages of CL detection and its potential when combined with the high separation ability offered by CE, research in this area has significantly increased. However, difficulties encountered when coupling the separation device with the CL detector still have to be dealt with.

In the present chapter an overview is presented about CL-based detection in CE, reviewing advances in the development of new detectors, the various CL-based reactions employed, and the applicability and usefulness of analysis to a wide range of samples.

2. CHARACTERISTICS OF CAPILLARY ELECTROPHORESIS

Tiselius originally introduced the conventional electrophoresis separation technique in 1937 [13], describing research on the separation of a protein mixture placed between buffer solutions in a tube. He observed that when applying an electric field, the sample components migrated in a direction and at a rate as determined by their charge and mobility. In this pioneering traditional work, Tiselius, who was awarded a Nobel Prize for his innovative research, recognized the separation advantages of smaller-diameter electrophoresis channels, but mentioned that detection considerations in narrow structures made the use of the technique unpractical.

Modern high-performance capillary electrophoresis (HPCE) is an instrumental approach to electrophoresis in which the components of a sample placed between two buffer solutions are separated in an open capillary tube with an inside diameter ranging from 2 to 200 μm and a length usually between 10 and 100 cm. The separation is based on the electrophoretic mobility of the analyte species induced by the large potential applied across the capillary (generally 10–30 kV). Separation efficiency in free solution, as performed initially, was limited by thermal diffusion and convection phenomena. To overcome these problems, various traditional stabilizing media were used, such as polyacrylamide, cellulose powder, glass wool, paper, or silica gel. Likewise, the use of tubes with small internal diameters showed numerous advantages because these narrow capillaries are themselves anticonvective allowing the performance of free-solution (or open-tube) electrophoresis, the gel media not being essential for this purpose. Moreover, their high electrical resistance allows the use of very high electrical fields with only minimal heat generation and efficient heat dissipation due to the large surface-area-to-volume ratio.

The separation of small ions, neutral molecules, and large biopolymers in modern CE can be carried out using different modes of operation [14, 15], including free (capillary) zone electrophoresis (CZE), isotachophoresis (CITP), isoelectric focusing (CIEF), micellar electrokinetic chromatography (MEKC), and gel electrophoresis (CGE). CZE is the most commonly used technique owing to the simplicity of operation and versatility. The separation mechanism is based on differences in solute size and charge at a given pH using fused silica capillaries only filled with buffer, the latter chosen based on UV transparency, electrical inertia, and durability properties. These capillaries contain surface silanol groups that may become ionized in the presence of the electrophoretic buffer, showing three layers in the interface between the fused silica capillary wall and the running buffer: the negatively charged silica surface due to the anionic form of silanol groups (at neutral or alkaline pH), an immobile layer (Stern layer), and the diffuse layer of solvated cations in the surface of the silica built up to maintain charge balance and creating a potential difference very similar to the wall capillary, called zeta potential. When the high voltage is applied across the capillary, this layer of cations migrates toward the cathode producing a migration bulk flow of liquid through the capillary called electroosmotic or endoosmotic flow (EOF). As the EOF in fused silica capillaries is normally greater than the electrophoretic mobility rates of the individual analytes in the sample, the different charged species are moved in the same direction (usually from the anode to the cathode), being then separated in the same run. In function of the highest charge/mass ratio, the migration of the different species is produced, being cations migrating fastest, neutral species moving at the velocity of the EOF but not being separated from each other, and anions with the greatest electrophoretic mobilities migrating last. Oncolumn or postcolumn detectors can be used, the "retention time" being the time required for a solute to migrate to the point where detection occurs based on different modes, depending on specific solute properties [16].

By controlling the pH of the buffer medium it is possible to change the cited charge/mass ratio of the analyte, affecting ionization and electrophoretic mobility. The electroosmotic velocity can be adjusted by means of an adequate selection of several parameters inherent to the buffer, such as pH (if more silanol groups are ionized, the bulk flow is increased), viscosity (as viscosity increases the velocity decreases), the concentration and ionic strength (decreasing the zeta potential and EOF when increased), the intensity of the electric field (flow increasing proportionally to voltage), and the dielectric constant. According to the charge, hydrophobicity, size, and stereochemical configuration of the analytes, an increasing range of applications can be carried out by CE, such as inorganic determinations, peptide and protein separations, environmental and food analysis, DNA sequencing, oligosaccharide separation, and single-cell assays. Several books and specific reviews include the fundamentals and applications of this technique [17–30].

Successfully validated CE methods are now routinely applied in many pharmaceutical quality control laboratories where applications include purity testing, quantitative assays, separation of enantiomers, and the determination of drug stoichiometry [31, 32]. Also, there has been a significant increase in publications dealing with the application of CE to clinical diagnosis and with biological sample preparations prior to CE analysis [33–38]. The application of CE in forensic sciences shows its capability of providing information about a wide range of chemical species and matrices, including gunshot and explosive residues, drug and DNA identification, to name a few. Due to the demonstrated superiority and taking into account that CE methodology preserves the requirement of legal systems based on the use of minute sample sizes only, CE has become a technique rigorously applied to analyze evidence and to help the criminal justice system to find correct conclusions [39].

Owing to the ultrasmall sample volumes introduced in the system and because of the small internal-diameter requirements in CE, together with the fact that in general the analytes of interest are present in low concentrations, poor detection limits are encountered, limiting the usefulness of the technique. For this reason, relatively concentrated analytical solutions, online preconcentration methods, and a variety of sample injection techniques for the preconcentration of analytes have been developed to improve sensitivity [38, 40]. Researchers have developed specific and nonspecific bioaffinity and molecular recognition CE methods for preconcentrating and characterizing analytes present in a wide concentration range in diluted liquids, fluids, or complex matrices. The methods combine the low as well as the high binding selectivities of the sorbing molecules with the efficient resolution abilities of CE. Detection limits of parts-per-trillion or -quadrillion can be reached, which allow a significant impact in forensic, environmental, biomedical, clinical, food, and pharmaceutical analysis [41, 42].

The selection of detection techniques capable of providing detection improvements has been a principal issue of research. A wide range of methods applied to meet detection limitations in CE have been taken mainly from liquid chromatographic techniques with only minor modifications, including ultraviolet (UV) absorption, fluorescence, mass spectrometry, conductivity, and electrochemistry principles.

It is clear that the main performance criteria that must be taken into account when selecting a detector for a particular determination are selectivity and sensitivity, followed by linearity of signal response, linear detection range, and reproducibility. These performance criteria are related to the independence of the detector response of buffer composition and of the physical devices (cells, joints, fittings, connectors, etc.), which must not contribute to extracolumn zone broadening, being compatible with the separation conditions.

Depending on the architecture of the detection cell, on-column or off-column detection can be carried out. Oncolumn detection is most commonly used as in this case the detection cell makes part of the electrophoretic capillary, thus eliminating the broadening effect mentioned above and producing high separation efficiencies. However, using offcolumn detectors, the band broadening is generally increased.

Another classification of detectors is based on the specificity of the detection principle, being divided into "bulk-property" and "specific-property" series [43]. The former evaluate differences between a physical property of the solute relative to the buffer alone, such as the refractive index, conductivity, and the application of indirect methods. In spite of their universal character, lower sensitivities and dynamic ranges are obtained. Specific-property detectors measure inherent physicochemical properties of species, for example UV absorption, fluorescence emission, and mass spectral behavior. In this sense, only the analytes showing these properties are detected, minimizing background signals and increasing sensitivity and width of linear ranges for these determinations. Detection may occur in the migration process (UV, fluorescence, conductivity, and refractive index) or as the components elute from the capillary (postcolumn derivatization before detection, and electrochemical and mass spectrometric methods).

UV (and much less frequent, visible) absorbance detection is the most widely used detection technique for CE owing to its ability to detect nearly all species without derivatization and to the easy adaptability of UV detectors originally designed for HPLC work, though several types of commercially available CE equipment offer some basic types of absorbance detectors [15]. High-quality fused-silica capillaries having a cutoff of approximately 170 nm are suitable for the UV detection of a wide range of compounds by applying fixed- or variable-wavelength instruments. The easiest way to perform oncolumn UV or luminescence detection is by making a window in the polyamide coating of the fused silica capillary, removing a small section (<1 cm) by burning off the polyamide coating, although alkaline etching or mechanical scraping can also be used. The main limitations of UV detection are its relatively low sensitivity and dynamic ranges. The concentration limits of detection (typically 0.1–1 µM) are usually limited by the short pathlength of oncolumn detection systems (inner diameter of 25 µm or greater) and the limited time available to observe the sample as it passes the detector. Various methods have been employed to increase the pathlength for optical detection in small capillaries, including the use of axial illumination, Z-shaped flow cells, multireflection cells, and the use of tubing with noncircular cross-section. Diode-array detectors provide additional spectral infor-

Table 1 General Overview of Detection Methods in CE

Method[a]	Mass detection limit (moles)	Concentration detection limit (molar)[b]	Characteristics
UV-Vis absorption	$10^{-13}-10^{-14}$	$10^{-5}-10^{-6}$	Universal
			Diode array offers spectral information
Fluorescence	$10^{-15}-10^{-17}$	$10^{-7}-10^{-9}$	Sensitive
			Usually requires sample derivatization
Laser-induced fluorescence	$10^{-18}-10^{-21}$	$10^{-14}-10^{-17}$	Extremely sensitive
			Usually requires sample derivatization
			Expensive
Amperometry	$10^{-18}-10^{-19}$	$10^{-10}-10^{-11}$	Sensitive
			Selective but useful only for electroactive analytes
			Requires special electronics and capillary modification
Conductivity	$10^{-15}-10^{-16}$	$10^{-7}-10^{-8}$	Universal
			Requires special electronics and capillary modification
Mass spectrometry	$10^{-16}-10^{-17}$	$10^{-8}-10^{-9}$	Sensitive and offers structural information
			Interface between CE and MS complicated
Indirect UV. fluorescence, amperometry	10–100 times larger (= poorer) than direct method	—	Universal
			Lower sensitivity than direct methods

[a] Other methods include radioactivity, thermal lens, refractive index, circular dichroism, and Raman spectroscopy.
[b] Assume 10 nL injection volume.
Source: Ref. 64.

mation from the separated analytes that can be used to assist in peak purity assessment, analyte identification, and prerun screening to determine the wavelength setting for optimum detection sensitivity.

Fluorescence detection is the second mode quite frequently applied in CE analysis because of the low detection limits—not strictly pathlength dependent—and the easy adaptability of fluorescence detectors to the smaller diameter capillary [44–46]. The high sensitivity obtained is due to the fluorescence emission inherently at higher wavelengths than the excitation wavelength, yielding low background signals. Moreover, a single fluorescent analyte may emit multiple photons. As with UV detection, depending upon the needs of each specific application, direct and indirect detection methods can be considered. Also, a wide variety of reagents exist for pre-, post-, and oncolumn derivatization in CE to convert analytes into products with more favorable detection characteristics [47].

Laser-induced fluorescence (LIF) has also been utilized as a highly sensitive detection principle for CE [48–51]. However, while the LIF detector is now able to achieve zeptomole (10^{-21}) detection limits, conventional derivatization techniques are inefficient at these exceptional levels [52]. Also, CE has successfully been coupled with mass spectrometry (MS) [53], nuclear magnetic resonance (NMR) [54, 55], near-infrared fluorescence (NIRF) [56, 57], radiometric [58], flame photometric [59], absorption imaging [60], and electrochemical (conductivity, amperometric, and potentiometry) [61–63] detectors. A general overview of the main detection methods is shown is Table 1 [64].

A most powerful detection mode under actual investigation in CE is CL. Because CL detection does not require a light source for excitation—the required energy being produced by a suitable chemical reaction—problems in baseline stability limiting detection limits are overcome, providing excellent sensitivities due to the low background noise. Recently, some review articles have been produced in this field [46, 65–68].

3. CHEMILUMINESCENT SYSTEMS FOR DETECTION IN CAPILLARY ELECTROPHORESIS

For CL purposes, only minimal instrumentation is required and because no external light source is needed, the optical system is quite simple. Hence strong background light levels are excluded, as occurring in absorption spectroscopy, reducing the background signal, and leading to improved detection limits. For this reason, CL has been defined as a ''dark-field technique'' as this technique produces a signal against a dark background, making it easier to detect and thus to acquire the CL signal by a photomultiplier tube (PMT), which must obviously be sufficiently sensitive in the spectral region of interest.

However, several problems that may limit the application of CL techniques should be considered. First, as a principle any fluorescing substance can be measured after suitable chemical excitation, which implies that a CL reagent is not limited to just one unique analyte. This lack of selectivity is thus brought along fundamentally in FIA applications, where selective reactors must be incorporated before the CL reaction can occur, which for HPLC purposes is not a real problem. Another disadvantage is the dependence of the CL emission on several environmental factors mentioned above, which should be dealt with during the HPLC or CE separation procedure as well as during FIA analysis. Hence a compromise between the required and optimized separation and detection conditions should be worked out for each analytical procedure. Finally, since CL emission is not constant but varies with time (light flash composed of a signal increase after reagent mixing, passing through a maximum, then declining to the baseline), and this emission-versus-time profile can widely vary in different CL systems, care must be taken to detect the signal in the flowing stream at strictly defined periods.

In the absence of analyte, many CL systems show a low emission background level. Hence, in flow systems, as the CL intensity is proportional to the analyte concentration, the emission appears as a sharp peak superimposed on a low constant blank signal, which is measured when the mixture of analyte and CL reagents passes through the detector cell. Because only a small portion of CL emission is measured from this time profile, nonlinear calibration curves may be obtained for reactions with complex kinetics [1].

It is clear that the need to obtain improved detection technology is now related to the general trend in analytical chemistry to reduce the waste volumes of organic solvents by using more aqueous systems and to study smaller samples at increasingly lower concentrations. As the CL technique may provide improvement in these areas, in terms of low detection limits, wide dynamic range, and high sensitivity, the instrumentation for the measurement of CL and the coupling with CE devices is being progressively developed using different indirect and direct CL systems. All the CL systems used in CE that will be described in this section are summarized in Table 2.

3.1 Chemiluminescence Reactions with Peroxyoxalates

Peroxyoxalate-based CL reactions are related to the hydrogen peroxide oxidation of an aryl oxalate ester, producing a high-energy intermediate. This intermediate (1,2-dioxetane-3,4-dione) forms, in the presence of a fluorophore, a charge transfer complex that dissociates to yield an excited-state fluorophore, which then emits. This type of CL reaction can be used to determine hydrogen peroxide or fluorophores including polycyclic aromatic hydrocarbons, dansyl- or fluorescamine-labeled analytes, or, indirectly, nonfluorescers that are easily oxidized (e.g., sulfite, nitrite) and quench the emission. The most widely used oxalate

Table 2 Applications of Chemiluminescence Detection in Capillary Electrophoresis

Analyte	CL system/migration buffer	Detection system	LOD	Ref.
Proteins	TCPO-H_2O_2-dyestuff (EY) Phosphate buffer (pH 3.5)	Online CL detector	—	69
Proteins	TCPO-H_2O_2-dyestuff (Rose bengal) Phosphate buffer (pH 3.5)	Online CL detector	5.10^{-7}–10^{-4} M (for BSA)	70
Proteins	TCPO-H_2O_2-dyestuff (EY) Phosphate buffer (pH 3.5)	Online CL detector	6.10^{-8} M (for BSA)	71
Proteins	TCPO-H_2O_2-dyestuff (RITC) Imidazole buffer (pH 6.0)	Online CL detector	5.10^{-8} M (for BSA)	72
Proteins	TCPO-H_2O_2-dyestuff (TRITC) Carbonate buffer (pH 9.0)	Online CL detector	6.10^{-9} M (for BSA)	73
Proteins	TCPO-H_2O_2-FR Borate buffer (pH 9.0)	Online CL detector	1.10^{-7} M (for BSA)	74
Protein	Heme protein-catalyzed Luminol- H_2O_2 Carbonate buffer (pH 10.0)	Online CL detector	10^{-10} M (for hemoglobin)	91
Liposomes	TCPO-H_2O_2-dyestuff (EY or Rhodamine) Carbonate buffer (pH 9.0)	Online CL detector	—	75, 76
Amino acids	TCPO-H_2O_2-Dansyl chloride Borate buffer (pH 8.9)	Postcolumn CL detector	1.2 fmol (for L-arginine)	78
Amino acids	Microperoxidase-catalyzed Isoluminol- H_2O_2 Borate buffer (pH 9.3)	Postcolumn sheath flow cuvette	500 amol (for valine)	83
Amino acids and peptides	Cu(II)-catalyzed luminol- H_2O_2 Carbonate buffer (pH 10.0)	Online indirect CL detection	100 fmol (for leucine) 300 fmol (for aspartic acid)	86, 87
Amino acids	Cu(II)-catalyzed luminol- H_2O_2 Phosphate buffer (pH 9.2)	Online indirect CL detection	9–250 fmol	90
Amino acids	ECL Ru(bpy)$_3^{3+}$ Borate buffer (pH 9.0)	In situ cell for ECL detection	0.7 μM valine 0.2 μM proline	98

Analyte	CL reaction / buffer	Detection	Detection limit	Ref.
Amino acids	ECL Ru(bpy)$_3^{3+}$ Borate buffer (pH 9.8)	End-column ECL detection	5.8 μM valine	97
Amino acids	Microperoxidase-catalyzed ILITC-H$_2$O$_2$ Phosphate buffer (pH 10.8)	Compact detection cell using optical fiber	8.1.10^{-8} for glycine 1.1.10^{-7} for glycylglycine	110
Amino acids ABEI	ABEI-BaO$_2$ Borate buffer (pH 8.3)	Online solid-phase CL detector using BaO$_2$	—	112
Acridinium esters	Acridinium- H$_2$O$_2$ Tartrate buffer (pH 2.8)	Postcolumn CL detector	—	79
Acridinium ester (Lucigenin)	Acridinium- BaO$_2$ Borate buffer (pH 8.3)	Online solid-phase CL detector using BaO$_2$	5.10^{-8} M	112
Peptides	Acridinium- H$_2$O$_2$ Citric acid buffer (pH 2.7)	Postcolumn CL detector	—	81
Luminol and ABEI	Luminol- H$_2$O$_2$ Phosphate buffer (pH 9.9)	Postcolumn CL detector	3.10^{-9} M luminol 7.10^{-9} M ABEI	82
Luminol and ABEI	Luminol- H$_2$O$_2$	End-column CL detector	500 amol luminol	84
Luminol and ABEI	Luminol-BaO$_2$ Borate buffer (pH 8.3)	Online solid-phase CL detector using BaO$_2$	1.10^{-8} M Luminol 7.10^{-8}M ABEI	112
Catecholamines and catechol	Cu(II)-catalyzed luminol- H$_2$O$_2$ Phosphate buffer (pH 9.2)	Online indirect CL detection	22 fmol norepinephrine 87 fmol cathecol	90
Catecholamines catechol and serotonin	KMnO$_4$ in H$_2$SO$_4$ Borate buffer (pH 9.5)	Off-column CL detector	100 μM	99
Amines	Luminol- H$_2$O$_2$ Borate buffer (pH 11.0)	ECL detection	5.6.10^{-7} M octylamine 2.7.10^{-7} M n-propylamine	85
β-blockers	ECL Ru(bpy)$_3^{3+}$ Borate buffer (pH 8.8)	Online ECL detection	0.6 ppb oxprenolol	94
Amines	ECL Ru(bpy)$_3^{3+}$ Borate buffer (pH 9.0)	Postcolumn ECL detection	3.7.10^{-5} M tripropylamine 1.5.10^{-5} M proline	96

Table 2 Continued

Analyte	CL system/migration buffer	Detection system	LOD	Ref.
Amines	ECL Ru(bpy)$_3^{3+}$ Borate buffer (pH 9.8)	End-column ECL detection	120 nM TEA	97
Cu(II)	ECL Ru(bpy)$_3^{3+}$ (emetine dithiocarbamate Cu(II) complex) Phosphate buffer (pH 3.5)	End-column ECL detection	1.10^{-10} M	95
Metal ions	Metals ions-catalyzed luminol- H_2O_2 Citric acid buffer (pH 4.5)	Online CL detection	20 zmol Co(II) 100 amol Mn(II)	88
Rare earth ions and catechol	Co(II)-catalyzed luminol- H_2O_2 Phosphate buffer (pH 4.5)	Online indirect CL detection	22 fmol La (III) 50 fmol Nd(III)	89
ATP	Luciferin-luciferase Phosphate buffer (pH 7.8)	End-column CL detection	5 nM	84
Creatine kinase	Luciferin-luciferase TAPS/AMPD (pH 7.8)	EMMA indirect CL detection	10^{-11}M	100
Horseradish peroxidase	HRP-catalyzed Luminol- H_2O_2 Borate buffer (pH 8.5)	Microchip-based CE coupled with CL detection	7–35 nM (for HRP)	107
Luminol	Microperoxidase-catalyzed ILITC- H_2O_2 Phosphate buffer (pH 10.8)	Compact detection cell using optical fiber	14 amol (for luminol)	110

ABEI, *N*-(4-aminobutyl)-*N*-ethylisoluminol; BSA, bovine serum albumin; CL, chemiluminescence; DNPO, *bis*-(2,4-dinitrophenyl)oxalate; ECL, electrogenerated chemiluminescence; EMMA, electrophoretically mediated microanalysis; EY, eosine Y; FR, fluorescamine; HRP, horseradish peroxidase; ILITC, isoluminol isothiocyanate; LOD, limit of detection; RITC, rhodamine B isothiocyanate; TCPO, *bis*-(2,4,6-trichlorophenyl)oxalate; TEA, triethylamine; TRITC, tetramethylrhodamine isothiocyanate.

esters are *bis*-(2,4,6-trichlorophenyl)oxalate (TCPO) and *bis*-(2,4-dinitrophenyl)oxalate (DNPO). The advantage of the oxalate reaction is the wide pH range to carry out the oxidation, generally occurring near neutral values, the main limitation being the need for an organic solvent based on solubility, stability, and efficiency considerations.

Using the peroxyoxalate reaction, Hara et al. reported in 1991 [69] for the first time on the combination of CZE and CL detection. In their work, the authors found that the dyestuff Eosine Y (EY) comigrates with the proteins as a supramolecular binary or ternary complex if molybdate, tungstate, silver(I), or mercury (II) is present, in a pH 3.5 phosphate buffer using a 50-μm i.d. fused silica capillary. Similar results were obtained using Bromopyrogallol Red. The use of EY provided the possibility to overcome several problems encountered in the separation of biopolymers owing to the appearance of plural peaks in the fluorescence spectra of the protein labeled with a fluorophore and adsorption of the protein onto the inner wall of the capillary tube. Quite an improved determination in terms of sensitivity was obtained by measuring EY in the supramolecular compound mentioned above. The experiment was carried out using a standard CE apparatus, with a UV-visible, fluorescence, or CL detector coupled to the capillary by burning off the polymer coating (Fig. 1). For CL detection, the EY complex with the protein was determined by measuring the CL intensity of the TCPO-H_2O_2-EY system using an interface between CZE and CL detector. The CL reagent obtained by mixing 1 mM TCPO and 75 mM hydrogen peroxide in acetonitrile and the migrating buffer solution were fed into a four-way joint by the use of two pumps. The protein-EY complex passing through the capillary tube from the upper to the lower parts was mixed with the feeding solution.

As the use of EY did not achieve the adequate sensitivity for protein detection, other fluorescent xanthene dyestuffs including Rose Bengal, Erythrosine B, Phloxin B, etc., were studied in later work [70] using the same experimental device and CL detector as previously proposed [69]. The addition of a fluorinated cationic surfactant (FC-135) to the migration buffer improved CE resolution by forming a bilayer via a hydrophobic interaction between the apolar chains, resulting in a reversal of the EOF and avoiding, in this way, the above-mentioned inconvenience encountered in the separation of proteins and other biological compounds. When labeling the proteins with the dyestuff by mixing both species, Rose Bengal was found most sensitive and gave the highest CL intensity. Using this dye, 5.10^{-7}–10^{-4} mol/L of bovine serum albumin (BSA) was determined in about 20 min with a detection limit of 4 fmol.

However, the former apparatus, proposed by Hara et al. [69], in which two streams comprising a buffer solution and a CL reagent solution are applied, presented some disadvantages such as a lowering of CL intensity due to the buffer solution and the difficulty of maintaining steady-state mixing. To overcome these problems, this research group developed a new, simple, and inexpensive online

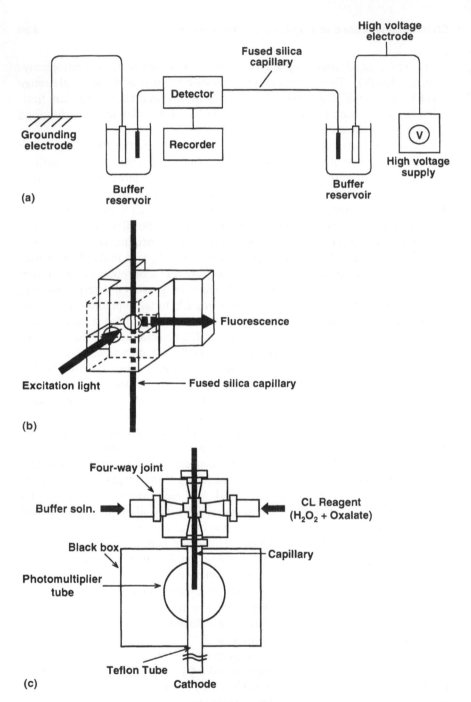

Figure 1 Schematic diagram of an HPCE apparatus. (a) Main part; (b) holder for FL detector; (c) CL detector. (From Ref. 69, with permission.)

CL detector [71]. In the latter, the stream of solution buffer is eliminated as a definitive migration current and a good resolved electropherogram for the separation of proteins is obtained using a stationary flow of fluorosurfactant. The complicated mechanical operation for preparing the cited four-way joint is avoided, using a three-way joint, commercially available, held at 10 cm below the horizontal surface of a reservoir solution to prevent any backflow of the CL reagent into the capillary tube. Using this online CL detector, an improved detection limit of 1.7 fmol for BSA was obtained by measuring the CL intensity of the complex with EY, in comparison with the CL device previously reported.

Consequently, the research work of Hara's group continued focusing on the improvement of protein determination using CE combined with online CL detection. By replacing EY by the Rhodamine B isothiocyanate (RITC) dye in the binary complexes formed with the proteins BSA or human serum albumin (HSA) and using a different imidazole buffer solution of pH 6, the sensitivity was increased [72]. However, best detection limits for these determinations were found employing the tetramethylrhodamine isothiocyanate isomer (TRITC) dye, left for 4 h with a standard solution of BSA in acetonitrile followed by introduction into the capillary. For BSA, a detection limit of 6 nM was reached [73].

Using the online CZE-CL method for the determination of proteins previously proposed [71], this research group introduced application of the fluorogenic reagent fluorescamine to the TCPO-H_2O_2 system, achieving high sensitivity and very short labeling time, as well as an undisturbed electropherogram in which a possible peak due to excess of labeling reagent did not appear as the excess of fluorescamine is quickly hydrolyzed in aqueous solution [74]. The structure and labeling reaction of fluorescamine are schematically represented in Figure 2. The reagent reacts readily and rapidly with primary amines under alkaline conditions

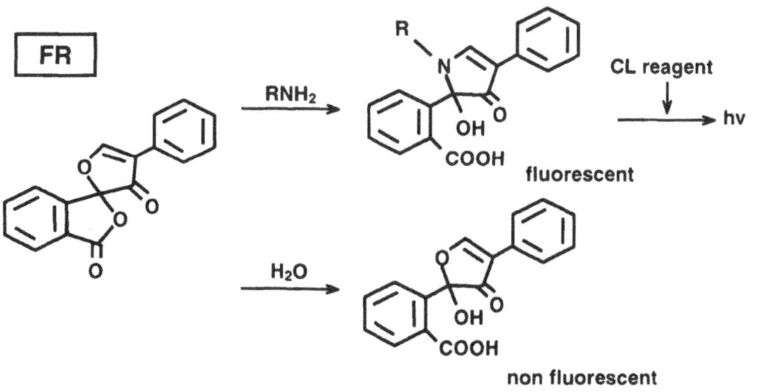

Figure 2 Structure of fluorescamine (FR) and its fluorogenic reaction with primary amines.

to form fluorescent substances, providing the basis for a rapid assay of amino acids, peptides, proteins, and other primary amines. Also, a new improvement in the former apparatus was proposed with respect to the attachment of the grounding electrode, which was directly immersed in the solution of the buffer reservoir, as can be seen in Figure 3, avoiding the formation of bubbles that could interfere with the CL measurements, thus providing a simpler and more efficient device.

More recently, this research group has focused its interest in the study of the migration and detection of an Eosin Y–containing liposome solution by CZE with CL detection using a peroxyoxalate system. Liposome, a synthetic lipid bilayer vesicle, has been used for model biomembranes, as drug carrier, and for other purposes such as immunoassays, in which dyestuff-containing liposomes play an important role as a probe and are used for characterizing liposomes in terms of homogeneity, capture volume, stability, permeability, etc. However, the methods for measuring the probe require tedious treatments and, in general, provide only indirect information. In this sense, Tsukagoshi et al. examined for the first time the electrophoretic behavior in a capillary [75, 76]: when a solution of Eosin Y–containing liposome was subjected to the CZE-CL method previously described for proteins, two peaks, due to Eosin Y entrapped in liposome and free Eosin Y in the bulk solution, were successfully observed in the electropherogram, from which various type of information about the liposomes can be provided rapidly, being only effective CL detection due to its high sensitivity. In these

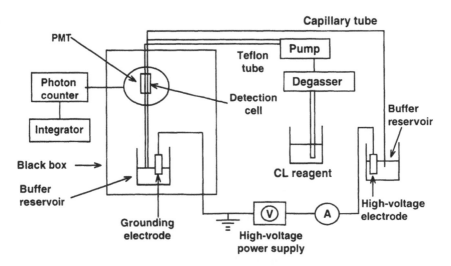

Figure 3 Schematic diagram of the CZE-CL apparatus. (From Ref. 74, with permission.)

studies, the Eosin Y–containing liposome migrated to the cathode by EOF in the capillary and was mixed with the CL reagent (TCPO + H_2O_2) at the tip of the capillary, where the liposome structure was broken by an organic solvent (aceto-nitrile) containing the reagent to induce a stable and reproducible CL intensity over all liposome concentrations. Under an alkaline condition the free Eosin Y in the bulk solution migrated slower than the liposome owing to the negative charge. The changes of retention times and ratios of the two peaks offered useful information as to the permeability and surface charge of the liposome membranes. Other dyestuffs such as Rhodamine B have been efficiently used [77], offering promising perspectives of this CZE-CL technique as a means of separating and detecting, e.g., in immunoassays using liposomes.

The peroxyoxalate CL (PO-CL) reaction has been most widely used for postcolumn detection in HPLC due to its high quantum efficiency and its ability to excite a wide range of fluorophores in comparison with other CL systems. The low solubility and instability of oxalate derivatives in aqueous solution, however, as cited above, require the use of organic solvents, which present several problems when applying the reaction for CE purposes. The organic solvents may influence the migration behavior of the analytes and their mobility in the aqueous electrophoretic buffer. Moreover, the stability of the PO-CL reagents can be affected by the high electric field strength inherent to CE systems.

Using a homemade CE apparatus and by means of a two-step approach for CE separation and dynamic elution (elution under pressure), Wu and Huie [78] avoided the disadvantages related to the incompatibilities between mixed aqueous-organic solvents and electrically driven systems by switching off the CE power supply at an appropriate time and connecting the CE capillary to a syringe pump to carry out dynamic elution. The employed postcolumn detector was equipped with various fused-silica capillaries held within a stainless-steel tee and a detection cell, as shown in the diagram of Figure 4. By burning off a 2-mm length of the polyimide coating, a detection window was created to detect CL emission generated within the postcolumn mixing region. The light was collected via one end of an optical fiber bundle situated directly above the detection window and the other end interfacing to the detection system using a PMT. The proposed system was used for measurement of three dansylated amino acids, showing a significant improvement in detection limits when compared to UV absorption detection. More investigations are still needed, nevertheless, to optimize the different experimental factors influencing detection and separation processes using the present methodology, such as pH, temperature, addition of catalysts, and organic modifiers. Likewise, the dynamic flow rate, the use of a deuterated running buffer to minimize the resolution loss due to slower rates of analyte diffusion within the higher-viscosity buffer solution, and the volume or geometry of the detection cell and connecting hardware, which affect the dispersion and/or the degree of mixing between analytes and reagents, will all need to be rigorously considered.

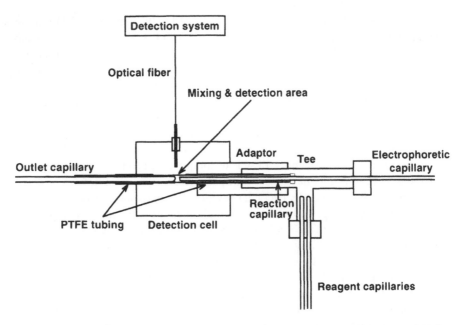

Figure 4 Schematic diagram of the postcolumn reactor developed by Wu and Huie. One arm of the tee contains the electrophoretic capillary, which is inserted in the reaction capillary (10 cm × 200 μm id × 400 μm od) situated at the opposite arm of the tee. The tee is connected to the detection cell via an adaptator and both the electrophoretic and reaction capillaries are inserted into the detection cell through the inner core of a PTFE tubing (400 μm id × 1.5 mm od). Two reagent capillaries (15 cm × 75 μm id × 144 μm od) inserted into the central arm of the tee are used to deliver the TCPO and H_2O_2 reagents into the mixing area through the small gaps that exist between the outer surface of the electrophoretic capillary and the inner surface of the reaction capillary. (From Ref. 78, with permission.)

3.2 Chemiluminescence Reactions with Acridinium Esters

A new detection interface designed for the addition of postcolumn reagents to evaluate the efficacy of CL as a detection method for CE was proposed in 1992 by Ruberto and Grayeski [79]. They used the acridinium CL reaction for CE purposes, which is based on the oxidation of an acridinium ester by hydrogen peroxide in alkaline medium to produce N-methylacridone in the excited state, which, upon relaxation, emits a photon, making it suitable as a derivatizing agent for amino acids, peptides, and proteins in CE analysis. This reaction has a high efficiency, yielding improved detectability. Its rate can be adjusted for measurements in flowing systems that require complexation reactions in a few seconds to minimize overlapping bands. Moreover, the acridinium esters used present some advantages; for example, they can be easily modified to include functional groups suitable for the derivatization of biomolecules. Also, the positive charge of the quaternary nitrogen atom in the ring structure provides greater mobility in the applied electric fields. The experimental configuration used is shown in Figure 5a. The detection interface uses a coaxial reactor consisting of two concentric fused silica capillaries, similar to the one proposed by Rose and Jorgenson for postcolumn reaction of amines with o-phthaldialdehyde followed by fluorescence detection [80], in which the smaller-diameter 50-cm-long electrophoretic capil-

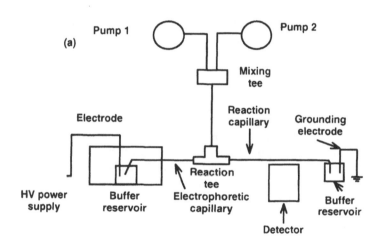

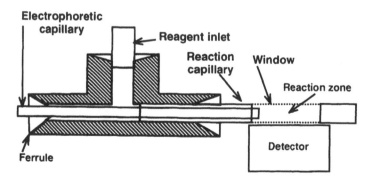

Figure 5 CE-CL system as proposed by Grayeski's group. (a) Experimental configuration. The hydrogen peroxide is introduced by pump 1 and pump 2 provides base. (b) Cross-sectional scheme of the CL detection interface. (From Ref. 79, with permission.)

lary is inserted into the larger-diameter 35-cm-long reaction capillary as illustrated in Figure 5b. The CL reagents enter the reaction tee and flow as a sheath around the electrophoretic capillary and its effluents. The hydrogen peroxide delivered by pump 1 is combined with the base from pump 2 by the mixing tee, the outlet of which leads to the reaction tee of the detection interface. By diffusion and radial migration, the reagents are mixed with the acridinium ester in a specific section of the reaction capillary called reaction zone. This zone is placed in front of the detector PMT at a distance of 1 cm and the photons emitted from the CL reaction are detected by the latter PMT. The end portion of the reaction capillary exits the detector and enters a buffer reservoir to complete the circuit. Several factors influencing the detector response have been taken into account, such as the flow rate of the postcolumn reagents, which requires an exhaustive control to procure adequate mixing of the reagents and a completed reaction in the time interval of the analyte being present in the proximity of the detector. A good separation of different acridinium esters at the optimum experimental conditions could be reached. The possible hydrolysis presented by the acridinium esters above pH 3 limits the working pH range, because this reaction is one of the competing processes to the photon-generating mechanism, decreasing the CL signal by more than 99% if the pH is increased up to 4. Nevertheless, biological species such as amino acids and proteins can be separated under these conditions.

A synthetic acridinium ester, 4-(2-succinimidyloxycarbonylethyl)phenyl-10-methylacridinium-9-carboxylate fluorosulfonate (acridinium NHS) can be used to label unhindered primary amine functionalities (Fig. 6), and using this interface for CL detection, it was later satisfactorily applied for performing trace peptide CE separation with CL detection [81]. In this case, the acridinium labeling of the peptides is done in a precolumn mode, prior to injection. The tagging reaction is run at pH 8, and is determined to reach completion in 15 min by

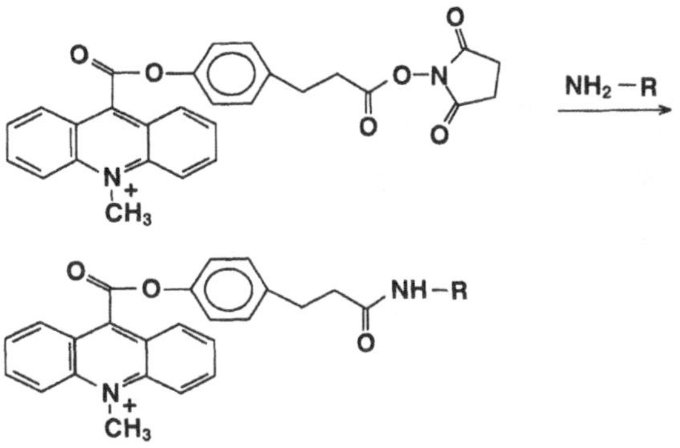

Figure 6   Acridinium-tagging reaction employing acridinium NHS.

monitoring the reaction progress with CE. To study the effect of acridinium tagging on migration time, a protein digest was injected into two different CE systems, both of which employed the same length of capillary. Into the first CE system, which contained the interface for CL detection, an acridinium-labeled tryptic digest was introduced and into the other, which was a conventional CE system with UV detection, unlabeled tryptic digest was injected. The protein used was β-casein and the enzyme used for the digestion was trypsin, which cleaves at the C-terminal side of lysine and arginine. The electropherogram produced in both cases is shown in Figure 7, where it can be seen that for CL detection, the run time is longer than the CE of untagged tryptic digest, while better resolution

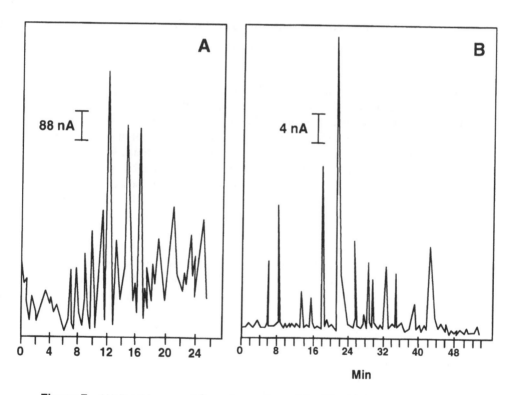

Figure 7 (A) Peptide map of β-caseine performed by CE with UV detection. Conditions: length from injection to detection 85 cm; detection at 200 nm. (B) Peptide map of β-caseine performed by CE with CL detection. The tryptic digestion was performed on 3 pmol of β-caseine, and 300 amol of the tagged trypic digest was injected into the CE capillary. Conditions: electrophoretic capillary 85 cm; reaction capillary 35 cm; operating buffer 50 mM citric acid and 20 mM γ-cyclodextrin (pH 2.7); operating voltage 25 kV. (From Ref. 81, with permission.)

is also obtained; the peaks seen from the CE run with CL detection do not show important band broadening even though the separation time is almost double that of the unlabeled tryptic digest separation. This type of phenomenon, called "chemical band narrowing," has been well documented in the literature for CL detection and is based on the nature of kinetics in CL [79]. The CL kinetics allow for an intense, rapidly decaying signal to be produced when the acridinium-tagged analyte enters the reaction zone of the reactor. Once the acridinium CL is complete, the signal stops, thus decreasing the effective volume of the flow cell because a measurement is not made for the entire residence time of the analyte in the flow cell.

3.3 Chemiluminescence Reactions with Luminol

3.3.1 Determination of Luminol, Derivatives, and Labeled Compounds

A successful study involving luminol CL detection was reported by Dadoo et al. in 1992 [82]. Luminol (5-amino-2,3-dihydro-1,4-phthalazinedione) reacts with an oxidant in the presence of a catalyst in alkaline medium to be oxidized to 3-aminophthalate emitting light with a wavelength in the interval of 425–435 nm, avoiding the inconvenience typical for peroxyoxalate reactions, which require the use of organic solvents. Although a variety of oxidants can be used, including permanganate, hypochlorite, and iodine, the most commonly used agent is hydrogen peroxide. The CL emission intensity is directly proportional to the concentration of luminol, H_2O_2, and catalyst, and for this reason measurements of CL intensity can be used to quantitate any of these species as well as species labeled with the catalyst, peroxide, or species that may be converted into peroxide, luminol, or species labeled with luminol. For example, carboxylic acids and amines can be labeled with luminol or its derivatives [isoluminol and N-(4-aminobutyl)-N-ethylisoluminol (ABEI)], being separated and then detected, as in HPLC, after postcolumn reagent addition.

The electrophoretic apparatus used as depicted in Figure 8 is provided with an electrophoretic capillary tube (48 cm × 75 μm id × 375 μm od), a reagent capillary (75 cm × 200 μm × 375 μm od), and a reaction (outlet) capillary (65 cm × 150 μm × 375 μm od), held in place by a tee connector. A 3–4-cm section at the end of the electrophoretic capillary is etched to an outer diameter of approximately 100–120 μm by placing it in concentrated hydrofluoric acid while purging the capillary with helium. The detection window, which is made on the reaction capillary by burning off the polyimide coating, is placed at the focal point of a parabolic mirror to collimate the light emitted, and subsequently focused on a PMT connected to a photon-counting system. To decrease the dark current of

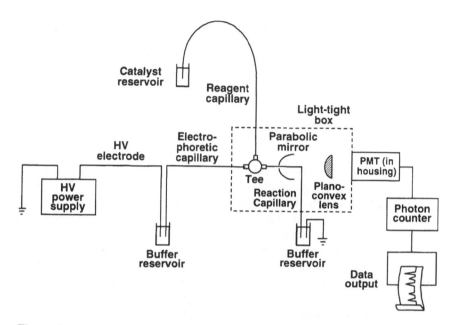

Figure 8 Schematic diagram of the CE-CL detector used by Dadoo et al. (From Ref. 82, with permission.)

the PMT, the latter was cooled down to $-20°C$ by means of a thermoelectric cooler.

This preliminary work demonstrated well the use of CL as a highly sensitive and selective detection method in CE by its application to the separation of luminol and ABEI using the same experimental conditions cited previously for the CL reaction of luminol in HPLC. Detection limits (S/N = 3) of 100 amol and 400 amol were obtained for the compounds mentioned, respectively, achieving an improvement in sensitivity of 2–3 orders of magnitude with respect to the ones obtained using UV absorption for detection.

Obviously, the main purpose for the introduction of CL detection coupled to CE separations is inherent to the development and improvement of sensitive and uncomplicated devices to achieve a decrease of the band broadening caused by turbulence at the column end, together with the attractive separation efficiency of CE setups. With this purpose in mind, Zhao et al. [83] designed a postcolumn reactor for CL detection in the capillary electrophoretic separation of isoluminol thiocarbamyl derivatives of amino acids, because, like other isothiocyanates, isoluminol isothiocyanate has potential applications in the protein-sequencing area.

The separation was performed in a 50 μm id × 190 μm od quartz capillary using a postcolumn sheath flow cuvette as the mixing chamber and a sheath stream from a syringe pump to carry the analytes away from the detection zone once emerging from the column. Mixing of the labeled analytes with hydrogen peroxide and microperoxidase as a catalyst is required because if these three components are introduced as separated streams, the peroxide will be destroyed by the catalyst before ever reacting with the analyte and hence no CL emission would be observed. In contrast to the experiments from Zare's group [82], who added peroxide to the separation buffer and mixed the catalyst in the detection chamber, Zhao et al. [83] added microperoxidase directly to the separation buffer, producing an intimate contact between the catalyst and analyte when mixed with peroxide, avoiding the problem of bubble formation in the separation capillary, which could perturb the separations. The sensitivity of the CL detector was influenced by several parameters, rigorously controlled in the experimental work, such as microperoxidase and hydrogen peroxide concentrations and mixing distance (distance downstream from the capillary to the center of the PMT). Nevertheless, separation of labeled amino acids was improved by the addition of a low concentration of the anionic surfactant sodium dodecyl sulfate (SDS) to the running buffer, taking into account that an increase in the concentration of the micellar medium may denature the microperoxidase, destroying the catalyst and inhibiting the reaction. Using this detector, the volumetric flow of peroxide added was much larger than the one used in Dadoo's design [82] for the same reaction. The optimal concentration of this reagent, however, was three orders of magnitude inferior.

The elimination of turbulent mixing in the flow chamber and the short residence time of the reaction mixture in the detection chamber provided a high separation efficiency of 100,000 theoretical plates for labeled amino acids, obtaining good resolution in comparison with previous CL detectors [79, 82]. For the isoluminol thiocarbamyl derivative of valine, a detection limit of 500 amol was reported; however, it was not possible to separate all 20 isoluminol-derivatized amino acids.

To facilitate the implementation of a CL-based detector in the CE system, Dadoo et al. focused their efforts in 1994 on the design of a simpler detector interface in which the signal is generated at the column outlet [84]. Based on the same CE system previously used [82] and removing a 1–3-mm section of the polyimide at the outlet end of the capillary by burning, the modified CL apparatus was coupled to the CE setup as shown in Figure 9. The outlet end of the separation capillary is immersed in the reservoir containing the electrolyte and the reagent for the CL reaction. When the analytes emerge from the column they react with the CL reagents in a reservoir producing visible light, which is transported by a fiberoptic (perpendicular to the capillary and with an end immersing in the solution) to a PMT tube. Immersing a platinum wire for the grounding electrode in the reservoir completes the CE electrical circuit. As in previous work, the same

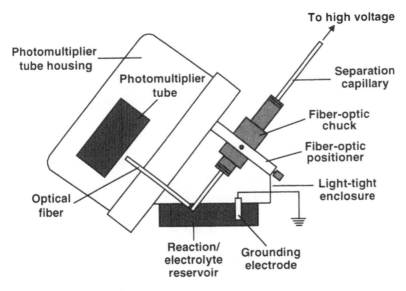

Figure 9 End-column CL detector for CE proposed by Zare's group. (From Ref. 84, with permission.)

reaction was used by the authors for separation of luminol and ABEI and also for separation of arginine and glycine derivatized with ABEI. In this separation of luminol and ABEI, the efficiency obtained was between 10,000 and 20,000 theoretical plates, indicating an improvement with respect to the one obtained for the previous design but with a detection limit for luminol (500 amol) lower than the one previously reported. The factors responsible for the decrease in sensitivity include the inefficient light collection as obtained using the fiberoptic in comparison with the application of the parabolic reflector and, also, the absence of cooling of the PMT. Although this end-column detector produces band broadening as a result of the mixing at the column outlet, and in spite of the slow CL reaction kinetics and a large detection zone, which provides relatively low numbers of theoretical plates, the simplicity and the ability to obtain detection limits in the nanomolar range make the setup suitable for routine work. The authors propose future modifications for achieving increased sensitivity, including cooling of the PMT tube to lower the dark current and the use of additional fiberoptics to increase light collection.

Based on the reaction of luminol and hydrogen peroxide, detection by electrogenerated CL (ECL) was also applied in CE [85]. In this detection technique, which has been used until now in LC and in FIA, the production of light is followed by an oxidation or reduction reaction at an electrode that serves the purpose of the catalyst used in luminol-based CL detection, offering the advantage of generating luminescence in a defined position of the electrode surface. This means of photon detection emitted from ECL provides an extreme sensitivity, which implies a close control of the placement of the electrode due to the confinement of generation of CL to a specific area as defined by the position of the electrode. Once luminol is separated using a conventional apparatus, it electrophoretically migrates into the detection cell containing hydrogen peroxide. A microelectrode is positioned immediately outside the bore of a fused silica capillary and when a potential is applied, the light produced upon electrochemical oxidation of luminol and hydrogen peroxide is generated at the microelectrode. This small size of electrode employed allows an easy alignment with the separation capillary and an optimum isolation of ECL generated in a reduced area. Subsequently, two optical fibers positioned 180 degrees apart collect the generated light, which is detected at the PMT. In this way, the postcolumn device for addition of a reactant in the CL reaction, as is the case with the CL detectors previously proposed, can be suppressed.

The electrogenerated luminescence response of luminol is strongly influenced by two factors: the hydrogen peroxide concentration in the detection buffer reservoir, which is dependent on the type of electrode material used in the electrooxidation process, and also the applied voltage at the microelectrodes because the latter regulates the reaction rate, which is correlated with the intensity of the emitted electrogenerated intensity. Throughout the experimental work it was deduced that using carbon microelectrodes the response for the detection of luminol is more stable; with platinum microelectrodes, however, the most sensitive response is obtained. The exact reason for this different behavior of ECL response depending on the electrode material cannot yet be explained. The efficacy of this methodology was proved in the analysis of amines derivatized with ABEI coupled to N,N-disuccinimidylcarbonate (ABEI-DSC), providing detection limits of 2.0 fmol and of 0.96 fmol for n-octylamine and n-propylamine, respectively (Fig. 10). Also, ABEI-DSC was used successfully to label the tripeptide Val-Tyr-Val and, applying MEKC with ECL detection, the separation of labeled amines was achieved. The main advantages of this ECL detection are the highly enhanced sensitivity by reducing interferences from solution impurities due to replacement of the catalyst, added in the luminol reaction, by the electrode and the elimination of complicated postcolumn reactors needed when CL detection is combined with CE.

3.3.2   Detection of Catalysts and Inhibitors: Ions, Amino Acids, Neurotransmitters, and Heme Proteins

Another detection mode, commonly used in LC and in FIA and recently adapted to CE separations, is indirect detection, based on the detection of a nonchemilumi-

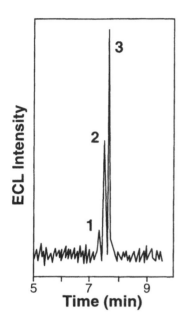

Figure 10 Separation of ABEI-DSC-derivatized *n*-octylamine (2) and *n*-propylamine (3) with ECL detection. Remaining ABEI (1) and ABEI DSC (not shown) are also detected. Conditions: 20% methanol in 5 mM sodium borate separation buffer, pH 10.9; 5-s injection at 25 kV, 1.0×10^{-6} for each labeled amine; 25-kV separation potential; 10-mm platinum wire electrode. (From Ref. 85, with permission.)

nescently active analyte that produces interference or suppression of a given CL reaction, the analyte being detected indirectly as an inverted peak, where the CL intensity decreases from a normally high background level. Liao et al. [86, 87] demonstrated for the first time the feasibility of this indirect CL detection technique in CE. Their contribution shows the utility of the Cu(II)-catalyzed luminol CL system for determination of five amino acids without previous pre- or postcolumn derivatization, making possible the detection of a wide range of biomolecules able to strongly and rapidly complex with Cu(II) as amines, catechol, catecholamines, and proteins. To carry out the separation and detection, the interface used is slightly modified with respect to the one proposed by Dadoo et al. [82]. Figure 11 illustrates the resolution obtained in the separation of five amino acids. Cu(II) catalyzes the luminol CL reaction, the CL emission intensity being proportional to the concentration of free Cu(II). In the presence of amino acids, the catalytic activity of free Cu(II) is decreased owing to the postcapillary formation of Cu(II)–amino acid complexes, and the CL intensity is considerably reduced. Good resolution efficiency is achieved by choosing an adequate electrophoretic

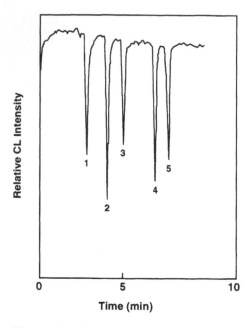

Figure 11   Electropherogram of a mixture of five amino acids using indirect CL detection. Conditions: 21-kV separation voltage, and 2 s at 21 kV for sample injection; sample concentration 0.5 mM of each amino acid. Peak identities: (1) arginine; (2) leucine; (3) serine; (4) cysteine; (5) aspartic acid. (From Ref. 86, with permission.)

buffer compatible with the detector reaction. In this case, to avoid the formation of microparticulates of Cu(II) hydroxide or carbonate, which will interfere with the CL detection, a small amount of tartaric acid must be added to the Cu(II) solution because it provides complexation with Cu(II) until the amino acid is introduced. This new detection system in CE is simpler than direct CL detection and reaches higher sensitivities, showing detection limits for the tested amino acids in the range 100–400 fmol, two orders of magnitude higher than obtained with CL detection in CE for labeled amino acids, as described in previous work [83].

Other studies in this specific area are also based on the catalytic effect of a variety of metal ions such as copper (II), cobalt (II), nickel (II), iron (III), and manganese (II) on the luminol–hydrogen peroxide reaction providing a rapid and efficient detection mode for these five ions, when an online CL detector is used before separation by CE [88]. This contribution combines capillary ion analysis (CIA) and CL detection by means of a postcapillary reactor similar to the one originally developed by Rose and Jorgenson [80] and finally modified by Wu and Huie [78] for CL detection. The reaction capillary is situated just in front of the PMT without the use of optical fibers to transport light, avoiding light loss by 10% and increasing the detection sensitivity due to the use of a sample injection volume lower than 20 nL. The most effective mixing mode to improve the sensitivity of this determination is by using luminol as one of the electrophoretic components and introducing only hydrogen peroxide solution in a postcapillary way. Thus, luminol, hydrogen peroxide, and the metal ions are placed at the detector window simultaneously, and the fast kinetic rate will produce a light response detected at the same time. It is clear that the pH must be conveniently selected because the EOF and the speciation of the metal ions are strongly influenced by this parameter. In the separation process, the net mobility of the ions is the sum of EOF and electrophoretic mobility, this last component being very similar for the transition metals and lanthanides, which do not allow an adequate separation. To overcome this problem, which impedes a good resolution, it is necessary to selectively alter the mobilities by means of a complexation process with a weak chelating reagent such as 8-hydroxyquinoline-5-sulfonic acid (HQS) or α-hydroxyisobutyric acid (HIBA). The net electrophoretic mobilities will be the weighted averages of the mobility of each free metal ion and its complexes and logically depend on the degree of complex formation obtained, using an optimal concentration of chelating agent and a pH value of 4.54. The latter represents a compromise between the optimum pH for the separation of metal ions and the required acidic medium to avoid hydrolysis effects, making it then possible to carry out the CL reaction. The detection limits of Co(II), Cu(II), Ni(II), Fe(III), and Mn(II) were 20 zmol, 2 amol, 80 amol, 740 amol, and 100 amol, respectively, obtaining a sensitivity considerably improved in comparison with common techniques for detection in CIA such as UV absorption and electrochemical detection.

Recently, this group has carried out for the first time CL detection with CE for rare-earth metal ion analysis [89]. Employing CL detection with CE as an oncolumn analysis method, dual effects of rare ions on the CL reaction of luminol with H_2O_2 were first reported. Under static conditions, rare-earth ions can complex with luminol and inhibit CL emission but on the other hand, they could catalyze the CL reaction second transformation from luminol free radical to aminophthalate and enhance CL emission in CE with online CL detection. Using the CE-CL detection system with a coaxial reactor similar to that shown in Figure 5 [79], in which the separation capillary is filled with electrophoretic buffer (10 mM phosphate buffer, pH 4.5), CL reagents (H_2O_2 and luminol) and Co^{2+} are siphoned into the tee and flowed into the detection window in incessant stream, and hence there frequently existed fresh reagent to react. The sample is injected by electroinjection for 10 s at a voltage of 10 kV. The background CL emission is constantly produced. As the background CL tends to be stable, a constant concentration of luminol free radical is present. When a rare-earth ion migrates into the detection window from the separation capillary, it reacts with a luminol free radical and generates a second enhanced CL signal. The stronger the background CL, the more the luminol free radical exists and hence the induced response of the rare earth is more intense. The proposed mechanism is depicted in Figure 12. In this case, the concentration of complexing agent HIBA in the electrophoretic solution can affect the separation of lanthanides, because several different lanthanide-HIBA complexes can exist simultaneously, leading to broadened peaks. The increase in HIBA concentration allows formation of a complex with high complexation degree, reducing the high charge of metal ion and leading to a decrease in EOF due to the decrease in ion strength. Both effects result in slower electrophoretic mobilities and a longer analysis time. On the other hand, the concentration of HIBA also has an influence on the background CL, decreasing this value for its complexation interaction with Co^{2+}. As free Co^{2+} in the electrophoretic solution is essential to obtain a stable CL background (but its

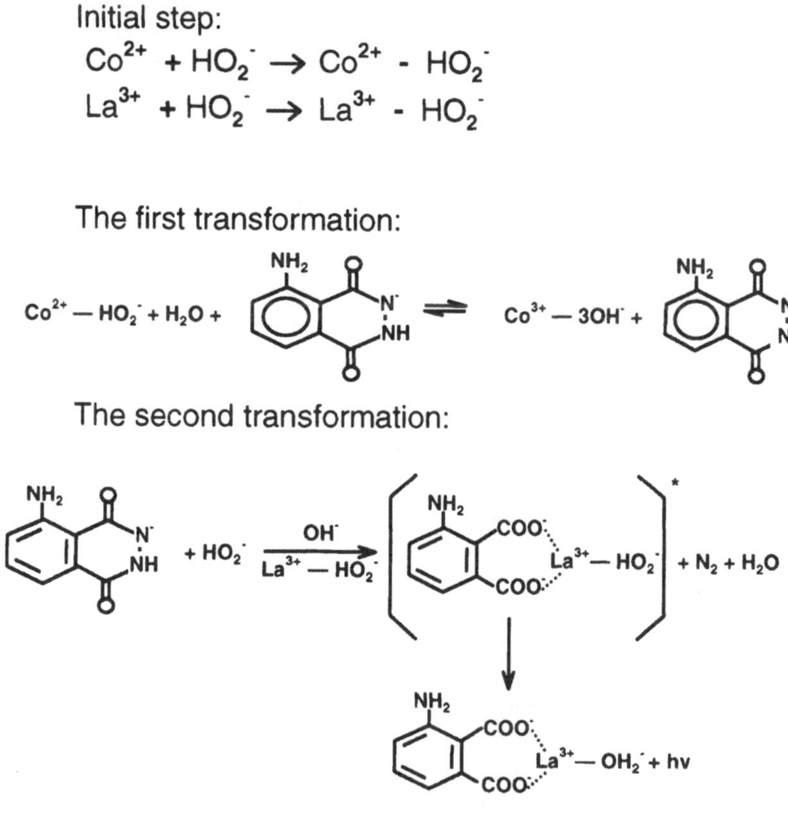
Initial step:
Co2+ + HO2- → Co2+ - HO2-
La3+ + HO2- → La3+ - HO2-
The first transformation:
Co2+ — HO2- + H2O +
Co3+ — 3OH- +
NH2
N-
NH
The second transformation:
+ HO2-
OH-
La3+ — HO2-
COO-
La3+ — HO2-
+ N2 + H2O
La3+ — OH2- + hv

Figure 12 Proposed mechanism for CL enhancement.

concentration may not be too high or else the baseline noise produced by the CL background becomes unstable), the optimum concentration ratio of Co^{2+} to HIBA was found to be 8.10^{-3} M. La^{3+}, Ce^{4+}, Pr^{3+}, and Nd^{3+} were successfully separated and detected providing detection limits of 33, 27, 42, and 50 fmol, respectively.

Zhang et al. have first used online indirect detection for analysis of catecholamines (CAS) and catechol (CAT) [90] using the CE-CL system reported above [88]. In this case, the high and constant CL background is obtained by the CL reaction of luminol enhanced with Co(II). The analytes complex with Co(II) and reduce the free Co(II) concentration, and thus the CL intensity decreases. The degree of CL suppression is a measure of the analyte concentration. The authors propose a new mixing mode of the analytes with the CL reagent in which luminol is used as a component of the electrophoretic carrier; H_2O_2 and Co(II) are introduced by postcapillary. In this way, luminol, H_2O_2, and Co(II) meet just at the detection window simultaneously. Against the mixing mode previously reported by Liao et al. [86], who used luminol and H_2O_2 as electrophoretic carrier and catalyst solution of a carbonate buffer containing copper sulfate, the formation of bubbles produced by H_2O_2 in the presence of base is impeded, which prevents the electrophoretic current and CL background from being unsteady, an increase in the noise, and a decrease in the separation efficiency. Sodium dodecyl sulfate was used in the separation of CAT, epinephrine, norepinephrine, and dopamine, obtaining detection limits of 87, 51, 22, and 38 fmol, respectively. Six amino acids such as Arg, Hyp, Lys, His, Glu, and Asp were detected, yielding better limits of detection than the ones reported by Liao et al. in their first study.

Tsukagoshi's group developed a new CE apparatus with an online CL detection using the luminol-H_2O_2 system for analyzing heme proteins [91]. It was found that iron (III), sulfate, hematin as an iron (III) porphyrin complex, and various heme proteins migrated and could be detected. They used the same three-way joint for mixing a CL reagent solution with an eluate from the capillary and the cell structure for detecting a low CL signal previously described [70, 74]. However, treatment of the CL solution (luminol + H_2O_2) was considered due to the drastic change of the CL intensity of the solution upon standing. In this case, the CL intensity quickly decreased and became about one-tenth of that of the initial solution within 1 h; an almost constant CL intensity was observed after about 12 h. For this reason, a CL reagent solution after being left for more than one night was used in this study because a fresh CL reagent solution provided high baseline noise levels and hence no reproducible results. They used a fresh capillary tube (50 μm id, 70 cm length) treated with 1 mol/L sodium hydroxide for 30 min and washed with distilled water and a migration buffer solution 10 mmol/L carbonate, pH 10. After a capillary tube was filled with the buffer solution in advance, a CL reagent solution (5 mmol/L luminol and 25 mmol/L H_2O_2 aqueous solution) was fed at a rate of 0.6 μL/min by a pump until a definitive electric current was obtained. Then, only the high voltage was removed, and a sample solution prepared by the carbonate buffer was introduced into a capillary tube having a positive electrode side for 10 s from a height of 30 cm by siphoning. After introduction of the sample solution, a voltage of 0–20 kV was gradually applied for 60 s. Monitoring was started just after the voltage reached 20 kV.

Though CE has shown excellent performances so far for the separation of many compounds having various molecular weights, it is not always satisfactory for the separation of biopolymers, such as proteins, glycoproteins, and lipoproteins due to the adsorption of proteins onto the inner wall of a capillary tube (and the low sensitivity in the detection of protein), which promote the absorption phenomenon due to the high concentration of the protein sample. However, in this case considerable sharp and symmetrical peaks were observed for all protein samples, in spite of turbulent mixing of the analyte with a CL solution at the end of the separation capillary. These satisfactory results are due to several reasons: first, since all protein samples were migrated at pH 10, which is higher than or equal to the isoelectric point values of the protein, the interaction between protein surfaces and negative charges due to silanol groups on the inner wall of capillary must be either very small or negligible, and second, the concentration of protein samples, which is much lower than that used for ordinary spectrophotometric and fluorimetric detection, was subjected to the CE-CL method. This CE-CL method was about 10^4 times as sensitive as the conventional CE-absorption detection system for the detection of hemoglobin.

3.4 Chemiluminescence Reaction with *Tris*(2,2′-bipyridine)ruthenium (II)

In the ruthenium *tris*-bipyridine system, an orange emission at 610 nm arises when the excited stated [Ru(bpy)$_3^{2+}$] decays to the ground state. Ru(bpy)$_3^{2+}$ is the stable species in the solution and the reactive species—Ru(bpy)$_3^{3+}$—can be generated from Ru(bpy)$_3^{2+}$ on the electrode surface by oxidation at about +1.3 V. Adding Ru(bpy)$_3^{2+}$ to the electrolyte and using an end-column electrode to convert the Ru(bpy)$_3^{2+}$ into the active Ru(bpy)$_3^{3+}$ form allow a simple and sensitive ECL detection mode. The reaction lends itself to electrochemical control due to the electrochemically induced interconversion of the key oxidation states:

$$Ru(bpy)_3^{2+} \rightarrow Ru(bpy)_3^{3+} + e^-$$

$$Ru(bpy)_3^{3+} + reductant \rightarrow product + [Ru(bpy)_3^{2+}]*$$

$$[Ru(bpy)_3^{2+}]* \rightarrow Ru(bpy)_3^{2+} + h\nu$$

Advantages include: in this CL system the reagent is regenerated and it can be recycled, and derivatization is not required for many classes of compounds. Many aliphatic amines such as alkylamines, amino acids, proteins, antibiotics such as erythromycin and clindamycin, and NADH, among others can participate in this reaction, which is compatible with FIA and HPLC solvent systems [92]. ECL employing $Ru(bpy)_3^{2+}$ offers an attractive detection scheme for CE because of the solubility and stability of the reagents in aqueous media [93]. Moreover, the efficiency of $Ru(bpy)_3^{2+}$ ECL is high over a wide pH range of pH making it compatible with most buffers systems commonly used in CE.

In 1997, Nieman's group used this system for the first time in CE introducing inactive $Ru(bpy)_3^{2+}$ into the electrophoretic buffer and generating electrochemically, online the active $Ru(bpy)_3^{3+}$ species just inside the outlet of the capillary [94]. This was done by applying -1.25 V versus Ag/AgCl to a Pt wire inserted 3 mm into the outlet of the capillary. In this way, analyte bands exiting the separation capillary react with the $Ru(bpy)_3^{3+}$ and produce light. The outlet is placed within a parabolic mirror that directs the emitted photons to a photon counting PMT. Concentration of $Ru(bpy)_3^{2+}$ in the electrolyte is an important parameter to be optimized because of its large impact on the background CL signal and dynamic range. The performance of the system was demonstrated using a series of β-blockers, a class of amine compounds that block the effect of norepinephrine on the adrenergic receptors. Detection limit for oxprenolol was 0.6 µg/mL with a separation efficiency of 15,000 plates. In the same period, Tsukasoshi et al. [95] found that the emetine dithiocarbamate Cu(II) complex, prepared from the emetine alkaloid, carbon disulfide, and Cu(II), showed a sensitive response on a $Ru(bpy)_3^{2+}$ ECL system, developing a CE-CL detection method for the analysis of emetine. They used an outline CE apparatus with an ECL detector. $Ru(bpy)_3^{2+}$ was fed at a rate of 40 mL/min by a pump, being oxidized using an electrolytic current of 100 mA in an electrochemical reactor and then mixed with the eluate at the tip of the capillary tube. A sample solution was introduced into the capillary tube having a positive electrode side for 20 min from 15-cm height by siphoning and a voltage of 0–20 kV was gradually applied for 60 s. A photon counter measured the CL signal at the tip of capillary. A combination of the dithiocarbamate complex formation of transition metal ions and their CL response to $Ru(bpy)_3^{2+}$ ECL is expected to be useful for the analysis of transition metal ions.

Another simple apparatus for postcolumn ECL detection [96] employs a conductive joint to isolate the separation field from the potential need to drive the ECL due to the electric currents generated in capillaries with id's greater than 25 µm, which greatly affects the faradaic currents at microelectrodes. This is accomplished by constructing a porous joint in the separation capillary (100 µm id, 360 µm od, 60-cm length). A 3-mm segment of the polyimide coating of the fused silica capillary, 5 cm from the capillary's end, is removed with hot concentrated sulfuric acid and the capillary is rigidly fixed to a Plexiglas plate, which serves to reinforce the capillary at the position where the polyimide coating has been removed. Teflon tubing provides a contact surface between the capillary and the Plexiglas plate. Next the entire capillary is placed in the cap of a Nalgene bottle, which later serves as a reservoir for the ground end of the separation capillary. The ends of the capillary are passed through the holes drilled in the Nalgene cap, which are then filled with epoxy. A porous joint is then etched in the capillary (where the polyimide coating has been removed) by immersing the capillary in 40% BF for 1–2 h. Capillaries are then filled with the running buffer and a potential of 10 kV is applied across the capillary. The ECL cell that houses the capillary, ECL reagent, Pt working, Pt auxiliary, and Ag/AgCl reference electrodes are fabricated from a Nalgene bottle cap and filled with 1 nM Ru(bpy)$_3^{2+}$ and Na$_2$HPO$_4$ (pH 9). The ECL signal is detected by a PMT positioned directly above the working electrode. The device is shown in Figure 9, Chapter 9 of this book. In this case, postcolumn and precolumn reagent addition were compared, showing that although the addition of the CL reagent to the running buffer and on-column detection ideally would lead to higher efficiency separations than the postcolumn arrangement, differences in migration between analytes and Ru(bpy)$_3^{2+}$ lead to zone broadening. Moreover the absorption of Ru(bpy)$_3^{2+}$ onto the silica walls of the separation capillary with an equilibration time of several hours and the perturbation in equilibrium produced when the capillary is flushed with dilute NaOH, water and finally buffer, hinder the use of on-column detection, suggesting the postcolumn addition of CL reagent to overcome these problems.

In this sense, a postcapillary reservoir of Ru(bpy)$_3^{2+}$ has been used for in situ generation of Ru(bpy)$_3^{3+}$ [97]. Ru(bpy)$_3^{2+}$ is added postcapillary as a small reservoir ($\approx$100 μL) at the interface of the separation capillary and the detection electrochemical cell and is then converted to Ru(bpy)$_3^{3+}$ at a carbon microfiber for reaction with eluting amines or amino acids. This detection approach has been found to provide a reproducible electrophoretic separation compatible with the nanoliter detection volumes required to maintain CE separation efficiencies, the major advantage being that the electrophoresis will not be inhibited by the presence of Ru(bpy)$_3^{2+}$ in the running buffer. Figure 13 shows the CE separation of triethylamine (TEA), proline, valine, and serine at pH 9.5 using CL detection. Detection limits range from approximately 100 nM for TEA and proline, the most efficient luminescent species, to approximately 100 μM for serine.

One problem associated with this design is that the Ru(bpy)$_3^{2+}$ reservoir evaporates over time and the Ru(bpy)$_3^{3+}$ concentration changes as the CE capillary effluent dilutes it, which affect both sensitivity and reproducibility of the CL response. To overcome this problem, recently a new in situ–generated Ru(bpy)$_3^{3+}$ CL cell has been proposed [98]. In this design, Ru(bpy)$_3^{2+}$ is continuously delivered to the cell and Ru(bpy)$_3^{3+}$ is then generated at the interface of the separation capillary and the working electrode. Electrochemical control of the production of Ru(bpy)$_3^{3+}$ at the distal end of the separation capillary without interference from the CE current is provided and finally the ECL process is cou-

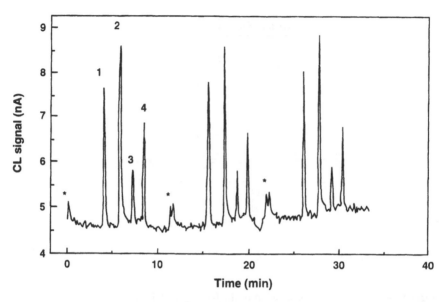

Figure 13 Electropherogram of selected amino acids with end-column addition of 1 mM Ru (bpy)$_3^{2+}$. Separation conditions: 20 kV with injection of analytes for 8 s at 20 kV. Capillary, 75 μm id, 62 cm long with a 4-cm detection capillary. Buffer 15 mM borate, pH 9.5. The electrode used for in situ generation of Ru(bpy)$_3^{3+}$ was a 35-μm-diameter carbon fiber, 3 mm long held at 1.15 V versus a saturated calomel electrode. The PMT was biased at 900 V. Peak identification: (1) 100 fmol TEA, (2) 70 fmol proline; (3) 1.6 pmol valine, (4) 50 pmol serine.* Injection points. (From Ref. 97, with permission.)

pled to an optical system for monitoring light emission. The authors used a CE homemade apparatus to perform electrophoretic separation and electrokinetic injections. A 2–3-mm detection window at the end of the separation capillary was formed by thermally removing the polyimide coating and this separation capillary was inserted into the reaction tube, Ru(bpy)$_3^{3+}$ reagent was delivered by a syringe pump to the reaction cell at a flow rate of 10 μL/min. This detection cell is schematically represented in Figure 14. For precise electrochemical control of the conversion of Ru(bpy)$_3^{2+}$ to Ru(bpy)$_3^{3+}$, the CE current is isolated from the detection capillary using an on-column fracture. The efficacy of this approach was shown in the detection of different amino acids. Further work is expected on the development of end-column detection, noise reduction strategies involving more sensitive PMTs cooled to minimize dark noise, the use of other capillary detection modes, and the application to a wide range of amino acids including derivatives that may be produced as a consequence of protein sequence analysis.

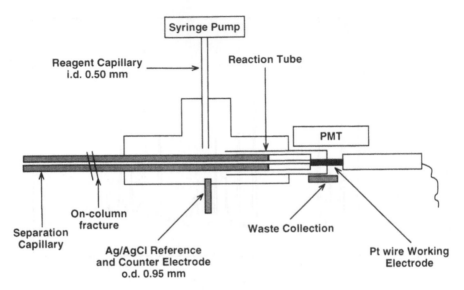

Figure 14 Schematic representation of the Ru(bpy)$_3^{3+}$ CL in situ detection cell. (From Ref. 98, with permission.)

3.5 Chemiluminescence Reaction with Potassium Permanganate in Acidic Medium

Oxidation of catecholamines by potassium permanganate in an acidic medium is known to produce CL, being used for the first time in 1997 in CE [99]. During the preliminary investigation for the analysis of catecholamines, the conventional end-column detection mode used in previous work [87] was first attempted by the authors. The end of a separation capillary was directly inserted into a short piece of a fused-silica reaction/detection capillary, and a tee connector was used to join the separation capillary, oxidant capillary, and reaction/detection capillary together. Acidic permanganate solution was fed through the oxidant capillary and mixed with analytes at the end of the separation capillary, which was inside the reaction/detection capillary. The downstream end of the reaction/detection capillary was dipped in a grounded buffer reservoir to complete the CE electric circuit. A 5-mm window was formed on the reaction/detection capillary (starting at the point where the inner separation capillary terminated) by burning off the polyimide coating. However, CL emission of catecholamines was not observed and the colorless buffer solution in the anodic reservoir gradually turned pink with successive CE experiments. This occurs because the oxidizing agent (permanganate anion) added at the column end migrates electrophoretically toward the anode and preoxidizes the analytes inside the capillary; therefore, no CL reaction occurs at the column end. To prevent the permanganate from the backstream migrating into the capillary, the reaction/detection zone at the column end was separated from the high-voltage electric field, developing for the first time an off-column CL detection device in CE. A porous polymer joint is easily constructed by fracturing the capillary followed by covering the fracture with a thin layer of cellulose acetate membrane. This porous joint, rather than the end of the capillary, was submerged in a buffer reservoir along the cathode (Fig. 15). The applied voltage was dropped across the capillary prior to the porous joint and the resulting EOF acted as a pump to push the analytes through the short section of capillary after the joint. The analytes mixed with permanganate emitted CL in a field-free region at the column outlet. The feasibility of this off-column CL detection mode was demonstrated in the CE of serotonin, catecholamines, and catechol. This detection mode is useful in cases where the CL reagent or catalyst added at the column end must not stream back into the separation capillary and degradation and decomposition of analytes may occur inside the capillary before they reach the column end. However, the only limitation is that at least some EOF is needed to push the analytes past the grounded joint. Further research is expected about the optimization of this system to enhance the sensitivity and efficiency.

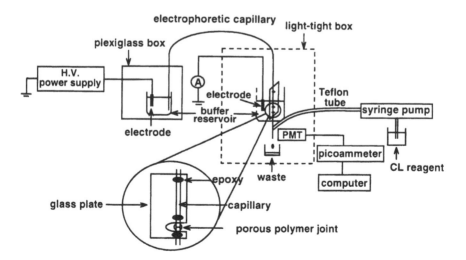

Figure 15 Schematic diagram of the CE with off-column CL detection system. (From Ref. 99, with permission.)

3.6 Chemiluminescence Reaction with Firefly Luciferase

Recent applications in the field of biochemical analysis have been developed based on the highly efficient firefly luciferin-luciferase reaction, a bioluminescence reaction in which two steps can be considered:

Luciferin + ATP → Adenyl-luciferin + PP_i

Adenyl-luciferin + O_2 → Oxyluciferin + AMP + CO_2 + light

the emission maximum occurring at 562 nm.

In this reaction, the most important analyte is adenosine 5'-triphosphate (ATP), appearing either directly or coupled with other enzymatic systems involving ATP as a reactant or product.

With the newly proposed detector, Dadoo et al. [84] adapted this bioluminescence reaction to determine ATP. A selective and sensitive determination is achieved because the use of CE as a separation technique minimizes the effect of several interfering substances such as some anions (e.g., SCN^-, I^-) that inhibit the reaction decreasing the luminescence emission, and even some nucleotides that generate light in this reaction but with lower intensity. A detection limit of 5 nM, approximately 3 orders of magnitude lower than using UV detection, was obtained.

The same reaction was recently proposed to detect creatine kinase (CK), an enzyme of high clinical significance in relation to the investigation of skeletal muscle disease and the diagnosis of myocardial infarct or cerebrovascular accidents. As ATP is a reaction product obtained from the reaction of ADP with creatine phosphate catalyzed by CK, this enzyme can be indirectly measured by the CL intensity read from the subsequent reaction of ATP with luciferin. Using the technique of electrophoretically mediated microanalysis (EMMA), it is possible to detect the enzyme using nanoliter volumes of biological sample with an improved speed and simplicity with respect to a conventional colorimetric method [100].

By application of EMMA Regehr and Regnier developed several assays for enzymes that produce (galactose oxidase and glucose oxidase) or consume (catalase) hydrogen peroxide. Unlabeled enzymes were determined in the femtomole mass range, while detection limits of less than 10,000 molecules were reported for catalase [101].

4. RECENT ADVANCES AND FUTURE PERSPECTIVES

4.1 Micromachining Techniques

Recent trends are focused on the use of micromaching techniques to form miniaturized capillary geometries in planar microdevices. These superminiaturized systems are being considered to overcome the problem of unsatisfactory detection limits characteristic of standard CE setups. Although in the last five years an important number of publications have been reported in this area, several devices are still being perfected offering great promise for the rapid and efficient processing of various types of samples, fundamentally in biological and clinical assays. A recent trend in miniaturization is the application of micromachining techniques (photolithography and chemical etching) in the fabrication of a complex manifold of flow channels on a microchip, capable of sample injection, pretreatment, and separation [102, 103]. The nature of electrokinetically driven systems as shown by CE makes it suitable for integration on a planar device, giving highly efficient separations in short capillaries together with a considerable reduction of analysis time and not requiring high-pressure pumps or gas supply. As electroosmotic flow velocity and electrophoretic migration depend only on the strength of the applied field, the separation efficiency is exclusively related to the voltage installed across the separation capillary and not to its length. Several materials such as planar glass, fused silica wafers, and quartz have been used to construct devices with different geometry and sizes. Since the first contribution by Manz's group in 1991 [104], introducing the concept of the CETAS system (capillary electrophoresis micro-total analysis system), further advances have been achieved and revised, reducing considerably the microchip size and extending the field of applications [105, 106].

Recently CL detection based on the horseradish peroxidase (HRP)-catalyzed reaction of luminol with peroxide has been investigated as a postseparation detection scheme for microchip-based CE [107]. Evaluation of CL detection on microchips was performed using the luminol reaction with various forms of HRP as the enzyme catalyst for oxidation of luminol by hydrogen peroxide [108]. In this contribution, an integrated injector, separator, and postseparation reactor were fabricated on planar glass wafer. The fluorescein conjugate of HRP (HRP-F1) was used as a sample for optimization of the CL detector response. The schematic layout of the microchip is shown in Figure 16. Devices consisted of two pieces of 1.95-mm thick glass, one with etched channels and the other with drilled access holes, thermally bonded together. For some devices, aluminum mirrors were sputter-deposited to 1000 Å thickness on the bottom plate after bonding. A shadow mask was formed with tape on the chip to define the 1-cm-x-1-cm-square mirrored region during deposition. When present, the mirror was centered on the Y-shaped reaction zone junction where the sample and peroxide stream met. Design PCRD1 was used for optimizing luminol and peroxide concentrations and evaluating the difference between double- and single-T injection modes and design PCRD2 was used for optimizing PMT bias, PMT operating temperature, reaction pH, comparing lens numerical apertures, and evaluating the effect of channel depth. This design, with an integrated mirror, was used in immunoassay applications. The sample is placed in reservoir A, luminol in B, and H_2O_2 in D. Reser-

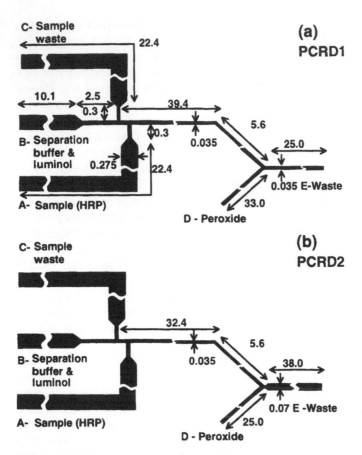

Figure 16 Schematic layout of microchip designs for (a) PCRD1 and (b) PCRD2; dimensions in mm. Letters are referred to in the text and identify the solution introduction reservoirs and points where potentials were applied. Indicated channels dimensions are for 10-μm-deep devices, and are not repeated in (b) except where differing from those in (a). (From Ref. 107, with permission.)

voirs C and E contain buffer with added luminol. A potential of −5 kV is applied to E, with A and B at ground and D at −1.1 kV. This produces a continuous stream of CL product flowing past the detector. The chip is translated in the x and y directions and the microscope in the z direction to maximize signal. The optimum occurs with the lens centered 0.1–1.0 mm from the Y intersection. The double-T injection is performed for 30 s, with 1.2 kV applied between A (at ground) and B, forming a geometrically defined sample plug of about 60 pL. Much larger plugs are formed with a single-T injection, which uses the same first step as the double-T, followed by 0.5–5 s with −3 to −6 kV at E, −1.2 kV at C, and A and B at ground. In both cases, separation uses −6 kV between B and E, with D at 1.3 kV. The detection limit obtained, 7–35 nM for HRP F1 for onchip CL, is about 50- to 100 fold lower than could be achieved for absorbance detection [109]. Using this microchip, separation and CL detection of the products of an immunological reaction of a fragment of the HRP conjugate of goat anti-mouse immunoglobulin G (IgG) with mouse IgG were performed.

4.2 Compact Detection Cells

Although research on the coupling of CE with CL detection has increased in recent years, the technique remains problematic. Most of the CE-CL detectors reported have involved variations of a postcapillary reactor to mix the reagents. The reactor requires insertion of the separation capillary into the reaction/detection capillary. These procedures are manually intensive and it is difficult to reproducibly control reagent concentrations at the reactor. To overcome these problems, a novel compact CL detection cell, made of PTFE, has been recently designed for CE [110]. This detection cell is equipped with an optical fiber, a fused-silica capillary, and a grounding electrode and it could easily be combined with CE equipment without any complex construction, expensive implements, tedious procedures, or special techniques (Fig. 17). The CL light generated at

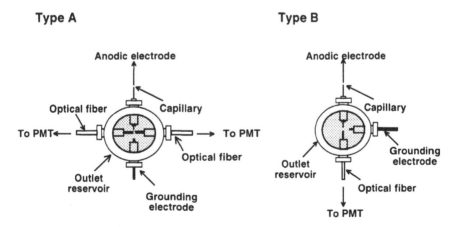

Figure 17 Schematic diagram of the CL detection cell. (From Ref. 110, with permission.)

the capillary outlet was transported by the optical fiber to a PMT. The luminol CL system was adapted for the use of this cell, carrying out detailed optimization of the concentration of hydrogen peroxide and catalyst. Using the cell, authors have obtained a detection limit for luminol of 5.10^{-10} M, which is the most sensitive result reported to date. Also a mixture of glycine, glycylglycine, and glycylglycylglycine, which was labeled with isoluminol isothiocyanate (ILITC), was sensitively detected and baseline-separated. Recently, the same group has developed a simpler, more convenient batch-type cell without an optical fiber for CE, to which only a fused-silica capillary and a grounding electrode were inserted [111]. In this sense, the cell also works as an outlet reservoir including the migration buffer and it is placed just in front of a photosensor module that captures directly the light generated at the capillary. Luminol, ILITC-labeled compounds, H_2O_2, and the catalyst are dissolved in a 10-mM phosphate buffer that is used as electrophoretic buffer. The catalyst is added to the inlet reservoir and H_2O_2 to the outlet reservoir (detection cell). Sample injections are performed by gravity for 10 s at a height of 20 cm. The sample migrates in the migration solution toward the CL detection cell and mixes with reagents generating the CL emission, which is captured by the detector. These CE-CL systems are expected to become a practical CL detection system for CE.

4.3   Online Solid-Phase CL Detector in CE

Recently, a novel online solid-phase CL detector has been designed for CE based on the strong CL signal observed when adding trace amounts of luminol, ABEI, or lucigenin in solution to milligram amounts of BaO_2 solid powder [112]. The oxidation reactions of luminol, ABEI, or lucigenin are relatively fast and the CL mechanism is considered to involve reactive oxygen species (O_2^-) existing on the surface of BaO_2 particles. The environment of the BaO_2 particles influenced the CL intensity significantly. Separations were carried out in 0.075 mm id $\times$ 0.375 mm od fused silica capillaries with a total length of 27 cm. The detection section was 0.3 cm in length. The preparation of this online CE-CL was finished in two steps, packing the BaO_2 powder into the detection window section and filling the electrolyte solution into the capillary using a microsyringe. Sample introduction was achieved by electroinjection method at a constant voltage (200 V/cm) for a fixed period (5–10 s). The light generated was collected onto a PMT connected to the recorder to produce the electropherogram. To avoid the BaO_2 particles moving from the detection window, the flow rate of the buffer solution must be controlled at lower than 1 cm/mm. Using a pH 8.3 0.05 M sodium borate solution as running buffer with 5% of acetonitrile, luminol, lucigenin, and ABEI were detected online after separation, providing detection limits of 1×10^{-8} M, 5×10^{-8} M, and 7×10^{-8} M, respectively.

5. CONCLUSIONS

The versatility and the robustness of CE separation in conjunction with the extreme sensitivity inherent to CL-based reactions make a combination of both techniques promising for application in a wide range of fields, including environmental analysis, biomedicine, and biological research and practice. Obviously, in comparison with other detection modes widely incorporated in CE, CL detection is a slowly evolving technique and advances should focus on the development of new detectors that are instrumentally simpler than existing systems and that offer the ability to detect various types of analytes at trace levels.

With the advancing automatization and computerization of CE instruments, the application of micromachining techniques, and the improvement of the devices for coupling CE with CL detection, it is hoped that both techniques may be incorporated in the future as suitable methodology in routine laboratories, being complementary to classical techniques such as HPLC and offering new alternatives to the analytical chemist.

REFERENCES

1. TA Nieman. In: WRG Baeyens, D De Keukeleire, K Korkidis, eds. Luminescence Techniques in Chemical and Biochemical Analysis. New York: Marcel Dekker, 1995, pp 523–561.
2. AC Calokerinos, NT Deftereos, WRG Baeyens. J Pharm Biomed Anal 13:1063–1071, 1995.
3. MJ Navas, AM Jiménez. Food Chem 55:7–15, 1996.
4. AM Jiménez, MJ Navas. Crit Rev Anal Chem 27:291–305, 1997.
5. PJM Kwakman, UATh Brinkman. Anal Chim Acta 266:175–192, 1992.
6. K Robards, PJ Worsfold. Anal Chim Acta 266:147–173, 1992.
7. SW Lewis, D Price, PJ Worsfold. J Biolum Chemilum 8:183–199, 1993.
8. AR Bowie, G Sander, PJ Worsfold. J Biolum Chemilum 11:61–90, 1996.
9. MG Sander, KN Andrew, PJ Worsfold. Anal Commun 34:13H–14H, 1997.
10. LP Palilis, AC Calokerinos, WRG Baeyens, Y Zhao, K Imai. Biomed Chromatogr 11:85–86, 1997.
11. WRG Baeyens, SG Schulman, AC Calokerinos, Y, Zhao, AM García-Campaña, K Nakashima, D De Keukeleire. J Pharm Biomed Anal 17:941–953, 1998.
12. AM García-Campaña, WRG Baeyens, XR Zhang, E Smet, G Van der Weken, K Nakashima, AC Calokerinos. Biomed Chromatogr 14:166–172, 2000.
13. A Tiselius. Trans Faraday Soc 33:524–530, 1937.
14. R Kuhn, S Hoffstetter-Kuhn. Capillary Electrophoresis: Principles and Practice. Berlin: Springer-Verlag, 1993, pp 161–232.
15. T Blanc, DE Schaufelberger, NA Guzman. In: GW Ewing, ed. Analytical Instrumentation Handbook, 2nd ed. New York: Marcel Dekker, 1997, pp 1351–1431.

16. RL St. Claire. Anal Chem 68:569R–586R, 1996.
17. SFY Li, ed. Capillary Electrophoresis: Principles, Practice and Applications, Vol. 52, Journal of Chromatography Library, Amsterdam: Elsevier, 1992.
18. PD Grossman, JC Colburn, eds. Capillary Electrophoresis: Theory and Practice; New York: Academic Press, 1992.
19. Z Deyl, F Tagliaro, I Mikski. J Chromatogr B 656:3–27, 1994.
20. NA Guzman, ed. Capillary Electrophoresis Technology. New York: Dekker, 1993.
21. R Weinberger. Practical Capillary Electrophoresis. San Diego: Academic, 1993.
22. JP Landers, ed. Handbook of Capillary Electrophoresis. Boca Ratón: CRC, 1994.
23. RA Hartwick. Introduction to Capillary Electrophoresis. Boca Ratón: CRC, 1994.
24. KD Altria. Capillary Electrophoresis Guidebook. London: Chapman & Hall, 1995.
25. DR Baker. Capillary Electrophoresis. New York: Wiley, 1995.
26. PG Righetti, ed. Capillary Electrophoresis in Analytical Biotechnology. Boca Ratón: CRC, 1996.
27. SC Beale. Anal Chem 70:279R–300R, 1998.
28. MG Khaledi, ed. High-Performance Capillary Electrophoresis. Theory, Techniques and Applications. New York: Wiley, 1998.
29. C Cruces Blanco. Electroforesis Capilar; Almería: Servicio de publicaciones Univ. de Almería, 1998.
30. MV Dabrio, GP Blanch, A Cifuentes, JC Diez Masa, M de Frutos, M Herraiz, I Martinez de Castro, J Sanz Perucha. Cromatografia y Electroforesis en Columna. Barcelona: Springer, 2000.
31. SR Rabel, JF Stobaugh. Pharm Res 10:171–186, 1993.
32. CA Morning, RT Kennedy. Anal Chem 66:280R–314R, 1994.
33. D Perret, G Ross. Trends Anal Chem 11:156–163, 1992.
34. E Jellum. J Cap Elec 1:97–105, 1994.
35. DJ Anderson, F Van Lente. Anal Chem 67:377R–524R, 1995.
36. JP Landers. Clin Chem 41:495–509, 1995.
37. MA Jenkins, MD Guerin. J Chromatogr B 682:23–34, 1996.
38. NA Guzman, SS Park, DE Schaufelberger, L Hernández, X Páez, P Rada, AJ Tomlinson, S Naylor. J Chromatogr B 697:37–66, 1997.
39. CA Kuffner, E Marchi, JM Morgado, CR Rubio. Anal Chem 68:241A–246A, 1996.
40. AJ Tomlison, NA Guzman, S Naylor. J Cap Elec 6:247–266, 1995.
41. NHH Heegaard, S Nilsson, NA Guzman. J Chromatogr B 715:29–54, 1998.
42. NA Guzman. LC-GC 17:16–27, 1999.
43. P Jandik, G Bonn, Capillary Electrophoresis of Small Molecules and Ions. New York: VHC, 1993, Chapter 3.
44. LN Amankwa, M Albin, WG Kuhr. Trends Anal Chem 11:114–120, 1992.
45. AT Timperman, JV Sweedler. Analyst 121:45R–52R, 1996.
46. R Zhu, WTh Kok. J Pharm Biomed Anal 17:985–999, 1998.
47. HA Bardelmeijer, H Lingeman, C de Ruiter, WJM Underberg. J Chromatogr 807: 3–26, 1998.
48. L Hernández, N Joshi, J Escalona, NA Guzman. J Chromatogr 559:183–196, 1991.
49. DY Chen, NJ Dovichi. J Chromatogr B 657:265–269, 1994.
50. AT Timperman, K Khatib, JV Sweedler. Anal Chem 67:139–144, 1995.
51. E González, JJ Laserna. Quím Anal 16:3–15, 1997.

52. SL Pentoney Jr, JV Sweedler. In: JP Landers, ed. Handbook of Capillary Electrophoresis. Boca Ratón: CRC, 1994.
53. J Cai. JD Henion. J Chromatogr A 703:667–692, 1995.
54. N Wu, TL Peck, AG Webb, RL Magin, JV Sweedler. J Am Chem Soc 116:7929–7930, 1994.
55. K Albert. J Chromatogr 703:123–147, 1995.
56. JH Flanagan Jr, BL Legendre Jr, RP Hammer, SA Soper. Anal Chem 67:341–347, 1995.
57. DC William, SA Soper. Anal Chem 67:3427–3432, 1995.
58. S Tracht, V Toma, JV Sweedler. Anal Chem 66:2382–2389, 1994.
59. SV de Griend, CE Kientz, UAT Brinkman. J Chromatogr A 673:299–302, 1994.
60. J Wu, J Pawliszyn. J Chromatogr B 657:327–332, 1994.
61. PL Weber, SM Lunte. Electrophoresis 17:302–309, 1996.
62. AM Fermier, ML Gostkowski, LA Colón. Anal Chem 68:1661–1664, 1996.
63. M Zhong, SM Lunte. Anal Chem 68:2488–2493, 1996.
64. AG Ewing, RA Wallinford, TM Olefirowicz. Anal Chem 61:292A–303A, 1989.
65. WRG Baeyens, B Lin Ling, K Imai, AC Calokerinos, SG Schulman. J Microcol Sep 6:195–206, 1994.
66. AM García-Campaña, WRG Baeyens, Y Zhao. Anal Chem 69: 83A–88A, 1997.
67. TD Staller, MJ Sepaniak. Electrophoresis 18:2291–2296, 1997.
68. AM García-Campaña, WRG Baeyens, NA Guzman. Biomed Chromatogr 12:172–176: 1998.
69. T Hara, S Okamura, J Kato, J Yokogi, R Nakajima. Anal Sci 7:261–264, 1991.
70. T Hara, J Yokogi, S Okamura, S Kato, R Nakajima. J Chromatogr A 652:361–367, 1993.
71. T Hara, S Kayama, H Nishida, R Nakajima. Anal Sci 10:223–225, 1994.
72. T Hara, H Nishida, S Kayama, R Nakajima. Bull Chem Soc Jpn 67:1193–1195, 1994.
73. T Hara, H Nishida, R Nakajima. Anal Sci 10:823–825, 1994.
74. K Tsukagoshi, A Tanaka, R Nakajima, T Hara. Anal Sci 12:525–528, 1996.
75. K Tsukagoshi, H Akasaka, R Nakajima, T Hara. Chem Lett 467–468, 1996.
76. K Tsukagoshi, Y Okumura, H Akasaka, R Nakajima, T Hara. Anal Sci 12:869–874, 1996.
77. K Tsukagoshi, Y Okumura, R Nakajima. J Chromatogr A 813:402–407, 1998.
78. N Wu, CW Huie. J Chomatogr 634:309–315, 1993.
79. MA Ruberto, ML Grayeski. Anal Chem 64:2758–2762, 1992.
80. DJ Rose, IW Jorgenson. J Chromatogr 447:117–131, 1988.
81. MA Ruberto, ML Grayeski, J Microcol Sep 6:545–550, 1994.
82. R Dadoo, LA Colón, RN Zare. J High Resol Chromatogr 15:133–135, 1992.
83. JY Zhao, J Labbe, NJ Dovichi. J Microcol Sep 5:331–339, 1993.
84. R Dadoo, AG Seto, LA Colón, RN Zare. Anal Chem 66:303–306, 1994.
85. SD Gilman, CE Silverman, AG Ewing. J Microcol Sep 6:97–106, 1994.
86. SY Liao, YC Chao, CW Whang. J High Resol Chromatogr 18:667–669, 1995.
87. SY Liao, CW Whang. J Chromatogr A 736:247–254, 1996.
88. B Huang, J Li, L Zhang, J Cheng. Anal Chem 68:2366–2369, 1996.
89. Y Zhang, J Cheng. J Chromatogr A 813:361–368, 1998.

90.　Y Zhang, B Huang, J Cheng. Anal Chim Acta 363:157–163, 1998.
91.　K Tsukagoshi, S Fujimura, R Nakajima. Anal Sci 13:279–281, 1997.
92.　LL Shultz, JS Stoyanoff, T. Nieman. Anal Chem 68:349–354, 1996.
93.　A Knight. Trends Anal Chem 18:47–62, 1999.
94.　GA Forbes, TA Nieman, JV Sweedler. Anal Chim Acta 347:289–293, 1997.
95.　K Tsukagoshi, K Miyamoto, E Saiko, R Nakajima, T Hara, K Fujinaga. Anal Sci 13:639–642, 1997.
96.　JA Dickson, MM Ferris, RE Milofsky. J High Resol Chromatogr 20:643–646, 1997.
97.　DR Bobbit, WA Jackson, HP Hendrickson. Talanta 46:565–572, 1998.
98.　X Wang, DR Bobbitt. Anal Chim Acta 383:213–220, 1999.
99.　YY Lee, CW Whang. J Chromatogr A 771:379–384, 1997.
100.　Y Yao, TA Taylor, M Fuchs. Unpublished data. PerSeptive Biosystems. Framingham, MA, 1996.
101.　MF Regehr. FE Regnier. J Capillary Electrophoresis 3:117–124, 1996.
102.　DJ Harrison, PG Glavina. A Manz Sens Actuators B 10:107–116, 1993.
103.　GTA Kovac, K Petersen, M Albin. Anal Chem 68:407A–412A, 1996.
104.　A Manz, JC Fettinger, EMJ Verpoorte, H Lüdi, DJ Harrison, HM Widmer, DJ Harrison. Trends Anal Chem 10:144–149, 1991.
105.　AM García-Campaña, WRG Baeyens, HY Aboul-Enein, X Zhang. J Microcol Sep 10:339–355, 1998.
106.　MD Luque de Castro, L Gámiz-Gracia. Anal Chim Acta 351:23–40, 1997.
107.　SD Mangru, DJ Harrison. Electrophoresis 19:2301–2307, 1998.
108.　TA Nieman. In: A Townshend, ed. Encyclopedia of Analytical Science. New York: Academic Press, 1995, pp 608–621.
109.　Z Liang, N Chiem, G Ocvirk, T Tang, K Fluri, DJ Harrison. Anal Chem 68:1040–1046, 1996.
110.　M Hashimoto, K Tsukagoshi, R Nakajima, K Kondo. J Chromatogr A 832:191–202, 1999.
111.　K Tsukagoshi, T Nakamura, M Hashimoto, R Nakajima. Anal Sci 15:1047–1048, 1999.
112.　JM Lin, H Goto, M Yamada. J Chromatogr A 844:341–348, 1999.

16

Bioanalytical Applications of Chemiluminescent Imaging

Aldo Roda, Patrizia Pasini, Monica Musiani, Mario Baraldini, Massimo Guardigli, Mara Mirasoli, and Carmela Russo
University of Bologna, Bologna, Italy

1. INTRODUCTION

Chemiluminescence (CL) is the light emission produced by a chemical reaction in which chemically excited molecules decay to the ground state and emit photons.

Bioluminescence (BL) is a type of CL that occurs naturally in living organisms and is also used in vitro. Measurement of light from a chemical reaction is very useful from an analytical point of view, because in appropriate experimental conditions the light output intensity is directly related to the analyte concentration, thus allowing precise and accurate quantitative analysis. In addition, the kinetics of light emission is usually a steady-state glow type, which simplifies sample handling and measurement procedures.

CL as an analytical tool has several advantages over other analytical techniques that involve light (mainly absorption spectroscopy and fluorometry): high detectability, high selectivity, wide dynamic range, and relatively inexpensive instrumentation.

The superior detectability of CL and BL measurements is partly due to a low background. In luminescence measurements two components of light reach the detector: the first one (i.e., the net analytical signal) is proportional to the analyte concentration, while the second component (i.e., the background) is an approximately constant light level due to various factors such as the phosphorescence of plastics, impurities in the reagents, emission from other sample components, detector dark current. Warmup and drift of light source and detector and interference from light scattering present in absorption and fluorescence methods are absent in CL, making the background light component much lower and thus achieving a significant gain in sensitivity.

Selectivity derives from the fact that the analyte of interest generates its signal in the presence of compounds that normally interfere in fluorescence measurement and that do not themselves produce light when the chemiluminescent reagents are mixed together.

Wide dynamic ranges allow samples to be measured across decades of concentrations without dilution or modification of the sample cell. This is due to the way the chemiluminescent signal is generated and measured, i.e., using no excitation source for light production and a phototransducer with an inherent wide range of response for light detection. The light emitted from chemi- and bioluminescent reactions is typically measured using a luminometer. Luminometers are simple, relatively inexpensive instruments designed to measure sample light output, generally by integrating light emission for a given period. All luminometers basically consist of a sample chamber, a detector, and a signal-processing apparatus, and are used to measure emission from different sample formats (the most common being single tubes and microtiter plates).

Photodiodes and photomultiplier tubes (PMTs) are the detection devices commonly found in commercial luminometers. Even if improvements in photodiodes have made them suitable for some applications, PMTs are still the detectors of choice for measuring extremely low levels of light.

A relatively recent advancement in light-detection technology for analytical purposes is represented by low-light imaging devices based on intensified Vidicon tubes or high-sensitivity charge-coupled devices (CCDs). These luminescence imaging instruments, also known as luminographs, allow not only the measurement of light intensity at the single-photon level, but also the spatial distribution of the light emission on a target surface to be evaluated.

CL is utilized in various analytical techniques in which small amounts of analytes are detected and quantitated by measurement of the light emission [1–3]. Chemiluminescent reaction systems often involve enzymes, such as alkaline phosphatase (AP) and horseradish peroxidase (HRP), and suitable CL substrates that allow for the detection of enzymes with very high efficiency [4–7]. These enzymes are widely used as labels in the development of immunoassays [8, 9], blotting [10, 11], and gene probe assays [12, 13]. Coupled enzymatic bio- and chemiluminescent analytical methods have also been developed: in particular, ATP-involving reactions (kinases) have been coupled with the firefly luciferin-luciferase system, NAD(P)H-producing or -consuming enzymatic reactions (dehydrogenases) with bacterial luciferases, and the luminol/H_2O_2/HRP system has been coupled with oxidase enzymes [14–18]. Advances in molecular biology and the increasing need for ultrasensitive assays have led to the development of novel luminescence systems for a wide variety of applications in genetic research, food technology, environmental monitoring, and clinical chemistry, so that CL-based immunoassay and gene detection kits became commercially available and routinely used [8, 9, 12, 13].

CL imaging also represents a promising detection system that is increasingly used for ultrasensitive quantitation and localization of analytes in a wide range of applications [19, 20]. CL imaging is suitable for filter membrane biospecific reactions, such as the southern, northern, or western blot tests, and dot blot hybridization reactions. In these techniques nucleic acids or proteins are either blotted on filter membrane after separation by gel electrophoresis, or directly dotted on the membrane. Nucleic acids are then hybridized with a complementary gene probe labeled with a hapten and detected by antihapten antibody conjugated with an enzyme and CL substrate; proteins are detected by specific antibody followed by antiantibody conjugated with an enzyme and CL substrate. The main advantage with respect to other detection systems (i.e., colorimetric or even chemiluminescent with photographic detection) is the direct and rapid quantitative evaluation of the signal over a wide dynamic range. It can be used for the measurement of CL signals in microtiter plates, with the advantage of a one-step measurement of the emission from the whole plate (differently from conventional luminometers, which usually read the emission well by well or strip by strip). This may be an important advantage in analytical methods relying on the kinetic behavior of the CL emission. CL imaging is also used to detect BL and CL in whole organs or plants, and on any kind of surface. It should be noted that CL imaging devices do not possess the very wide dynamic range characteristic of photomultiplier tube-based instruments. However, this limitation can be partly overcome by taking advantage of the intrinsic wide dynamic range of CL as detection principle.

When coupled with an optical microscope, CL imaging is a potent analytical tool for the development of ultrasensitive enzymatic, immunohistochemistry (IHC) and in situ hybridization (ISH) assays, allowing spatial localization and semiquantitative evaluation of the distribution of the labeled probe in tissue sections or single cells to be performed.

In this chapter we report recent analytical applications of CL imaging for the detection of biospecific reactions in macrosamples such as microtiter plates of different format (96 or 384 wells), filter membranes and irregular surfaces represented by specimens related to the cultural heritage, and results obtained when the CCD detector is coupled with optical microscopy for enzyme localization, immunohistochemical reactions, and complementary DNA (cDNA) detection (Table 1).

Table 1 Summary of the Applications of BL/CL Imaging Described in the Present Chapter

Sample formats	BL/CL systems	Applications
Filter membranes	HRP/H_2O_2/luminol AP/dioxetanes	Detection of nucleic acids and proteins
Microtiter plates	HRP/H_2O_2/luminol AP/dioxetanes Firefly luciferin/luciferase Bacterial luciferin/luciferase	Detection of enzymes and metabolites by direct or coupled enzyme reactions
		Determination of antioxidant and enzyme inhibitory activities
		Immunoassay
Irregular and flat surfaces	Firefly luciferin/luciferase HRP/H_2O_2/luminol AP/dioxetanes Firefly luciferin/luciferase Bacterial luciferin/luciferase	Detection of ATP as an indicator of microbial contamination
		Evaluation of the spatial distribution of immobilized biomolecules
Tissue sections and single cells	HRP/H_2O_2/luminol AP/dioxetanes Firefly luciferin/luciferase Bacterial luciferin/luciferase	Localization of enzyme activities, antigens by immunohistochemical techniques, and nucleic acids by in situ hybridization techniques

2. INSTRUMENTATION

The basic instrumentation for CL imaging includes an ultrasensitive video camera, an optical system, and appropriate software for image analysis [21].

The video camera must be characterized by high sensitivity and very low instrument noise to detect the weak CL emission. In the last few years high-performance CCD cameras have substituted for older devices based on Vidicons. Cooled, back-illuminated CCD cameras provide very high detection sensitivity owing to their high quantum efficiency and low background noise and allow a quantification of emitted light at a single-photon level [22–26]. Moreover, these devices do not require an image intensification step, which could negatively affect the image quality and the signal-to-noise ratio [19, 20, 23, 27]. However, relevant limitations of CCD-based imaging, in particular in quantitative analysis, are the narrow dynamic range (2–3 decades lower) and the lower sensitivity (5–10 times lower) with respect to photomultiplier-based detection [24].

For macrosample analysis, the CCD device is connected to standard or custom camera optics and enclosed in a lighttight box. In this configuration the instrument can be used for reading CL emission in 96- or 384-well microtiter plates (luminographs able to read four standard plates simultaneously are commercially available), on target surfaces such as gels, thin-layer chromatography plates, or dot blot membranes [28], and in other kinds of samples. Several commercial instruments can be used not only for detecting CL emission, but also for fluorescence (using additional UV sources) or to perform densitometric measurements using a transilluminator. In this way even fluorescence or colorimetric measurements can take advantage of the high sensitivity of the CL detector and the easy acquisition and quantification of the signal.

In conjunction with an optical microscope, a CCD camera can be used to localize the light emission from tissues or cells, and to obtain semiquantitative information on the localization of the probed species [27, 29–35]. Also in this case, to avoid interference from ambient light, a lighttight box is required. Owing to the very low light intensities involved in CL measurements, the potential loss of light in the optical system should be minimized to achieve the maximum analytical detectability. Therefore, the microscope should have a lens coupling system that is as simple as possible and objectives with the highest numerical aperture compatible with focal aberration and depth of field.

In CL measurements many factors that influence the intensity of the CL signal should be taken into account. The CL signal may depend on the geometry of the sample. Internal refraction and reflection at the air-solution interfaces are important factors in determining the measured CL intensity, and should be taken into account, for example, when a CL cocktail is placed over a sample. The effect of sample geometry can be evaluated using model systems, such as enzymes chemically immobilized on calibrated nylon net, controlled-porosity glass particles, and macroporous acrylic beads [25, 36]. In general samples and references should be analyzed employing the same experimental conditions to minimize any kind of interference due to the experimental setup.

The CL signal measured in a white microtiter plate is several times higher than that obtained in a black one, owing to the higher reflectivity of the well surfaces in the white plate. In preliminary experiments, we observed that when the CL signal is measured in a 384-well microtiter plate using a luminograph the values obtained for the wells lying in the peripheral zone of the plate are affected by serious underestimation (up to 30%), owing to the system geometry [37]. Since this effect is reproducible, depending only on the position and the volume content of the wells, the use of a correction factor, which multiplies the measured CL signal of each well depending on its position, made it possible to equalize the signal over the entire microtiter plate area.

When calculating the intensity of the light emitted by a sample using the data obtained with an imaging device, the light collection angle must be considered [23]. The light collection solid angle (Ω) of an optical device is related to the aperture of the collection lens by the relationship

$$\Omega = 2\pi(1 - \cos\alpha)$$

where α is one-half of the aperture angle (i.e., one-half of the light collection angle). The overall collection efficiency of the lens system (η) is given by the ratio between the light collection solid angle and the full solid angle (4π). Therefore,

$$\eta = (1 - \cos\alpha)/2$$

The maximum light collection efficiency ($\eta = 0.5$) is achieved when the emitting sample is in close contact with the detector. When the detector is coupled with a lens, the light collection efficiency may be much lower.

It should be pointed out that the quality of a CL image (and therefore the precision of a quantitative measurement or the sharpness of a CL probe localization) depends not only on the absolute value of the CL signal but also on its signal-to-noise ratio. Since most of the low-light imaging devices are integrating devices, the signal-to-noise ratio can easily be improved by increasing the exposure time, which means accumulating the CL signal for a longer period [23]. In the ideal case, an increase in the exposure time by a factor N determines an increase of the signal-to-noise ratio of a factor $\sqrt{N}$.

In our studies the detection and analysis of chemiluminescent signals were performed using two different high-performance, low-light-level imaging apparatuses, the Luminographs LB 980 and LB 981 (EG&G Berthold, Bad Wildbad, Germany), which permit emitted light measurement at the single-photon level.

The instrument setup and CL imaging processing are quite similar for the two devices, which differ as concerns the video systems. The LB 980 is provided with a 1-in. Saticon, high-dynamic-range pickup tube (which is a Vidicon-type tube with Se-As-Tl light target photoconductor) linked to an image intensifier, by high-transmission lenses, and also to a videoamplifier. The LB 981 lumino-graph, based on a back-illuminated, slow-scan, cooled CCD without intensifica-tion stages, was used when higher detectability was required. In both cases the video system is connected to a PC for quantitative image analysis, and a sample dark box is provided to prevent contact with external light. The videocamera can be connected to a Model BH-2 Optical Microscope (Olympus Optical, Tokyo, Japan) also enclosed in a dark box. The system operates in the following consecu-tive steps: (1) samples are recorded as transmitted light; (2) the luminescent signal is measured with an optimized photon accumulation lasting 1 min; and (3) after a computer elaboration of the luminescent signal with pseudocolors correspond-ing to the light intensity, an overlay of the images on the screen provided by the transmitted light and by the luminescent signal allows the spatial distribution of the target analytes to be localized and evaluated. The light emission from the sample is quantified by defining a fixed area and counting the number of photon fluxes from within that area.

3. CHEMILUMINESCENCE REAGENTS IN IMAGING

Labeling with enzymes is generally preferred to labeling with CL substances because of the possibility to obtain a higher sensitivity owing to amplification of the CL signal in the presence of an excess of the CL substrate. Moreover, with enzyme labels glow-type light emission kinetics can be usually obtained (e.g., the CL emission reaches a steady-state intensity), which permits both easy handling and standardization of the experimental conditions and quantitation of the labeled probe under investigation, because the steady-state light intensity is directly related to the enzyme activity.

AP and HRP are probably the most widely used CL labels, owing to the availability of many different CL substrates that originate glow-type emis-sion kinetics and permit the detection of very small amounts (in the order of 10^{-18}–10^{-21} mol) of CL label. The most sensitive and widely investigated CL substrates for AP are based on adamantyl-1,2-dioxetane aryl phosphate deriva-tives [5, 13], while commercial substrates for HRP are mainly luminol-based reagents (Fig. 1). Adamantyl 1,2-dioxetane aryl phosphates are dephosphorylated by AP to give an unstable intermediate; decomposition of this intermediate pro-duces an excited-state aryl ester that emits light. Luminol and its derivatives are oxidized by peroxides (usually hydrogen peroxide) giving light emission. This

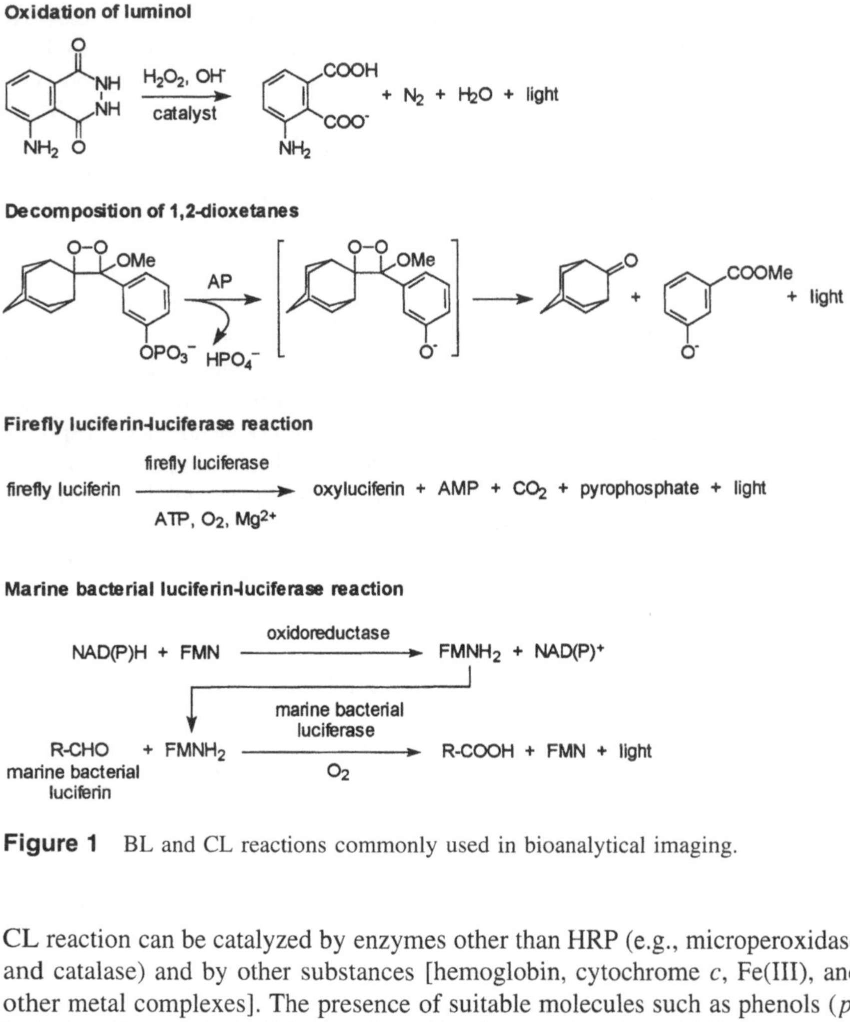

Oxidation of luminol

luminol $\xrightarrow[\text{catalyst}]{H_2O_2,\ OH^-}$ (aminophthalate) $+\ N_2\ +\ H_2O\ +\ \text{light}$

Decomposition of 1,2-dioxetanes

dioxetane phosphate $\xrightarrow[HPO_4^-]{AP}$ [dioxetane anion] $\longrightarrow$ ketone $+$ methyl ester $+\ \text{light}$

Firefly luciferin-luciferase reaction

firefly luciferin $\xrightarrow[\text{ATP, } O_2,\ Mg^{2+}]{\text{firefly luciferase}}$ oxyluciferin $+$ AMP $+$ CO_2 $+$ pyrophosphate $+$ light

Marine bacterial luciferin-luciferase reaction

NAD(P)H $+$ FMN $\xrightarrow{\text{oxidoreductase}}$ FMNH$_2$ $+$ NAD(P)$^+$

R-CHO $+$ FMNH$_2$ $\xrightarrow[O_2]{\text{marine bacterial luciferase}}$ R-COOH $+$ FMN $+$ light
(marine bacterial luciferin)

Figure 1 BL and CL reactions commonly used in bioanalytical imaging.

CL reaction can be catalyzed by enzymes other than HRP (e.g., microperoxidase and catalase) and by other substances [hemoglobin, cytochrome c, Fe(III), and other metal complexes]. The presence of suitable molecules such as phenols (p-iodophenol), naphthols (1-bromo-2-naphthol), or amines (p-anisidine) increases the light production deriving from the HRP-catalyzed oxidation of luminol and produces glow-type kinetics [6, 7]. The use of other enzymes, such as glucose-6-phosphate dehydrogenase [38–41], β-galactosidase [42], and xanthine oxidase [43–46], as CL labels has been reported.

Bioluminescent reactions are also employed for imaging purposes, in particular the firefly and the bacterial luciferin/luciferase ones (Fig. 1). The firefly luciferin/luciferase reaction requires ATP, magnesium ions, and oxygen. Many different luciferins and mutant luciferases have been investigated to optimize the reaction performance. Bioluminescent bacterial systems involve bacterial lucifer-ases that catalyze the oxidation of $FMNH_2$ and bacterial luciferin (a long-chain aldehyde) in the presence of oxygen. In vivo, the luciferase is coupled to an oxidoreductase that catalyzes the oxidation of NAD(P)H leading to the formation of $FMNH_2$. In vitro, $FMNH_2$ can also be obtained by various chemical means.

Most of the commercially available CL substrates can, in principle, be used for imaging purposes. However, in general neither the substrates nor the experi-mental conditions are optimized for imaging. A crucial factor in determining the applicability of a CL substrate in imaging, particularly as concerns their use in the imaging of microsamples through an optical microscope [47], is the diffusion of the excited species responsible for the CL emission. In fact, to achieve a sharp localization of the labeled probe, the light emission should occur as near as possi-ble to the site of the primary biospecific recognition. The use of CL substrates that produce excited species with quite long half-life results in poor spatial resolu-tion in localization of the labeled probe because the excited species diffuse in the solution before light emission occurs. CL substrates suitable for imaging should therefore be characterized by very short-lived excited species to make diffusion of the emitting species negligible. Resolution can also be improved by increasing the viscosity of the chemiluminescent cocktail solution by adding substances like gelatin, glycerol, or polyvinylpyrrolidone at suitable concentrations.

The problem related to the diffusion of the species involved in the CL reaction is even greater when CL derives from a chain of enzymatic reactions, which is a rather common situation in enzyme activity determinations. In this case several diffusion processes, involving both the emitting species and the inter-mediate reactive species (or enzymes), could contribute to the decrease of spatial resolution in localization of the labeled probe. Therefore, careful optimization of the experimental conditions and of the concentrations of all reagents is required. The 3α-hydroxysteroid dehydrogenase (3α-HSD) enzyme, immobilized on a ny-lon net, was localized by means of coupled enzymatic reactions involving the bacterial enzymes FMN-oxidoreductase and luciferase. Localization of 3α-HSD was very poor with the two bioluminescent enzymes free in solution (Fig. 2a). It improved using a nylon net on which FMN-oxidoreductase was also immobi-lized (Fig. 2b), and became quite good with all the enzymes coimmobilized (Fig. 2c) [25, 48]. The image quality can be improved by increasing the viscosity of the solution and the concentration of the indicator bioluminescent enzymes.

The analytical detectability applying a CL method should, in principle, be comparable to that obtained using radioactive labels, without all the disadvan-tages related to the use of isotopic labeling. In fact, assuming reasonable values for the quantum efficiency of the chemiluminescent reaction ($\Phi_{CL} \sim 0.01$), for the overall photon collection efficiency of the optical system-CCD camera assem-bly ($\eta \sim 0.01\%$), and for the intensity of the lowest detectable CL signal (about

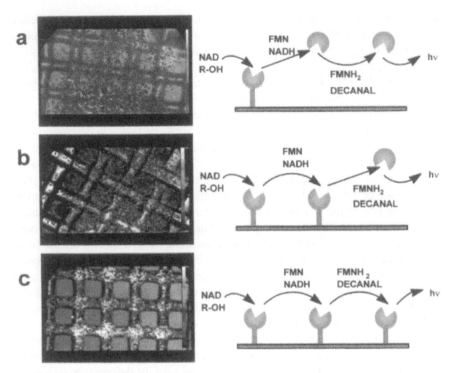

Figure 2 Effect of enzyme immobilization on luminescent image spatial resolution evaluated using coupled enzymatic reactions on nylon net as a model system. (a) Immobilized 3α-hydroxysteroid dehydrogenase; (b) immobilized 3α-hydroxysteroid dehydrogenase and FMN-NADH oxidoreductase; (c) immobilized 3α-hydroxysteroid dehydrogenase, FMN-NADH oxidoreductase, and bacterial luciferase. (From Ref. 47. Copyright John Wiley & Sons Ltd. Reproduced with permission.)

1 photon/s/pixel), the detection limit using a CL label may be of the order of 10^{-18} mol. If an enzyme is used as a label instead of a CL substrate, amplification factors of 10^3–10^4 can easily be obtained; thus the detection limit can be as low as 10^{-21} mol of enzyme. In practice, the detection limit depends not only on the absolute intensity of the CL signal, but also on the signal-to-noise ratio. Therefore, among different CL systems the maximum sensitivity is not necessarily obtained using the CL system that gives the maximum emission intensity, because even if it is desirable to have an intense absolute signal it is more important to have a high signal-to-noise ratio to improve the image quality and to allow sensitive and specific quantification of the analyte.

4. CHEMILUMINESCENT ANALYSIS OF MACROSAMPLES

4.1 Filter Membranes

As stated earlier, the main advantage of CL imaging with respect to other detection systems is the possibility to perform a direct and rapid quantitative evaluation of the signal over a relatively wide dynamic range [34]. The analytical performance of CL imaging is then comparable or superior to that of systems using different principles to detect immunological or genetic reactions, such as radioisotopes or color-producing substrates [11, 49], or even of CL with photographic detection. Moreover, CL images can be permanently recorded for archiving or further elaboration and easily exchanged with other laboratories.

Nucleic acid hybridization techniques are able to detect viral genomes directly in clinical samples, allowing rapid and sensitive diagnosis of viral infections, especially for viruses that do not grow in cell cultures or have a long replication cycle. In CL dot blot hybridization reactions the specimens are dotted on a membrane, hybridized with a specific gene probe labeled with a hapten, and then the hybrid is detected using an enzyme-conjugated antihapten antibody and a suitable CL substrate [50–52]. The analytical performances of several CL substrates for HRP or AP were compared in dot blot hybridization assays for the detection of B19 parvovirus DNA [32, 53]. The assays used digoxigenin-labeled DNA probes that were immunoenzymatically revealed using antidigoxigenin Fabs (antigen-binding fragments) conjugated with HRP or AP. The detection limits were between 0.5 and 2 pg of target homologous DNA using HRP as label, and between 10 and 50 fg of DNA using AP as label. Since the detection limits for colorimetric methods are about 5 pg and 100 fg when HRP and AP are used as labels, respectively, the chemiluminescent method was superior to colorimetry.

4.2 Microtiter Plate Format

Chemiluminescent imaging can be used to measure CL emission in microtiter plates, even if the performance of imaging devices is not comparable to that of the standard luminometers used for this purpose. In particular, both the sensitivity and the dynamic range of imaging devices are lower than those of photomultiplier-based microtiter plate readers. However, commercial imaging devices make it possible to measure simultaneously, in a minute or less, up to four microtiter plates (corresponding to up to 4×384 microtiter wells). This time is comparable to or shorter than that required by standard microtiter plate readers, making this technique interesting in the development of high-throughput-screening (HTS) analytical methods based on CL. Such methods could be used, for example, in screening the identity and biological activity of new compounds synthesized using combinatorial chemical synthesis. Another potential application field for imaging devices are CL analyses based on the kinetics of the CL emission [54]. In fact, standard luminometers, which read the CL emission well by well or strip by strip, would not be able to follow relatively fast-emission kinetics across an entire microtiter plate. Despite these potential advantages of imaging techniques, as far as we know no data have been reported in the literature on this type of application.

A CL method has been used to evaluate the antioxidant activity of natural compounds, in comparison to that of a reference compound, in a 384-well microtiter plate [37, 55]. In fact, antioxidant compounds are able to temporarily interrupt the light output of a luminol-based CL system because they act as radical scavengers and block the formation of the luminol radicals that drive the CL reaction. Light emission is restored after an interval that is directly proportional to the amount and activity of the antioxidant added to the CL solution. The same analytical format has been used for determination of the activity of acetylcholinesterase (AChE) inhibitors [37, 56]. For this determination, AChE was involved with choline oxidase and HRP in a chain of enzymatic reactions leading to light emission. Under suitable experimental conditions (i.e., excess of both choline oxidase and HRP) the intensity of the CL emission is proportional to the activity of AChE. Therefore, the addition of AChE inhibitors to the system results in a reduction of the light output proportional to the enzyme inhibition. Both methods were proven to be suitable for analysis of the biological activity of a large number of samples (about 200 for each 384-well microtiter plate) in a short period; the results obtained for the AChE inhibitors were comparable to those obtained using conventional color-producing AChE substrates to evaluate the inhibitory activity.

We have developed chemiluminescent immunoenzymatic assays for β-agonist drugs in the 96-well-microtiter-plate format. Such competitive assays have been used for determination of clenbuterol and of the overall content of β-agonist drugs in the sample. They matched the standard requirements of precision and accuracy, and were more sensitive compared to the conventional colorimetric methods. Moreover, CL detection was very rapid, making these assays suitable for screening analysis.

4.3 Irregular Surfaces

A rapid, nondestructive method based on determination of the spatial distribution of ATP, as a potential bioindicator of microbial presence and activity on monuments, artworks, and other samples related to the cultural heritage, was developed [57]. After cell lysis, ATP was detected using the bioluminescent firefly luciferin-luciferase system and the method was tested on different kinds of surfaces and matrices. Figure 3 reports the localization of biodeteriogen agents on a marble specimen. Sample geometry is a critical point especially when a quantitative analysis has to be performed; however, the developed method showed that with opti-

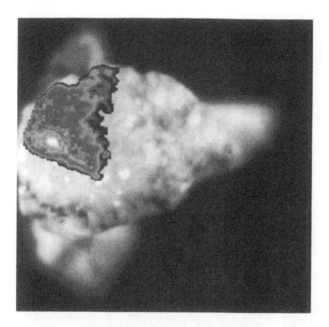

Figure 3 Localization of biodeteriogen agents on a marble sample by luminescent ATP detection using the firefly luciferin-luciferase system. (Courtesy of Dr. G. Ranalli, University of Molise, Campobasso, Italy.)

mized experimental conditions an accurate evaluation of the spatial distribution can be achieved. Along with rapidity, nondestructivity, and sensitivity, the most interesting feature of such a technique is the potential use as a rapid diagnostic tool for in situ applications.

4.4 Electrodes

CL imaging can also be a tool for evaluating the distribution of a protein (or another biomolecule) immobilized on a given solid support. It can be used, for example, to study the immobilization of a protein on the gold surface of a quartz microbalance electrode via disulfide bond formation with the activated metal surface. The quartz microbalance makes it possible to easily determine the amount of protein on the electrode, but does not give information on its distribution on the electrode surface. By using a suitably labeled protein, CL imaging could allow comparison of different immobilization procedures in terms of both the amount of protein immobilized and its distribution on the solid support. A protein A-HRP conjugate was immobilized on a quartz microbalance electrode (Fig. 4a)

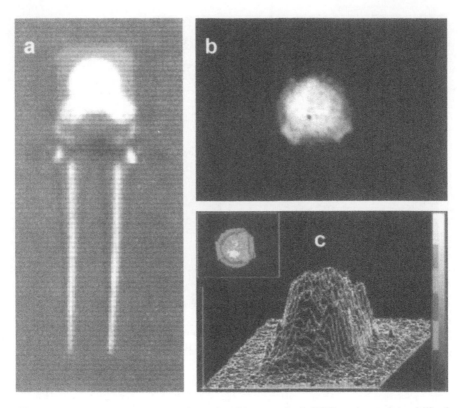

Figure 4 Quartz microbalance electrode with a protein A-HRP conjugate immobilized on the gold surface. (a) Transmitted light image; (b) chemiluminescent signal after addition of CL substrate; (c) 3-D display of the light signal spatial distribution.

and the light signal was obtained after the addition of a CL substrate for HRP (Fig. 4b). Figure 4c shows the pseudocolor-processed CL image and the 3D-display of the light signal, which allow a better evaluation of the immobilized protein distribution.

5. CHEMILUMINESCENT ANALYSIS OF MICROSAMPLES

The analytical performance of CL microscopy imaging in terms of detectability, precision, accuracy, and spatial resolution was previously reported by us [25]. The system allowed for the detection of 400 amol of enzymes such as HRP or AP, with a spatial resolution as low as 1 μm and very low background. CL coupled with optical microscopy thus represents a useful tool for enzyme, antigen, and DNA probe localization [29, 30] and is particularly suitable for those samples that require high detectability and, at the same time, quantitative information. This method can be applied to any kind of specimen such as fixed cells, tissue cryosections, and paraffin-embedded sections. The specimens are prepared following the same protocols developed for enzymatic, immunohistochemical, and in situ hybridization techniques with other detection systems. The resolving power of the CL image obtained using a videocamera connected to an optical microscope is comparable to that obtained with the optical microscope and a transmitted light image, making it possible to localize a CL signal in a tissue section or within a single cell, which is particularly suitable for analysis at the subcellular level [31–33, 35, 58].

CL mapping of the analytes in the specimen requires three steps: first, transmitted light image collection; second, addition of the CL cocktail and acquisition of the emitted light; third, after computer processing of the luminescent signal with pseudocolors corresponding to the light intensity, superimposition of the CL image with the live image to correctly localize the analyte. The multistep procedure can be facilitated by the use of an optical microscope equipped with a computer-assisted device able to precisely locate the same sample spots. With an automated system the detection of two or more reactions in the same specimen can also be easily performed, thus increasing the diagnostic power of such techniques.

Semiquantitative analysis can be performed once the system is optimized taking into account the nonspecific chemical and electronic background signal [25, 59]. Since quantification could be imprecise or inaccurate owing to factors such as sample shape, light-scattering phenomena, and internal reflectance in the reaction medium, negative and positive control experiments with the same type of specimens under the same experimental conditions must be performed and the results expressed as a relative value [47]. Therefore, a positive signal in CL microscopy imaging is generally considered to be above a predetermined threshold value (usually, the background value plus five standard deviations). An objective evaluation of the results can be achieved without any microscopic training thus minimizing uncertainties about positive or negative results. When tissue sections are analyzed, their thickness must be as reproducible as possible because there is a linear relationship between light emission and section thickness [35]. The analyte concentration in cell or tissue samples could be quantified using a calibration curve obtained by immobilizing different amounts of analyte on a target surface such as activated oxirane acrylic beads or nylon net, as previously reported [25, 36].

5.1 Enzyme Activity

Endogenous AP activity was detected in rabbit intestine cryosections by simply adding a few drops of CL substrate [25]. AP was localized in the epithelial cells with good resolution and sensitivity, the aspecific signal being very low (Fig. 5a). Diffusion of the chemiluminescent products was minimized by removing excess solution from over the cryosection, after which the chemiluminescent substrate was still in excess and embedded in the tissue. This method is superior to conventional histochemical colorimetric detection of AP, since experiments performed on the same specimen, first analyzed with CL and then by a conventional colorimetric technique, showed that CL is more sensitive and also permits a more accurate quantification of the enzyme [60].

Endogenous AChE activity was detected in rat coronal brain slices by using coupled enzymatic reactions terminating with light emission (AChE/choline oxidase/peroxidase) [56]. The developed CL reagent solution for AChE proved to be suitable for analyte localization on a target surface: it allowed direct localization of the spatial distribution of the enzyme in rat brain sections, with a sharp CL signal localized in cholinergic neurons but with very low background emission. The relative concentrations of the components of the CL reagent solution were optimized to avoid the diffusion of the light-emitting species in the surrounding solution. This imaging system could be a useful tool to study both the pathophysiological role of AChE distribution in brain and the effect of in vivo administration of enzyme-inhibiting drugs, providing a system that is more predictive than in vitro assay of inhibition activity.

5.2   Immunohistochemistry

IHC techniques provide important tools for the localization of specific antigens within individual cells. In CL IHC the probes used are highly specific antibodies that bind to antigens such as proteins, enzymes, and viral or bacterial products. The bound specific antibody is revealed indirectly by species-specific or class-specific secondary antibodies conjugated to CL enzymes.

A CL IHC technique was developed to localize interleukin 8 (IL-8) in gastric mucosa biopsy specimens, using monoclonal mouse anti-IL-8 and AP-labeled goat anti-mouse antibodies [47]. The CL IHC technique effectively localized IL-8 in the gastric mucosa (Fig. 5b), and an increased IL-8 content was observed in epithelial cells in the presence of *Helicobacter pylori* infection, thus confirming earlier results achieved with an immunofluorescence technique [61]. Alternatively, the immunochemiluminescent approach could be used to identify bacterial phenotype using antisera against known strain-specific virulence determinants. Such an approach could allow identification of bacterial phenotype in situ without microbial culture. A future development, consisting of the use of double immunochemiluminescence for quantifying epithelial expression of mediators in relation to luminal pathogenic agents, has been explored.

The main advantage of using immunochemiluminescence techniques is that they allow more sensitive detection and improved localization of infectious

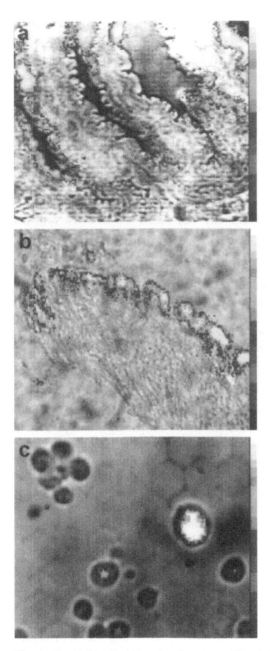

Figure 5 (a) Localization of endogenous AP activity in rabbit intestine cryosection by
addition of CL substrate; (b) epithelial localization of interleukin 8 in gastric mucosa
cryosection infected with *Helicobacter pylori* by immunohistochemistry; (c) localization
of cytomegalovirus DNA in infected human fibroblasts by in situ hybridization. [(a) From
Ref. 46. Copyright Academic Press. Reproduced with permission. (b) Courtesy of Dr.
J. E. Crabtree, St. James's University Hospital, Leeds, England.]

pathogens, and permit quantification of tissue antigens. Another advantage of the technique is that it allows examination of the immunopathogenic aspects of infections: it can be used to investigate host-pathogen interactions, in particular, inflammatory mediators in epithelial cells that are upregulated in response to infectious agents.

Immunocytochemical methods were developed to detect specific antibodies in sera from patients affected by different infectious diseases. P3-HR1 cells, which express Epstein-Barr-virus-induced virus capsid antigens (VCA), were used to search for specific human IgM (class M immunoglobulins) to VCA in infectious mononucleosis patients. After treatment of cells with serial dilutions of sera, HRP-conjugated anti-IgM antibody was added and detected with CL substrate [25].

Human embryo lung fibroblasts infected with a reference laboratory strain of herpes simplex virus (HSV) type 2 were used to detect antibody to HSV type 2 in serum samples. After treatment of cells with serial dilutions of sera, HRP-labeled immunoglobulins to human IgG (class G immunoglobulins) were added and detected with CL substrate [36]. In both cases a sharp detection of the specific antibodies was achieved with chemiluminescent assays, which proved more sensitive than the colorimetric immunoperoxidase assays.

5.3 In Situ Hybridization

ISH is a suitable method for the localization of specific nucleic acids inside individual cells with preservation of cellular and tissue morphology, thus permitting a simultaneous assessment of the morphological alterations associated with the lesion [62]. Moreover, the possibility of simultaneous detection of two or more CL probes can improve the diagnostic significance of such technique [63]. In recent years there has been a growing interest in the application of ISH for the rapid, specific, and reliable diagnosis of viral diseases, especially for those viruses that cannot be diagnosed by isolation procedures.

Different CL ISH assays for the detection of viral DNAs in tissue sections or single cells have been developed in our laboratory, taking advantage of the rapidity and sensitivity of CL detection and obtaining a reliable semiquantitative evaluation of the viral DNA presence.

A CL ISH assay for the detection of human papillomavirus (HPV) DNA was developed, in which the hybridization reaction was performed using either digoxigenin-, biotin-, or fluorescein-labeled probes [64]. The hybrids were visualized using AP as the enzyme label and a highly sensitive 1,2-dioxetane phosphate as chemiluminescent substrate. This assay was applied to biopsy specimens from different pathologies associated with HPV, which had previously proved positive for HPV DNA by polymerase chain reaction (PCR). The analytical sensitivity was assessed using samples of HeLa and CaSki cell lines, whose content in HPV DNA is known (10–50 copies of HPV 18 DNA in HeLa cells and 400–600 copies of HPV 16 DNA in CaSki cells). The CL ISH assay proved sensitive and specific using either digoxigenin-, biotin-, or fluorescein-labeled probes and provided an objective evaluation of the results. CL detection was more sensitive than colorimetry, being able to detect a positive signal in HeLa cells that was negative in the colorimetric ISH assay.

B19 parvovirus DNA was detected in bone marrow cells employing digoxigenin-labeled B19 DNA probes, antidigoxigenin Fab fragment conjugated with AP, and a 1,2-dioxetane phosphate CL substrate [32]. The CL ISH was applied to samples that had previously been tested for B19 DNA content using ISH with colorimetric detection, dot blot hybridization, and nested PCR. The CL assay proved specific and showed an increased sensitivity in detecting B19 DNA when compared to ISH with colorimetric detection, being able to find a higher number of positive cells per 100 counted cells with highly statistically significant difference.

Cytomegalovirus (CMV) DNA in cultured CMV-infected cells and in different clinical samples (tissue sections and cellular smears) was detected using digoxigenin-labeled probes, antidigoxigenin Fab fragments labeled with AP, and the adamantyl 1,2-dioxetane phenyl phosphate CL substrate [33]. The presence of hybridized CMV DNA was observed in infected cells fixed at various times after infection, and it was possible to measure increasing light emission values thus following the CMV replication cycle (Fig. 5c). When the assay was performed on clinical samples from patients with acute CMV infections, CMV DNA was detected in all the positive samples tested, in both cellular samples and tissue sections.

HSV in human fibroblasts was detected using biotin-labeled HSV DNA probes, streptavidin-HRP complex, and enhanced CL substrate reagent for HRP [56]. The presence of HSV DNA was observed in cells infected with clinical samples known to contain the HSV virus fixed at 48 h postinfection, with a sharp topographical localization and a good preservation of cellular morphology.

Chemiluminescent ISH can be used for the simultaneous detection of different DNA sequences in the same sample, by means of a double hybridization reaction in which probes to different DNA targets are labeled and detected using different enzymatic systems. The two different chemiluminescent detections have to be made sequentially, provided the first CL substrate is removed before the second CL analysis is performed by adding the other substrate. Therefore, in comparison with other existing ISH methods to detect two targets (for example, those using two different fluorescent labels), CL ISH has the important limitation of requiring sequential staining and signal visualization.

A CL ISH assay for simultaneous detection of two different viral DNAs (HSV and CMV DNAs) was developed utilizing both HRP and AP as reporter enzymes [63]. A biotinylated HSV DNA probe and a digoxigenin-labeled CMV DNA probe were cohybridized with samples; then CL detection of the two probes was performed. The HSV DNA was revealed using a streptavidin-HRP complex amplified with biotinyl tyramide and a luminol-based CL substrate for HRP, while CMV DNA was detected by means of antidigoxigenin Fab fragments conjugated with AP and a dioxetane phosphate derivative as CL substrate. Both HSV and CMV DNAs were found in infected cells with distinct localization and absence of cross-reactions. In the double CL ISH, the enzymatic reaction of HRP is usually performed before that of AP. This is because the HRP/luminol reaction rapidly reaches a steady-state light output that is maintained for a relatively short time, while the kinetics of the AP/dioxetane reaction is slower and the CL emission lasts for a relatively long time. This detection sequence requires shorter washing to remove the HRP substrate, to avoid any interfering photon emission during the AP reaction, making the assay more rapid. Consistent results were obtained when the substrate reaction sequence was reversed; however, longer washing was required.

In conclusion, CL ISH has proved more sensitive than ISH followed by colorimetric detection and can be as sensitive as ISH using radioactive labels [31, 32, 64, 65]. When compared with fluorometric detection, CL showed an improved signal-to-noise ratio, thus a higher specific detectability; furthermore, once standardized, it provided more precise and accurate quantitative results.

6. FUTURE PERSPECTIVES

A significant improvement in the CL technique would be enhancement of the transmitted light image quality, to facilitate morphological localization of the analyte. Cryosections, generally 5–10 μm thick, can be directly observed through the microscope with a phase-contrast system. A more significant improvement in image quality could also be obtained by staining the samples with appropriate dyes either after or before CL measurement, provided the staining does not interfere with the CL reaction.

Since CL imaging offers the possibility of measuring not only the light output but also its topographic distribution on a given area of the sample, it enlarges the application of biospecific reactions to simultaneous multianalyte detection by developing multiarray devices where different antigens, antibodies, or gene probes are immobilized in different areas. Miniaturized multiarray devices could be set up using chip microfabrication technology, the CL signal being evaluated by an imaging device coupled with an optical microscope or suitable optics.

An array or a matrix of nucleic acid probes immobilized at discrete locations on a silicon or glass surface provides a convenient means to simultaneously probe a sample for the presence of many different target sequences. Microarray biochip scanning devices, mostly based on fluorescent labels, are now currently available, and could also be used with CL labels to take advantage of the higher sensitivity of this detection principle.

A sensor array where different haptens are immobilized at well-defined areas on a plain glass surface has been developed [66]. Using an automated flow injection system it was possible to incubate all areas on the chip with analytes, specific antibodies, secondary HRP-labeled antibodies, and CL substrate. Measurement of the light output via imaging performed with a CCD device allowed determination of the analytes present in the sample on the basis of the spatial localization of the CL signal.

Imaging can also be useful for multiprobe detection, for example using fluorescent probes together with CL reactions. Potentially it is possible to detect first a fluorescent probe, and second detect the CL probe by adding the CL substrate. Probes marked with different CL labels, usually enzymes requiring different substrates, can also be used at the same time, provided the first CL substrate is removed before a second CL analysis is performed by adding another substrate.

REFERENCES

1. M DeLuca, ed.. Methods Enzymol. Vol 57. New York: Academic Press, 1978.
2. LJ Kricka, PE Stanley, GHG Thorpe, TP Whitehead. Analytical Applications of Bioluminescence and Chemiluminescence. London: Academic Press, 1984.
3. WRG Baeyens, D De Keukeleire, K Korkidis. Luminescence Techniques in Chemical and Biochemical Analysis. New York: Marcel Dekker, 1990.
4. I Bronstein, B Edwards, JC Voyta. J Biolumin Chemilumin 4:99–111, 1989.
5. S Beck, H Köster. Anal Chem 62:2258–2270, 1990.
6. GHG Thorpe, LJ Kricka. Methods Enzymol 133:311–354, 1986.
7. GHG Thorpe, LJ Kricka. In: J Scholmerich, R Andreesen, A Kapp, M Ernst, WG Woods, eds. Bioluminescence and Chemiluminescence: New Perspectives. Chichester: Wiley, 1987, pp 199–208.
8. LJ Kricka, RAW Stott, GHG Thorpe. In: WP Collins, ed. Complementary Immunoassays. Chichester: Wiley, 1988, pp 169–179.
9. Y Ashihara, H Saruta, S Ando, Y Kikuchi, Y Kasahara. In: AK Campbell, LJ Kricka, PE Stanley, eds. Bioluminescence and Chemiluminescence: Fundamentals and Applied Aspects. Chichester: Wiley, 1994, pp 321–324.
10. R Schneppenheim, P Rautenberg. Eur J Clin Microbiol 6:49–51, 1987.
11. I Bronstein, JC Voyta, KG Lazzari, O Murphy, B Edwards, LJ Kricka. BioTechniques 8:310–314, 1990.
12. JA Matthews, A Batki, C Hynds, LJ Kricka. Anal Biochem 151:205–209, 1985.
13. I Bronstein, JC Voyta, B Edwards. Anal Biochem 180:95–98, 1989.
14. A Roda, S Girotti, S Ghini, G Carrea. J Biolumin Chemilumin 4:423–435, 1989.
15. G Carrea, R Bovara, G Mazzola, S Girotti, A Roda, S Ghini. Anal Chem 58:331–333, 1986.
16. M DeLuca, WD McElroy. Methods Enzymol 133:331–584, 1986.
17. DC Vellom, LJ Kricka. Methods Enzymol 133:229–237, 1986.
18. PJ Worsfold, A Nabi. Anal Chim Acta 179:307–313, 1986.

19. JC Nicolas. J Biolumin Chemilumin 9:139–144, 1994.
20. CE Hooper, RE Ansorge, JG Rushbrooke. J Biolumin Chemilumin 9:113–122, 1994.
21. PE Stanley. J Biolumin Chemilumin 12:61–78, 1997, and references therein.
22. RA Wick. BioTechniques 7:262–268, 1989.
23. R Bräuer, B Lübbe, R Ochs, H Helma, J Hofmann. In: AA Szalay, LJ Kricka, PE Stanley, eds. Bioluminescence and Chemiluminescence: Status Report. Chichester: Wiley, 1993, pp 13–17.
24. CE Hooper, RE Ansorge. In: PE Stanley, LJ Kricka, eds. Bioluminescence and Chemiluminescence: Current Status. Chichester: Wiley, 1991, pp 337–344.
25. A Roda, P Pasini, M Musiani, S Girotti, M Baraldini, G Carrea, A Suozzi. Anal Chem 68:1073–1080, 1996.
26. CE Hooper, RE Ansorge, HM Browne, P Tomkins. J Biolumin Chemilumin 5:123–130, 1990.
27. Y Hiraoka, JW Sedat, DA Agard. Science 238:36–41, 1987.
28. A Roda, P Pasini, M Musiani, M Baraldini. Methods Enzymol 305:120–132, 2000.
29. W Muller-Klieser, S Walenta, W Paschen. J Natl Canc Inst 80:842–848, 1988.
30. E Hawkins, R Cumming. J Histochem Cytochem 38:415–419, 1990.
31. P Lorimier, L Lamarcq, F Labat-Moleur, C Guillermet, R Berthier, P Stoebner. J Histochem Cytochem 41:1591–1597, 1993.
32. M Musiani, A Roda, M Zerbini, G Gentilomi, P Pasini, G Gallinella, S Venturoli. J Clin Microbiol 34:1313–1316, 1996.
33. M Musiani, A Roda, M Zerbini, P Pasini, G Gentilomi, G Gallinella, S Venturoli. Am J Pathol 148:1105–1112, 1996.
34. CS Martin, I Bronstein. J Biolumin Chemilumin 9:145–153, 1994.
35. W Muller-Klieser, S Walenta. Histochem J 25:407–420, 1993.
36. A Roda, P Pasini, M Baraldini, M Musiani, G Gentilomi, C Robert. Anal Biochem 257:53–62, 1998.
37. M Mirasoli, P Pasini, C Russo, M Lotierzo, P Valenti, M Guardigli, A Roda. In: A Roda, M Pazzagli, LJ Kricka, PE Stanley, eds. Bioluminescence and Chemiluminescence: Perspectives for the 21st Century. Wiley: Chichester, 1999, pp 524–527.
38. JC Nicolas, P Balaguer, B Térouanne, MA Villebrun, AM Boussioux. In: LJ Kricka, ed. Nonisotopic Probing, Blotting and Sequencing. San Diego: Academic Press, 1995, pp 237–260.
39. B Térouanne, ML Carrie, JC Nicolas, A Crastes de Paulet. Anal Biochem 154:118–125, 1986.
40. LE Morrison, TC Halder, LM Stols. Anal Biochem 183:231–244, 1989.
41. YM Lo, WZ Mehla, KA Fleming. Nucleic Acids Res 16:8719, 1988.
42. CS Martin, CEM Olesen, B Liu, JC Voyta, JL Shumway, RR Juo, I Bronstein. In: JW Hastings, LJ Kricka, PE Stanley, eds. Bioluminescence and Chemiluminescence: Molecular Reporting with Photons. Wiley: Chichester, 1997, pp 525–528.
43. A Baret. In: LJ Kricka, ed. Nonisotopic Probing, Blotting and Sequencing. San Diego: Academic Press, 1995, pp 261–269.
44. A Baret, V Fert. J Biolumin Chemilumin 4:149–153, 1989.
45. V Fert, A Baret. J Immunol Methods 131:237–247, 1990.
46. K Withby, J Garson, A Baret. A microtiter format quantitative PCR assay for HCV RNA employing xanthine oxidase generated chemiluminescence. Proceedings of

Federation of European Microbiological Societies: Symposium on Hepatitis C and Its Infection, Istanbul, 1993, p 137.

47. A Roda, M Musiani, P Pasini, M Baraldini, JE Crabtree. Methods Enzymol 305: 577–590, 2000.

48. A Roda, P Pasini, M Musiani, C Robert, M Baraldini, G Carrea. In: JW Hastings, LJ Kricka, PE Stanley, eds. Bioluminescence and Chemiluminescence: Molecular Reporting with Photons. Wiley: Chichester, 1997, pp 307–310.

49. S Girotti, M Musiani, P Pasini, E Ferri, G Gallinella, ML Zerbini, A Roda, G Gentilomi, S Venturoli. Clin Chem 41:1693–1697, 1995.

50. S Chou. Rev Infect Dis 12:727–736, 1990.

51. G Gentilomi, M Musiani, ML Zerbini, G Gallinella, D Gibellini, M La Placa. J Immunol Methods 125:177–183, 1989.

52. M Musiani, M Zerbini, G Gentilomi, G Gallinella, S Venturoli, D Gibellini, M La Placa. J Clin Microbiol 28:2101–2103, 1990.

53. S Girotti, M Musiani, E Ferri, G Gallinella, ML Zerbini, A Roda, G Gentilomi, S Venturoli. Anal Biochem 236:290–295, 1996.

54. TP Whitehead, GHG Thorpe, SRJ Maxwell. Anal Chim Acta 266:265–277, 1992.

55. E Speroni, MC Guerra, A Rossetti, L Pozzetti, A Sapone, M Paolini, G Cantelli-Forti, P Pasini, A Roda. Phytother Res 10:S95–S97, 1996.

56. P Pasini, M Musiani, C Russo, P Valenti, G Aicardi, JE Crabtree, M Baraldini, A Roda. J Pharm Biomed Anal 18:555–564, 1998.

57. G Ranalli, P Pasini, A Roda. In: A Roda, M Pazzagli, LJ Kricka, PE Stanley, eds. Bioluminescence and Chemiluminescence: Perspectives for the 21st Century. Wiley: Chichester, 1999, pp 153–156.

58. M Musiani, P Pasini, ML Zerbini, A Roda, G Gentilomi, G Gallinella, S Venturoli, E Manaresi. Histol Histopathol 13:243–248, 1998.

59. L Lamarcq, P Lorimier, A Negoescu, F Labat Moler, I Durrant, E Brambilla. J Biolumin Chemilumin 10:247–256, 1995.

60. A Roda, M Musiani, P Pasini, A Suozzi, M Baraldini, S Girotti, S Venturoli, C Polimeni. In: AK Campbell, LJ Kricka, PE Stanley, eds. Bioluminescence and Chemiluminescence: Fundamentals and Applied Aspects. Chichester: Wiley, 1994, pp 625–628.

61. JE Crabtree. In: A Lee, F Mégraud, eds. *Helicobacter pylori*: Techniques for Clinical Diagnosis and Basic Research. London: WB Saunders, 1996, pp 235–244.

62. JH Wilcox. J Histochem Cytochem 41:1725–1733, 1993.

63. G Gentilomi, M Musiani, A Roda, P Pasini, M Zerbini, G Gallinella, M Baraldini, S Venturoli, E Manaresi. BioTechniques 23:1076–1083, 1997.

64. M Musiani, ML Zerbini, S Venturoli, G Gentilomi, G Gallinella, E Manaresi, M La Placa, A D'Antuono, A Roda, P Pasini. J Histochem Cytochem 45:729–735, 1997.

65. P Lorimier, L Lamarcq, C Negoescu, C Robert, F Labat Moleur, F Gras Chappuis, I Durrant, E Brambilla. J Histochem Cytochem 44:665–671, 1996.

66. AJ Schüetz, M Winklmair, MG Weller. In: A Roda, M Pazzagli, LJ Kricka, PE Stanley, eds. Bioluminescence and Chemiluminescence: Perspectives for the 21st Century. Chichester: Wiley, 1999, pp 67–70.

17

Photosensitized Chemiluminescence

Its Medical and Industrial Applications for Antioxidizability Tests

Igor Popov and Gudrun Lewin
Research Institute for Antioxidant Therapy, Berlin, Germany

1. ANTIOXIDIZABILITY AND ITS QUANTIFICATION

Antioxidizability and its control are relevant for various areas in medicine and industry. Atherosclerosis, cardiac infarction, malignant growth, and aging are consequences of uncontrolled oxidation. Currently, oxidizability and antioxidants are also actual problems for alternative and complementary therapies like phyto-, helio-, and aero-ion therapy.

Among others, in the development of suntan lotions or in the production of beer, the oxidation of ingredients during storage is a known cause of reduced product quality.

A principal characteristic of living organisms is their capability to actively protect themselves against uncontrolled oxidation. Although all organisms are subject to the permanent influence of oxygen and other oxidatively active causes (UV sun irradiation, atmospheric noxae, natural and artificial radiation, etc.), they maintain their integrity due to the effect of a special antioxidative system that developed in the course of phylogenesis [1, 2]. This is in contrast to avital compounds the oxidizability of which depends only on their chemical composition. Thus, antioxidizability is a characteristic of living organisms.

Antioxidative protection mechanisms can be classified into at least four categories [2], i.e., compartmentation, detoxification, repair, and utilization; the first two have a direct relationship to antioxidizability.

Compartmentation means both spatial separation of potentially harmful but essential compounds (e.g., storage of iron in ferritin) and cell- and tissue-specific distribution of antioxidative compounds, and it serves to prevent uncontrolled oxidation.

Detoxification of pro-oxidatively active molecules (radicals, peroxides) is ensured by enzymatic and nonenzymatic compounds.

In addition to the antioxidative detoxificating enzymes, superoxide dismutase, catalase, and glutathione peroxidase with primarily intracellular occurrence that protect the cells from the destructive side effects of the physiological metabolism by prevention of initiation and branching of free-radical chain reactions, nonenzymatic antioxidants are of essential relevance. In contrast to enzymes, they are more or less equally distributed in all compartments of an organism. Exogenous stimuli, like X-rays, UV and ionizing radiation, lesions, inflammations, etc., can cause oxidative stress anywhere in the organism. Therefore, easily oxidizable structures are protected by a permanent homeostatically controlled antioxidant influx. The maintenance of antioxidative homeostasis is ensured by the antioxidative system with all its mechanisms such as release from depots, new synthesis, and control of excretion. In ex vivo studies, these mechanisms cannot be detected; the same applies to most industrial applications.

The antioxidative state of the organism can be defined by the antioxidizability and the oxidation state of blood components. The investigation includes measurement of the antioxidative capacity of the blood plasma and the degree of existing oxidative damage to lipids, proteins, and nucleic acids.

A large number of nonenzymatic compounds, including tocopherols, carotinoids, vitamins C and D, steroids, ubiquinones, thiols, uric acid, bilirubin, inosine, taurine, pyruvate, CRP, and so on, demonstrate qualitative antioxidant properties under experimental conditions. However, the quantitative relevance of most findings remains unclear.

Owing to phenomena such as synergism, antagonism, competition, potentiation, mimicry, sparing, pseudoactivity, etc., it does not seem relevant to perform an assessment of the antioxidative state of the organism merely by selective determination of the physiologically most relevant antioxidants in blood plasma, vitamins C and E, carotinoids, and the compounds uric acid and bilirubin synthetizable in the organism.

Measurement of the total antioxidative overall efficacy in an oxidant-generating test system is considered a physiologically relevant alternative to selective determination of individual components.

A system for determination of antioxidizability consists of two components:

1. A generator of pro-oxidatively acting species, e.g., free radicals
2. Their detector allowing quantification of the generated species and indicating changes in the measured signal as a response to the presence of antioxidative compounds

In the case of free radicals, all generation systems can be classified into physical (radiolysis, photolysis, electrolysis, etc.), physicochemical (thermic decomposition of nitrogen compounds, photosensitized generation), chemical (Fe^{2+}/H_2O_2 system, KO_2 decomposition), and biochemical systems of varying complexity including individual enzymes (e.g., xanthine oxidase), subcellular fractions (NADPH-consuming microsomes), and tissue homogenates (e.g., brain homogenate).

Detection of free radicals can be performed using light absorption, luminescence, oxygen consumption, electrical conductivity, and enzyme activity measurements. A number of examples of relatively common systems are given in Table 1.

Despite the diversity of analytical procedures, no valid selection of solely "right methods" exists. Depending on the problem, the nature of the free radicals themselves or the physicochemical properties of the test system such as composition and hydrophobicity of the medium, oxygen partial pressure, pH value, etc., are important as well [15–17]. An adequate test system is characterized by the fact that with regard to size, lipophilicity, and reactivity the biological target substrate reacts with the most relevant radicals.

Table 1 Some Principal Systems for Antioxidant Assay

Author, year	Generator	Detector	Duration	Ref.
Emanuel et al., 1961	Methyloleate + O_2	Peroxide number	12–16 h	3
Stocks et al., 1974	Brain homogenate + O_2	O_2 consumption	1 h	4
Frank et al., 1982	Oil + O_2	El. conductivity	1–3 h	5
Wayner et al., 1985	ABAP	O_2 consumption	30–60 min	6
Popov et al., 1985	Luminol + hv	Chemiluminescence	30–100 s	7
Niki et al., 1985	ABAP	O_2 consumption	30–60 min	8
Klebanov et al., 1988	Egg yolk + Fe^{2+}	Chemiluminescence	10–20 min	9
Miller et al., 1993	ABTS + peroxidase + H_2O_2	Light absorption	5 min	10
Nakano et al., 1994	Meth-Hb	Luminescence, O_2	20–40 min	11
Ghiselli et al., 1995	ABAP	Fluorescence	20–40 min	12
Saramet et al., 1996	Luminol + H_2O_2	Chemiluminescence	10–20 min	13
Abella et al., 1996	AAPH	LDH activity	20–40 min	14

As the superoxide radical is a precursor of the other reactive oxygen species and interacts with blood plasma components under physiological and pathological conditions as well, systems related to its generation are biologically relevant. It should be noted, however, that with respect to the initiation of lipid peroxidation as one of the main causes of oxidative cell damage, its own reactivity is very weak and that only in protonized form is its toxicity comparable to that of lipid peroxyl radicals [18].

Systems generating only hydroxyl radicals are of minor relevance because of the high reactivity of these radicals resulting in oxidation of all types of molecules at the site of origin. This approach makes it possible to declare almost every compound an "antioxidant," although this thesis will not stand in vivo examination. According to Halliwell and Gutteridge [19], who proposed a practicable definition, an antioxidant is "any substance that, when present at low concentrations compared to those of an oxidizable substrate, significantly delays or prevents oxidation of that substrate."

According to this principle, all antioxidant detection methods can be categorized in two groups, i.e., undefined and defined. The first group includes those with free radicals of unknown characteristics, e.g., the oxidation of a suitable object (unsaturated fatty acids, tissue homogenate of a lab animal, egg yolk) in vitro by treatment with air or pure oxygen. The time when the antioxidants are used up ("lag phase") is recognizable and measurable by the accelerated generation of oxidation products and increased oxygen consumption. The antioxidative effect of a test substance can be estimated after repeated oxidation of the sample in the presence of this substance followed by calculation of the difference between the first and second lag phase. In addition to shortcomings of chemical nature, particularly in the use of egg yolk or tissue homogenate, the major disadvantage of the method is its long duration. It takes several hours to perform one single test. Another shortcoming is due to the fact that the nature of the free radicals involved remains insufficiently defined. The methods based on systems, in which the nature of the generated radicals is known, for example thermic decomposition of nitrogen compounds, photosensitized generation, KO_2 decomposition, etc., are free of this disadvantage.

Below several useful methods are discussed in greater detail.

1.1 Autoxidation of Unsaturated Fatty Acids

Approximately 10 mL of oil are fumigated with oxygen in a closed system, so that the gas stream from the reactor cell is directed into the measuring cell containing distilled water. After the antioxidants are used up, the lipid peroxidation is initiated, and the volatile reaction products get into the measuring cell improving its electrical conductivity. This is recorded graphically as the lag phase. According to this principle, the antioxidizability of oils is investigated with the Rancimat device (manufacturer: Metrohm, Switzerland) [20]. A single measurement with this instrument takes several minutes to hours, and it is suitable only for investigation of fat-soluble compounds.

1.2 Enzymatic Radical Generation

The reaction of peroxidase (metmyoglobin) with hydrogen peroxide leads to the generation of a green-blue radical from a colorless compound 2,2'-azino-*bis*(3-ethylbenzothiazoline-6-sulfonic acid) (ABTS). It is slowed down in the presence of an antioxidant, an effect that is used for its quantitation in the Total Antioxidant Status Kit (manufacturer: Randox, UK) [10]. Problems associated with this method are due to potential interference of the reaction compound H_2O_2 with components of the sample to be investigated. No investigation of fat-soluble compounds is possible.

1.3 Thermally Induced Decomposition of Nitrogen Compounds

A number of water- and fat-soluble nitrogen compounds, e.g., 2,2'-azo-*bis*(2-amidinopropane) dihydrochloride (ABAP), 2,2'-azo-*bis*(2,4-dimethylvaleronitrile) (AMVN), and 2,2'-azo-*bis*(2-cyanopropane) (ABCP), form free radicals during decomposition that in the sample to be investigated initiate lipid peroxidation [16]:

$$R{-}N{=}N{-}R \rightarrow 2R^{\cdot} + N_2$$
$$R^{\cdot} + O_2 \rightarrow ROO^{\cdot}$$
$$ROO^{\cdot} + L{-}H \rightarrow ROOH + L^{\cdot}$$

As in the first example, the time when the antioxidants are used up can be determined by measurement of the lag phase of oxygen consumption or of the generation of oxidation products. Although the measuring time is shorter than it is with the other variants mentioned, it still takes many minutes to obtain a result, obviously too long a period for routine tests.

Pronounced improvement of quantity and quality in the determination of antioxidizability is obtained by the photosensibilized chemiluminescence (PCL) method, realized in the Photochem device (manufacturer: Analytik Jena AG, Germany).

2. PHYSICOCHEMICAL BASIS OF PCL

The main feature of the PCL measuring method is combination of the simple and reliable photochemical generation of free radicals with their very sensitive chemiluminometric detection. Compared to standard conditions, the oxidative reaction is accelerated by a factor of 1000. This results in a reduction in measuring time by a factor of 10–1000 compared to other methods.

2.1 Photosensitized Generation of Free Radicals

This is obtained by optical excitation of the photosensitizer (S) and followed by oxygen reduction.

There are two possible initial steps of photosensitized reactions, leading to the formation of superoxide radical:

I. $S + h\nu \rightarrow S^* + R(RH) \rightarrow R^{+\cdot} + S^{\cdot-}$

$$R^{\cdot} + {}^{\cdot}SH$$

$$S^{\cdot-}({}^{\cdot}SH) + {}^{3}O_2 \rightarrow O_2^{\cdot-}(HO_2^{\cdot}) + S$$

II. $S + h\nu \rightarrow S^* + {}^{3}O_2 \rightarrow S + {}^{1}O_2$ \hfill (a)

$$S^{+\cdot} + O_2^{\cdot-} \hfill \text{(b)}$$

where S* is the photosensitizer in triplet state, and R is the reducing substance.

Various dyes can be used as photosensitizers, including methylene blue, riboflavine, and hematoporphyrin derivative. The selection of the photosensitizer should be in favor of a compound that exclusively leads to Reaction (b), so that a clear interpretation of the results is possible.

2.2 Chemiluminometric Detection of Free Radicals

The dismutation (disproportioning) of two free radicals is accompanied by release of a portion of reaction energy as a light quantum. As the quantum yield of such a process is extremely low, the detection of this type of chemiluminescence is technically complicated. Several compounds like lucigenin and luminol have a high quantum yield after reaction with peroxide radicals. Therefore, they are widely used for the detection of these radicals, particularly in the examination of phagocyting cells.

The design of a device that would unify the photochemical method, i.e., the generation of free radicals, and chemiluminometric detection conflicts with controversial requirements—irradiation of the photosensitizer-containing solution with high-intensity light and the need to completely darken the environment during registration of the chemiluminescence signal.

The solution of this problem was to design a device with circular sample transfer. Irradiation and chemiluminescence measurement are spatially separated;

a peristaltic pump transfers the irradiated solution to the chemiluminometer measuring cell, immediately after measurement back to the irradiation cell. This spatial separation results in a delay in detection; therefore, the radical dismutation is slowed down by increasing the pH value of the solution to 10–11.

Figure 1 shows the graphs of the PCL that were recorded with riboflavin as the photosensitizer and luminol as the detector for free radicals [21]. The course of the PCL reaction has two maxima at approximately 30 s and 3 min after the start of irradiation. It has been demonstrated by analysis of kinetics after addition of the reactants at varying times that the first maximum is riboflavin-dependent. Luminol is needed only for visualization of the superoxide radicals.

The process of the superoxide-dependent PCL that can be inhibited by enzyme superoxide dismutase (SOD) is shown in Figure 2. Luminol can be replaced by lucigenin. In this case, only the first maximum is detected. This variant of the system is useful for SOD activity measurements. The system is very sensitive and rugged; therefore, it is even possible to perform the enzyme determination in whole blood [22].

The second maximum is riboflavin-independent (Fig. 1). In this case, luminol obviously plays a double role; it is the chemiluminogenous detection compound for free radicals and photosensitizer as well. It is a remarkable characteristic of this system that the signal intensity decreases only very slowly, giving an opportunity for detection of nonenzymatic antioxidants.

At high pH values the luminol (LH_2) exists in two forms: LH^- and L^{2-}; hence the following reactions are possible after the absorption of light:

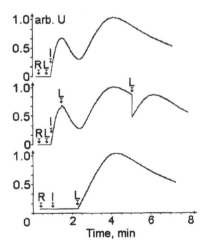

Figure 1 PCL graphs with riboflavin (R) as the photosensitizer and luminol (L) as the detector for free radicals. I = start of irradiation. (From Ref. 21.)

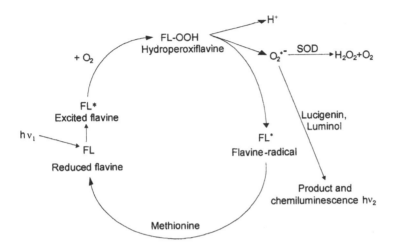

Figure 2 Chemical mechanisms of superoxide-dependent PCL with riboflavin as photo-sensitizer. (From Ref. 22.)

$$^1LH^- + h\nu \rightarrow {}^3LH^- \rightarrow {}^3LH^- \cdots O_2 \rightarrow {}^1LH^{\textbf{·}} + O_2^{\textbf{·}-}$$
$$\downarrow -H^+$$

$$^1L^{2-} + h\nu \rightarrow {}^3L^{2-} \rightarrow {}^3L^{2-} \cdots O_2 \rightarrow {}^1L^{\textbf{·}-} + O_2^{\textbf{·}-}$$

Thereafter the free radicals can react with one another and with the initial molecules. The reaction $LH^{\textbf{·}} + O_2^{\textbf{·}-} \rightarrow O_2 + LH^-$ is not accompanied by chemiluminescence [23].

The reaction $L^{\textbf{·}-} + O_2^{\textbf{·}-} \rightarrow LHOO^-$ leads, after some intermediate stages, to generation of electronically excited compound AP* (aminophthalate anion) [23, 24]. In these reactions, let $k_1 = 31k_2$, but at pH = 9.2 both proceed equally fast.

Independent of the precise mechanism of the chemiluminescence-producing reaction, the process of luminol-dependent PCL can be subdivided into two stages:

I. $L + h\nu_1 \rightarrow L^{\textbf{·}-} + O_2^{\textbf{·}-}$ (generation of free radicals)

II. $L^{\textbf{·}-} + O_2^{\textbf{·}-} \rightarrow N_2 + AP^{2-} + h\nu_2$ (chemiluminescence).

The reaction partners for antiradical substances are products of the first reaction. The SOD reacts selectively with $O_2^{\textbf{·}-}$; the nonenzymic antioxidants can react with both superoxide and luminol radicals. Theoretically, the carbonate radicals can also be involved in the PCL [25, 26].

3. THE "PHOTOCHEM" DEVICE FOR PCL MEASUREMENT

The measuring principle of photosensitized chemiluminescence is applied in the Photochem. The principal functional units of the instrument are shown in Figure 3.

Irradiation of the photosensitizer solution takes place in a vessel (**1**) by means of a lamp (**2**). During measurements, a continuously irradiated solution is transferred with a peristaltic pump (**5**) from the vessel (**1**) into the measuring cell (**6**) of a chemiluminometer (**7**). Control of the unit, signal recording, and data processing are performed by a computer (**8**). The tube (**4**) serves for injection of various components during measurement.

Data acquisition software (Fig. 4) makes possible:

Recording and plotting of PCL signals
Evaluation of curves with and without calibration
Archiving and printing of curves and tables of results

3.1 Testing of Water-Soluble Compounds

A typical time course of PCL with luminol as the photosensitizer is shown in Figure 5, as "blank." The presence of a water-soluble antioxidant leads to dose-dependent temporary inhibition of PCL. "ACW" (antioxidant capacity of water-soluble compounds) represents the effect of human blood plasma (2 μL) on PCL; all tested antioxidants, such as ascorbic acid, uric acid, Trolox, taurine, bilirubin, ceruloplasmin, etc., produced the same effects.

An evaluation parameter of the graphs is the duration of the lag phase L of the PCL ($L = L_1 - L_0$, where L_1 is the lag phase of a sample with an antioxidant and L_0 is the same of a blank sample). All compounds tested demonstrate

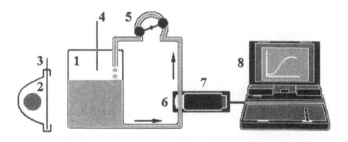

Figure 3 Scheme of the apparatus for measurement of PCL. 1, Vessel for irradiation of a photosensitizer-containing solution; 2, UV lamp; 3, shutter; 4, injection tube; 5, peristaltic pump; 6, measuring cell; 7, photomultiplier; 8, computer. (From Ref. 45.)

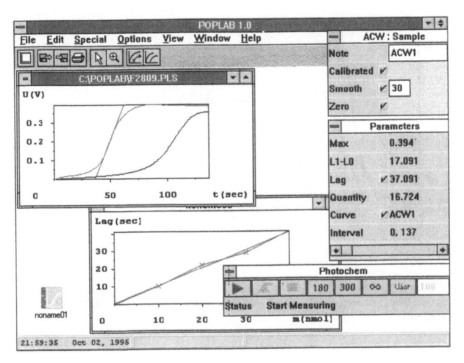

Figure 4   Poplab software.

a linear dependence of the lag phase on the substance quantity. Figure 6 shows the effects of increasing concentrations of Trolox on ACW.

"SOD" in Figure 5 demonstrates the effect of superoxide dismutase on PCL. Its effect is characterized by an inhibition of the chemiluminescence intensity without having an impact on the lag phase. According to McCord and Fridovich [27] (50% test signal inhibition), 1 activity unit in this system refers to approximately 100 ng of the enzyme preparation of the Sigma Co.

3.2   Testing of Lipid-Soluble Compounds

A modification of the measuring system concerns the composition of the assay mixture, in which a part of the water (90%) is replaced by methanol to provide for solubility of hydrophobous substances.

Figure 7 shows an example of the measurements of the antioxidant capacity of lipid-soluble compounds (ACL). In this figure, recording 1 corresponds to "blank," while recordings 2–5 demonstrate the effect of adding lipid extracts from equivalently 20, 40, 60, and 80 μL of blood plasma from a healthy volunteer

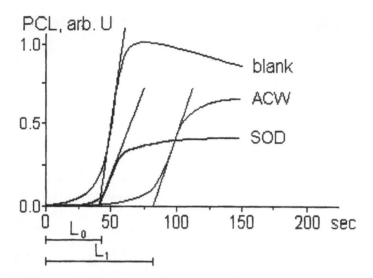

Figure 5 Time course of PCL of luminol. Blank, without additions; ACW, effect of human blood plasma (2 µL); SOD, effect of superoxide dismutase (150 ng). (From Ref. 34.)

to the assay mixture. The evaluation parameter of measurement is inhibition of PCL (I), which can be calculated as: $I = 1 - S_1/S_0$. Here S_0 is the integral of PCL intensity recorded during 2 min, S_1 is the same parameter in the presence of an antioxidant.

A comparison of the PCL method with HPLC gave the results shown in Figure 8. The measurements were performed with the both methods on the same

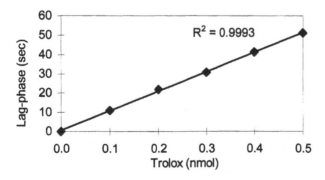

Figure 6 ACW calibration curve for Trolox.

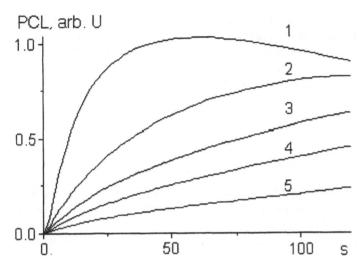

Figure 7 PCL recordings in ACL system. 1, blank; 2–5, effects of lipid extracts from, respectively, 20, 40, 60, and 80 μL of blood plasma from a healthy blood donor. (From Ref. 28.)

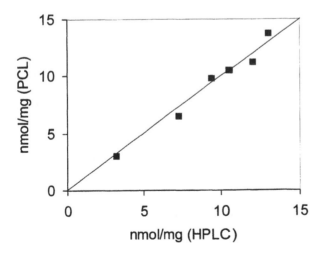

Figure 8 Effect of LDL supplementation with α-tocopherol in vitro determined by HPLC and PCL techniques. The surplus in comparison to the initial value is represented for each parameter. (From Ref. 28.)

lipid extracts from LDL, supplemented in vitro with α-tocopherol [28]. Supplementation of LDL was made before isolation. Plasma was enriched with increasing concentrations of α-tocopherol in dimethylsulfoxide (10 μL DMSO/1 mL plasma). The LDL was isolated after an incubation time of 3 h at 37°C using the microultracentrifuge HITACHI CS 120 as described in Ref. 28, covered with nitrogen, and stored in the refrigerator at 4°C until the lipid extraction. A good quantitative coincidence of both methods was established with $r = 0.998$, $p < 0.001$.

3.3 Specialized Analytical Opportunities

In the context of the theoretically existing problems, a need exists to investigate the antiradical efficacy of compounds independent of pH value or polarity of the solvent, respectively. These problems are particularly relevant in the design of new pharmaceutical antioxidant preparations for specialized therapeutic applications.

It is well known that under pathological conditions the pH value in tissues may significantly differ from the physiological pH value of 7.4. No direct measurement with the Photochem is possible at various pH values, including particularly the acidic range. The reason is the pronounced decrease in chemiluminescence intensity caused by spontaneous dismutation of the superoxide radicals during transfer of the irradiated solution into the measuring cell, according to the following equation:

$$2O_2^- + 2H^+ \rightarrow H_2O_2 + O_2$$

Already at a pH value of 8.0, the signal amplitude is so low that no sufficient accuracy will be obtained during measurement. In the physiological pH value range, the signal equals zero. In such cases, a special pH jump method is a valuable tool. Analogously to measurement of the optical spectrum of chemiluminescence using cutoff filters, the total graph expected in PCL is separated into sections. Each measurement starts with the preferred pH value, while the measuring curve tends to the zero line. At a predefined time, the pH value is changed in a substantial step by the injection of buffer solution, and starting from this time, a signal can be detected. It has been demonstrated by suitable experiments that the change in pH value has no impact on photochemical radical generation.

4.   MEDICAL APPLICATIONS OF PCL: CHARACTERIZATION OF ANTIOXIDATIVE HOMEOSTASIS

4.1   Parameters Assayed in Blood Plasma

A large number of reports on the biological activity of antioxidants in blood plasma and cell membranes verify that vitamin E (VE) is the most significant antioxidant in the lipid phase and vitamin C in the aqueous phase. Vitamin C causes vitamin E to change from the oxidized into the reduced state (vitamin E changes into the oxidized state while performing its main function, namely inhibition of free-radical-mediated lipid peroxidation).

Aside from these two vitamins, uric acid has also been assigned an important role as antioxidant [29]. It has the ability to protect vitamin C from oxidizing [30]. Under strong oxidative stress, bilirubin also seems to be important [31, 32].

4.1.1 ACW—Integral Antiradical Capacity of Water-Soluble Compounds

During determination of the integral antiradical capacity of blood plasma (ACW) by the PCL method, the above-mentioned substances, among others, will be detected primarily.

Basic procedure (ACW kit): Mix 1500 µL of ACW reagent 1 (diluter) with 1000 µL of ACW reagent 2 (buffer) and 25 µL of photosensitizer reagent (luminol based). Start measurement after brief vortexing. Assayed solution (or control) is added before addition of photosensitizer reagent. Volume of ACW reagent 1 is reduced by the volume of assayed plasma sample. Standard substance: ascorbic acid. Duration of measurement: 2–3 min. Measured parameter: effective lag phase = lag-phase sample − lag-phase blank. Assayed amount of human blood plasma: 2 µL.

Tests performed on humans and animals under normal and pathological conditions showed that under normal conditions the ACW is a stable parameter specific to species and age. On the contrary it is very sensitive to any stress factors (bacterial and sterile inflammations, surgical and psychoemotional stress) and to the application of β-agonists [33].

These and other findings relating to the importance of ACW justified the concept of antioxidant homeostasis in organisms and postulate the ACW as a central and evidently regulated parameter (comparable with pH or body temperature) [2].

4.1.2 ACL—Integral Antiradical Capacity of Lipid-Soluble Compounds

Basic procedure (ACL kit): Mix 2400 µL of ACL reagent 1 (diluter) with 100 µL of ACL reagent 2 (buffer) and 25 µL of photosensitizer reagent (luminol based). Start measurement after brief vortexing. Assayed solution (lipid extract) is added before addition of photosensitizer reagent. Volume of ACL reagent 1 is reduced by the volume of assayed solution. Standard substance: α-tocopherol or Trolox. Duration of measurement: 1 min. Measured parameter: integral (area under the kinetic curve of PCL).

Lipid extraction: 200 µL of plasma sample is mixed by brief vortexing with 200 µL of ethanol followed 1 min of vortexing with 800 µL hexane. After centrifugation for 5 min at 5000 rpm, 400 µL of upper (hexane) phase is transferred to a glass tube with screwed cap, dried under nitrogen, and kept at $-60°C$ until PCL analysis. For PCL analysis the sample is dissolved in 400 µL of MeOH by 30 s of vortexing, and 100 µL aliquots are taken for ACL assay.

Similar to ACW corresponding to water-soluble compounds, ACL is a sum parameter including various fat-soluble compounds, the majority of which belong to the vitamin E group. The scatterplot diagram in Figure 9 shows the results of PCL (ACL) and HPLC (VE) investigations of antioxidants in 142 plasma samples from healthy donors. It illustrates a good linear correlation between VE as a sum of α- and γ-tocopherols and ACL: $r = 0.811$, $p < 0.001$. In the average belongs to the VE (23.54 ± 8.78 µmol/L) a share of about 84% in the ACL (28.03 ± 8.02 µmol/L).

Figure 10 presents the results of assay of VE and ACL in blood plasma of rabbits treated with probucol and two other synthetic antioxidants (S-1, S-2) for 4 weeks. In addition to an improvement in antioxidative blood plasma protection, in the case of compound S-2 a statistically significant ($p < 0.01$) decrease in vitamin E content was detected, a finding considered physiologically unfavorable.

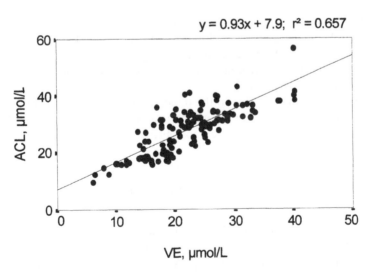

Figure 9 Antiradical capacity in the lipid phase of blood plasma (ACL) determined with the PCL method versus vitamin E (VE) as a sum of α- and γ-tocopherols determined with HPLC. (From Ref. 28.)

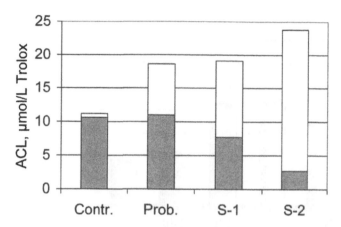

Figure 10 Lipid-soluble antioxidants in blood plasma in rabbits fed 4 weeks with probucol and two newly synthesized compounds. All products were administered in equal amounts. The VE portion is the dark gray portion of the columns. Contr., control group; Prob., group fed with probucol; S-1 and S-2, groups fed with new antioxidants.

4.1.3 ACU—Urate-Independent Component of ACW

Using suitable oxidases it is possible to analyze the proportion of individual ACW compounds.

Modified procedure (ACU kit): The plasma sample is preincubated with uricase (urate oxidase) and then quantified in the ACW assay. The method requires 10 µL of plasma. It is also possible to determine the uric acid (UA) portion in ACW: $UA = ACW - ACU$.

The main value of ACU measurement is that it makes it possible to give a fast estimate of vitamin C deficiency because in healthy subjects ascorbic acid is about 90% of total ACU [33]. A strong deficit of ACU, possibly due to vitamin C deficiency, was found in patients with breast tumors (Fig. 11).

These results do not have any diagnostic meaning. The dynamics of the ACU values after an operative intervention could be taken into account regarding the existence of metastases. As demonstrated in Figure 12, prognostic relevance can be ascribed to ACU. The rise in ACU value is caused in this case by increase of the bilirubin level under the conditions of the life-threatening inflammatory processes [1].

4.1.4 ASC—Vitamin C

Modified procedure (ASC kit): Ascorbic acid is isolated from plasma proteins and uric acid in a single-step liquid gel chromatography procedure and its amount

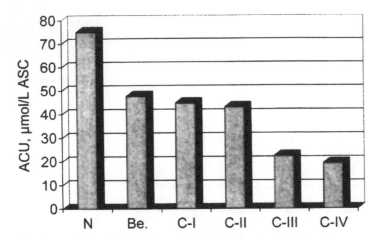

Figure 11 Blood plasma ACU in patients with mammary tumors. N, healthy women ($n = 24$); Be., benign mastopathies ($n = 40$); C, carcinomas in stages I ($n = 14$), II ($n = 13$), III ($n = 9$), and IV ($n = 7$). Compared to healthy subjects, $p < 0.001$ for all carcinomas.

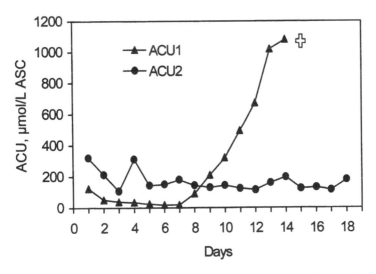

Figure 12 ACU course in the blood plasma of two intensive-care patients with peritonitis. Lethal outcome in patient 1. Normal ACU value, determined in 144 healthy blood donors: 70 ± 27 µmol/L ASC.

(based on its antiradical capacity) is measured in a modified ACW assay. The method requires 500 μL of plasma; its sensitivity is 5 μM [34].

An example of the prognostic meaning of vitamin C measurements is given in Figure 13 [33]. In this study on neurosurgical patients drastic reduction of the vitamin C concentration in blood during the operation was associated with postoperative brain edema.

4.1.5 ACP—Antiradical Capacity of Proteins

Modified procedure (ACP kit): Total plasma protein is isolated from low-molecular-weight plasma antioxidants in a single-step liquid gel chromatography procedure and its antiradical capacity ACP is measured in the ACW assay.

The antiradical capacity of proteins is thought to be an important component of the total antioxidant capacity of blood plasma. Analysis of the ACW of blood plasma showed, that under normal conditions, its main components are the UA and ASC. The rest of the total antiradical capacity (ACR), which can be

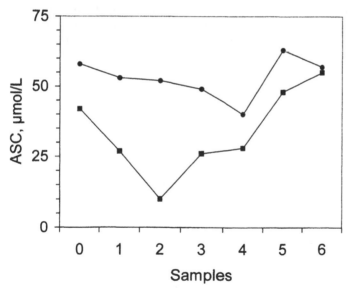

Figure 13 Dynamics of ASC in the blood plasma of neurosurgical patients. Conditions of blood taking referring to the sample number: 0, 1 day before operation; 1, before the first cut; 2 and 3, during the operation, in 30–40-min intervals; 4, at the end of the operation; 5, the next day; 6, at discharge from the hospital. Mean values from $n = 14$ in the group without complications (circles) and $n = 6$ in the group with brain edema (squares). $p < 0.05$ for 1st and 3rd samples, and $p < 0.01$ for 2nd sample. (From Ref. 33.)

calculated as: ACR = ACW − (UA + ASC), was assumed to be the sum of minor antioxidants, including serum albumin, tocopherols, carotenoids, vitamin D, steroids, ubiquinones, thiols, bilirubin, inosine, taurine, pyruvate, C-reactive protein, etc. According to the literature, these substances have antioxidant properties. The practical quantitative relevance of these findings for ACR, however, remains unknown.

In investigations of healthy persons under normal conditions and after UV irradiation of the whole body we have found that ASC correlates significantly ($p < 0.001$) negatively with ACR [35]. A plausible explanation was that some of the above antioxidants can replace ASC in case of its increased consumption during an oxidative stress.

Investigations of the effects of UV- and hypochlorite-induced oxidative modification of 20 amino acids and human serum albumin (HSA) on their antiradical properties showed unexpected results [36]. Seven amino acids (cystine, histidine, methionine, phenylalanine, serine, tryptophan, and tyrosine) and HSA developed ACW following oxidation (see examples in Fig. 14). The fresh (produced in 1998) HSA from Serva had no antiradical capacity, but it acquired this quality during irradiation. The out-of-date HSA sample (Dessau, GDR, 1987, expiration date 7/1/1992) showed a remarkable ACW even in an unirradiated state.

Figure 15 presents a qualitative comparison of different oxidation types on histidine ACW. Hence, the ACP cannot be seen to be characteristic for antioxidant defense, but more likely it is a feature of prehistory connected with the free-radical processes, reflecting the degree of oxidative stress.

A check of the results of the parameter ACP of human blood plasma after removal of low-molecular-weight antioxidants by means of gel filtration was positive and showed a clear difference between the results in healthy donors and cancer patients, as can be seen in Figure 16 [35].

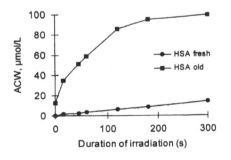 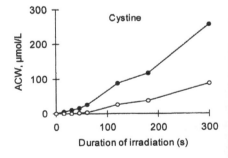

Figure 14   Time course of ACW of HSA (60 g/L) and cystine (2 mmol/L) during (filled circles) and 24 h after UV irradiation (open circles). (From Ref. 36.)

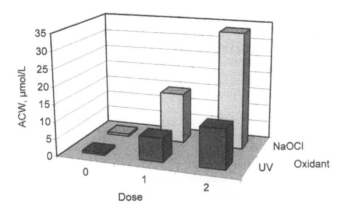

Figure 15 Antiradical capacity of histidine (2 mmol/L) undergoing chemical (NaOCl) and physical (UV, 254 nm) oxidation in equivalent concentrations of ascorbic acid used as calibrator. For UV: dose 1 = 60 s, dose 2 = 120 s. For NaOCl: after 45 min of incubation with 16 (dose 1) or 32 (dose 2) mg/L NaOCl.

Due to the fact that cancer patients are subjected to increased oxidative stress, these results suggest that the degree of oxidative stress in living organisms can be judged by measuring of the antiradical qualities of high-molecular-weight-protein-containing components of blood plasma (HSA and lipoproteins).

If this hypothesis is confirmed in further investigations, measuring the parameter ACP will provide a possibility of recognition of premorbid states caused

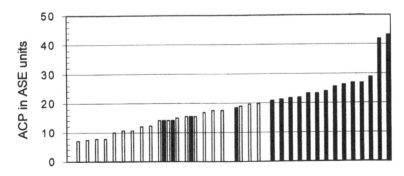

Figure 16 Results of ACP measurements on healthy persons (blank columns) and cancer patients (filled columns). ASE = ASC equivalent. 1 ASE = 10 pmol ASC/mg protein. Mean values: 13.5 and 21.6 ASE; $p < 0.0005$. (From Ref. 35.)

by environmental pollutants or professional noxious factors, and diagnosis of the seriousness as well as control of the therapy efficacy of illnesses like atherosclerosis and heart attack, cancer, diabetes, asthma, allergies, and others.

Analysis of antioxidants (antioxidant status) for characterization of the antioxidative homeostasis in organisms by selective measurement of the ACP can be very meaningful for efficient supervision of antioxidant therapy as well.

4.2 Parameters Assayed for Plasma LDL

Oxidized low-density lipoprotein (LDL) may play a key role in the initiation and progression of atherosclerosis. Risk factors for elevated levels of oxidized LDL are not well established and may be important in identifying individuals who may benefit from antioxidant supplementation or interventions to reduce oxidant stress.

4.2.1 ACW, ACL, and ACL_0 for Isolated LDL

Basic procedure: LDL is isolated from fresh EDTA(Na_2) plasma by sequential ultracentrifugation in KBr step gradient at 120,000 rpm (microultracentrifuge HITACHI CS 120) and a temperature of 10°C for 2 × 2 h. The LDL was covered with nitrogen in a screw-cap tube and stored in the refrigerator at 4°C until further use.

Prior to application, the LDL proceeded through the gel filtration desalting procedure and its cholesterol content was adjusted to 0.4 mM.

LDL isolation is used for measurement of its oxidizability under the influence of various factors in a model system using $CuCl_2$ (final concentration 3.3 μM) as the initiator of lipid peroxidation. Oxidation of LDL was followed by changes in optical density at 234 nm (conjugated dienes formation assay) [37].

ACL_0 is a simplified variation of the ACL assay with direct LDL investigation in the ACL system without prior lipid extraction. The ACL_0 value correlates with the ACL value but is lower by approximately 25–30%.

In Figure 17 the changes in ACW and ACL_0 during Cu^{2+}-initiated LDL oxidation are shown. It should be noted that the native LDL has negligible ACW and that the ACL_0 kinetic course is composed of two distinct phases. The same processes such as UV-induced oxidation of proteins obviously take place during the relatively slow Cu^{2+}-initiated LDL oxidation. It was demonstrated that approximately 80% of the antioxidant capacity of LDL can be accounted for by α-tocopherol [38, 39]. It was assumed that after all antioxidant capacity is exhausted, the oxidative stress imposed by copper ions or by the peroxidase-catalyzed reaction results in lipid peroxidation [40, 41]. The formation of conjugated dienes is conditioned by exhaustion of endogenous antioxidants, but the increase

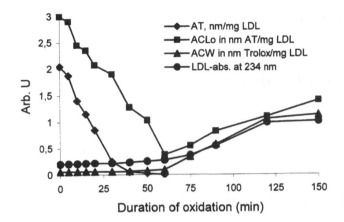

Figure 17 Time course of the antiradical parameters ACL_0 and ACW of LDL measured by the PCL method; its α-tocopherol content (AT) measured by the HPLC technique and conjugated dienes (LDL-abs. at 234 nm) during Cu^{2+}-initiated oxidation in vitro. (From Ref. 36.)

of the ACW and ACL_0 is obviously the consequence of acquisition of antiradical activity by proteins, possibly apolipoprotein B [42].

In contrast to lipid peroxidation the products of which typically appear after a lag phase, protein damage by reactive oxygen species takes place directly and immediately [43, 44]. In contrast to single amino acids and HSA, the ACW of LDL was not elevated until all lipid-soluble antioxidants were exhausted. Hence the first points of the ACL_0 curve in Figure 17 represent the sum of antioxidants without a contribution from protein-dependent ACW. The late was detected at the same time as the formation of conjugated dienes showing the possible protective effect of endogenous antioxidants in LDL on free-radical-induced modification of apolipoprotein B.

4.2.2   ACW, ACL, and ACL_0 for Nonisolated LDL

As isolation of LDL for determination of its antiradical properties is a tedious task, it was alternatively selectively removed from the blood plasma.

Modified procedure: HDL plasma—plasma depleted with apo B–containing lipoproteins by pretreatment with dextran sulfate/Mg^{2+} HDL reagent (Sigma, procedure 352-3). Plasma LDL-bound antiradical parameters can be calculated as the difference between ACW/ACL/ACL_0 for whole plasma and for HDL plasma as well.

An example is shown in Figure 21 (see below). Further investigations are in progress by our group.

5.   INDUSTRIAL APPLICATIONS OF PCL

In many industrial areas, antioxidants play a significant role in the context of stability improvement of easily oxidizable compounds. There is a considerable interest in antioxidants as bioactive components of food and as nutritional agents with a role in the maintenance of health and in disease prevention.

5.1   Detection of Consequences of Food Irradiation

The objective of irradiation of food with γ-rays is elimination of parasitizing insects, fungi, and bacteria to prevent premature spoiling of the food and the outbreak of diseases. In addition, retardation of aging and ripening of fruits and vegetables can be achieved. In the Federal Republic of Germany, irradiation of food with the exception of spices is not permissible.

The examination was performed with the apple variety Jonagold. Apple pieces weighting 0.3–0.5 g without skin were homogenized in a Potter-type glass homogenizer and then centrifuged. Before examination, the supernatant was diluted 1:10 with double distilled water. PCL measurements were performed with 2 μL of the diluted supernatant. Apple juice contains mainly vitamin C and carotinoids as principal ACW components.

A typical change in ACW as measured in various apples 3 days after irradiation is shown in Figure 18. The results are in agreement with the expectation that ACW is reduced due to an interaction of the free-radical scavengers with the free radicals originating during irradiation.

5.2   Antiradical Properties of Wines

People in France eat a lot of fatty foods but suffer less from fatal heart strokes than people in the northern regions of Europe or in North America, where wine is not consumed on a regular basis ("French paradox"). There is an increased favorable effect from red wine. The unique cardioprotective properties of red wine are due to the action of flavonoids, which are minimal in white wine. The best-researched flavonoids are resveratrol and quercetin, which confer antioxidant properties more potent than α-tocopherol.

Some structural changes of the native flavonoids occur during wine conservation, and one of the most studied of those changes concerns red wine color evolution, called "wine aging." It has been demonstrated that as a wine ages, the initially present grape pigments slowly turn into new, more stable red pig-

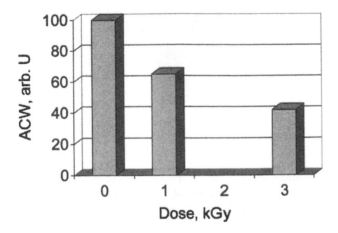

Figure 18 Dosage-dependent decrease in ACW detected in apple juice after γ-irradiation of apples.

ments. That phenomenon continues for many weeks and years. The main questions to be answered in PCL investigations of wines are:

Can antiradical properties of wines be quantified by means of PCL?
Which subpopulations of blood plasma lipoproteins bind the wine-derived antioxidants?
Do the results of oxidizability tests of LDL in the Cu^{2+} system correlate with the results of PCL examinations after treatment with wine?
Can technological processes of wine production be followed using the PCL method?

Figure 19 presents results of PCL examination of wines, grape skins, and grape pips. The very low antioxidant capacity of white wine is obviously related to the technology of its manufacturing: it is prepared from pure juce without grape skins and pips, in contrast to red wine. The dependency of ACW on the storage time of red wine is depicted in Figure 20. White wine is stable in this sense.

Three different types of white wines and four red wines were compared after PCL testing in an ex vivo model system of the influence of oxidation resistance on LDL.

Results of PCL measurements and LDL investigations are depicted in Table 2 and Figure 21. Dr. V. Ivanov (Linus Pauling Institute, Oregon State University), a recipient of a grant from the German Federal Ministry for Research and Technology (AiF FKP 0034904B7B), was involved in this part of the study.

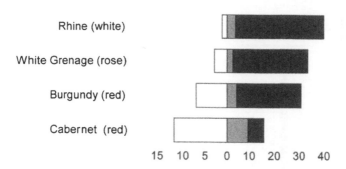

Rhine (white)

White Grenage (rose)

Burgundy (red)

Cabernet (red)

15 10 5 0 10 20 30 40

Figure 19 Antioxidative qualities of wines (white rectangles) in comparison to watery extracts from solid grape constituents: skins (gray rectangles) and pips (black rectangles) in equivalent concentrations of ascorbic acid (mmol/L, mmol/g).

The results of the LDL protection determined with both methods correlate very well with each other. The linear regression between LDL oxidation assay by conjugated dienes formation and PCL (ACW_{LDL}) for all wines was: $y = 0.2505x - 16.563$, $R^2 = 0.9348$.

Figure 21 shows the distribution of wine-derived antioxidants between apo B–containing lipoproteins and other high-molecular-weight components of

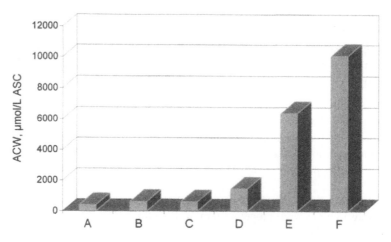

Figure 20 Comparison of ACW values of human blood plasma (A) with different wine types: Riesling Spätlese 1989 (B), Riesling Kabinett 1996 (C), Spätburgunder 1996 (D), Spätburgunder 1994 (E), Spätburgunder 1990 (F). B and C, white wines; D, E, F, red wines.

Table 2 Antiradical Activity of Red and White Wines (ACW$_w$) and Their Antioxidative Protective Effectiveness in the Test System of the Cu^{2+}-Initiated LDL Oxidation

No.	Wine types: red (1–4), white (5–7)	ACW$_w$, mM ASC	LDL protection, Δ(Lag), %	LDL protection, Δ(ACW$_{LDL}$), %
1	Burgundy Cabinett, 1997	7.22	221.7	1289
2	Burgundy Qba, 1997	7.70	143.4	721
3	Burgundy Qba, 1995	7.23	265.5	1064
4	Burgundy Qba, 1989	9.32	>500	1826
5	Riesling Cabinett, medium dry, 1995	1.07	−4.6	−26
6	Riesling, medium dry, 1996	0.86	−6.2	−7
7	Riesling, dry, 1996	0.52	—	−14

Parameters of protection: Δ(Lag), change of the lag phase of conjugated dienes formation, Δ(ACW$_{LDL}$), change of the ACW of LDL after its preincubation with wine, both expressed in percentage of the initial value.

plasma, mainly HDL and serum albumin. After preincubation of plasma with wine in vitro the apo B–containing lipoproteins were removed by pretreatment with dextran sulfate/Mg^{2+} HDL Sigma reagent.

The plasma LDL/VLDL-bound antiradical parameter was calculated as the difference between ACP for whole plasma and for HDL plasma as well.

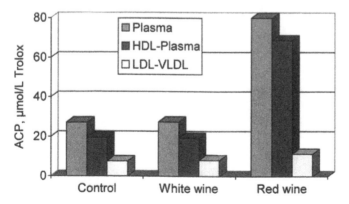

Figure 21 Ex vivo effects of wines on ACP of plasma, HDL plasma, and LDL-VLDL fraction.

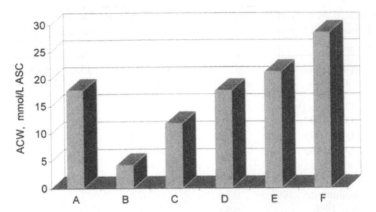

Figure 22 Meaning of technological variants for the gain of antioxidant substances in red wine. A, Heating of the mash for 2 h at 60°C; B, immediately pressed; different duration of mash storage: C, 2 days; D, 3 days; E, 7 days; F, 10 days.

Figure 22 shows an example of the impact of a modified mash treatment (heating, storage) on antiradical properties of resulting wines (the authors thanks Dr. Voigt for support of these investigations).

Obviously the amount of antiradical-effective substances in wine can be raised by optimizing the technology under PCL monitoring.

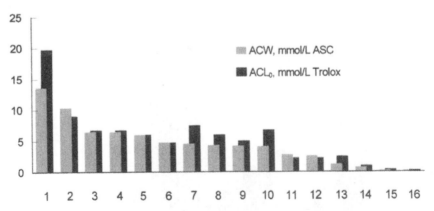

Figure 23 Comparison of antiradical properties of different tea types. 1, Sage tea; 2, green tea; 3, black tea; 4, peppermint tea; 5, bladder tea; 6, dandelion tea; 7, Saint-John's-wort tea; 8, tea for cough; 9, nettle tea; 10, rosehip tea; 11, chamomile tea; 12, horsetail tea; 13, linden flowers tea; 14, marigold tea; 15, fennel tea; 16, cumin tea.

5.3 Antiradical Properties of Different Tea Types

Tea also contains phenolic substances and carotenoids. The long-living Caucasian population regularly consume wine and tea. The aim of the studies was to compare the antiradical properties of different tea types. The results of measurements of ACW and ACL_0 for 16 various types are shown in Figure 23.

In the investigated teas the lowest content of antioxidants is in cumin tea, the highest in sage tea, followed by green and black tea.

6. CONCLUSIONS

The method of photosensitized chemiluminescence is suitable for antioxidizability testing of biological fluids. In addition to measurement of the integral antiradical capacity of blood plasma [45], the PCL method allows selective determination of uric acid and of vitamin C [34] and the degree of oxidative modification of proteins as well, therefore offering the possibility of screening tests of new artificial, herbal, and gene technologically obtained compounds in vitro and in vivo.

The significant benefits of the PCL method include speed, precision, and the possibility to investigate water- and lipid-soluble compounds in a single system and to alter the type of radicals originated by selection of the respective photosensitizer. In contrast to other methods currently in use, the method discussed here is not restricted to a limited pH or temperature range.

Potential PCL applications in medicine (these investigations are in progress) include characterization of disturbances in antioxidative homeostasis, in the following areas:

Surgery (diagnosis and prevention of ischemic damage to tissues)

Internal medicine (characterization of the oxidation state of LDL relevant for atherosclerosis and cardiac infarction)

Anesthesiology and intensive care medicine (control of blood damage in a heart-lung machine)

Obstetrics and gynecology (diagnosis of pregnancy pathology)

Oncology (assessment of antioxidative homeostasis in tumor patients)

Transfusion (quality control of stored blood)

Sports medicine (antioxidative vitamins and cell damage in competitive sportsmen)

Medical genetics (diagnostics of Down's syndrome and familial hypercholesterinemia)

Naturopathy (objectivization of the efficacy of helio-, phyto-, oxygen, and other kinds of unconventional therapy)

Public health (studies on antioxidative homeostasis in various segments of the population the meaning of which could potentially unravel the relevance of environmental factors, life style, East-West gap, etc., on health, fitness for work and life-span)

Possible industrial applications include screening of substances with anti-radical activity, quality testing of raw materials, pharmaceuticals, cosmetic products, fruit juices, wines, beers, edible oils, detection of food irradiation, and many more.

REFERENCES

1. G Lewin, I Popov. Entwicklung und experimentell-klinische Nutzung der Photo-chemolumineszenz-Methode für die Beurteilung der antioxidativen Homöostase im Organismus. Berlin: Humboldt-Universität, 1990.
2. G Lewin, I Popov. Med Hypoth 42:269–275, 1994.
3. NM Emanuel, YN Ljaskovskaja. Inhibition of Fat Oxidation. Moscow: Pistsheprom-isdat, 1961 (in Russian).
4. J Stocks, JMS Gutteridge, RJ Sharp, T Dormandy. Clin Sci 47:215–221, 1974.
5. J Frank, JV Geil, R Freaso. Food Technol 36:71, 1982.
6. DDM Wayner, GW Burton, KU Ingold, S Locke. FEBS Lett 187:33–37, 1985.
7. I Popov, J Hörnig, R von Baehr. Z med Labor-Diagn 26:417–421, 1985.
8. E Niki, Y Yamamoto, Y Kamiya. Chem Lett 1267–1270, 1985.
9. GI Klebanov, IV Babenkova, YuO Teslkin, OS Komarov, YuA Vladimirov. Lab Delo 1:59–62, 1988 (in Russian).
10. NJ Miller, C Rice-Evans, MJ Davies, V Gopinathan, A Milner. Clin. Sci 84:407–412, 1993.
11. M Nakano, T Ito, T Arimoto, Y Ushijima, K Kamiya. Biochem Biophys Res Commun 202:940–946, 1994.
12. A Ghiselli, M Serafini, G Maiani, E Azzini, A Ferro-Luzzi. Free Rad Biol Med 18:29–36, 1995.
13. A Saramet, G Danila, I Paduraru, D Petrariu, R Olinescu. Arzneimittelforschung 46:501–504, 1996.
14. A Abella, C Messaoudi, D Laurent, D Marot, J Chalas, J Breux, C Claise, A Lin-denbaum. Br J Clin Pharmacol 42:737–741, 1996.
15. P Palozza, C Luberto, GM Bartoli. Free Rad Biol Med 18:943–948, 1995.
16. MC Hanlon, DW Seybert. Free Rad Biol Med 23:712–719, 1997.
17. L Landi, D Fiorentini, M Lucarini, E Marchesi, GF Pedulli. Effect of the medium on the antioxidant activity of dipyridamole. In: SFRR Europe Summer Meeting, June 26–28, 1997, Abano Terme. Book of Abstracts, p 265.
18. E Niki. In: G Scott, ed. Atmospheric Oxidation and Antioxidants. Amsterdam: Elsevier, 1993, Vol III, pp 1–31.
19. B Halliwell, JMC Gutteridge. Arch Biochem Biophys 280:1–8, 1990.
20. SP Kochhar. In: G Scott, ed. Atmospheric Oxidation and Antioxidants. Amsterdam: Elsevier, 1995, Vol II, pp 71–131.
21. I Popov, P Falck, R von Baehr. Studia Biophys 115:39–44, 1986.

22. DV Magin, G Lewin, I Popov. Vopr Med Khim 45:70–79, 1999 (in Russian).
23. BA Rusin, In: B Tarusov, ed. Biochemiluminescence. Moscow: Nauka, 1983, pp 69–117 (in Russian).
24. G Merenyi, J Lind, TE Eriksen. J Am Chem Soc 108:7716–7726, 1986.
25. R Radi, TP Cosgrove, JS Beckman, BA Freeman. Biochem J 290 (Pt 1):51–57, 1993.
26. GN Semenkova, TM Novikova, SN Cherenkevich, AI Drapeza. Lab Delo 11:13–15, 1991 (in Russian)
27. JM McCord, I Fridovich. J Biol Chem 244:6049–6055, 1969.
28. I Popov, G Lewin. J Biochem Biophys Methods 31:1–8, 1996.
29. M Ryan, L Grayson, DJ Clarke. Ann Clin Biochem 34(Pt 6):688–689, 1997.
30. A Sevanian, KJA Davies, P Hochstein. J Free Rad Biol Med 1:263–271, 1985.
31. K Bernhard. Helv Chim Acta 38:306–333, 1954.
32. CR Krishna Murthy. Proc Ind Nat Acad. Sci B48:163–175, 1982.
33. I Popov, G Lewin. In: L Packer, AN Glazer, eds. Methods in Enzymology. New York: Academic Press, 1999, Vol 300, pp 437–456.
34. G Lewin, I Popov. J Biochem Biophys Methods 28:277–282, 1994.
35. G Lewin, G Kühn, U Deuse, M Bühring, B Ramsauer, I Popov. Paradox of the antiradical capacity of blood plasma proteins. In: SFRR Europe Summer Meeting, July 2–5, 1999, Dresden, Germany. Book of Abstracts, p 029.
36. I Popov, G Lewin. Luminescence 14:1–6, 1999.
37. G Lewin, I Popov. Drug Res 44:604–607, 1994.
38. H Esterbauer, M Dieber-Rotheneder, G Waeg, G Striegl. Chem Res Toxicol 3:77–92, 1990.
39. D Smith, VJ O'Leary, VM Darley-Usmar. Biochem Pharmacol 45:2195–2201, 1993.
40. H Esterbauer, M Dieber-Rotheneder, G Striegl, G Waeg. J Clin Nutr 53:314–321, 1991.
41. SR Maxwell, O Wiklund, G Bondjers. Atherosclerosis 111:79–89, 1994.
42. B Kalyanaraman, V Darley-Usmar, A Struck, N Hogg, S Parthasarathy. J Lipid Res 36:1037–1045, 1995.
43. RE Pacifici, KJA Devis. In: L Packer, AN Glazer, eds. Methods in Enzymology. New York: Academic Press, 1990, 186 B, pp 485–502.
44. ER Stadtman. Annu Rev Biochem 62:797–821, 1993.
45. I Popov, G Lewin. Free Rad Biol Med 17:267–271, 1994.

18

Application of Novel Acridan Esters as Chemiluminogenic Signal Reagents in Immunoassay

Gijsbert Zomer and Marjorie Jacquemijns
National Institute of Public Health and the Environment, Bilthoven, The Netherlands

This chapter is dedicated to the memory of J. F. C. (Hans) Stavenuiter, a long-term friend and mentor.

1. INTRODUCTION

Bioluminescence, the phenomenon of biological light emission, has fascinated mankind for many centuries. Scientists have been intrigued by it, and for many years have tried to answer "simple" questions like how and why certain animals and bacteria bioluminesce. This research has led to new insights in (molecular) biology and biochemistry. For example, in the 1960s, McCapra studied the chemical mechanisms of bioluminescence and devised a model for firefly luciferin: the acridan esters (Fig. 1) [1–5].

In the firefly, the carboxyl group of luciferin is activated. Key features of both compounds include an easily autoxidizable CH group, an activated carbonyl group in juxtaposition, and an oxidation product that is very fluorescent. Studies of the chemistry of the model acridan ester have led to a better understanding of firefly bioluminescence including correct prediction of the structure of the oxidized luciferin product.

One of the spinoffs of this work has been the appreciation that bio- or chemiluminescence could be used as an aid in analytical chemistry. This has led to the successful applications of acridinum esters as labels in immunoassays [6] and DNA probes [7, 8].

Using acridinium esters (Fig. 2) as chemiluminogenic labels, very sensitive immunoassays were developed [6]. Also commercially, immunoassays using acridinium ester labels have proven to be successful. The sensitivity of detection

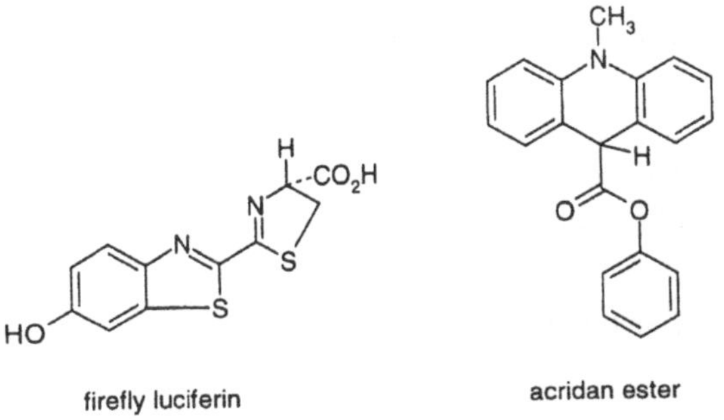

firefly luciferin   acridan ester

Figure 1   Structures of firefly luciferin and acridan ester.

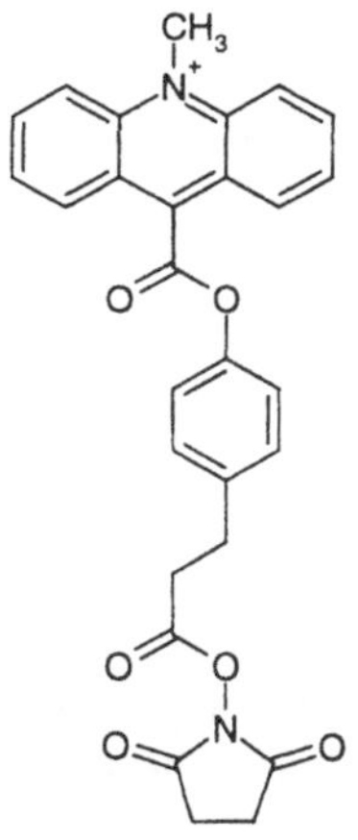

Figure 2 Structure of acridinium ester.

of acridinium ester chemiluminescence is in the order of 10 amol. This is considerably better than the sensitivity of radioactive labels like ^{3}H and ^{125}I. Even greater sensitivities can be reached by using an enzyme as the label: in this case, because of enzyme turnover, many chemiluminescent molecules can be produced from one enzyme molecule. Examples include the luminol-enhanced chemiluminescence [9–11] catalyzed by horseradish peroxidase (HRP) and certain luminol analogs [12] reported to be more efficient than luminol.

Recently, it was found that certain acridan esters could be oxidized to acridinium esters in a reaction catalyzed by HRP [13, 14]. Like the luminol case it was found that the light yield of the reaction could be increased when certain additives (enhancers) were added.

1.1 Immunoassay

Immunoassays are based on the ability of the immune system to produce a virtually unlimited variety of antibodies each with a high affinity for foreign compounds (immunogens like viruses, bacteria, proteins, and haptens). Analytically, this phenomenon can be exploited by detection of this immunoreaction using labeled antibodies or antigens (i.e., compounds that can be bound by antibodies). The equilibrium between antibody (Ab), antigen (Ag), and immune complex (Ab-Ag) may be expressed as:

$$\text{Ab} + \text{Ag} \underset{k_d}{\overset{k_a}{\rightleftarrows}} \text{Ab} - \text{Ag} \tag{1}$$

where k_a and k_d represent the association and dissociation rate constants, respectively. The equilibrium (affinity) constant K is defined as:

$$K = \frac{k_a}{k_d}$$

And, according to the law of mass action:

$$k_a[Ab][Ag] = k_d[Ab - Ag]$$

This becomes:

$$k = \frac{[Ab - Ag]}{[Ab]\,[Ag]}$$

In a so-called competitive immunoassay format the antigen competes with a labeled antigen for a limited number of antibody-binding sites. It can be shown that in this case the ultimate sensitivity of the assay (when the [Ab] approaches zero) is dependent on the equilibrium constant K and the reliability of the signal measurement of the bound fraction at zero dose [15].

An example of an immunoassay format is shown in Figure 3. This immunoassay format relies on partial saturation of the solid-phase antibody (Ab) by the antigen (Ag) and on its competition with the labeled antigen (Ag-L) for the available antibody sites. At low antibody and tracer concentrations the sensitivity of

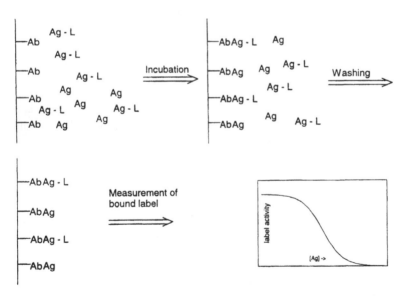

Figure 3 Layout of competitive solid-phase antibody immunoassay format; in an incubation step antigen (Ag) and labeled antigen (Ag-L) compete for the solid-phase antibody-binding sites (Ab); after the solid-phase antibodies are washed, the activity of the bound labeled antigen is measured, and a calibration curve constructed. The measured signal is inversely proportional to the antigen concentration.

this type of assay increases since at low concentrations a variation in the amount of competing antigen has a larger influence on the change of signal that is ultimately measured. When using low concentrations of antibody and tracer, the rate of immunocomplex formation is reduced, resulting in relatively long incubation times. It can be shown that in this case the ultimate sensitivity of the assay (when the [Ab] approaches zero) is dependent on the equilibrium constant K and the reliability of the signal measurement of the bound fraction at zero dose (B0) [15]. For example, an antibody-antigen system with a K of 10^{10} M^{-1} and a relative error of 5% can be detected at $5 * 10^{-12}$ M. With a low-molecular-weight compound (mw 200) this corresponds to a maximal attainable sensitivity of 10^{-9} g/L or 1 ppt.

1.2 Enzyme Immunoassay

1.2.1 Introduction

Using enzymes as labels in immunoassays potentially affords much more sensitive tracer (i.e., enzyme-labeled antigen or antibody) detection in comparison with radiolabels because of the great catalytic power resulting in generation of many product molecules from one enzyme molecule through turnover. A major advantage of radiolabels is that their measurement is not prone to chemical interference. This is not the case with enzymes. Measurement of enzyme activity is strongly influenced by suboptimal temperature, pH, salt concentration, etc. However, the ease and low price of enzyme detection, together with the increased costs of using radiolabels, has made enzymes popular labels in immunoassay. The most widely used enzyme for immunoassay purposes is horseradish peroxidase (HRP)

1.2.2 Enzyme Detection

Although HRP is often detected and quantified using a chromogenic substrate (actually, hydrogen peroxide is the substrate for HRP, generating an oxidized form of HRP, which is subsequently reduced in two steps converting a hydrogen donor DH to an oxidized donor D; in this case the chromogenic substrate is colorless DH, which is transformed to colored product D), there are other possibilities to measure HRP. One of the most sensitive and convenient is use of chemiluminogenic substrates.

1.3 Practical Aspects of Enzyme Immunoassay

An enzyme-linked immunosorbent assay (ELISA) has two major components. The first is the immunological reaction that occurs between an antigen and antibody. This reaction is crucial and needs careful optimization. The second component is the surface to which antigens or antibodies are immobilized. This aspect too needs consideration with respect to the effect of the solid phase on, e.g., reproducibility of antigen or antibody coating, nonspecific binding, sensitivity of the assay, etc. Furthermore, parameters like temperature, pH, composition of coating and assay buffer, incubation times, etc., play important roles with respect to assay sensitivity and reliability. In the next sections some of these factors are discussed in detail, taking as an example the assay format described in Figure 3.

1.3.1   Coating of the Solid-Phase with Antibody

Although other assay containers can be used (e.g., coated tubes or beads), the most popular one is the 96-well microtiter plate. ELISAs that are detected in a microplate luminometer must be performed in opaque white or black plates.

Biomolecules like antibodies attach to surfaces via a variety of mechanisms. This attachment phenomenon is controlled by the chemical properties of the surface, but can be influenced by factors such as pH and temperature. In the case of antibody coating to a solid support the use of so-called medium-binding plates is to be recommended. Coating conditions can be optimized by performing a checkerboard titration (in the following example the optimal coating antibody concentration is determined):

A.   Antibody is serially diluted across the plate (starting concentration of approximately 10 µg/mL in coating buffer (PBS or carbonate buffer), allowed to coat to the surface (overnight at 4°C), and excess is washed away.

B.   The surface is then blocked (PBS containing 0.02% Tween 20 and 1% BSA).

C.   The antigen is serially diluted down the plate, allowed to react with the immobilized antibody, and excess is washed away (PBS containg 0.1% Tween 20, PBST).

D.   Enzyme-labeled antigen is used at a constant (and excess) concentration of approximately 10 µg/mL (such that it is not the limiting factor in the assay), allowed to react, and excess is washed away (PBST).

E.   Signal reagent is added to each well.

F.   Following signal detection, select the optimal reading (this reading is normally chosen based on the following: readings will increase from the bottom right corner to the upper left corner of the plate; the optimal reading lies just prior to the set of wells exhibiting the highest readings).

G.   From the well corresponding to the optimal reading, derive the optimal concentration of coating antibody.

This procedure can be repeated for other coating conditions (time, temperature, pH, buffer compositions, blocking steps, wash buffers, etc.).

1.3.2 Determining the Optimal Concentration of Labeled Antigen

The optimal concentration of labeled antigen can also be determined via a checkerboard titration protocol:

A. The antigen is serially diluted down the plate (optimized with respect to antibody concentration as described in Sec. 1.3.1.), allowed to react with the immobilized antibody, and excess is washed away (PBS containing 0.1% Tween 20, PBST).

B. Enzyme-labeled antigen is serially diluted across the plate, allowed to react, and excess is washed away (PBST).

C. Signal reagent is added to each well.

D. Following signal detection, select the optimal set of column wells (this will be dependent on the goal of the assay: if, for instance, a screening assay is required with a certain cutoff concentration, e.g., 0.1 ng/mL for pesticides in water, then the labeled antigen concentration that yields a B/B0 of 50% at 0.1 ng/mL will be ideal).

E. From the wells corresponding to the optimal reading, derive the optimal concentration of labeled antigen.

2. CHEMILUMINOGENIC SIGNAL REAGENTS FOR HRP

2.1 Enhanced Chemiluminescence Using Luminol and Analogs

In the early 1980s a serendipitous discovery was made when to a mixture of HRP, luminol, and hydrogen peroxide in buffer, firefly luciferin was accidentally added. The light yield of the HRP-catalyzed peroxidation of luminol was greatly enhanced. This remarkable discovery marked the beginning of a very successful analytical tool for immunoassay and all kinds of blotting (protein, DNA, and RNA) applications [16–18].

2.2 Enhanced Chemiluminescence Using Acridan Esters

In 1994 at the 8th International Symposium on Bioluminescence and Chemiluminescence, Schaap and Akhavan-Tafti presented a new technology for the detection of HRP [13, 19, 20]. They proposed the use of certain aromatic acridan esters (see Fig. 4 for the structures) as part of a signal reagent for very sensitive detection

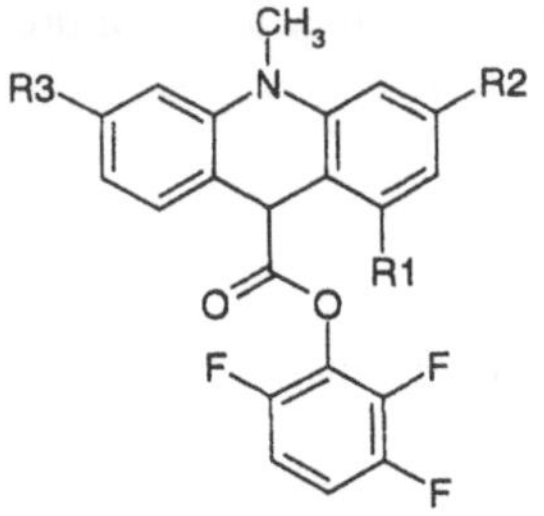

Figure 4 Structures of Lumigen acridan esters: PS-1: R1=R3=H, R2=OCH$_3$, PS-2: R1=R3=OCH$_3$, R2=H, PS-3: R1=R2=R3=H.

of HRP at the 10^{-19}–10^{-15} mol range. This constitutes a major improvement over the commercial luminol-based enhanced chemiluminescence reagent.

2.3   Mechanism of Enhanced Chemiluminescence

Little is known about the role of the enhancer in the HRP-catalyzed chemiluminescent oxidation of acridan esters. From the work of Akhavan-Tafti and Schaap's group the intermediacy of the corresponding acridinium ester can be inferred [14], which in a subsequent reaction with hydrogen peroxide affords *N*-methylacridone with the concomitant production of light. This last reaction is well known from McCapra [2]. In the corresponding HRP-catalyzed chemiluminescent oxidation of luminol, the enhancer is assumed to rapidly react with the peroxidase reactive intermediates, Compound I (CI) and Compound II (CII), accelerating enzyme turnover and producing enhancer radicals [21]. These enhancer radicals react rapidly with luminol, acting as a redox mediator between HRP intermediates and luminol [22]. Assuming a similar role for the enhancer in acridan ester oxidation, a simplified reaction scheme incorporating some of the hypothesized reactions can be devised (Fig. 5).

As in the luminol case, the main role of the enhancer (EnH) seems to be related to turnover of the enzyme, generating enhancer radicals (En rad) in the process that are capable of oxidizing the acridan ester (AcH). The structure of the enhancer obviously is very important. To accelerate HRP turnover, the enhancer must on the one hand be able to rapidly react with the reactive HRP intermediates CI and especially CII (k_2 and k_3 large). On the other hand, the oxidized enhancer intermediate (radical or radical cation) must be able to oxidize the acridan ester (light-generating step). This last reaction also depends on the structure of the acridan ester: in a very unfavorable case, adding an "enhancer" for enzyme turnover could actually diminish the light production if $k_{-4} > k_4$ (Fig. 5), i.e., if the enhancer radical would not be able to oxidize the acridan ester.

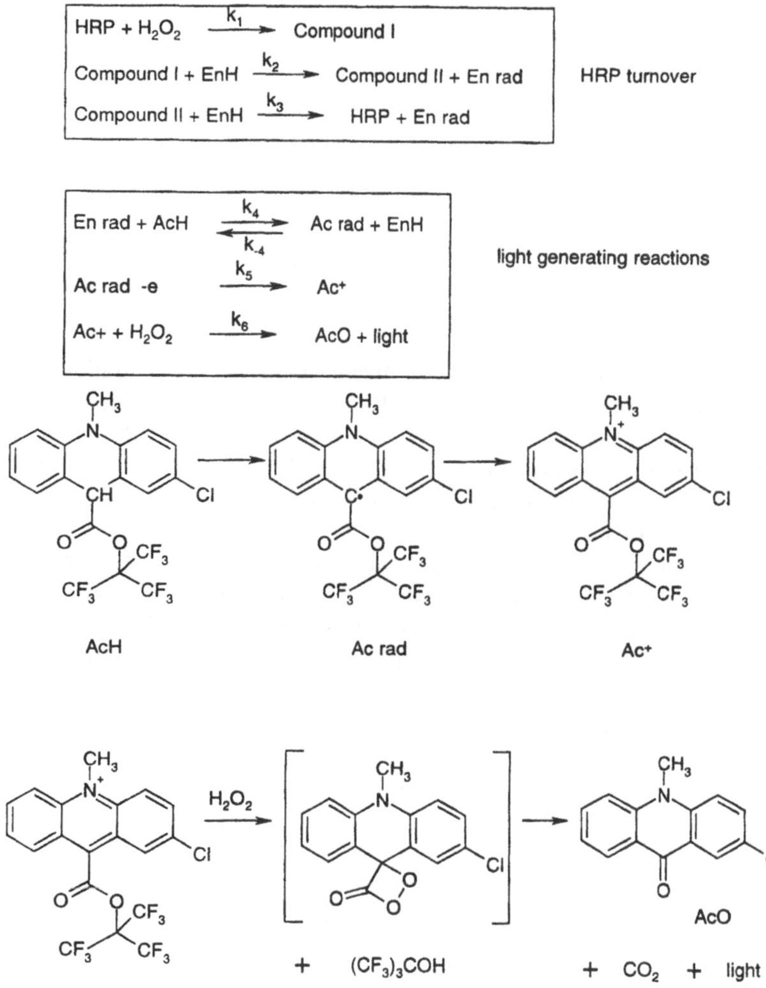

Figure 5 Proposed reactions involved in the HRP-catalyzed chemiluminescent peroxidation of acridan esters.

Further complicating factors in the choice of an enhancer include degradation of HRP by enhancer radicals [23], pH effects [24] on reduction and oxidation potentials for enhancer and acridan ester, inactivation of enhancer radicals because of dimerization or other reactions, etc. All these, and other, effects of the structures (and because of the kinetics also the concentrations) of enhancer and acridan ester may cause erratic results when optimization studies are conducted. When

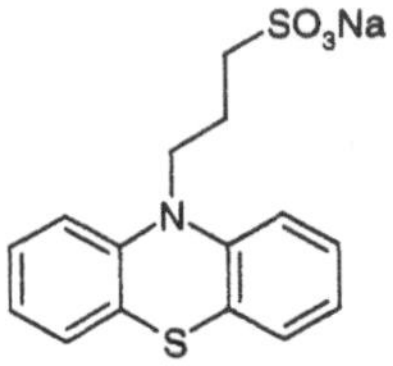

phenothiazine-N-propanesulfonate

Figure 6   Structure of the enhancer as incorporated in the Pierce Supersignal family.

testing new enhancers it is advisable to optimize the enhancer concentration with respect to a complete calibration line for HRP. It is possible that a certain enhancer concentration will give good results at the high end of the HRP calibration curve while performing worse than optimal at the low end of the calibration curve. Studies that show only enhancement factors for one HRP concentration should be treated with suspicion.

Although substituted phenols (e.g., para-iodophenol, para-phenylphenol, firefly luciferin, coumaric acid) are popular enhancers, in both luminol and acridan ester oxidation, enhancers with other functional groups [24], e.g., phenylboronic acids [25–28], phenothiazines [29], are also useful. As an example the structure of the phenothiazine enhancer used in the Supersignal substrate family is shown in Figure 6.

Other additives to the signal reagent for enhanced chemiluminescence signal reagents include Tween20, protein, e.g., bovine serum albumin, and ionene [30] (their main effect seeming to be stabilization of the HRP). In preliminary experiments a stabilizing effect of ionene on the activity of HRP could be shown when phenylphenol was used as the enhancer but not with other enhancers (Zomer, unpublished results). Also, the salts to prepare the buffer solutions have effects on the intensity and duration of the signal [31].

2.4   Characteristics of the GZ-11 Acridan Ester Signal Reagent

In the ideal case the kinetics of the HRP-catalyzed oxidation of acridan ester is solely dependent on the enzyme concentration; i.e., the signal is proportional to the HRP concentration irrespective of the point in time. To achieve such a situation the mechanism of the light-generating reaction must be the same for all HRP concentrations. This implies that the concentrations of hydrogen peroxide and acridan ester should be large enough to cover the whole HRP concentration range without affecting the mechanism. For convenient measurement of the signal the plateau of steady-state light emission is chosen. This steady state of light emission results from the equal rates of formation and breakdown of the intermediate acridinium ester. This allows for starting the chemiluminescence reaction outside the light-measuring device. To approach this ideal situation a careful optimization has to be performed. Important parameters relating to this optimization include the concentration of hydrogen peroxide, the pH, the nature and concentration of the enhancer, and the temperature at which the reaction is performed.

Hydrogen peroxide has at least two functions: on the one hand it is the substrate for the enzyme, on the other hand it is involved in the light-emitting step. At high concentrations hydrogen peroxide deactivates the enzyme and also causes the background signal, i.e., the light emission in the absence of enzyme, to rise.

The pH optimum of HRP is around pH 5. Therefore, this would be the pH of choice. Unfortunately this is not the optimal pH for the light-generating reaction (the general base-catalyzed reaction of acridinium ester with hydrogen peroxide). An acridan ester like GZ-11 with a leaving group of low pK_a (perfluoro-*tert*-butanol has a pK_a below 6) is clearly advantageous.

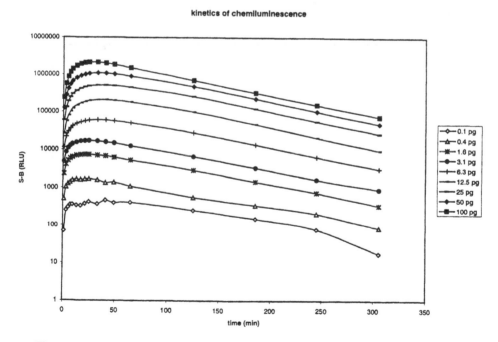

Figure 7 Kinetics of the chemiluminescent HRP-catalyzed peroxidation of GZ-11. The course of the light output, corrected for background (S-B), is plotted versus time for different amounts of HRP (0.1–100 pg).

Although the signal reagent can be used over a wide pH range, the optimal pH was found to be 5–7. Most of the results described in this chapter were obtained using a pH of 5.4 [phosphate-buffered saline (PBS), 10 mM].

From the simplified mechanism of enhanced chemiluminescence it might be concluded that the enzyme turnover benefits from a high concentration of enhancer. After all, the rate of enzyme turnover is governed by the rate-limiting step, the reduction of compound II to the resting enzyme, which is proportional to the enhancer concentration. However, when enhancer concentrations are plotted against enhancement factors (at constant HRP concentration) a bell-shaped curve is obtained [21]. Furthermore, the optimal enhancer concentration is not necessarily equal for all HRP concentrations. This implies that secondary reactions of enhancer (or of impurities present in the enhancer [32]) take place. The characteristics of the chemiluminescence signal from the HRP-catalyzed peroxidation of acridan ester GZ-11 are shown in Figures 7 (kinetics of the reaction) and 8 (HRP-calibration curve).

As can be seen from Figure 7, the signal reaches a plateau within 10 min and lasts for a relatively long time (even after 5 h a useful signal can be obtained). Furthermore, the signal reagent allows measurement of HRP over a wide range (Fig. 8).

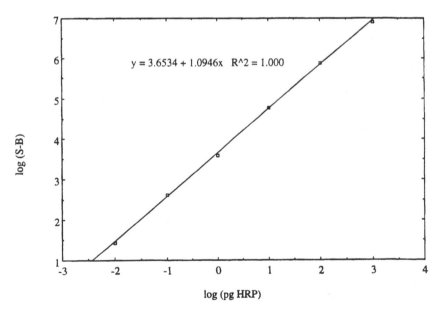

Figure 8   HRP calibration curve (log-log plot of HRP dose vs. chemiluminescence signal, corrected for background) using signal reagent incorporating GZ-11.

3. APPLICATIONS

To test our new signal reagent based on GZ-11 the detection system was applied to two competitive-format immunoassays. These two assays (for atrazine and clenbuterol) normally use chromogenic detection systems. While these colorimetric assays may be adequate for laboratory use, chemiluminescent detection offers potential advantages in sensitivity and on site screening applications [33].

3.1 Atrazine Immunoassay

3.1.1 Introduction

Monitoring water samples for the presence of atrazine is normally done using conventional techniques involving chromatographic separation followed by detection (UV, MS). Recently an enzyme immunoassay for atrazine was reported. This assay uses enhanced luminol chemiluminescence detection of an HRP conjugate and results in an assay with a theoretical sensitivity of 0.05 ppb using 2 h of incubation in an ice-water bath [34]. Using a signal reagent containing GZ-11 we decided to develop an enzyme immunoassay for atrazine.

3.1.2 Assay Development

The assay conditions were optimized with respect to antibody and tracer concentration to obtain maximal sensitivity at 0.1 ppb, the current maximum admissible concentration (MAC) in drinking water.

The normal protocol using monoclonal antibody (mAb) K4E7 [35] involves the following steps: Coating of plates using goat anti-mouse antibody (0.25 mL, 1:5000 in 50 mM carbonate buffer, pH 9.6) was performed overnight at 4°C. After being washed with washing buffer (PBS, pH 7.6; 4 mM, 0.05% Tween 20), the plates were incubated with 0.2 mL of mAb K4E7 (dilution 1:100,000) in PBS (pH 7.6, 40 mM) for 2 h at room temperature. The plates were washed with washing buffer. In the competition step 0.15 mL of atrazine standards (0–50 μg/L) and samples were incubated in the wells with 0.05 mL of atrazine-HRP conjugate (dilution 1:50,000) for 1 h at room temperature. After the plate was washed with PBST, the substrate reaction using the chromogen tetramethylbenzidine (TMB) was performed.

The assay conditions involving the GZ-11 signal reagent were optimized with respect to mAb and tracer dilutions. This resulted in the following protocol: The atrazine assay was performed in coated Lumacuvettes (overnight, room temperature, pH 9.6, goat-anti mouse 1/5000) using antiatrazine monoclonal antibodies (1/160,000 in PBS, pH 7.4, 40 mM), with 2 h of incubation at room temperature. The antibody-coated tubes were incubated with atrazine-HRP conjugate

(1/200,000) and standards and samples. After a 1-h incubation at room temperature, the tubes were washed, and signal reagent was added to all tubes. The chemiluminescence was measured at different points in time. The results are shown in Figure 9.

3.1.3   Results

From Figure 9 it can be concluded that the sensitivity of the assay is adequate to screen water samples for the presence of atrazine at a MAC value of 0.1 ppb. In a preliminary experiment 10 river water samples were screened for the presence of atrazine. In eight out of 10 an immunological response corresponding to atrazine levels greater than 0.1 ppb could be confirmed by GC-MS analysis. The other two samples contained atrazine at levels below 0.1 ppb (both with the enzyme immunoassay as with the GC-MS analysis).

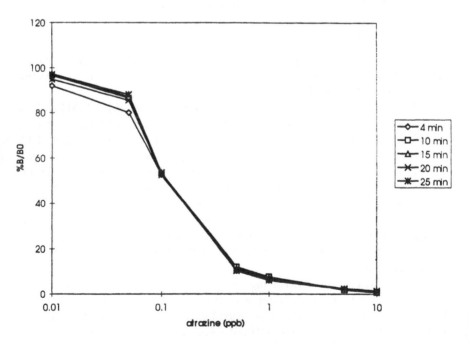

Figure 9   Atrazine calibration curve, plotted as percentage binding (B) over binding at zero dose (B0) against dose, as detected after different signal-developing times.

3.2 Clenbuterol Immunoassay

3.2.1 Introduction

β-Agonists like clenbuterol are widely used, not only for the treatment of respiratory diseases but also to improve carcass characteristics and growth rates of farm animals. The use of β-agonists as anabolic agents is illegal in the European Community and in North America. Within the framework of Residue Control Directive 86/469/EEC the European Member States try to gain control of the growing black market for β-agonists. To do so a large number of biological samples (urine, hair, eyes, etc.) are screened annually for the presence of clenbuterol and other β-agonists. Enzyme immunoassays are frequently used for these examinations. However, in a recent evaluation of nine commercially available ELISA kits for the detection of clenbuterol in bovine urine it was concluded that none met the requirement of the official residue control plans of the EU Member States to detect β-agonists at the 1-ppb level [36]. The above ELISA methods are suitable for laboratory use only.

Ideally, one prefers to perform a first screening at the place of sampling (farm- or slaughterhouses). To do this a tube enzyme immunoassay for β-agonists was developed [37]. This test was capable of detecting clenbuterol at a level of 3–4 ppb when performed in fivefold diluted urine. Here we describe the adaptation of an enzyme immunoassay capable of detecting clenbuterol and other β-agonists at the sub-ppb level in bovine urine using HRP as the label and the chemiluminescence signal reagent based on GZ-11 acridan ester. The immunoassay can be performed in microtiterplates with the signal being detected either in a plate luminometer or on photographic materials (X-ray and instant film).

3.2.2 Materials and Methods

Chemiluminescence measurements were performed on X-ray film or Polaroid 20,000 ASA film using a camera luminometer (Tropix, Inc., Bedford, MA). The microplate luminometer we use is the Lucy 1 (Anthos Labtec Instruments, Wals, Austria). Black, white, and transparent microtiterplates and strips were from Corning Costar (Badhoevedorp, The Netherlands).

Immunoassays were performed in microtiter plates or eight-well strips. Briefly, plates were coated with a suitable antibody dilution (1:3000) overnight at 4°C in carbonate buffer (50 mM, pH 9.6). The coated strips were washed four times with PBST (40 mM, pH 7.4 containing 0.1% Tween 20). The strips were incubated with 10-fold diluted urine samples or standards (200 μL) and tracer dilution 1:4000 (10 μL). After standing for 1–2 h the strips were washed with PBST. To each well 200 μL of signal reagent containing GZ-11 was added. The glowing plate was either analyzed in the microplate luminometer or exposed to

X-ray film or the strips were placed in the mask of the camera luminometer loaded with the high-speed Polaroid film. After the strips were exposed to the photographic material for 10–15 min, the film was pulled from the film holder and after 45 s (processing time) the film backing was removed to reveal the developed image.

Optimization of the immunoassay was performed with respect to tracer and antibody concentrations to obtain the required sensitivity. These conditions differed depending on the detection system used: photographic detection required higher antibody and tracer concentrations than when the plate luminometer was used. A further complication arose from the very low affinity of the tracer for the antibody: when using an antibody dilution of 1:3000 and a tracer dilution of 1:4000 less than 1% of the tracer was bound after a 2-h incubation. This means that the antibody, in the absence of clenbuterol, binds less than 10 pg of the

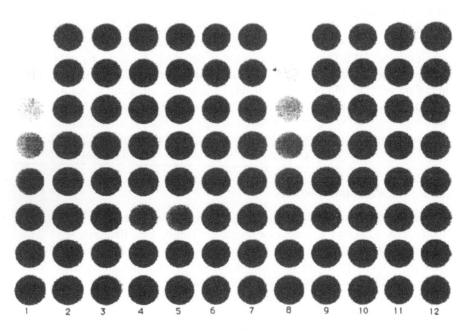

Figure 10 Image obtained after performance of 40 clenbuterol assays on 10-fold diluted bovine urine samples (in duplicate), as detected on X-ray film: Column 1 and 8 contain standards 0, 0.3, 0.6, 1.2, 2.5, 5, 10, and 20 ppb, from bottom to top, respectively. Columns 2, 4, 6, 9, and 11 contain samples 1–40, with their duplicates in columns 3, 5, 7, 10, and 12. Sample 14 (columns 4 and 5, third from bottom) was later determined using GC-MS to contain 1 ppb of clenbuterol. In all other samples no clenbuterol or other β-agonists could be detected.

tracer. Consequently, the signal reagent must be capable of generating enough light to be detectable on instant film in the low-pg range of the tracer.

3.2.3 Results

An example of a clenbuterol assay of 10-fold diluted bovine urine samples as detected on X-ray film is shown in Figure 10. This image results from an incubation of standards (column 1 and 8200 µL) or urine samples (columns 2–7 and 9–12) and tracer (1:8000, 10 µL) in PBS (100 mM, pH 7.2) in the wells of an antibody-coated (1:3000) microtiter plate for 2 h at room temperature, followed by incubation with signal reagent. After a 10-min incubation, the glowing plate was covered with X-ray film, and exposed for 13 min to the film. After development, the photographic image was scanned into a computer and inverted. From Figure 10 it can be concluded that it should be possible to screen 10-fold-diluted urine samples for the presence of clenbuterol below ppb levels.

4. SYNTHESIS AND CHARACTERIZATION OF ACRIDAN ESTERS

Acridan esters were synthesized according to the general synthetic scheme depicted in Figure 11.

4.1 Acetanilide

Aromatic amine (20 g), acetic acid anhydride (20 mL), acetic acid (glacial, 20 mL), and zinc powder (0.1 g) were mixed and refluxed. After 30 min the solution was poured into 500 mL of ice water and the precipitate was filtered and dried to obtain the acetanilide.

4.2 4-Chlorophenylphenyl Amine

Acetanilide (13.5 g), (substituted) aromatic bromide (25 g), potassium carbonate (13.2 g), and copper iodide (1.9 g) were heated (190°C) and stirred overnight. After cooling to room temperature toluene was added and the precipitate filtered. The solution was concentrated and the excess of bromide removed by distillation under reduced pressure. The residue was dissolved in ethanol (200 mL), potassium hydroxide (10.3 g) was added, and the mixture refluxed overnight. Ethanol was evaporated, the residue dissolved in dichloromethane, and washed with brine. The organic layer was dried over $MgSO_4$ and concentrated to obtain the crude diphenylamine.

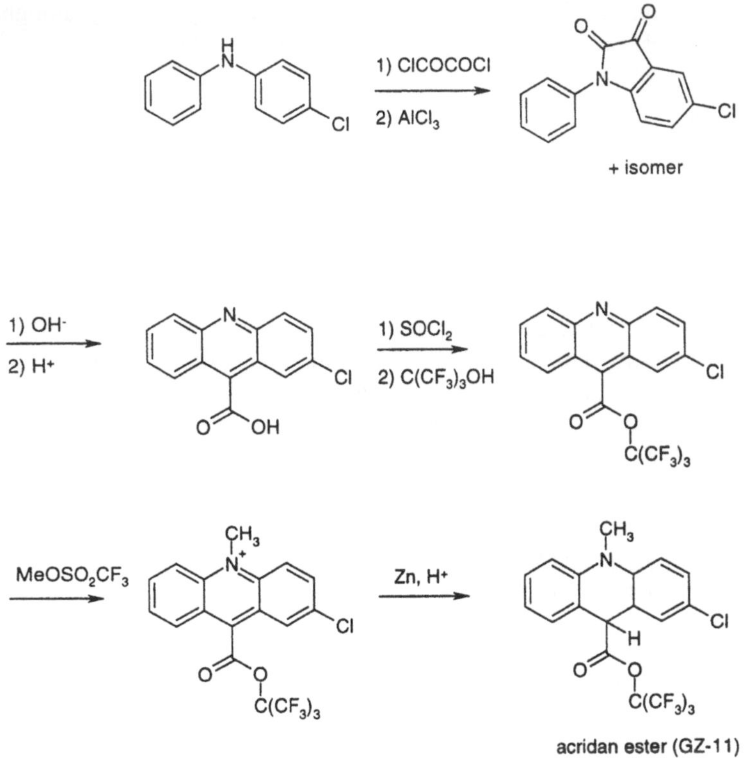

Figure 11   Synthetic scheme for the synthesis of acridan ester GZ-11.

4.3   2-Chloro-Acridine-9-Carboxylic Acid

To a solution of oxalic chloride (5 g) in dichloromethane, a solution of diphenyl amine (5 g) in dichloromethane was added dropwise and refluxed for 30 min. The solution was concentrated (50%) and aluminum trichloride (8 g) added in portions. The mixture was refluxed for 45 min and the solvent evaporated. To this residue hydrochloric acid in ice water (1 M) was added and the red-colored precipitate filtered. The precipitate was dissolved in potassium hydroxide (10% in water), refluxed overnight, and poured into hydrochloric acid in ice water (5 M). The yellow acridine-9-carboxylic acid was filtered, washed with water, and dried.

4.4   Perfluoro-*t*-Butyl 2-Chloroacridine-9-Carboxylate

Acridine-9-carboxylic acid (1 g) was mixed with thionyl chloride (20 mL) and refluxed until a clear solution was obtained. The solution was concentrated and the excess thionyl chloride coevaporated with toluene. To the residue pyridine (20 mL), 4-dimethylaminopyridine (1 g) and an excess of perfluoro-*tert*-butanol were added. The mixture was stirred overnight at room temperature, poured into hydrochloric acid in ice water (1 M), and extracted with dichloromethane. The organic layer was washed with water, dried over $MgSO_4$, and concentrated. The crude product was purified by flash chromatography (66% hexane/ethyl acetate).

^{1}H NMR (CDCl$_3$) δ 7.6–7.9 (m, 5H), 8.2–8.3 (m, 2H).

4.5 Perfluoro-*t*-Butyl 2-Chloro-10-Methylacridinium-9-Carboxylate Trifluoromethanesulfonate

Acridine-9-carboxylic ester (100 mg) was dissolved in dichloromethane (5 mL). To this solution an excess of methyl trifluoromethanesulfonate was added. The mixture was stirred overnight at room temperature. Diethyl ether was added and the obtained precipitate was filtered, washed with diethyl ether, and dried.

^{1}H NMR (acetone-d$_6$) δ 5.3 (s, 3H), 8.2–9.3 (m, 7H).

4.6 Perfluoro-*t*-Butyl 2-Chloro-10-Methylacridane-9-Carboxylate (GZ-11)

10-Methylacridinium-9-carboxylic ester trifluoromethanesulfonate (100 mg) was dissolved in dichloromethane (100 mL). To this solution perchloric acid (20 drops) and zinc powder (6 g) were added. The mixture was placed in an ultrasonic bath for 3 h and filtered. The solution was washed with water (1×) and hydrochloric acid (2×, 0.1 M), dried over $MgSO_4$, and concentrated.

^{1}H NMR (CDCl$_3$) δ 3.4 (s, 3H), 5.1 (s, 1H), 6.9–7.4 (m, 7H).

^{13}C NMR (CDCl$_3$) δ 163.9, 141.7, 140.8, 129.3, 128.9, 128.5, 125.7, 121.4, 121.2, 119.7, 117.8, 117.1, 114.0, 113.0, 49.5, 33.1

REFERENCES

1. F McCapra. Acc Chem Res 9:201–208, 1976.
2. F McCapra. Prog Org Chem 8:231–277, 1974.
3. F McCapra. Biochem Soc Trans 7:1239–1246, 1979.
4. F McCapra. Endeavour 32:139–145, 1973.
5. F McCapra. Proc R Soc Lond B Biol Sci 215:247–272, 1982.
6. I Weeks, I Beheshti, F McCapra, AK Campbell, JS Woodhead. Clin Chem 29:1474–1479, 1983.
7. LJJ Arnold, PW Hammond, WA Wiese, NC Nelson. Clin Chem 35:1588–1594, 1989.
8. M Septak. J Biolumin Chemilumin 4:351–356, 1989.
9. GH Thorpe, LJ Kricka. Methods Enzymol 133:331–353, 1986.

10. GH Thorpe, LJ Kricka, E Gillespie, S Moseley, R Amess, N Baggett, TP Whitehead. Anal Biochem 145:96–100, 1985.
11. GH Thorpe, LJ Kricka, SB Moseley, TP Whitehead. Clin Chem 31:1335–1341, 1985.
12. M Ii, H Yoshida, Y Aramaki, H Masuya, T Hada, M Terada, M Hatanaka, Y Ichimori. Biochem Biophys Res Commun 193:540–545, 1993.
13. H Akhavan-Tafti, R DeSilva, Z Arghavani, RA Eickholt, R Handley, AP Schaap. In: AK Campbell, LJ Kricka, PE Stanley, eds. Bioluminescence and Chemiluminescence: Fundamentals and Applied Aspects. Chichester: Wiley, 1994, pp 199–202.
14. H Akhavan-Tafti, R DeSilva, Z Arghavani, RA Eickholt, RS Handley, BA Schoenfelner, Y Sugioka, AP Schaap. J Org Chem 68:930–937, 1998.
15. R Ekins. Pure Appl Chem 62:1286–1290, 1991.
16. GH Thorpe, LJ Kricka. In: J Schoelmerich, R Andreesen, A Kapp, M Ernst, WG Woods, eds. Bioluminescence and Chemiluminescence: New Perspectives. Chichester: Wiley, 1987, pp 199–208.
17. EHJM Jansen, CH Van Peteghem, G Zomer. In: WRG Baeyens, D De Keukeleire, K Korkidis, eds. Luminescence Techniques in Chemical and Biochemical Analysis. New York: Marcel Dekker, 1991, pp 477–504.
18. LJ Kricka, RA Stott, GH Thorpe. In: WRG Baeyens, D De Keukeleire, K Korkidis, eds. Luminescence Techniques in Chemical and Biochemical Analysis. New York: Marcl Dekker, 1991, pp 599–636.
19. H Akhavan-Tafti, Z Arghavani, R DeSilva, BA Schoenfelner, AP Schaap. In: AK Campbell, LJ Kricka, PE Stanley, eds. Bioluminescence and Chemiluminescence: Fundamentals and Applied Aspects. Chichester: Wiley, 1994, pp 309–312.
20. H Akhavan-Tafti, R DeSilva, Z Arghavani, K Sugioka, Y Sugioka, AP Schaap. In: AK Campbell, LJ Kricka, PE Stanley, eds. Bioluminescence and Chemiluminescence: Fundamentals and Applied Aspects. Chichester: Wiley, 1994, pp 313–316.
21. PM Easton, AC Simmonds, A Rakishev, AM Egorov, LP Candeias. J Am Chem Soc 118:6619–6624, 1996.
22. A Lundin, LOB Hallander. In: J Schoelmerich, R Andreesen, A Kapp, M Ernst, WG Woods, eds. Bioluminescence and Chemiluminescence. New Perspectives. Chichester: Wiley, 1987, pp 555–558.
23. KJ Baynton, JK Bewtra, N Biswas, KE Taylor. Biochim Biophys Acta 1206:272–278, 1994.
24. F Garcia Sanchez, A Navas Diaz, JAG Garcia. J Photochem Photobiol 105:11–14, 1997.
25. X Ji, K Kondo, Y Aramaki, LJ Kricka. J Biolumin Chemilumin 11:1–7, 1996.
26. LJ Kricka, M Cooper, X Ji. Anal Biochem 240:119–125, 1996.
27. LJ Kricka, X Ji. J Biolumin Chemilumin 10:49–54, 1995.
28. O Nozaki, X Ji, LJ Kricka. J Biolumin Chemilumin 10:151–156, 1995.
29. KD Feather-Henigan, I Horney, CJ Clothier, KK Hines. In: A Roda, M Pazzagli, LJ Kricka, PE Stanley, eds. Bioluminescence and Chemiluminescence. Perspectives for the 21st Century. Chichester: Wiley, 1999, pp 103–106.
30. YL Kapeluich, MY Rubtsova, AM Egorov. J Biolumin Chemilumin 12:299–308, 1997.

31. B Cercek, RJ Obremski, CS Oh. In: PE Stanley, LJ Kricka, eds. Bioluminescence and Chemiluminescence; Current Status. Chichester: Wiley, 1990, pp 1995–1998.
32. PD Davis, K Feather-Henigan, K Hines. PCT/US97/06422.
33. G Zomer, M Jacquemijns. In: A Roda, M Pazzagli, LJ Kricka, PE Stanley, eds. Bioluminescence and Chemiluminescence. Perspectives for the 21st century. Chichester: Wiley, 1999, pp 87–90.
34. GW Aherne, A Hardcastle, P Saleem, N England. In: PE Stanley, LJ Kricka, eds. Bioluminescence & Chemiluminescence. Current status. Chichester: Wiley, 1991, pp 91–98.
35. T Giersch. J Agric Food Chem 41:1006–1011, 1993.
36. S Hahnau, B Jnlicher. Food Addit Contam 13:259–274, 1996.
37. W Haasnoot, L Streppel, G Cazemier, M Salden, P Stouten, M Essers, P van Wichen. Analyst 121:1111–1114, 1996.

19

Chemiluminescence and Bioluminescence in DNA Analysis

Masaaki Kai, Kazuko Ohta, Naotaka Kuroda, and Kenichiro Nakashima
Nagasaki University, Nagasaki, Japan

1. INTRODUCTION

Over the last decade, various nonradiochemical methods utilizing chemiluminescent or bioluminescent reactions have been developed to increase the sensitivity and speed of detecting DNA probes or DNA itself. This has permitted the devel-

551

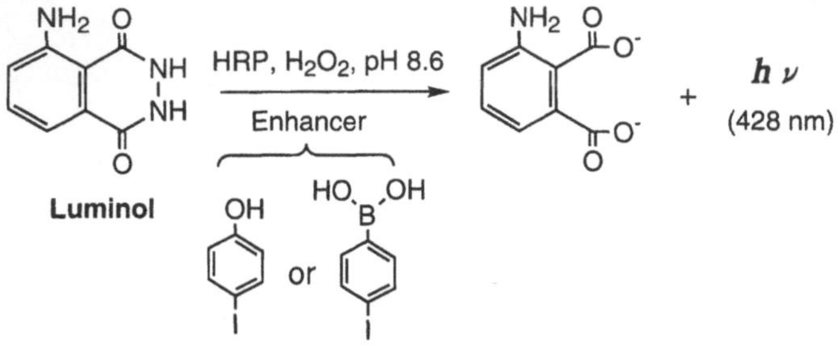

Figure 1 HRP-catalyzed chemiluminescent reaction of luminol.

opment of powerful analytical techniques for obtaining information on gene structure and function.

Recently, two major enzyme-catalyzed chemiluminescent reactions have become popular. These use either luminol as a substrate of peroxidase or 3-(2'-spiroadamantane)-4-methoxy-4-(3''-phosphoryloxy)phenyl-1,2-dioxetane (AMPPD) as a substrate of alkaline phosphatase (ALP).

Luminol is the most popular chemiluminescent compound. Chemiluminescent detection based on horseradish peroxidase (HRP)-catalyzed oxidation of luminol in the presence of hydrogen peroxide requires a halogenated phenol such as 4-iodophenol or 4-iodophenylboronic acid [1] as a potent enhancer (Fig. 1). This method is often called enhanced chemiluminescent detection. In the enhanced reaction, the light emission is increased over 100-fold, permitting detection of HRP at subfemtomole levels. The light has a maximum wavelength of 428 nm and can be captured with high efficiency by blue-light-sensitive X-ray film.

AMPPD is the best chemiluminescent substrate for detecting an ALP-labeled probe [2, 3]. The enhanced sensitivity of the chemiluminescence based on the reaction of AMPPD with ALP depends on the enzymatic reaction time (Fig. 2), because the slow kinetics of the signal decay result in the accumulation

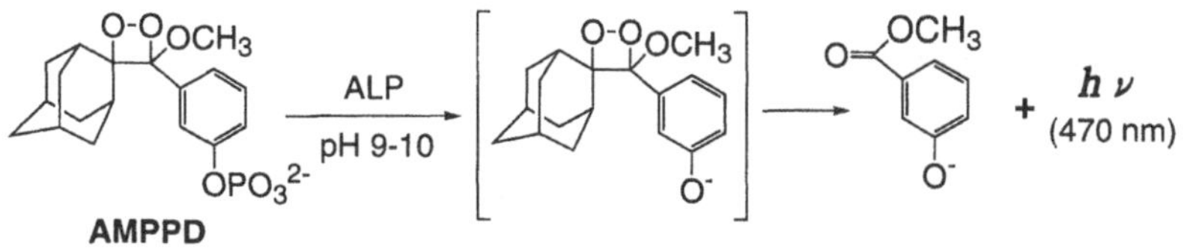

Figure 2 ALP-catalyzed chemiluminescent reaction of AMPPD.

of light (maximum wavelength of 470 nm). This protocol permits highly sensitive detection of enzyme molecules at subattomole levels [2].

Both ALP and HRP are commonly used as signal markers for chemiluminescent detection of complementary deoxyribonucleic acid (cDNA) probes. Recently, marine bacterial luciferase and firely luciferase have been utilized as the marker enzymes of cDNA probes [4, 5]. The light emission by reaction of luciferase with luciferin in the presence of ATP, magnesium(II) ion, and oxygen is called bioluminescence since this reaction is a naturally occurring chemiluminescent reaction (Fig. 3A) [6]. Luciferase also mediates an oxidative reaction of an aldehyde compound in the presence of the reduced type ($FMNH_2$) of flavin mononucleotide (FMN), and then emits a strong and stable light (Fig. 3B) [6, 7]. These enzymatic chemiluminescent methods permit reliable and sensitive nonisotopic detection in immobilized hybridization assays, as well as in common membrane hybridization assays.

5-Bromo-4-chloro-3-indolyl (BCI) substrates can also be used with several commercially available marker enzymes, including ALP, β-D-galactosidase, and β-glucosidase (Fig. 4) [8]. The enzymes hydrolyze the BCI substrate, producing the corresponding chromogenic indigo dye and hydrogen peroxide stoichiometrically in the reaction mixture. Therefore, the enhanced luminol-HRP detection system can detect the enzyme used as a signal marker when high sensitivity is required in hybridization assays.

Acridinium esters have also been utilized for chemiluminescent detection of cDNA probes (Fig. 5) [9–11]. The hydrolysis rate is much faster when the ester is conjugated to single-stranded DNA, rather than to double-stranded DNA. This means that the chemiluminescence from unhybridized acridinium ester–labeled probe is rapidly lost, whereas the chemiluminescence from the hybridized probe is minimally affected. This permits discrimination between hybridized and unhybridized acridinium ester–labeled DNA probes without separation steps.

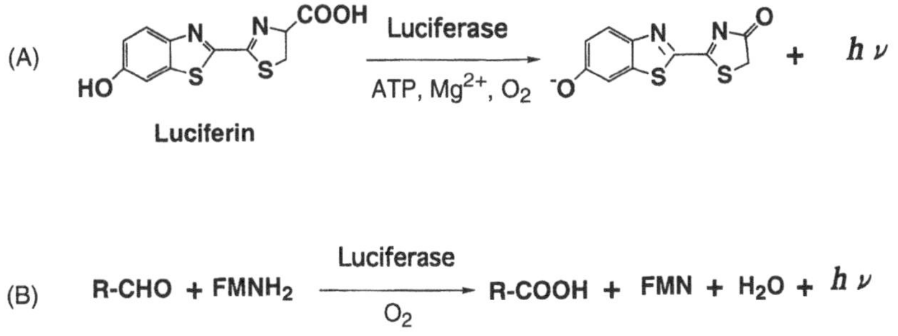

Figure 3 Luciferase-catalyzed bioluminescent reaction with (A) luciferin and (B) an aldehyde compound and $FMNH_2$.

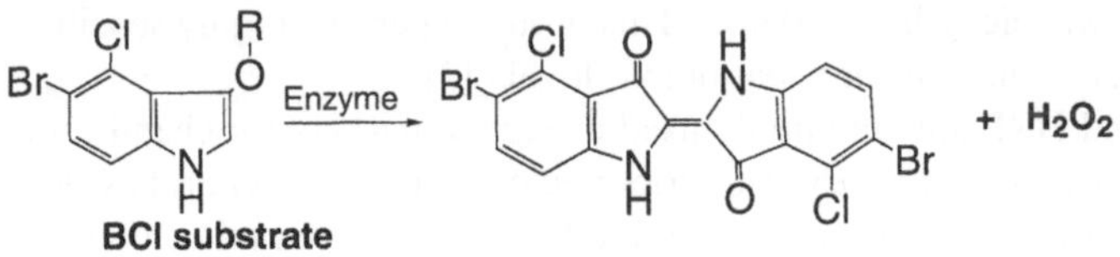

BCI substrate

$$H_2O_2 + \text{Luminol} \xrightarrow{\text{HRP}} h\nu$$

R of substrate	Enzyme
Phosphate	ALP
D-Galactopyranoside	β-Galactosidase
D-Glucoside	β-Glucosidase

Figure 4 Enzymatic reactions of BCI substrates and their chemiluminescent detection by the luminol reaction.

Therefore, chemiluminescent methods using an acridinium ester–labeled cDNA probe allow the discrimination of a mismatched DNA sequence in a homogeneous assay.

On the other hand, few reagents react directly with the target DNA for chemiluminescent detection. Recently, a unique chemical derivatization reagent, 3′,4′,5′-trimethoxyphenylglyoxal (TMPG), has been developed [12–15]. This reagent reacts specifically with guanine bases in nucleic acids to produce a chemiluminescent, fluorescent derivative quickly, under mild reaction conditions (Fig. 6). The derivative emits chemiluminescence (maximum wavelength of 510 nm) un-

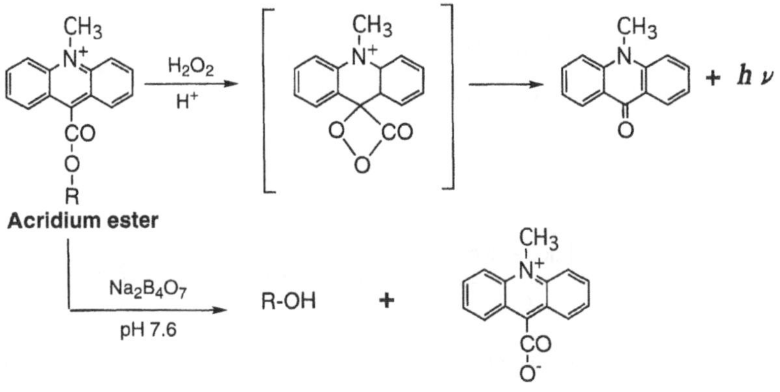

Figure 5 Chemiluminescent reaction of acridinium ester and its hydrolysis.

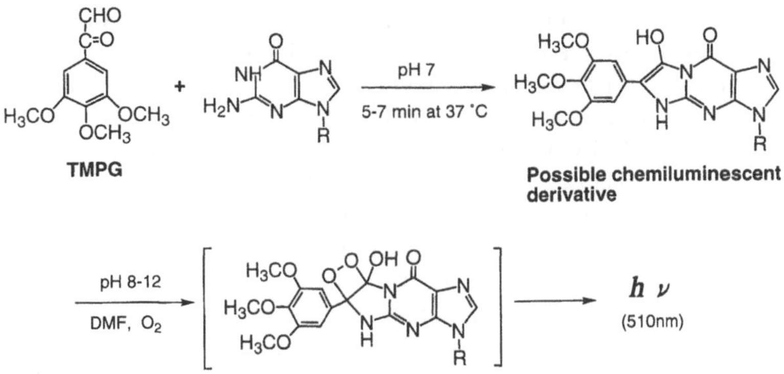

Figure 6 Derivatization and chemiluminescent reactions of guanine-containing compounds with TMPG.

der weakly alkaline conditions in the presence of dimethylformamide (DMF) at room temperature. The TMPG reaction can be used to detect DNA or polydeoxyguanylic acid (d(G)n), both on nylon membranes and in aqueous solution.

This chapter outlines the chemical principles for luminescent detection of target DNA in hybridization and quantitative assays that utilize the above-mentioned chemiluminogenic and bioluminogenic reagents.

2. ENZYMATIC CHEMILUMINESCENT AND BIOLUMINESCENT DETECTIONS

Enzyme-labeled probes have been employed for sensitive chemiluminescent or bioluminescent detection of hybridized target DNA that is bound to a membrane or otherwise immobilized. For these assays, HRP, ALP, and luciferase have been used as the marker enzyme predominantly. These enzymes are far bulkier than isotopic labels, and are also less thermally stable. For labeling, a small molecule such as digoxigenin or biotin is employed to conjugate the enzymes to a DNA hybrid duplex by means of either bioaffinity interaction of avidin or immunochemical interaction of antidigoxigenin antibody [16]. The primary advantage of using biotin or digoxigenin probes is that they are suitable for indirect labeling of many unstable enzymes to cDNA probes.

2.1 Membrane Hybridization

Hybridization assays of membrane-bound DNA are important for characterizing or searching for cloned genes related to a genetic disease, and for identifying

DNA fragments in restriction fragment length polymorphism analyses [16]. These assays usually involve electrophoretic separation of the DNA fragments produced using appropriate restricted enzymes, transferring the resulting band pattern onto a membrane in a process called Southern blotting, and subsequent hybridization with a cDNA probe to detect only the target DNA fragment. In this assay system, a higher detection sensitivity for the hybrid duplex is desirable because less target DNA is necessary for the assay [6, 16].

After hybridization with the target DNA, the enzyme-conjugated complex can be sensitively detected using chemiluminescent reactions with either AMPPD-ALP [17–24] or HRP-enhanced luminol [25, 26] that involves the incorporation of digoxigenin (Fig. 7A) or biotin (Fig. 7B) into cDNA probes. Biotin- or digoxigenin-modified probes can be generated with the polymerase chain reaction (PCR) using digoxigenin- or biotin-labeled nucleotides in the PCR medium. Several modified nucleotide analogs, antidigoxigenin antibody-conjugated enzymes, and biotin-conjugated enzymes are available from commercial suppliers.

These chemiluminescent detection methods have the advantages of high sensitivity and safety. The protocols use a conventional X-ray-film-reading device or an optical instrument such as a cooled charge-coupled device (CCD) camera [27, 28] for acquiring images. If the light emission is increased, the exposure time required for detecting light in methods based on enzymatic amplification can be reduced considerably.

2.2 Immobilized Hybridization

PCR is a technique for in vitro amplification of DNA sequences that involves repeated cycles of denaturation, oligonucleotide annealing, and DNA polymerase extension [29]. The amplified products following PCR cycles contain double-stranded DNA fragments of discrete length. These DNAs are copies of the template DNA that are bounded at the 5′-terminus by the oligonucleotide primer for the sequence extension with a heat-resistant DNA polymerase. In quantitative assays of PCR products, therefore, nonspecific products interfere with the assay.

A target PCR product incorporating biotin or digoxigenin deoxynucleotides can be immobilized to a solid phase such as magnetic beads, Sepharose polymer, or a microtiter well by bioaffinity binding between avidin and biotin or digoxigenin and antidigoxigenin antibody, or by covalently binding a spacer compound to the cDNA probe. The target is hybridized with biotin- or digoxigenin-labeled cDNA probe to capture the marker enzyme that produces the luminescent signal (Fig. 8A and 8B). [30–32]. The sensitivity of immobilization-based hybridization assays of a target PCR product can be increased markedly by utilizing the luminol-HRP or AMPPD-ALP methods.

The protocol shown in Figure 8C is based on a multienzymatic channeling reaction leading to light emission [4]. This method requires three enzymes: glu-

(A)

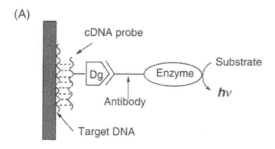

(B)

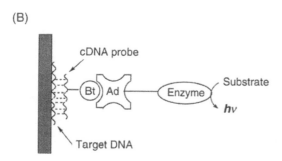

Figure 7 Chemiluminescent detections using (A) digoxigenin- and (B) biotin-binding probes for membrane hybridizations of DNA. Dg, digoxigenin; Bt, biotin; Ad, avidin. Procedures A and B [20]: Membrane hybridization of DNA involves the use of nylon membranes as a solid support onto which various sample DNAs are adsorbed. Of the adsorbed DNAs, the target DNA only can be hybridized to its cDNA probe (12 ng/mL) in which Dg-11-dUTP for protocol A or Bt-11-dUTP for protocol B is incorporated in advance by PCR. After hybridization at 65°C for 16 h, the membrane is washed with an appropriate buffer to remove excess cDNA probe and nonspecifically bound probe, and rinsed with a blocking solution to avoid binding nonspecific antibody or avidin proteins. For membranes using Dg-binding probe, anti-Dg Fab fragment containing ALP is conjugated to the probe for 30 min at 60–70°C. For membranes using Bt-binding probe, a streptavidin-ALP conjugate is used similarly. After conjugate treatment, membranes are rinsed with blocking reagent and washed four times. Each membrane is equilibrated for 5 min in a pH 9.5 substrate buffer, and reacted for 10 min in 0.26 mM AMPPD in fresh substrate buffer. The membrane is then sealed in a polyester/polyethylene bag, and exposed to X-ray film at ambient temperature.

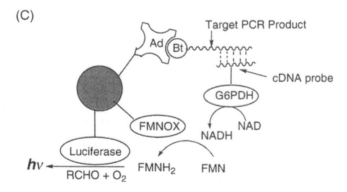

(A)

(B)

(C)

Figure 8 Chemiluminescent (A and B) and bioluminescent (C) detections for immobilized hybridizations of PCR product. Dg, digoxigenin; Bt, biotin; Ad, avidin. Procedure A [30]: Biotin moiety is incorporated into PCR products during the amplification reaction, using one 5'-biotinylated primer. The product is hybridized with a Dg-labeled probe and is immobilized on streptavidin-coated magnetic beads. This capture reaction is carried out for 30 min at 37°C. A permanent magnet is used to sediment the beads during washing to remove unbound DNA. By incubation with the washed beads for 45 min at 37°C, anti-Dg antibody conjugated to HRP enzyme is bound to the Dg-labeled probe, and luminol reaction is performed for CL detection. Procedure B [31]: Wells of a polystyrene microtiter plate are activated with 1-ethyl-3-(3-dimethylaminopropyl)-carbodiimide, and then coated with a labeled cDNA probe complementary to an internal region of the target DNA.

cose-6-phosphate dehydrogenase (G6PDH), flavin mononucleotide oxidoreductase (FMNOX), and luciferase. FMNOX and luciferase are immobilized to a suitable solid phase. The hybrid duplex of the target PCR product with a cDNA probe labeled with G6PDH is formed by the interaction of avidin and biotin on a bioluminescent adsorbent. In the assay, NADH produced by G6PDH attached to the cDNA probe bound to the target is used directly by the FMNOX reaction to reduce FMN. The immobilized luciferase produces stable light emission in the presence of $FMNH_2$ and an aldehyde compound. The luminescence signals are proportional to the amount of bound label.

The immobilized-hybridization methods provide simple detection systems that do not require separation of the target PCR product from contaminating DNA. These protocols are also applicable to the detection of an unmodified DNA target [5, 33, 34].

In the method shown in Figure 9A, a biotin-labeled cDNA probe is first immobilized to a polyvinylchloride microtiter plate well that is coated with biotinylated-bovine serum albumin [33]. The target DNA is hybridized in the liquid-phase with a digoxigenin-labeled probe, so that the biotin-labeled probe can capture a marker enzyme. An antibody-conjugated enzyme is then added, followed by a chemiluminescent substrate.

In the method shown in Figure 9B, a firefly luciferase gene is introduced for sensitive bioluminescent detection of target DNA [5]. The luciferase-coding DNA requires no posttranslational modification, and the activity of the luciferase produced can be readily measured in the transcription/translation mixture without prior purification. In this assay system, the digoxigenin-labeled probe is first immobilized to polystyrene wells coated with antidigoxigenin antibody. The target

---

Dg-labeled PCR products are added to the wells, and hybridized for 4 h at 65°C. The hybridized wells are washed and rinsed with a blocking agent. For chemiluminescent detection, an anti-Dg antibody conjugated to ALP and AMPPD are added to the wells at ambient temperature. Procedure C [4]: PCR product contains biotin moiety by the PCR reaction in the presence of Bt-dUTP. The target PCR product and its cDNA probe, which is labeled covalently with G6PDH by using 4-(N-maleimidomethyl)cyclohexane-1-carbonate and dithiobis(propionate N-hydrosuccinimide ester), are hybridized for 2 h at 37°C. In the solid phase, the enzyme produces NADH from NAD. The produced NADH is directly used by the FMN oxidoreductase to reduce FMN. The immobilized luciferase uses the $FMNH_2$ plus an aldehyde to produce a stable light emission. In the supernatant, the unbound cDNA probe produces also NADH, which is oxidized by lactate dehydrogenase plus pyruvate that are previously added to the reaction mixture. This process does not require any separation step since the unbound enzyme does not lead to light emission. Luminescence values are related to the amount of the bound cDNA probe.

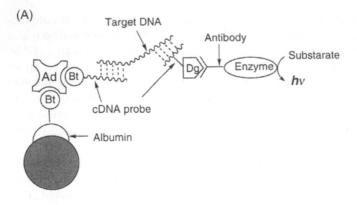

Figure 9 Chemiluminescent (A) and bioluminescent (B) detections for immobilized hybridizations of unmodified DNA target. Dg, digoxigenin; Bt, biotin; Ad, avidin. Procedure A [33]: A Bt-labeled cDNA probe is first bound to a polyvinylchloride microtiter plate well that was previously coated overnight at 4°C with a biotinylated-bovine serum albumin, and saturated with streptavidin for 1 h at 37°C. The test sample containing an unmodified target DNA is hybridized in liquid-phase at 55°C for 1 h with the Dg-labeled cDNA or cRNA probe, and then the mixture is added to the well to be captured at 55°C for 1 h by the Bt-labeled probe in the well. Anti-Dg Fab conjugated with alkaline phosphatase is then reacted at 37°C for 1 h. The well is washed sufficiently, and followed by AMPPD reaction for chemiluminescent detection. Procedure B [5]: An unmodified target DNA is hybridized with its two different sequencing cDNA probes for 1 h at 42°C. One of the probes is immobilized in a microtiter well, via Dg and anti-Dg antibody interaction that is physically adsorbed to the polystyrene well, and the other probe is biotinylated. After completion of the hybridization, the hybrids are reacted with streptavidin and a biotinylated luciferase-coding DNA (2.1 k base pair) for 20 min. After washing of the well, the solid-phase bound luciferase-coding DNA is expressed by adding the transcription-translation solution for 90 min at 30°C. The synthesized luciferase in the well is transferred to a tube, and mixed with the luciferin substrate to detect the luminescence.

DNA is then hybridized with the probe, and also another biotin-labeled cDNA probe is hybridized to capture the luciferase gene. The complex is prepared by mixing biotinylated luciferase-coding DNA with a large excess of avidin. The assay is completed by expressing the solid-phase bound DNA label and measuring the activity of the luciferase synthesized. Expressed luciferase, which mediates the luminescence reaction of luciferin as the substrate, has been exploited to detect target DNA in the hybridization assay. The chemiluminescent signal obtained depends on a linear relationship between the target DNA (5–5000 amol) and the activity of the luciferase produced.

3. CHEMILUMINESCENT DETECTION OF ACRIDINIUM ESTER–LABELED PROBE

The acridinium ester (AE) in an AE-labeled cDNA probe hybridized to target DNA is less likely to be hydrolyzed than in the unhybridized conformation (Fig. 10) [9–11]. Single-base mismatches in the duplex adjacent to the site of AE attachment disrupt this protection, resulting in rapid AE hydrolysis [11]. Hydrolysis by a weak base renders AE permanently nonchemiluminescent. After hydrolysis, it is possible to use the remaining chemiluminescence as a direct measure of the amount of hybrid present. This selective degradation process is a highly specific chemical hydrolysis reaction, which is sensitive to the local environment of the acridinium ester. The matched duplex can be detected and quantified readily, whereas the mismatched duplex produces a minimal signal.

Therefore, a homogeneous hybridization assay for detecting a DNA target not mismatched with its AE-labeled cDNA probe sequences can be performed. The AE-containing probe is synthesized using protected alkylamine linker arms at any location within the synthetic DNA probe. The method is very simple and sensitive, and is completed in 30–60 min. In this method, the background produced by unhybridized probe is sufficiently low. Therefore, this assay is useful for searching for genetic disorders [10, 11].

4. DERIVATIZATION-BASED CHEMILUMINESCENT DETECTION

Methods currently available for chemiluminescent detection of nucleic acids are not based on derivatization techniques that directly recognize one of the nucleic acid bases or nucleotides. For chemical derivatization-based chemiluminescent detection, the specific reactivity of alkyl glyoxals and arylglyoxals with adenine or guanine nucleotides has been investigated.

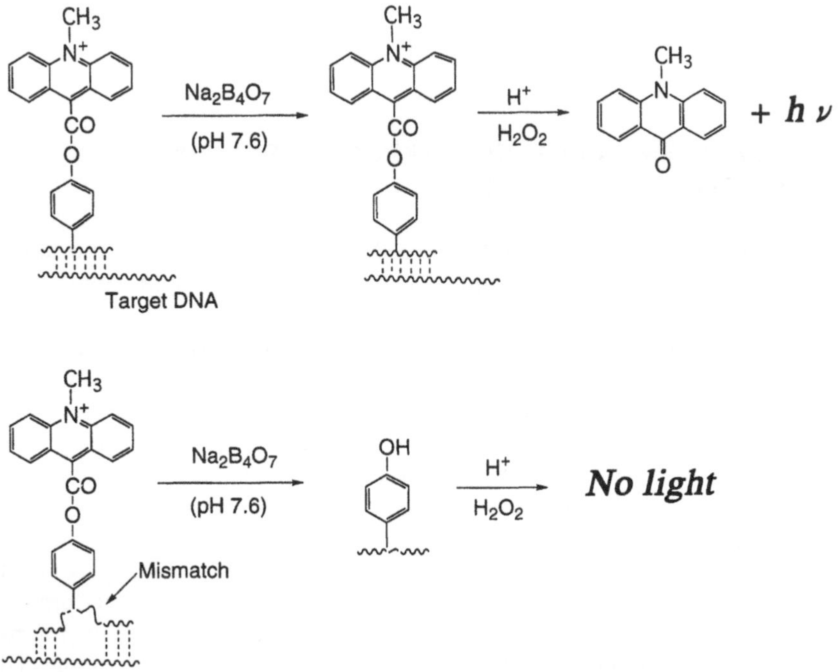

Figure 10 Mismatch detection by using a chemiluminescent AE-labeled cDNA probe. Procedure [9, 11]: Acridinium ester–labeled probes specific for either wild-type or mutant sequence corresponding to a target DNA are hybridized with the sample DNA for 1.0 h at 60°C in a hybridization buffer (pH 5.2). Hybridized and nonhybridized probes are discriminated by the hydrolysis reaction for 12 min at 62.5°C in the presence of $Na_2B_4O_7$ (pH 8.5) and Triton X-100. The chemiluminescence of each sample is then measured in a luminometer.

4.1 DNA Quantification

Phenylglyoxal and alkoxyphenylglyoxals react selectively with the guanine moiety of nucleosides and nucleotides in phosphate buffer (pH 7.0) at 37°C for 5–7 min to give the corresponding fluorescent derivatives [12–15], as shown in Figure 6. Other nucleic acid bases and nucleotides (e.g., adenine, cytosine, uracil, thymine, AMP, CMP) do not produce derivatives under such mild reaction conditions. The fluorescent derivative emits chemiluminescence on oxidation with dimethylformamide (DMF) and H_2O_2 at pH 8.0–12 [14, 15].

Of the alkoxyphenylglyoxals, 3′,4′,5′-trimethoxyphenylglyoxal (TMPG) produces the most intense chemiluminescence for both DNA and guanine nucleotides in aqueous solution [15]. TMPG also reacts readily with DNA absorbed to a nylon membrane at pH 9–10 for 30–60 s at room temperature, and the derivatives produced on the membrane emit chemiluminescence in the presence of DMF [15]. The sensitivity for detecting DNA samples bound to a nylon membrane is approximately 4 ng when reacted with TMPG and detected with a CCD camera cooled to $-25°C$. This is equivalent to a genome size of about 3×10^9 base pair or a molecular weight of about 2×10^{12}, and corresponds to zmol (10^{-21} mol) levels. This method is highly sensitive for nucleic acids containing a large amount of guanine.

On the other hand, glyoxals and phenylglyoxal also react with adenine and DNA under strongly acidic conditions when heated at 100°C for approximately 1 h [35–37]. The products then emit chemiluminescence under strong alkaline conditions. With the method using methylglyoxal, the detection limit for adenine-containing DNA is 5 ng. Under these drastic reaction conditions, however, guanine and guanine nucleotides do not produce any chemiluminescence.

4.2 Detection of Polydeoxyguanilic Acid (d(G)n)–Labeled Probe

The TMPG reagent can be used for chemiluminescent detection of d(G)n on a blotting membrane [15]. The chemiluminescence can be detected with a densitometric luminometer equipped with a cooled CCD camera. This method is used to detect a hybrid with a DNA probe containing a d(G)n oligomer to increase its chemiluminescence (Fig. 11). In Figure 11, the chemiluminescence produced by the hybrid of target DNA bound to the $d(G)_{30}$ probe is greater than that of the target DNA alone.

This protocol permits detection of 10 pmol of target DNA dotted on a nylon membrane after hybridization with the $d(G)_{30}$ probe [15]. The background chemiluminescence caused by nonspecific binding of the probe in the hybridization buffer to the membrane is negligible in this assay system. However, both the target DNA and the cDNA probe bound to the membrane are detected solely

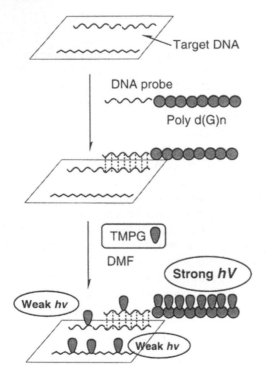

Figure 11 Chemiluminescent detection for membrane hybridization of unmodified DNA target by derivatization reaction with TMPG. Procedure [15]: A portion of the DNA solution is spotted on a nylon membrane. The target DNA is hybridized to its cDNA probe having a -(G)15TT(G)15TT at its 3′ terminus in a hybridization buffer (pH 7.0) at 42°C for 2 h. After washing, the membrane is moistened with sodium phosphate solution (pH 10) for a few seconds, and then immersed in 0.2 M TMPG dissolved with dimethyl sulfoxide for 0.5 min at ambient temperature. The moist membrane is then dipped in dimethylformamide for a few seconds, and the luminescence is detected for 0.5 min.

by the guanine they contain. This hybridization gives a positive, reproducible signal, with a lower detection limit of approximately 1.0 pmol of target DNA.

5. CONCLUSIONS

The advantages of detecting cDNA probes by chemiluminescence and bioluminescence include high sensitivity and simple protocols, using either manual film reading or sophisticated instrumentation such as CCD and photon-counting cameras for processing image data. In addition, the luminescent reagents are considerably more stable than radioisotopes, and adaptable for detecting hybridization on membranes or with immobilized DNA.

The chemiluminescent derivatizing method for detecting DNA and d(G)n on a nylon membrane, based on the chemical reaction with TMPG, has the advantages of speed and simplicity for the quantitative determination of nucleic acids and DNA probes containing d(G)n. Using a d(G)n probe enhances the chemiluminescence in proportion to the amount of guanylic acid in the probe molecule. The TMPG reaction is useful not only for in situ detection of guanine-rich regions of the genome, but also for immobilized hybridization assays.

Current methods for DNA detection usually require enzymatic amplification of the target DNA sequence prior to analysis. For example, the PCR technique selectively increases the concentration of the target sequence relative to unrelated sequences. PCR methods, however, introduce ambiguities resulting from contamination by different DNA sequence. Therefore, a definitive method is required for the analysis of a single, original DNA sequence. To achieve this objective, the sensitivity and speed of the chemiluminescent enhancement techniques described in this chapter must be improved.

REFERENCES

1. LJ Kricka, M Cooper, X Ji. Anal Biochem 240:119–125, 1996.
2. S Beck, H Koster. Anal Chem 62:2258–2270, 1990.
3. R Tizard, RL Cate, KL Ramachandran, M Wysk, JC Voyta, OJ Murrhy, I Bronstein. Proc Natl Acad Sci USA 87:4514–4518, 1990.
4. P Balaguer, B Terouanne, AM Boussioux, JC Nicolas. Anal Biochem 195:105–110, 1991.
5. NHL Chiu, TK Christopoulos. Anal Chem 68:2304–2308, 1996.
6. LJ Kricka. Clin Chem 37:1472–1481, 1991.
7. J Wannlund, M DeLuca. Anal Biochem 122:385–393, 1982.
8. H Arakawa, M Maeda, A Tsuji. Anal Biochem 199:238–242, 1991.
9. LJ Arnord Jr, PW Hammond, WA Wiese, NC Nelson. Clin Chem 35:1588–1594, 1989.
10. NC Nelson, AB Cheikh, E Matsuda, MM Becker. Biochemistry 35:8429–8438, 1996.
11. NC Nelson, PW Hammond, E Matsuda, AA Goud, MM Becker. Nucleic Acids Res 24:4998–5003, 1996.
12. M Kai, Y Ohkura, S Yonekura, M Iwasaki. Anal Chim Acta 207:243–249, 1988.
13. S Yonekura, M Iwasaki, M Kai, H Nohta, Y Ohkura. J Chromatogr 641:235–239, 1993.

14. M Kai, Y Ohkura, S Yonekura, M Iwasaki. Anal Chim Acta 287:75–91, 1994.

15. M Kai, S Kishida, K Sakai. Anal Chim Acta 381:155–163, 1999.

16. ES Mansfield, JM Worley, SE Mckenzie, S Surrey, E Rappaport, P Fortina. Mol Cell Prob 9:145–156, 1995.

17. ED Johnson, TM Kotowski. J Forens Sci 41:569–578, 1996.

18. I Bronstein, JC Voyta, B Edwards. Anal Biochem 180:95–98, 1989.

19. DP Knight, AC Simmonds, AP Schaap, H Akhavan, MAW Bray. Anal Biochem 185:353–358, 1990.

20. JJ Lanzillo. Anal Biochem 194:45–53, 1991.

21. M Musiani, M Zerbini, D Gibellini, G Gentilomi, S Venturoli, G Gallinella, E Ferri, S Girotti. J Clin Microbiol 29:2047–2050, 1991.

22. EH Fiss, FF Chehab, GF Brooks. J Clin Microbiol 30:1220–1224, 1992.

23. I Bronstein, JC Voyta, OJ Murphy, R Tizard, CW Ehrenfels, RL Cate. Methods Enzymol 217:398–414, 1993.

24. EL Durigon, DD Erdman, BC Anderson, BP Holloway, LJ Anderson. Mol Cell Probes 8:199–204, 1994.

25. T Stone, I Durrant. Mol Biotechnol 6:69–73, 1996.

26. P Wattre, A Dewilde, D Subtil, L Andreoletti, V Thirion. J Med Virol 54:140–144, 1998.

27. AE Karger, R Weiss, RF Gesteland. Nucleic Acids Res 20:6657–6665, 1992.

28. CS Martin, I Bronstein. J Biolumin Chemilumin 9:145–153, 1994.

29. RA Gibbs. Anal Chem 62:1202–1214, 1990.

30. AM Vlieger, AMJC Medenblik, RPM van Gijlswijk, HJ Tanke, M van der Ploeg, JW Gratama, AK Raap. Anal Biochem 205:1–7, 1992.

31. J Stevens, FS Yu, PM Hassoun, JJ Lanzillo. Mol Cell Prob 10:31–41, 1996.

32. A Erhardt, S Schaefer, N Athanassiou, M Kann, WH Gerlich. J Clin Microbiol 22:1885–1891, 1996.

33. EC Kim, EL Durigon, DD Erdman, LJ Anderson. J Virol Methods 50:349–354, 1994.

34. ML Collins, B Irvine, D Tyner, E Fine, C Zayati, C Chang, T Horn, D Ahle, J Detmer, LP Shen, J Kolberg, S Bushnell, MS Urdea, DD Ho. Nucleic Acids Res 25:2979–2984, 1997.

35. N Kuroda, K Nakashima, S Akiyama. Anal Chim Acta 278:275–278, 1993.

36. N Sato, K Shirakawa, K Sugihara, T Kanamori. Anal Sci 13:59–65, 1997.

37. N Kuroda, K Nakashima, S Akiyama, N Sato, N Imi, K Shirakawa, A Uemura. J Biolumin Chemilumin 13:25–29, 1998.

20

Recent Developments in Chemiluminescence Sensors

Xinrong Zhang
Tsinghua University, Beijing, P.R. China

Ana M. García-Campaña
University of Granada, Granada, Spain

Willy R. G. Baeyens
Ghent University, Ghent, Belgium

Raluca-Ioana Stefan and Jacobus F. van Staden
University of Pretoria, Pretoria, South Africa

Hassan Y. Aboul-Enein
King Faisal Specialist Hospital and Research Centre, Riyadh, Saudi Arabia

1. INTRODUCTION

Optical sensors belong to the most important types of chemical sensors that have
been extensively studied in recent years for the continuous and real-time monitoring of analytes. Depending on the origin of the optical signals, these types of
sensors may be roughly classified into absorbance and luminescence-based sensors, the latter mainly utilizing the principle of fluorescence and chemiluminescence (CL) detection. Although they offer many advantages such as high sensitivity, good selectivity, and fast response time, fluorescence-based sensors require
an excitation light source and spectral separation of exciting and emitted light,
leading to relatively sophisticated equipment producing high background signals.

However, these drawbacks do not occur in the case of CL detection since
the energy required for CL emission does not originate from an exciting light
beam but is produced in a chemical reaction. Significant advances in design and
applications of CL sensors and biosensors were recorded in the last few years
[1–3]. The most utilized type of sensor is the flow-through one [4]. The reliability
of this type of sensor made them suitable for utilization as detectors in flow
injection analysis (FIA) systems [5]. The first CL sensor for hydrogen peroxide
analysis was reported by Freeman and Seitz in 1978 [6] and since then various
types of CL-based sensors have been extensively studied for inorganic, organic,
and biological/pharmaceutical compounds. In this chapter, the development of
CL sensors is reviewed and the advantages and limitations when applying the
technique to routine analysis are discussed.

2. TYPES OF CL SENSORS

A CL sensor can be defined as an analytical device incorporating an active material with a transducer, with the purpose of detecting in a continuous, selective,

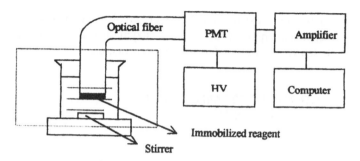

Figure 1 Schematic diagram of a fiberoptic sensor for batch assay. PMT, photomultiplier tube; HV, high-voltage supplier.

and reversible way, the concentration of chemicals in diverse kinds of sample, employing CL emission. There are several ways to classify CL sensors, based on different points of view. Relating to the manner of bringing the sample in contact with the sensing surface, there are batch and flow CL sensing systems. In the former, the sensing surface is immersed in the analyte solution and an optical fiber is sometimes used as light transducer. This type of CL sensor is also called a CL optrode. Figure 1 gives a schematic diagram of this kind of CL sensor. Flow CL sensors are more popular because a flow system allows the management of additional reagent solutions, and improves the analytical speed and repeatability, thus easing automation. Figure 2 shows a typical flow CL sensing system.

Another classification is based on the analytes that occur in gas or liquid phases. The CL gas sensors are applied to the analysis of compounds in gases or vapors, such as O_2, O_3, NO, NO_2, chloride, ammonia-containing compounds

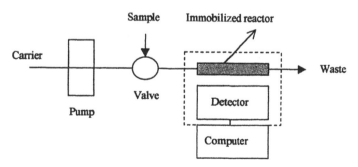

Figure 2 Schematic diagram of a flow-sensing system for online assay.

as well as organic vapors, etc. Most reports, however, involve the analysis of compounds using CL sensors in liquids, which include enzyme-based CL sensors, non-enzyme-based CL sensors as well as CL immunosensors, etc.

3. CL SENSORS FOR THE DETERMINATION OF ANALYTES IN AIR OR VAPORS

3.1 NO$_2$-Air Sensors

Different instrumental variations, based upon the oxidation of luminol-producing CL, have been developed for the detection of gas-phase oxidants. Maeda et al. [7] designed a CL sensor for the detection of NO$_2$ that comprised a pool of alkaline luminol solution directly below a photomultiplier tube (PMT) introducing the sampled airstream into the region above the solution. This system was very sensitive to movements of the compartment and had a relatively slow time response. An alternate design for CL detection of NO$_2$ was subsequently presented by Wendel et al. [8], wherein a length of filter paper was positioned adjacent to a PMT, and a flow of alkaline luminol solution was directed down the paper in a fine film. A simpler optrode for CL detection of NO$_2$ was reported by Yin et al. [9] wherein a piece of filter paper soaked in alkaline luminol solution was positioned to one end of an optical fiber. The CL signal transmitted through the optical fiber could be detected by a PMT positioned at another end of this fiber. A further variation of the instrument as described by Wendel et al. [8] led to the development of a commercially available instrument devoted to the measurement of NO$_2$; a detailed description of this instrument was presented by Schiff et al. [10]. The reaction cell consists of a fabric wick positioned in front of a PMT that is continually wetted with fresh luminol solution delivered via a peristaltic pump, and whose surface is exposed to a stream of the ambient air pumped through the cell. In recent years, this instrument has been utilized in combination with various pretreatment stages for trace detection of organic nitrates.

High sensitivities can be achieved by utilizing the CL biosensors proposed by Spicer et al. [11]: the first one is based on the reduction of NO$_2$ to NO followed by the detection of NO by the CL produced from its reaction with O$_3$ while the second one is based on the detection of CL produced from the reaction of NO$_2$ with luminol solution. The working concentration range for the O$_3$ CL sensor (0–800 µg/L NO$_2$) is larger than the working concentration range obtained using the luminol CL sensor (0–50 µg/L). The main disadvantage of these types of CL sensors is nonselectivity with respect to several substances (e.g., nitrous acid).

Another variation on the luminol CL sensor for NO$_2$ was introduced by Collins and Ross-Pehrsson [12] where a solid-phase reagent was positioned below a PMT, across which the air under test is pumped. Of the hydrogel or polymeric sorbents investigated, a Waterlock superabsorbing polymer (hydrogel)

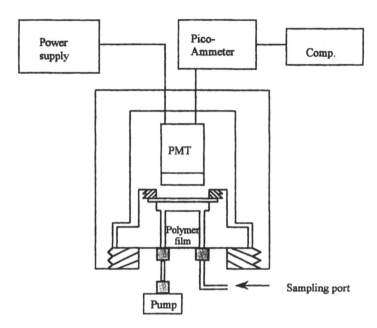

Figure 3 Schematic diagram of a solid-phase NO_2 sensor. The sensor consists of a small cell supporting the polymer-coated, glass substrate behind a glass window in full view of a PMT. The CL reagent is immobilized on the hydrogel substrate. The gel is sandwiched between the glass window and a Teflon PTFE membrane. The purpose of the Teflon membrane is to permit the diffusion of NO_2 from the airstream into the gel while preventing the loss of water from the hydrogel. Inlet and outlet tubes (PTFE) allow a vacuum pump to sample air (2 L/min) directly across the surface of the chemical sensor. (Adapted with permission from Ref. 12.)

incorporating luminol and Cu(II) proved most suitable for the selective determination of NO_2. The sensitivity was 0.46 ppb. The sensors showed drift during continuous exposure due to the irreversibility of the luminol reaction. Figure 3 shows the schematic diagram of a typical NO_2-air sensor.

3.2 O_2- and O_3-Air Sensors

The first CL sensor for oxygen analysis was reported by Freeman and Seitz in 1978 [6]. Collins and Ross-Pehrsson [12] investigated the effect of polymer type, pH, and metal catalyst incorporated within the film. Oxygen levels as low as 2.4 ppm in nitrogen have been detected using the oligomer fluoropolyol as the support matrix for immobilizing luminol, KOH, and the metal catalyst $Fe_2(SO_4)_3$. A sensor for monitoring atmospheric ozone was proposed by Schurath et al. [13] by coating a dye, Coumarin 47 on a thermostatized commercially available TLC plate of dry silica. Ozone was determined by the CL reaction with the surface-absorbed dye. The flow rate, pressure, temperature, and humidity dependence of the detector sensitivity were measured. The lightest balloon-borne version of the detector, including batteries, weighed 1.3 kg.

3.3 Sensor for Carbon Dioxide

Lan and Mottola [14] have presented two continuous-flow-sensing strategies for the determination of CO_2 in gas mixtures using a special reaction cell. Both approaches are based on the effect of the complex of Co(II) with phthalocyanine as a rate modifier of the CL emission generated by luminol in the absence of an added oxidant agent, which is enhanced by the presence of CO_2 in the system. This enhancement allows the fast and simple determination of carbon dioxide at ppm levels (v/v) in atmospheric air and in human breath. In the first case, a continuous monitoring system was applied; however, because the flow of expired gas is not constant, a discrete sample introduction approach was used in the analysis of CO_2 in breath.

3.4 Sensors for Chloride- and Ammonia-Containing Compounds

A sensor for organic chloride-containing compounds was constructed by immobilization of luminol or *tris*-(2,2'-bipyridyl)ruthenium(III) between a PMT and a poly(tetrafluoro)ethylene (PTFE) membrane [15], through which a stream of air was sampled by diffusion. A heated Pt filament incorporated in the gas line leading to the CL cell was used to oxidize the analytes prior to diffusion across the PTFE membrane. Detection limits for CCl_4, $CHCl_3$, and CH_2Cl_2 were 1.2–4 ppm. A similar device could also be used for the determination of hydrazine and its monomethyl and dimethyl derivatives or NH_3 vapor. The detection limit for hydrazine was only 0.42 ppb [16].

3.5 Sensors for Sulfur-Containing Compounds

Meng et al. [17] described a CL sensor based on *tris*(2,2'-bipyridylruthenium(II) permanganate for the SO_2 assay. SO_2 can be sampled if air is purged through a 0.1% triethanolamine absorbing solution. Furthermore, the slope of the calibration graph is constant for a given triethanolamine solution and it was stepwise linear from 1×10^{-7} to 1.25×10^{-5} mol/L of SO_2 in the triethanolamine solution. The recovery of SO_2 in air samples is between 94.6 and 105.4%.

A high-speed sensor for the assay of dimethyl sulfide in the marine troposphere based on its CL reaction with F_2 was recently reported [18]. Sample air and F_2 in He were introduced at opposite ends of a reaction cell with a window at one end. The production of vibrationally excited HF and electronically excited fluorohydrocarbon (FHC) produced CL emission in the wavelength range 450–650 nm, which was monitored via photon counting. Dimethyl sulfide could be determined in the 0–1200 pptv (parts per trillion by volume) concentration range, with a 4-pptv detection limit.

3.6 Organic Vapor Sensors

Ethanol and acetone in air could also be detected by a vapor sensor that was constructed from a sintered layer (0.5 mm) thick of γ-Al_2O_3 or an α-Al_2O_3 substrate on a ceramic substrate (3 × 1.5 mm) with a heater layer of Pt thin film [19, 20]. It was placed in a quartz tube (8.4 mm id) through which the gaseous sample was passed. The CL produced on oxidation of the sample was measured at 380–420 nm using the photocounting technique, and at 450–650°C with a flow rate of 400 mL/min. It was possible to determine up to 400 ppm of ethanol and 200 ppm of acetone in gaseous mixtures. A similar design was also reported for discriminating and determining constituents in mixed gases [21]. Ethanol and butanol spectra were similar, but they differed from propanol and butyric acid spectra. Detection limits were around 1 ppm for the vapors.

Recently "cataluminescence" has been used as detection technique in a gas sensor for recognizing organic vapors [22]. Cataluminescence is termed CL emitted during catalytic oxidation of a combustible gas. In this new approach, spectroscopic images of cataluminescence intensity, which reflect the type and concentration of organic vapor as function of wavelength and temperature, can be measured continuously. Alcohol and ketone or different kinds of alcohols could be discriminated and determined using this method.

4. CL SENSORS FOR THE DETERMINATION OF ANALYTES IN LIQUIDS

4.1 Enzyme-Based CL Sensors

4.1.1 Hydrogen Peroxide

Freeman and Seitz [6] developed one of the first enzyme-based CL sensors with convincing performance. They immobilized horseradish peroxidase (HRP) at the end of an optical fiber and achieved a detection limit of 2 × 10^{-6} mol/L H_2O_2. Preuschoff et al. [23] developed a fiberoptic flow cell for H_2O_2 detection with long-term stability, suitable for fast FIA. Different peroxidases were covalently immobilized on an affinity membrane and compared with respect to the catalytic luminol oxidation, achieving high sensitivity and best long-term stability with microbial peroxidase. The operational stability of the sensor is longer than 10 weeks and hydrogen peroxide can be determined in the range between 10^{-3} and 10^{-8} M. Higher sensitivity was obtained using a different approach for a H_2O_2 biosensor based on the luminol-peroxidase system [24]. CL intensity was increased by using a high-salt-concentration medium, such as 3 M KCl or NaCl, and the enhancement phenomenon was just as effective on soluble peroxidase and on immobilized peroxidase, its magnitude depending on the hydrogen peroxide concentration.

More recently a CL biosensor was developed using silicate glasses obtained by the sol-gel method [25], which represents an alternative for immobilization of biological entities owing to its low-temperature preparation, providing an adequate supporting matrix in which HRP is immobilized by microencapsulation. This sol-gel biosensor based on the CL reaction of the hydrogen peroxide-luminol-HRP system permits the determination of hydrogen peroxide in the range of 0.1–3.0 mM by measuring CL in a cuvette and through an optical fiber modified at its end with immobilized HPR gel, providing a detection limit of 6.7×10^{-4} M. The method was satisfactorily applied to hydrogen peroxide determinations in disinfectant solutions for contact lenses.

4.1.2 Glucose

Several types of CL biosensors are described for glucose assays. Some of them are based on the utilization of glucose oxidase for enzymatic reaction in the coupling with luminol for the CL reaction. Aizawa et al. [26] and Blum [27] developed the CL bienzyme sensors for glucose determination on the basis of the HRP-catalyzed H_2O_2/luminol reaction and coimmobilized glucose oxidase. A fiberoptic was incorporated in the FIA system for the analysis of glucose using the above-mentioned technique [28]. The working concentration range is of mM magnitude order, with a detection limit of 1 μM. High sensitivity was obtained in the analysis of glucose in plasma using the luminol reaction with H_2O_2 produced by immobilized pyranose oxidase within a flow-through cell containing immobilized peroxidase [29]. The enzyme catalyzes the reaction

$$\text{D-Glucose} + O_2 + H_2O \rightarrow \text{D-glucosone} + H_2O_2$$

This enzyme oxidizes α- and β-anomers of D-glucose to the same extent and shows excellent stability and sensitivity about twice that for the methods with immobilized glucose oxidase. In this approach, pyranose oxidase is immobilized on tresylate-poly(vinylalcohol) beads and packed into a stainless column and peroxidase is immobilized on tresylate-hydrophylic vinyl polymer beads and packed into a transparent PTFE tube that is used as the CL flow cell. The H_2O_2

produced in the reactor was detected by the CL emitted in the flow cell. The method allows the determination of glucose in plasma without any pretreatment procedure, except for dilution, offering a limit of detection of 3 nM (5 pg in 10-µL injection).

Common to most CL sensing techniques is that the analyte is being monitored in a dark box for exclusion of ambient light from the analytical signal. Preston and Nieman [30] developed an electrogenerated chemiluminescence (ECL) probe that does not require the use of a dark box. The probe is placed within the sample solution in a fashion similar to the use of a pH electrode. ECL systems including luminol and *tris*(2,2′-bipyridyl) ruthenium(III) appear to be suitable for the probe. By immobilization of glucose oxidase within the probe body, glucose determination could be performed from 3.0 to 1000 µM. The key design is the top piece, which is shown in Figure 4 with a detailed view of the probe.

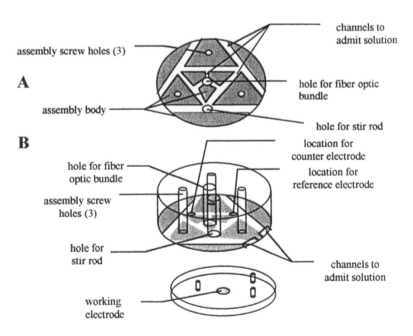

Figure 4 An ECL probe that does not require the use of a dark box. (A) Bottom view of the top piece. Shown are the patterns of channels leading to the fiberoptic bundle and the location of the stirring rod with respect to the fiberoptic bundle. (B) A detailed view of the probe. The working electrode is aligned directly under the fiberoptic bundle, and both the reference and counterelectrodes are inserted through the top of the top piece and access the solution from the side of the probe. (Adapted with permission from Ref. 30.)

Another type of sensor was based on the utilization of glucose dehydrogenase enzyme coupling with *tris*(2,2'-bipyridylruthenium(II) complex [31]. This sensor can be used in the 10–2500-μmol/L concentration range. Several interferences occur, like NADH, oxalate, proline, and tripropylamine. However, gluconic acid and NAD^+ do not interfere.

4.1.3 Amino Acids and Related Compounds

The luminol-H_2O_2 CL system was applied for the determination of glutamine by immobilization of glutaminase [32, 33]. An interesting design was made by Spohn et al. [34] (Fig. 5), who described an extension of the sensor concept by coimmobilization of microbial peroxidase with lysine oxidase, glutamate oxidase, or xanthine oxidase on the sensor membrane. Both a photomultiplier-based fiber optical setup and a photodiode with an integrated preamplifier were used as the signal transducer. On this basis a five-channel FIA system for the determination of glucose, lactate, glutamine, glutamate, and ammonia in an animal cell culture could be achieved [35].

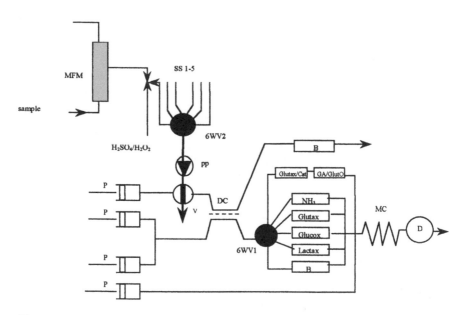

Figure 5 Five-channel enzyme sensor for the simultaneous determination of glucose, lactate, glutamate, glutamine, and ammonium. MFM, microfiltration module; WV, valves; P, pumps; DC, dialysis cell; B, blank reactors; MC, reactor; D, biosensor flow cell. (Adapted with permission from Ref. 34.)

Similar to the work described by Spohn et al. [34], a trienzyme sensor was developed recently for the determination of branched-chain amino acids (L-valine, L-leucine, and L-isoleucine). Leucine dehydrogenase, NADH oxidase, and peroxidase were coimmobilized covalently on tresylate-hydrophylic vinyl polymer beads and packed into a transparent PTFE tube (20 cm × 1.0 id), which was used as flow cell. The sensor was free of interferences from protein and NH_4^+ and it was stable for 2 weeks. The sensor system was applied to the determination of branched-chain amino acids in plasma with recoveries ranging from 98 to 100% [36].

A packed bed flow microreactor containing alanine aminotransferase and glutamate oxidase immobilized on sieved porous glass beads was combined with a CL detector for the generated H_2O_2 [37]. To catalyze the indicator reaction between luminol and H_2O_2, Co(II) and immobilized peroxidase from *Arthromyces ramosus* (ARP) were used in a fiberoptic detection cell. L-Alanine was determined from cell cultivation media in the 2–500-μM concentration range, with a limit of detection of 1 μM using Co(II), and in the 5–800-μM concentration range with a limit of detection of 2 μM when ARP was used. L-Glutamate and α-ketoglutarate were also determined with detection limits of 5 and 1 μM, respectively.

4.1.4 Choline and Acetylcholine

The detection of choline using a CL biosensor is based on the immobilization on a polymer [38] or nylon [39] of choline oxidase and fungal peroxidase. The calibration graphs were linear in the 0.1–1-μmol/L concentration range with a limit of detection of 1 μmol/L. Acetylcholine and choline were determined satisfactorily in human serum using a biosensor prepared by covalently coupling acetylcholinesterase (AchE) and choline oxidase (ChO) to the controlled-pore glass as an enzyme reactor to produce hydrogen peroxide [40]. Electrostatically immobilized luminol and copper ion exchange resin were used as a transduction element in an injection system. The analytes were injected into a continuous stream of simple medium flowing through a sequence of enzyme reactors in which hydrogen peroxide was produced. Luminol and Cu^{2+} were eluted and reacted with H_2O_2 to produce CL. The complete analysis was finished within 2 min with a detection limit of 500 fmol; the stability of the sensor was for 6 months.

4.1.5 Phosphate

A FIA system has been proposed for the CL detection of phosphate based on an enzymatic reaction and the application of a subsequent luminol reaction [41]. The system consists of an immobilized pyruvate oxidase column, a mixing chamber for the CL reaction, and a PMT. H_2O_2 is generated by the reaction of phosphate and pyruvate oxidase and then reacts with luminol and HRP, producing

CL emission. The system allows a simple determination of phosphate in 3 min with a linear range of 4.8–160 μM. Owing to its sensitivity, this method could be satisfactorily applied to the analysis of maximum permissible phosphate concentrations in natural waters [42–44]. Also, the maltose-phosphorylase, mutarose, and glucose oxidase (MP-MUT-GOD) reaction system combined with an ARP-luminol reaction system has been used in a highly sensitive CL-FIA sensor [45]. In this system, MP-MUT-GOD is immobilized on N-hydroxysuccinimide beads and packed in a column. A linear range of 10 nM–30 μM and a measuring time of 3 min were provided, yielding a limit of detection of 1.0 μM as well as a satisfactory application in the analysis of river water.

4.1.6   Carbamate and Organophosphorous Compounds

The above-mentioned system has also been used for the indirect CL determination of some carbamate and organophosphorous pesticides that inhibit acetylcholinesterase. Acetylcholinesterase in solution or immobilized on methacrylate beads is coupled to immobilized choline oxidase and peroxidase [46].

Acetylcholine

$\quad$ ↓ Acetylcholinesterase

Choline

$\quad$ ↓ Choline oxidase

$$\text{Betaine} + 2\,H_2O_2 \xrightarrow[\text{luminol} + OH^-]{\text{Peroxidase}} \text{aminophthalate anion} + N_2 + 3H_2O + \text{light}$$

In this system, choline formed by acetylcholinesterase is oxidized by choline oxidase and the hydrogen peroxide produced is determined using the luminol/peroxidase CL reaction. The sensor has been used for the analysis of Paraoxon and Aldicarb pesticides, with detection limits of 0.75 μg/L and 4 μg/L, respectively. Recoveries in the range of 81–108% in contaminated samples of soils and vegetables were obtained.

4.1.7   Xanthine and Hypoxanthine

A fiberoptic biosensor has been used for the determination of xanthine and hypoxanthine by immobilization of xanthine oxidase and peroxidase on different preactivated membranes, which were mounted onto the tip of the fiberoptic bundle [47]. The hydrogen peroxide generated was measured using the luminol reaction. A linear calibration curve of the sensors occurred in the range of 1–316 nM hypoxanthine and of 3.1–316 nM xanthine, respectively, with a detection limit of 0.55 nM.

4.1.8 L-Lactate

A bienzyme fiberoptic sensor for the CL-FIA of L-lactate was developed by immobilizing lactate oxidase and peroxidase covalently on preactivated polyamide membrane. Hydrogen peroxide generated by the lactate oxidase reaction in the presence of L-lactate was the substrate of the second reaction catalyzed by peroxidase, in which light was produced in the presence of luminol [48]. Compartmentalization of the enzyme layer was obtained by stacking a peroxidase membrane on a lactate oxidase membrane at the sensing tip of the fiberoptic sensor. The detection limit was 250 pmol and the method was satisfactorily applied to lactate determinations in reconstituted whey solutions. Enzyme-modified silica and graphite paste were used to construct another CL biosensor for L-lactate [49]. L-Lactate oxidase was coupled with luminol/Na_2CO_3 (pH = 9.2) to generate the CL. The system is very sensitive and selective when it is used in clinical analysis.

4.1.9 Oxalate

An oxalate sensor by immobilizing spinach tissue as the source of oxalate oxidase was developed by Li [50]. The sensor responds linearly to oxalate concentration in the range of 1.0–100 μM with a detection limit of 0.6 μM. The sensor was stable for 30 days when stored at 4°C and a complete analysis for the determination of oxalate could be performed in 1 min including sampling and washing. Considering the low cost of the plant tissue and simple procedure for plant tissue immobilization, this report is most valuable.

4.1.10 Nicotinamide Adenine Dinucleotide (NADH)

In recent years, dehydrogenase-based biosensors have been extensively reported. NAD^+ is required to catalyze the enzyme reaction of dehydrogenase. NADH produced by the enzyme reaction is related to the substrate concentration, and therefore, the substrate concentration can be determined by measuring NADH concentration. Usually, an electrochemical method is applied as sensing technique for NADH. Because high overpotential is necessary to oxidize NADH directly at an electrode, an electron transfer mediator is normally employed for electrochemical oxidation of NADH, such as ruthenium *tris*(2,2′-bipyridine). In this sense, a regenerable ECL biosensor for NADH based on dehydrogenase and *tris*(2,2′-bipyridyl) ruthenium(II) complex immobilized on Eastman AQ and Nafion polymer films was reported [51]. The biosensor design places an enzyme-loaded polymer film adjacent to a *tris*(2,2′-bipyridyl)ruthenium(II)-loaded polymer film covering a platinum electrode, which is used in a FIA system as an ECL detector. The CL response to samples containing enzyme substrate and cofactor NAD^+ results from the $Ru(bpy)_3^{2+}$ ECL reaction with NADH produced by the enzyme. Similar results were obtained with both kinds of films, providing a detection range of 0.01–100 µM. Another bioluminescence biosensor is based on the detection on NADH by coupling NADH-flavin mononucleotide (FMN) oxidoreductase and bacterial luciferase [52]. A bioactive layer associated with the transducer was designed using a commercial preactivated polyamide membrane to which bacterial luciferase and oxidoreductase were covalently bound. The calibration graph was linear in the 10 pmol/L–0.5 nmol/L concentration range.

4.1.11 Ethanol

Alcohol oxidase was used to generate H_2O_2 followed by its reaction with luminol in the presence of $K_3[Fe(CN)_6]$ as a catalyst [53]. The luminescence was transmitted from the flow cell to the detector via optical fibers. Ethanol can be determined in the 3–750-µmol/L concentration range, with a detection limit of 3 µmol/L. Also, using an immobilized alcohol dehydrogenase reactor in glass beads, a FIA sensor for a reduced form of NADH was constructed by the ECL using the abovementioned ruthenium *tris*(2,2′-biryridine) complex. The sensor was satisfactorily applied to the determination of ethanol concentration [54].

4.1.12 Cholesterol

Cholesterol oxidase was recently immobilized onto amine-modified silica gel via glutaraldehyde activation, and packed in a column [55]. The analytical reagents, including luminol and ferricyanide, were electrostatically coimmobilized on an anion-exchange column. Cholesterol was detected by the CL reaction between H_2O_2 generated in the enzymatic reaction and luminol and ferricyanide, which were released from a column by elution with immobilized reagents. The method was satisfactorily applied to the determination of cholesterol in human serum in a linear range of 5–100 ppm with a relative standard deviation less than 5%.

Some other typical examples of the enzyme-based CL sensors [56–62] are also included in Table 1.

4.2 Non-Enzyme-Based CL Sensors

4.2.1 Chlorine

CL sensors based on immobilization of nonenzyme reagents have been extensively studied in recent years. Nakagama et al. [63] developed a CL sensor for monitoring free chlorine in tap water. This sensor consisted of a Pyrex tube, packed with the uranine (fluoresceine disodium) complex immobilized on IRA-93 anion-exchange resin, and a PMT placed close to the Pyrex tube. It was used for monitoring the concentration of free chlorine (as HClO) in tap water, up to 1 mmol/L, with a detection limit of 2 µmol/L. The coefficient of variation ($n =$

Table 1 Some Typical Examples of Enzyme-Based CL Sensors

Analyte	Immobilized enzyme	CL reaction	Ref.
Alanine	Alanine aminotransferase, ARP	Luminol-H_2O_2-Co(II)	37
Acetylcholine	Acetylcholinesterase	Luminol-H_2O_2-Cu(II)	40
Alcohol	Alcohol oxidase	Luminol-H_2O_2-Fe(CN)$_6^{3-}$	53
Cholesterol	Cholesterol oxidase	Luminol-H_2O_2-Fe(CN)$_6$	55
	Cholesterol oxidase, HRP	Luminol-H_2O_2-HRP	56
Choline	Choline oxidase	Luminol-H_2O_2-Cu(II)	40
Glucose	Glucose oxidase, HRP	Luminol-H_2O_2-HRP	26, 27
	Pyranose oxidase, HRP	Luminol-H_2O_2-HRP	29
	Glucose oxidase	Luminol-H_2O_2-HRP	30
	Glucose oxidase	Tris(2,2′-bipyridyl)ruthenium(III)	30
	Glucose dehydrogenase	Tris(2,2′-bipyridyl)ruthenium(III)	31
	Glucose oxidase	TCPO-H_2O_2-perylene	57
Glutamine	Glutaminase	Luminol-H_2O_2-Fe(CN)$_6^{3-}$	32, 33
Glutamate	Glutamate oxidase, HRP	Luminol-H_2O_2-HRP	58
	Glutamate oxidase, ARP	Luminol-H_2O_2-Co	37
Hypoxanthine	Hypoxanthine oxidase, HRP	Luminol-H_2O_2-HRP	47
Lactate	Lactate oxidase, HRP	Luminol-H_2O_2-HRP	48, 62
Lysine	Lysine oxidase	Luminol-H_2O_2-HRP	59
Oxaloacetate	Malate deydrogenase, bacterial luciferase	FMA-NADH	60
Phosphate	Pyruvate oxidase	Luminol-H_2O_2-HRP	41
Sorbitol	Sorbitol dehydrogenase, bacterial luciferase	FMA-NADH	60
Uric acid	Uricase oxidase, HRP	Luminol-H_2O_2-HRP	61
Xanthine	Xanthine oxidase, HRP	Luminol-H_2O_2-HRP	47

HRP, horseradish peroxidase; ARP, *Arthromyces ramosus* peroxidase.

10) obtained for the free chlorine assay is 1.6%, for a concentration of 10 μmol/ L. The main disadvantage is the short lifetime of the sensor.

4.2.2 Copper

The copper flow-through CL sensor comprised an anion-exchange column having luminol and cyanide coimmobilized on the resin, while copper was temporarily retained by electrochemical preconcentration on a Au electrode placed in an anodic stripping voltammetric cell [64]. Injection of 0.1 mol/L NaOH through the column eluted the reagents, which then reacted with copper, stripped from the electrode to produce a CL signal. The response was linear in the 0.01–10-μg/L

Cu(II) concentration range in solution, with a detection limit of 8 ng/L. The RSD value at the 40 ng/L concentration level was 7.4%, for the assay of Cu(II) in natural waters and human serum.

4.2.3 Epinephrine

By immobilizing Mn(III)-tetrakis(4-sulfonatophenyl)-porphyrin on dioctadecyl-dimethyl ammonium chloride bilayer membranes incorporated into a PVC film, Kuniyoshi et al. [65] developed an epinephrine CL sensor, which allowed determination of epinephrine down to 3 μM with an RSD of 1.0% for 50 μM of this biological compound. Compared with the previously reported epinephrine CL sensor [66], the present authors noted that the alkaline carrier solution, at high concentration levels, caused gradual deterioration of the immobilized catalyst, and this problem could be solved by the use of immobilization techniques other than ion exchange, e.g., solubilization of the catalyst that has octadecyl groups in the bilayer molecules.

4.2.4 Hydrogen Peroxide, Ions, and Related Compounds

Zhang's group proposed a type of so-called bleeding CL sensors for determination of inorganic and some organic analytes. The sensors were prepared by electrostatically immobilizing the reagents, luminol and some catalysts such as metal ions Co(II), Cu(II), or $Fe(CN)_6^{3-}$, etc., on anion/cation-exchange columns. The analytes, e.g., H_2O_2, were sensed by the CL reaction of luminol and metal ions bleeding from the ion-exchange columns by hydrolysis. A flow injection system was used throughout for the measurements. The sensors could be used for determination of ClO^- [67]; CN^- [68]; Co^{2+} [69]; V(V) [70]; Fe^{2+} and Fe^{3+} [71]; Fe^{3+} [72]; SO_3^{2-} [73]; H_2O_2 [74] and its monitoring in rain water [75], etc.

By immobilization of luminol and Co(II) on a strongly basic anion-exchange resin and a weakly acid cation-exchange resin, H_2O_2 can be determined in the 40 nmol/L–10 μmol/L concentration range with a limit of detection of 12 nmol/L [76]. This system was combined with FIA and its high sensitivity made possible the analysis of hydrogen peroxide in water and the assay of glucose in serum by measuring the formation of H_2O_2 from a packed bed reactor with immobilized glucose oxidase. By in situ electrogenerated H_2O_2 on an electrode, a CL sensor for vitamin B_{12} determinations was developed, based on immobilizing luminol on an ion-exchange resin [77].

Also, a selective sensor for hydrogen peroxide based on FIA was proposed by Janasek et al. [78]. H_2O_2 can be detected in the presence of luminol and Co(II) and Cu(II) foils, obtaining linear determination ranges of 0.1–200 μmol/L and 5–200 μmol/L under FI conditions, respectively. To improve the selectivity, the CL detector was combined with a thin-layer gas dialysis cell.

4.2.5 Sulfur-Containing Compounds

A reagentless flow sensor for sulfite was developed by electrostatically immobilizing the oxidizer permanganate and the sensitizer riboflavin phosphate on an anion-exchange column, allowing a sensitive determination of sulfite in beverage [79]. In this sensor, both the oxidation and the subsequent energy transfer process proceed directly with the immobilized reagent and there is no need to supply an eluent, thus permitting the flow sensor to operate in a reagentless way. As compared to the use of continuously delivered reagents in conventional CL flow systems, this CL sensor shows some advantages in terms of operational convenience, instrumental simplification, reducing reagent consumption, and decreasing analyte dilution, achieving a lifetime much longer than with eluting reagent. Sulfite can be assayed in beer and wine in the 0.1–100-mg/L concentration range with a detection limit of 0.06 mg/L, and an RSD of 3.7% for 1.0 mg/L of sulfite. A sensitive CL sensor was recently proposed for the determination of sulfite with FIA [80]. It is based on the weak CL produced by auto-oxidation of sulfite in the presence of rhodamine 6G immobilized electrostatically on a cation exchange column. A strong enhancement of the weak CL signal was observed in the presence of micelles of the Tween 80 surfactant, showing a linear calibration range of 0.01–5 ppm with a detection limit of 0.01 ppm. Interfering metal ions coexisting in the sample solution can be eliminated online by an upstream cation exchanger.

Using ECL, an assay for $S_2O_8^{2-}$ was reported by application of a microring electrode [81]. A gold-coated fiber was polished to a flat surface, such that the gold formed a micro-ring electrode around the optical fiber. *Tris*(2,2'-bipyridyl)ruthenium(II) was used as reagent for CL generation. The detection limit was 4 nmol/L.

4.2.6 Ammonium Ion

Zhang's group [82] recently presented a novel CL sensor combined with FIA for ammonium ion determination. It is based on reaction between luminol, immobilized electrostatically on an anion-exchange column, and chlorine, electrochemically generated online via a Pt electrode from hydrochloric acid in a coulometric cell. Ammonium ion reacts with the chlorine and decreases the produced CL intensity. The system responds linearly to ammonium ion concentration in a range of 1.0–100 µM, with a detection limit of 0.4 µM. A complete analysis can be performed in 1 min, being satisfactorily applied to the analysis of rainwater.

4.2.7 Ascorbic Acid

Three sensors based on luminol and different cations immobilized on a resin are proposed for ascorbic acid assay, in the following way: (1) D-201 type anion-exchange resin containing luminol and permanganate immobilized [83]; (2) D-201 × 7 anion-exchange resin was used for immobilization of luminol and 732 cation-exchange resin (Na form) was used for Fe(II)-immobilization [84]; (3) Amberlite A-27 anion-exchange resin containing immobilized luminol and potassium ferricyanide [85]. Using these types of flow-through sensors, the CL signal produced by the reaction with luminol was decreased in the presence of ascorbic acid. This fact allows the indirect determination of ascorbic acid on: (1) 10 μg/L–4 mg/L; (2) 1 nmol/L–1 μmol/L; and (3) 0.01–0.8 μg/mL concentration ranges, with the following detection limits: (1) 5 μg/L; (2) 0.4 nmol/L; (3) 5.5 ng/mL, respectively. While the first proposed sensor is free of interferences, Cu(II), thiourea, uric acid, and vitamin B_1 seriously interfere with the third sensor.

4.2.8   Ethanol and Organic Molecules

CL sensors are described for the assay of ethanol and organic molecules in water, comprising a γ-Al_2O_3 layer as catalyst, which can be coated with a Pt thin film [86]. When a mixture of air and organic molecules (e.g., ethanol and acetone) vaporized from a solution flows around the sensor, CL is emitted during the catalytic oxidation. The CL spectra consist of subbands with peak wavelengths independent of the type of vapors, the CL intensity depending on the concentration of these organic compounds, and the temperature of the sensor. The limit of detection is in the mg/L-magnitude order.

4.2.9   Trichloroethylene

An optical-fiber CL sensor is reported for trichlorethylene assay [87]. The sensor consists of a glass fiber bundle and a transducer consisting of three components: (i) a gas-permeable membrane to separate trichlorethylene from water, (ii) H_2SO_4-$NaNO_3$ mixture as oxidizing agent, and (iii) a luminol solution. The assay of trichloroethylene can be done in the 0.05–0.6-μg/mL concentration range with a detection limit of 0.03 μg/mL.

4.2.10   Uric Acid

For the assay of uric acid, a sensor based on $KMnO_4$–octylphenyl polyglycol ether is proposed [88]. Uric acid can be assayed directly in urine in the 0.10–600-μg/mL concentration range with a detection limit of 55 ng/mL. The system is free of interferences.

4.2.11   Oxalate, Alkylamine, and NADH

The development of an ECL sensor based on $Ru(bpy)_3^{2+}$ immobilized in a Nafion film coated on an electrode was discussed by Downey and Nieman [89]. The sensor was used to determine oxalate, alkylamine, and NADH. Detection limits were of 1 µM, 10 nM, and 1 µM, respectively, with working ranges extending over four magnitudes in concentration. The sensor remains stable for several days under suitable storage conditions, being the first research regarding NADH detection using $Ru(bpy)_3^{2+}$ ECL. Using the same system, NADH detection was optimized in a FIA sensor, providing a linear concentration range of 10–250 µM [54].

The substrate selectivity of the $Ru(bpy)_3^{2+}$ ECL was changed by coating a Pt working electrode with $Ru(bpy)_3^{2+}$-modified chitosan membrane and successively with a silica gel membrane that was prepared by the sol-gel method using tetramethoxysilane as a precursor [90]. The double coating resulted in a high selectivity toward oxalic acid at a pH 6 allowing development of an ECL sensor for oxalate [91]. This high selectivity in the presence of trimethylamine can be explained partly on the basis of the electrostatic repulsion of the chitosan membrane having a positive charge with trimethylamine, as being supported by cyclic voltametry. A linear calibration range was established of 0.1–10 mM with a detection limit of 0.03 mM [92].

4.2.12 Primary Alcohols

Another ECL sensor suitable for primary alcohols based on ECL of hydroxyl compounds by cyclic square-wave electrolysis was developed by Egashira et al.

Table 2 Some Typical Examples of Non-Enzyme-Based CL Sensors

Analyte	Immobilized reagent	CL reaction	Ref.
ClO⁻	Uranine	ClO⁻-uranine	63
	Luminol	ClO⁻-luminol	67
CN⁻	Luminol, Cu(II)	Luminol-Cu(CN)$_4^{2-}$	68
Co(II)	Luminol, IO$_4^-$	Co(II)-luminol-IO$_4^-$	69
H$_2$O$_2$	Luminol, Co(II)	Co(II)-luminol-H$_2$O$_2$	76, 78
NH$_4^+$	Luminol	ClO⁻-luminol	82
Glucose	Luminol, Co(II), glucose oxidase	Co(II)-luminol-H$_2$O$_2$	76
Sulfite	Permanganate, riboflavin	Sulfite-permanganate-riboflavin	79
Vitamin C	Luminol, Fe(CN)$_6^{3-}$	Vitamin C-luminol-Fe(CN)$_6^{3-}$	85
Vitamin B$_{12}$	Luminol	Luminol-H$_2$O$_2$(EG)-vitamin B$_{12}$	93
Oxalate	Ru(bpy)$_3^{2+}$	Ru(bpy)$_3^{3+}$ (EG)-oxalate	89
Alkylamine	Ru(bpy)$_3^{2+}$	Ru(bpy)$_3^{3+}$ (EG)-oxalate	89
NADH	Ru(bpy)$_3^{2+}$	Ru(bpy)$_3^{3+}$ (EG)-oxalate	89
Epinephrine	Mn(III)-tetrakis (4-sulfonatophenyl)-porphyrin	Epinephrine magnesium-porphyrin	65

EG, electrogenerated.

[93]. The sensor was stable up to 2 weeks and could be easily regenerated in water making it suitable for use in fermentation industries.

Table 2 lists some typical examples of non-enzyme-based CL sensors.

5. CL IMMUNOSENSORS

In immunoassays the main advantages of CL as a detection principle are the high sensitivity and rapidity of measurement without lengthy incubations. These advantages are more than potential so as to design a CL immunosensor. Aizawa [94] reported on a fiberoptic CL immunosensor by using an aromatic hydrocarbon and enzyme as labeling agents. The antibody was not directly immobilized on an optical fiber. Hara et al. [95] developed a fiberoptic immunosensor by immobilizing the antibody on an optical fiber. A metal-complex compound iron(III)-2,9,16,23-tetrakis(chlorocarbonyl) phthalocyanine [TCCP-Fe(III)] was used as a CL catalyst. Competitive immunoassay was carried out by immersing the end of the optical fiber in a solution of human serum albumin (HSA) complexed with the metal complex. The flow rates for phosphate buffer, H_2O_2, and luminol were 1.50, 0.25, and, 0.25 mL/min, respectively. The HSA was determined in the 0.1–100-mg/L concentration range, with a detection limit of 40 pg. The selectivity can be improved by utilization of a monoclonal antibody. A schematic flow diagram of a fiberoptic immunosensor designed by Hara et al. [95] is illustrated in Figure 6.

Zhang et al. [96] developed a FIA-CL immunosensor by immobilizing antibody on controlled pore glass filled into a column (0.2 × 5 cm). With this method, the time required per assay was 20 min compared to 20 h with conventional enzyme-linked immunosorbent assay (ELISA). Moreover, the whole assay could be proceeded automatically by control through a microprocessor. The technique was used for determination of hepatitis B surface antigen (HBsAg) in serum. Gatto-Menking et al. [97] developed an immunomagnetic ECL sensor for sensitive detection of biotoxoid and bacterial spores based on the CL from the ruthenium(III) trisbipyridal chelate-labeled reporter antibodies induced by a potential of 650–800 V. Detection limits of the order of fg were achieved for botulinus A, cholera β subunit, ricin, and staphylococcal enterotoxoid B.

Starodub et al. [98] studied different constructions and biomedical applications of immunosensors based on fiberoptic and enhanced CL. They discussed three different approaches of immobilization of one of the immunocomponents on the fiberoptic surface. Good results could be achieved by the use of a special membrane closely connected to the fiberoptic, with sensitivities compared to that obtained by the ELISA method but with a faster rate of analysis. The sensor was

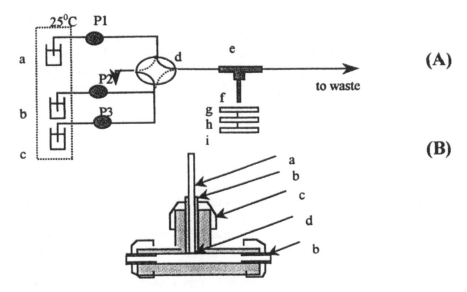

Figure 6 (A) Schematic flow diagram of a fiberoptic immunosensor. (a) Buffer solution; (b) H₂O₂ solution; (c) luminol solution; (d) four-way cock; (e) cell; (f) optical fiber; (g) PMT; (h) photon-counter; (I) integrator; (P1, P2, P3), pumps. (B) The cell set with an optical fiber. (a) optical fiber; (b) silicone tube (3 mm od × 1 mm id); (c) Teflon-made three-way joint (3 mm id); (d) immobilized antibody.

applied to determination of several antigens such as estradiol-17, α-2-interferon, chorionic gonadotropine, antibodies, and cells of *Salmonella* [99].

6. DNA SENSORS

A DNA optical sensor system was proposed based on the combination of sandwich solution hybridization, magnetic bead capture, FI, and CL for rapid detection of DNA hybridization [100]. Bacterial alkaline phosphatase (phoA) gene and hepatitis B virus (HBV) DNA were used as target DNA. A biotinylated DNA probe was used to capture the target gene onto the streptavidin-coated magnetic beads and a calf intestine alkaline phosphatase (CAP)-labeled DNA probe was used for subsequent enzymatic CL detection. The detection cycle was less than 30 min, excluding the DNA hybridization time, which was about 100 min. Both the phoA gene and HBV DNA could be detected at picogram or femtomole levels. Successive sample detection could be made by removing the magnetic field and using a washing step.

DNA probes have been covalently immobilized onto the distal end of the optical fiber bundle in the construction of a novel CL DNA bisensor [101]. Hybridization of HRP-labeled complementary nucleotides to the immobilized probes was detected by enhanced CL. This approach was specific, sensitive, and available for diagnostic application with a detection limit of 0.1 ppm.

7. CONCLUSIONS

Analytical techniques using CL reactions in air or in solution as detection methods have received much attention in various fields owing to their extremely high sensitivity along with extra advantages such as the simple instrumentation required, fast dynamic response properties, wide calibration ranges, the easy coupling to various separation techniques such as HPLC and capillary electrophoresis (CE), and suitability for miniaturization in analytical chemistry. Most of these techniques, however, require several types of reagents, which are continuously delivered and thus wasted in large quantities. This is undesirable not only in view of the simplification of the detection device, but also because of cost, apart from environmental and resource considerations. Some efforts so far have been devoted to reduce the consumption of reagents, including controlled release, cyclic use, immobilization, and regeneration. These proceedings may be directed toward the construction of CL sensors. Compared with chemical sensors based on other sensing principles, very few CL sensors have been reported so far. Although one reason might be the fact that the CL technique is unfamiliar to most scientists, the consumption of reagents accompanying CL reactions, resulting in deterioration of the indicator phases or of sensors on prolonged use, could also be an important factor. Further developments should focus on improvement of the lifetime of this type of sensor and on improvement of selectivity by careful selection of suitable CL reaction systems. Miniaturization of CL sensors for special applications should also be studied to extend this new technique to life sciences as well as to specific areas.

REFERENCES

1. ZJ Zhang, W Qin. Fenxi Kexue Xubao 13:72–77, 1997.
2. S Girotti, EN Ferri, F Fini, M Musiani, G Carrea, A Roda, P Rauch. Chem Listy 91:477–482, 1997.
3. X Zhang, WRG Baeyens, AM García-Campaña, J Ouyang. Trends Anal Chem 18: 384–391, 1999.
4. M Valcárcel, MD Luque de Castro. Analyst 118:593–600, 1993.
5. M Ishii, M Yamada. J Flow Injection Anal 11:154–168, 1994.

6. TM Freeman, WR Seitz. Anal Chem 50:1242–1246, 1978.
7. Y Maeda, K Aoki, M Munemori. Anal Chem 52:307–311, 1980.
8. GJ Wendel, DH Stedman, CA Cantrell, L Damrauer. Anal Chem 55:937–940, 1983.
9. J Yin, J Ouyang, H Yue. Daxue Huaxue 10:38–39, 1995 (Ch.).
10. H Schiff, GI Mackay, C Castledine, GW Harris, Q Tran. Water Air Soil Pollut 30: 105–114, 1986.
11. CW Spicer, DV Kenny, GF Ward, IH Billick, NP Leslie. Air Waste 44:163–168, 1994.
12. GE Collins, SL Ross-Pehrsson. Anal Chem 67:2224–2230, 1995.
13. U Schurath, W Speuser, R Schmidt. Fresenius J Anal Chem 340:544–547, 1991.
14. ZH Lan, HA Mottola. Anal Chim Acta 329:305–310, 1996.
15. GE Collins, SL Ross-Pehrsson. Sens Actuators B 34:506–510, 1996.
16. GE Collins. Sens Actuators B 35:202–206, 1996.
17. H Meng, F Wu, Z He, Y Zeng. Talanta 48:571–577, 1999.
18. AJ Hills, DH Lenschow, JW Birks. Anal Chem 70:1735–1742, 1998.
19. M Nakagawa, S Kawabata, K Nishiyama, K Utsunomiya, I Yamamoto, T Wada, Y Yamashita, N Yamashita. Sens Actuators B 34:334–338, 1996.
20. K Utsunomiya, M Nakagawa, N Sanari, M Kohota, T Tomiyama, I Yamamoto, T Wada, N Yamashita, Y Yamashita. Sens Actuators B 25:790–793, 1995.
21. M Nakagawa. Sens Actuators B 29:94–100, 1995.
22. M Nakagawa, T Okabayashi, T Fujimoto, K Utsunomiya, I Yamamoto, T Wada, Y, N Yamashita. Sens Actuators B 51:159–162, 1998.
23. F Preuschoff, U Spohn, G Blankenstein, K Mohr, M Kula. Fresenius J Anal Chem 346:924–929, 1993.
24. AB Collaudin, LJ Blum. Sens Actuators B 39:189–194, 1997.
25. A Navas Díaz, MC Ramos Peinado, MC Torijas Minguez. Anal Chim Acta 363: 221–227, 1998.
26. M Aizawa, Y Ikariyama, M Kuno. Anal Lett B 7:555–564, 1984.
27. LJ Blum. Enzyme Microb Technol 15:407–411, 1993.
28. AA Suleiman, RL Villarta, GG Guilbault. Anal Lett 26:1493–1503, 1993.
29. N Kiba, A Itagaki, S Fukumura, K Saegusa, M Furusawa. Anal Chim Acta 354: 205–210, 1997.
30. JP Preston, TA Nieman. Anal Chem 68:966–970, 1996.
31. AF Martin, TA Nieman. Anal Chim Acta 281:475–481, 1993.
32. MV Cattaneo, KB Males, JHT Luong. Biosens Bioelectron 7:569–574, 1992.
33. MV Cattaneo, JHT Luong. Biotechnol Bioeng 41:659–665, 1993.
34. U Spohn, F Preuschoff, G Blankenstein, D Janasek, MR Kula, A Hacker. Anal Chim Acta 303:109–120, 1995.
35. G Blankenstein, U Spohn, F Preuschoff, J Thommes, MR Kula. Biotech Appl Biochem 20:291–307, 1994.
36. N Kiba, M Tachibana, K Tani, T Miwa. Anal Chim Acta 375:65–70, 1998.
37. D Janasek, U Spohn. Biosens Bioelectron 14:123–129, 1999.
38. YK Zhou, XD Yuan, QL Hao, S Ren. Sens Actuators B 12:37–40, 1993.
39. H Lapp, U Spohn, D Janasek. Anal Lett 29:1–17, 1996.
40. ZH Song, ZJ Zhang, WZ Fan. Acta Chim Sinica 56:1207–1213, 1998.

41. K Ikebukuro, H Wakamura, I Karube, I Kubo, M Inagawa, T Sugawara, Y Arikawa, M Suzuki, T Takeuchi. Biosens Bioelectron 11:959–965, 1996.
42. K Ikebukuro, R Nishida, H Yamamoto, Y Arikawa, H Wakamura, M Suzuki, I Kubo, T Takeuchi, I Karube. J Biotechnol 48:67–72, 1996.
43. H Nakamura, K Ikebukuro, S Mcniven, I Karube, H Yamamoto, K Hayashi, M Suzuki, I Kubo. Biosens Bioelectron 12:959–966, 1997.
44. H Nakamura, H Tanaka, M Hasegawa, Y Masuda, Y Arikawa, Y Nomura, K Ikebukuro, I Karube. Talanta 50:799–807, 1999.
45. H Nakamura, M Hasegawa, Y Nomura, Y Arikawa, R Matsukawa, K Ikebukuro, I Karube. J Biotechnol 75:127–133, 1999.
46. A Roda, P Rauch, E Ferri, S Girotti, S Ghini, G Carrea, R Bovara. Anal Chim Acta 294:35–42, 1994.
47. J Hlavay, SD Haemerli, GG Guilbault. Biosens Bioelectron 9:189–195, 1994.
48. A Berger, LJ Blum. Enzyme Microbial Technol 16:979–984, 1994.
49. D Janasek, U Spohn. Sens Actuators B 39:291–294, 1997.
50. BX Li. Master's thesis, Department of Chemistry, Shaanxi Normal University, Xian, P.R. China, 1998.
51. AF Martin, TA Nieman. Biosens Bioelectron 12:479–489, 1997.
52. LJ Blum, SM Gautier, A Berger, PE Michel, PR Coulet. Sens Actuator B 29:1–9, 1995.
53. XF Xie, AA Suleiman, GG Guilbault, ZM Yang, ZA Sun. Anal Chim Acta 266:325–329, 1992.
54. K Yokoyama, S Sasaki, K Ikebukuro, T Takeuchi, I Karube, Y Tokitsu, Y Masuda. Talanta 41:1035–1040, 1994.
55. YM Huang, C Zhang, ZJ Zhang. Anal Sci 15:867–870, 1999.
56. ZJ Zhang, WB Ma. Gaodeng Xuexiao Huaxue Xuebao 14:1366–1369, 1993 (Ch.).
57. MS Abdel-Latif, GG Guilbault. Anal Chem 60:2671–2674, 1988.
58. G Blankenstein, F Preuschoff, U Spohn, KH Mohr, MR Kula. Anal Chim Acta 271:231–237, 1993.
59. F Preuschoff, U Spohn, E Weber, K Unverhau, KH Mohr. Anal Chim Acta 280:185–189, 1993.
60. SM Gautier, LJ Blum, PR Coulet. J Biolumin Chemilumin 5:57–63, 1990.
61. ZJ Zhang, WB Ma, ML Yang. Fenxi Huanxue 20:1048–1051, 1992 (Ch.).
62. ZJ Zhang, WB Ma. Kexue Tongbao 38:230–232, 1993 (Ch.).
63. T Nakagama, M Yamada, T Hobo. Anal Chim Acta 231:7–12, 1990.
64. W Qin, ZJ Zhang, HJ Liu. Anal Chem 70:3579–3584, 1998.
65. A Kuniyoshi, K Hatta, T Suzuki, A Masuda, M Yamada. Anal Lett 29:673–685, 1996.
66. F Yoshimura, T Suzuki, M Yamada, T Hobo. Bunseki Kagaku 41:191–196, 1992.
67. W Qin, ZJ Zhang, SN Liu. Anal Lett 30:11–19, 1997.
68. JZ Lu, W Qin, ZJ Zhang, ML Feng, YJ Wang. Anal Chim Acta 304:369–373, 1995.
69. JZ Lu, ZJ Zhang. Huaxue Xuebao 54:71–76, 1996 (Ch.).
70. W Qin, ZJ Zhang, CJ Zhang. Analyst 122:685–688, 1997.
71. W Qin, ZJ Zhang, FC Wang. Fresenius J Anal Chem 360:130–132, 1998.
72. W Qin, ZJ Zhang, CJ Zhang. Mikrochim Acta 129:97–101, 1998.

73. W Qin, ZJ Zhang, CJ Zhang. Fresenius J Anal Chem 361:824–826, 1998.
74. BX Li, W Qin, ZJ Zhang. Chin Chem Lett 9:471–472, 1998.
75. W Qin, ZJ Zhang, HH Chen. Int J Environ Anal Chem 66:191–200, 1997.
76. W Qin, ZJ Zhang, BX Li, SN Liu. Anal Chim Acta 372:357–363, 1998.
77. W Qin, ZJ Zhang, HJ Liu. Anal Chim Acta 357:127–132, 1997.
78. D Janasek, U Spohn, D Beckmann. Sens Actuators B 51:107–113, 1998.
79. W Qin, ZJ Zhang, CJ Zhang. Anal Chim Acta 361:201–203, 1998.
80. YM Huang, C Zhang, XR Zhang, ZJ Zhang. Anal Lett 32:1211–1224, 1999.
81. LS Kuhn, A Weber, SG Weber. Anal Chem 62:1631–1636, 1990.
82. W Qin, ZJ Zhang, BX Li, YY Peng. Talanta 48:225–229, 1999.
83. FC Wang, W Qin, ZJ Zhang. Fenxi Huaxue 25:1255–1258, 1997.
84. W Qin, ZJ Zhang, HH Chen. Fresenius J Anal Chem 358:861–863, 1997.
85. ZJ Zhang, W Qin. Talanta 43:119–124, 1996.
86. M Nakagawa, I Yamamoto, N Yamashita. Anal Sci 14:209–214, 1998.
87. K Bansho, H Tao, T Imagawa, A Miyazaki. Bunseki Kagaku 42:55–60, 1993.
88. Z Li, ML Feng, JR Lu. Mikrochem J 59:278–283, 1998.
89. TM Downey, TA Nieman. Anal Chem 64:261–268, 1992.
90. CZ Zhao, N Egashira, Y Kurauchi, K Ohga. Anal Sci 13:333–336, 1998.
91. CZ Zhao, N Egashira, Y Kurauchi, K Ohga. Anal Sci 14:439–441, 1998.
92. CZ Zhao, N Egashira, Y Kurauchi, K Ohga. Electrochim Acta 43:2167–2173, 1998.
93. N Egashira, Y Nabeyama, Y Kurauchi, K Ohga. Anal Sci 12:793–795, 1996.
94. M Aizawa. Protein, Nucleic Acid and Enzyme. Bessatsu 31:278–280, 1987.
95. T Hara, K Tsukagoshi, A Arai, Y Imashiro. Bull Chem Soc Jpn 62:2844–2848, 1989.
96. X Zhang, M Feng, J Lu, Z Zhang. Huaxue Xuebao 52:83–88, 1994 (Ch.).
97. DL Gatto-Menking, H Yu, JG Bruno, MT Goode, M Miller, AW Zulich. Biosens Bioelectron 10:501–507, 1995.
98. NF Starodub, PY Arenkov, AN Starodub, VA Berenzin. Optic Eng 33:2958–2963, 1994.
99. NF Starodub, PY Arenkov, AN Starodub, VA Berenzin. Sens Actuators B 18:161–165, 1994.
100. X Chen, XE Zhang, YQ Chai, WP Hu, ZP Zhang, XM Zhang, AEG Cass. Biosens Bioelectron 13:451–458, 1998.
101. GJ Zhang, YK Zhou, JW Yuan, S Ren. Anal Lett 32:2725–2736, 1999.

Appendix: Abbreviations

A/D	Analog to digital
AAS	Atomic absorption spectrometry
Ab	Antibody
ABAP	2,2'-Azo-*bis*(2-aminodipropane)dihydrochloride
ABCP	2,2'-Azo-*bis*(2-cyanopropane)
ABEI	*N*-(4-Aminobutyl)-*N*-ethylisoluminol
ABTS	2,2'-Azino-*bis*(3-ethylbenzothiazoline-6-sulfonic acid)
AchE	Acetylcholinesterase
ACL	Antioxidative capacity of lipid-soluble compounds
ACP	Antioxidative capacity of proteins
ACR	Rest of total antiradical capacity
ACU	Urate independent component of ACW
ACW	Antioxidative capacity of water-soluble compounds
ADP	Adenosine diphosphate
AE	Acridinium ester
AFP	α-Fetoprotein
Ag	Antigen
ALP (or AP)	Alkaline phosphatase
AMP	Adenosine monophosphate
6-AMP	6-Aminomethylphthalhydrazide
AMPPD	3-(2'-Spiroadamantane)-4-methoxy-4-(3''-phosphoryloxy)phenyl-1,2-dioxetane
AMVN	2,2'-Azo-*bis* (2,4-dimethylvaleronitrile)
ANS	8-Aniline-1-naphthalenesulfonic acid

AP	Aminophthalate anion
APAN	Atmospheric pressure active nitrogen
APP-CLS	Analyte pulse perturbation-chemiluminescence spectroscopy
ARP	*Arthromyces rasomus* peroxidase
ASC	Ascorbic acid
ATP	Adenosine triphosphate
AVP	Avalanche photodiode
BCI	5-Bromo-4-chloro-3-indolyl
BHT	2,6-Di-*tert*-butyl-4-methylphenol
BL	Bioluminescence
Brij-35	Polyoxyethylene (23) dodecanol
BSA	Bovine serum albumin
c.m.c.	Critical micelle concentration
CAP	Calf alkaline phosphatase
CAR	Continuous-addition-of-reagent
CAR-CLS	Continuous-addition-of-reagent chemiluminescence spectroscopy
CAS	Catecholamines
CAT	Catechol
CBI	Cyanobenz[*f*] isoindole
CCD	Charge coupled device
cDNA	Complementary deoxyribonucleic acid
CE	Capillary electrophoresis
CE-CL	Capillary electrophoresis-chemiluminescence
CEDAB	Cetylethyldimethylammonium bromide
CE-ECL	Capillary electrophoresis-electrogenerated chemiluminescence
CETAS	Capillary electrophoresis micrototal analysis system
CGE	Capillary gel electrophoresis
CGH	Comparative genomic hybridization
ChO	Choline oxidase
CIA	Capillary ion analysis
CID	Charge injection device
CIEEL	Chemically initiated electron exchange luminescence
CIEF	Capillary isoelectric focusing
CITP	Capillary isotachophoresis
CK	Creatine kinase
CL	Chemiluminescence
CMPI	2-Chloro-1-methylpyridinium iodide
CMV	Cytomegalovirus
CNN	Computational neural network
CPPO	2,4,5-Trichloro-6-carbopentoxyphenyl oxalate
CSTR	Continuous stirring tank reactor

CZE	Capillary zone electrophoresis
d(G)n	Polydeoxyguanylic acid
DBD-F	4-(N,N-Dimethylaminosulphonyl)-7-fluoro-2,1,3-benzoxadiazole
DBD-H	4-(N,N-Dimethylaminosulphonyl)-7-hydrazino-2,1,3-benzoxadiazole
DBPM	N-[4-(6-Dimethylamino-2-benzofuranyl)phenyl]maleimide
DCIA	7-Dimethylamino-3-{4-[(iodoacetyl)amino]phenyl}-4-methylcoumarin
DDDAB	Didodecyldimethylammonium bromide
DEA	Diethanolamine
DETBA	1,3-Diethyl-2-thiobarbituric acid
DFPO	*Bis*(2,6-difluorophenyl)oxalate
DIC	Differential interference contrast
DMF	N,N'-Dimethylformamide
DMQPH	6,7-Dimethoxy-1-methyl-2($1H$)-quinoxalinone-3-proprionylcarboxylic acid hydrazide
DMS	Dimethyl sulfide
DMSO	Dimethyl sulfoxide
DNA	Deoxyribonucleic acid
DNPO	*Bis*-(2,4-dinitrophenyl) oxalate
DNS-Cl	Dansyl chloride
DNS-H	Dansyl hydrazine
DODAB	Dioctadecyldimethylammonium bromide
DPA	9,10-Diphenylanthracene
DPH	4,5-Diaminophthalhydrazide
DPPA	4-[4,5-Di(2-pyridyl)-1 H-imidazol-2-yl]phenylboronic acid
DSC	N,N'-Disuccinimidyl carbodiimide
DTDCI	3,3'-Diethylthiadicarbocyanine iodide
DVS	Divinyl sulfone
EC	Electrochemical
ECL	Electrogenerated chemiluminescence
ELISA	Enzyme-linked immunosorbent assay
EMMA	Electrophoretically-mediated microanalysis
EOF	Electroosmotic flow
EY	Eosine Y
FCLD	Fluorine-induced chemiluminescence detector
FHC	Fluorohydrocarbon
FIA	Flow injection analysis
FID	Flame ionization detector
Fl	Fluorescein
FL	Fluorescence

FMN	Flavin monocleotide
FMNOX	Flavin monocleotide oxidoreductase
FPD	Flame photometric detector
FPS	Frame per second
FR	Fluorescamine
G6PDH	Glucose-6-phosphate dehydrogenase
GC	Gas chromatography
GFP	Green fluorescent protein
GOD	Glucose oxidase
HA	Hyaluronic acid
HACCP	Hazard analysis and critical control point
HBV	Hepatitis B virus
HCCD	High resolution charge coupled device
HCPI	2-(4-Hydrazinocarbonylphenyl)-4,5-diphenylimidazole
HDI	2-(4-Hydroxyphenyl)-4,5-diphenylimidazole
3α-HSD	3α-Hydroxysteroid dehydrogenase
HEG	Hydroxyethylglycine
HIBA	α-Hydroxyisobutyric acid
HPB	Hexadecylpyridinium bromide
HPCE	High performance capillary electrophoresis
HPD	Hematoporphyrine derivatives
HPI	2-(4-Hydroxyphenyl)-4,5-di(2-pyridyl)imidazole
HPLC	High performance liquid chromatography
HPV	Human papillomavirus
HQS	8-Hydroxyquinoline-5-sulphonic acid
HRP	Horseradish peroxidase
HSA	Human serum albumin
HSV	Herpes simplex virus
HTAB	Hexadecyltrimethylammonium bromide
HTAC	Hexadecyltrimethylammonium chloride
HTAH	Hexadecyltrimethylammonium hydroxide
HTS	High throughput screening
HV	High voltage
IAECL	Ion annihilation electrogenerated chemiluminescence
ICP-MS	Inductively coupled plasma-mass spectrometry
ICP-OES	Inductively coupled plasma-optical emission spectrometry
IDA	Iminodiacetic acid
IHC	Immunohistochemistry
IL8	Interleukin 8
ILITC	Isoluminol isothiocyanate (4-isothiocyanatophthalhydrazide)
IMER	Immobilized enzyme reactor
IPO	6-Isothiocyanatobenzo[g]phthalazine-1,4 (2H,3H)-dione

ISH	In situ hybridization
ITO	Indium tin oxide
LC	Liquid chromatography
LC-ECL	Liquid chromatography-electrogenerated chemiluminescence
LDA	Lithium diisopropylamide
LDL	Low-density lipoprotein
LIF	Laser-induced fluorescence
LOD	Limit of detection
mAb	Monoclonal antibody
MAC	Maximum admissible concentration
m-CED	1,2-*Bis*(3-chlorophenyl)ethylenediamine
MEKC	Micellar electrokinetic chromatography
METQ	4,4'-{Oxalyl *bis*[(trifluoromethylsulfonyl)imino]-ethylene} *bis*(4-methylmorpholinium)trifluoromethanesulfonate
MP	Maltose phosphorylase
MPP	Multipinned phase
MPTQ	3,3'-Oxalyl-*bis*[(trifluoromethylsulfonyl)imino]trimethylene-*bis*(*N*-methylmorpholinium) trifluoromethanesulfonate and 4,4'-{Oxalyl-*bis*[(trifluoromethylsulfonyl)imino]-trimethylene}*bis*(4-methylmorpholinium) trifluoromethanesulfonate
MS	Mass spectrometry
MTES	Metastable transfer emission spectroscopy
MUT	Mutarose
NAD	Naphthalene-2,3-dicarboxaldehyde
NADH	Nicotinamide adenine dinucleotide
NANA	*N*-Acetylneuraminic acid
NBD-F	4-Fluoro-7-nitrobenzoxadiazole
NDA	Naphthalene-2,3-dicarboxaldehyde
NDRO	Nondestructive readout
NIR	Near infrared
NIRF	Near infrared fluorescence
NMR	Nuclear magnetic resonance
2-NPO	2-Nitrophenyl oxalate
ODI	1,1'-Oxalyldiimidazole
PAH	Polycyclic aromatic hydrocarbon
PCA	Principal component analysis
PCL	Photosensitized chemiluminescence
PCPO	Pentachlorophenyl oxalate
PCR	Polymerase chain reaction
PETQ	2,2' Oxalyl-*bis*[(trifluoromethanesulfonyl)imino]ethylene-*bis*(*N*-methylpyridinium) trifluoromethanesulfonate

PFPD	Pulse flame photometric detector
PFPO	*Bis*(pentafluorophenyl)oxalate
PhoA	Alkaline phosphatase
PM-DIC	Polarization-modulated differential interference contrast
PMT	Photomultiplier tube
POCL	Peroxyoxalate chemiluminescence
ppbv	Parts per billion volume
ppmv	Parts per million volume
PSA	Prostate-specific antigen
PTAB	Pentadecyltrimetylammonium bromide
PTFE	Poly(tetrafluoro)ethylene
PVC	Polyvinylchloride
QE	Quantum efficiency
RCD	Redox chemiluminescence detector
RHB	Rhodamine B
RI	Refractive index
RITC	Rhodamine B isothiocyanate
RMS	Root mean square
RRD	Rotating ring-disk (electrode)
RSD	Relative standard deviation
RTP	Room temperature phosphorescence
RT-PCR	Reverse transcriptase- polymerase chain reaction
S/N	Signal to noise ratio
SB-12	*N*-Dodecyl-*N,N*-dimethylammonium-3-propane-1-sulfonate
SCD	Sulfur chemiluminescence detector
SDBS	Sodium dodecylbenzene sulfonate
SDS	Sodium dodecyl sulfate
SFA	Segmented flow analyzer
SFC	Supercritical fluid chromatography
SF-CLS	Stopped-flow-chemiluminescence spectrometry
SIA	Sequential injection analysis
SOD	Superoxide dismutase
SSCL	Solid surface chemiluminescence
SSF	Solid surface fluorescence
SSRTP	Solid surface room temperature phosphorescence
STAC	Stearyltrimethylammonium chloride
TCCP	2,9,16,23-*Tetrakis*(chlorocarbonyl) phthalocyanine
TCP	Trichlorophenol
TCPO	*Bis*(2,4,6-trichlorophenyl)oxalate
TDPO	*Bis*[2-(3,6,9-trioxadecyloxycarbonyl)-4-phenyl]oxalate
TEA	Thermal energy analysis
TEA	Triethylamine

TEP	Triethylphosphine
TL	Total luminescence
TLC	Thin layer chromatography
TMB	Tetramethylbenzidine
TMP	2,4,6,8-Tetrathiomorpholinopyrimido [5,4-*d*] pyrimidine
TMP	Trimethylphosphine
TMPG	3′,4′,5′-Trimethoxyphenylglyoxal
TNS	Potassium 2-*p*-toluidinylnaphthalene-6-sulfonate
TPB	Tetradecylpyridine bromide
TRIS	Tris (hydroxymethyl) aminomethane
TRITC	Tetramethylrhodamine isothiocyanate
TTAB	Tetradecyltrimethyl ammonium bromide
UA	Uric acid
USDA-FSIS	U.S. Department of Agriculture-Food Safety and Inspection Service
UV	Ultraviolet
VCA	Virus capsid antigens
VE	Vitamin E
YFP	Yellow fluorescent protein

Index